Thalassotherapy and Cosmeceuticals

Books in the Series

- Seabird Biodiversity and Human Activities
  Jaime A. Ramos and Leonel Pereira (eds.), 2022
- Thalassotherapy and Cosmeceuticals
  M. Lourdes Mourelle and Haresh S. Kalasariya (eds.), 2025

Series: Aquatic Sciences
Series Editor: Leonel Pereira

Volume 2

Thalassotherapy and Cosmeceuticals

Editors
M. Lourdes Mourelle
FA2 Research Group
Applied Physics Department
University of Vigo
Spain

Haresh S. Kalasariya
Microbiology Department
Sankalchand Patel University
Gujarat, India

CRC Press
Taylor & Francis Group
Boca Raton London New York

CRC Press is an imprint of the
Taylor & Francis Group, an **informa** business
A SCIENCE PUBLISHERS BOOK

Cover credit: Images reproduced by kind courtesy of Dr. Leonel Pereira and Dr. M. Lourdes Mourelle.

First edition published 2025
by CRC Press
2385 NW Executive Center Drive, Suite 320, Boca Raton FL 33431

and by CRC Press
4 Park Square, Milton Park, Abingdon, Oxon, OX14 4RN

CRC Press is an imprint of Taylor & Francis Group, LLC

Library of Congress Cataloging-in-Publication Data (applied for)

ISBN: 978-1-032-50613-5 (hbk)
ISBN: 978-1-032-50614-2 (pbk)
ISBN: 978-1-003-39928-5 (ebk)

DOI: 10.1201/9781003399285

Typeset in Times New Roman
by Prime Publishing Services

Preface to the Series

Aquatic Science is the multidisciplinary study of aquatic ecosystems, both freshwater and marine. Scientific investigations range in scale from the molecular level of contaminants to the stresses on entire ecosystems.

Some of the major fields of study within the aquatic sciences include: limnology (study of lakes, rivers, wetlands and groundwater); biogeochemistry; aquatic ecology; oceanography; Marine biology; ecology applied to marine macroalgae; shrimp farming; ecophysiology and behavior of fish; microalgae ecology; microbial aquatic ecology, nutrient cycling and limnology; ecology of marine benthos and cephalopods; zooplankton ecology; aquatic ecotoxicology; seabirds biodiversity; dynamics of populations of aquatic beings; fisheries biology; physical chemistry, marine pollution and coastal management; malacology; crustacean nutrition; oceanography; pathology of aquatic organisms; continental fish farming; aquaculture technology, mariculture and coastal management. In each of these fields, different volumes edited by renowned experts will be published.

The Series Editor
Leonel Pereira

Preface

Thalassotherapy is the combined use of marine elements (seawater, algae, mud, and climate), in a marine environment, for healing and well-being purposes. Seawater cures have been extensively developed in the last century, but there is a lack of scientific studies and books that allow support to this kind of therapy. The second part of the book focuses on cosmeceuticals. Cosmeceuticals, derived from marine macroalgae, particularly seaweeds, stand at the forefront of pioneering advancements in skincare science. Abundant in bioactive compounds like polysaccharides, polyphenols, and peptides, seaweeds offer a rich source of anti-aging, antioxidant, and moisturizing properties. The distinctive composition of seaweed extracts holds promise for stimulating collagen synthesis, preserving elastin, and fostering overall skin regeneration. Hence, harnessing the immense potential of seaweed and its bioactive compounds is essential. This text unveils their profound impact, seamlessly aligning with the core theme of advanced skincare at the heart of cosmeceutical exploration.

This book seeks to bridge the gap between the basic science of seawater and highly advanced published literature and is intended to serve as a comprehensive reference for the field of thalassotherapy and cosmeceuticals. It covers all the major aspects of sweater composition and properties, mechanisms of action, techniques, and cures linked to the environment, climate, and other marine resources such as mud and seaweeds. It is designed to provide extensive information on marine cures and their mechanisms of action, suitable therapies and techniques depending on the type of treatment, as well as an overview of complementary techniques.

The book is divided into eighteen chapters. The first nine chapters are related to seawater cure, and discuss the main techniques and mechanisms of action, while the second part of the book is concerned with cosmeceuticals, focusing on those linked to seaweed and seaweed-based biologically active compounds for skin benefits in cosmetics and cosmeceuticals.

Chapter 1 focuses on the concept of thalassotherapy and its history. Chapter 2 deals with the composition and properties of seawater with special regard to those linked to the therapies. Chapter 3 provides an overview of the mechanisms of action of seawater and its potential as a means of healing. Chapter 4 discusses the health benefits of marine cures, including the different diseases and ailments that can be improved through the use of seawater. Chapter 5 focuses on the different seawater techniques, including the so-called complementary or auxiliary ones. Chapter 6 discusses extensively the use of seaweeds in thalassotherapy but also in cosmeceuticals. Chapter 7 provides an overview of marine muds, a marine resource also used in cosmeceuticals as well as in thalasso cures. Chapter 8 deals with the use of the climate and marine environment in healing. Chapter 9 discusses the use of seawater as a nutrient, as well as for sport recovery and other nutritional fields.

The second part begins with an insightful Introduction to cosmeceuticals in Chapter 10. Chapters 11 and 12 delve into the biology of marine algae and cosmetic applications of marine algae, respectively while Chapter 13 magnifies the significance of algal lipids in the cosmeceutical sector. The exploration of carbohydrates and proteins in cosmetics unfolds in Chapter 14, followed by a dedicated focus on phlorotannins in Cosmeceuticals in Chapter 15. Chapter 16 navigates recent advancements in phenolic compounds derived from algae, leading to an in-depth analysis of the current status of seaweed utilization in cosmeceuticals in Chapter 17. Finally,

Chapter 18 looks at the future, addressing challenges and unveiling opportunities on utilization of seaweed in cosmeceuticals.

The wide variety of topics found in this book should benefit a large audience interested in thalassotherapy and cosmeceuticals, including doctors, nurses, graduate students, researchers, and those concerned with seawater therapies, such as technicians, aquatic therapists and others. The contemporary and expansive content of this book also holds appeal for non-specialist readers seeking a comprehensive understanding. Ultimately, our aspiration is for this book to stand as a definitive textbook, setting a standard for aspiring seawater therapists, medical practitioners, cosmetic scientists, and researchers of cosmetic science in the future.

Contents

CHAPTER 1

Introduction to the Thalassotherapy

Marie Perez Siscar

Introduction

The origin of the word "thalassotherapy" comes from the Greek, where "*thalasso*" means "sea" and "*therapia*" means "treatment" or "care".

Thalassotherapy is a therapeutic approach that utilizes the benefits of seawater to promote health and well-being. For millennia, ancient civilizations have recognized the healing properties of seawater and have used this natural resource to treat various conditions. Today, thalassotherapy has become a popular form of complementary medicine, offering a range of treatments that combine seawater, marine algae, mud, and sea salts to promote relaxation, regeneration, and healing of certain pathologies.

It can be said that thalassotherapy has emerged multiple times, in different places, and during various periods:

- As a separate discipline, encompassing extensive information on the history and major discoveries related to the marine environment, justifying the use of seawater and its derivatives.
- As a concept, which allowed various perceptions of thalassotherapy to be linked and organized. Initially, it was associated with medical and scientific practices in specialized medical centers called "*hélio marins*", and later became more focused on well-being and health within thalassotherapy institutes.

Thalassotherapy: its origin in history

The Egyptians were the first to recognize the therapeutic virtues of seawater and marine mud.

The Ebers Papyrus (See a page of the papyrus in Appendix 1) is one of the oldest known medical treatises, dating back to the 16th century BCE, during the reign of Amenhotep I. This text is a list of remedies with brief indications of the ailments to be treated. It is difficult to interpret, as some medical terms still remain enigmatic. Among the identifiable substances, there are numerous references to mud and seawater.

The Greeks and Romans were already familiar with and utilized the therapeutic benefits of the sea.

Présidente de Côté Thalasso.
Email: marie.perez@cote-thalasso.fr

Numerous examples illustrate the use of seawater by the ancient Greeks for therapeutic purposes. They attest to the recognition of the health benefits of seawater and its integration into medical practices of the time.

Hippocrates, considered the father of Western medicine, mentioned saltwater baths to alleviate joint pain, skin conditions, and gastrointestinal disorders (Klarić et al. 2007).

Herod of Alexandria and Asclepiades of Bithynia, Greek physicians of the 1st century BCE, respectively recommended saline baths for treating various female ailments and musculoskeletal disorders as well as respiratory problems.

Pausanias, a Greek traveler and geographer from the 2nd century CE, described thermal baths in Epidaurus and mentioned that seawater baths were frequented for their healing properties.

In ancient Rome, the use of seawater and its derivatives was also widespread for therapeutic and medicinal purposes.

Pliny the Elder, a Roman writer and naturalist from the 1st century CE, mentioned the use of seawater in his works "Natural History". He discussed seawater baths as remedies for various ailments, including muscular pain, skin problems, and respiratory disorders (Bluteau 1931).

The city of Baiae in the Gulf of Naples, Italy, was famous for its seawater therapy facilities during the Roman era and is considered the first major seawater spa. It attracted influential personalities and Roman emperors who came to enjoy seawater baths for their healing properties. The ruins of the seawater therapy facilities in Baiae bear witness to the medical use of seawater during that time.

Pompeii, an ancient Roman city buried by the eruption of Mount Vesuvius in 79 CE, has revealed Roman thermal baths with seawater bathing facilities. These baths were used both for hygiene and therapeutic purposes and included pools of seawater and hot and cold baths (Bluteau 1931).

Galen, a famous Greek physician who lived during the Roman era, also mentioned the use of seawater in his medical writings. He recommended seawater baths to relieve joint pain, skin conditions, and respiratory disorders.

From ancient Rome to the Middle Ages, the use of seawater and its derivatives in medical and therapeutic contexts continued and evolved.

After ancient Rome, Roman baths continued to be used, and seawater baths was considered beneficial for health and well-being.

Public baths, known as thermae, were popular, and Byzantine emperors supported their development. For example, Emperor Justinian I built the Baths of Zeuxippus in Constantinople, which included seawater bath facilities.

During the medieval period, knowledge of seawater use was transmitted by the Arabs, who have preserved and developed medical knowledge from antiquity. Arab physicians such as Avicenna wrote about the benefits of seawater in their medical treatises. This knowledge was later integrated into medieval European medicine.

Avicenna's famous "Qanun" was translated into Latin by Gerard of Cremona between 1150 and 1187 during his extended stay in Toledo, Spain, and was printed for the first time in Hebrew in Milan in 1473, then in Venice in 1527, and in Rome in 1593 in Italy (Bluteau 1931).

Medieval public baths, called "baños" in Spain and "St. Thomas Baths" in England, continued to use seawater in their facilities, especially those close to the coast. These baths were used for both hygiene and therapeutic purposes.

Medieval monasteries and abbeys, often located near the coast, utilized the benefits of seawater in their care practices. Monks and nuns practiced balneotherapy, consisting of seawater baths and treatments with sea salts and algae.

It is important to note that bathing practices varied across regions and cultures during the Middle Ages. Medical knowledge and hygiene practices could also differ from one period and society to another.

At the end of the Middle Ages, the Catholic Church had some control over public baths. Ecclesiastical authorities ensured that the baths adhered to hygiene rules and did not become places of debauchery or disorder. Beliefs in contagion and the spread of diseases led to increasing concerns about frequenting public baths.

Toward the end of the Middle Ages and the beginning of the Renaissance, attitudes toward hygiene began to change. Cleanliness and freshness were increasingly valued, and personal hygiene practices, including the use of baths, were encouraged.

Private baths became more common among the affluent classes. The homes of nobles and wealthy bourgeois started to be equipped with private baths, usually in specially dedicated rooms called "bathing rooms". These private baths were more luxurious and offered more privacy than public baths.

Italy, especially the cities of Florence and Venice, played a significant role in promoting hygiene practices and the use of baths. Influenced by the Roman heritage, Italians valued hygiene and regularly practiced bathing for both hygienic and social reasons.

During this period, seawater continued to be used for therapeutic purposes. Seawater baths, also known as 'salt baths', were popular among people suffering from health problems, including joint pain, skin conditions, and respiratory disorders. Coastal cities such as Bath in England and Trouville in France became favored destinations for therapeutic seawater baths.

From the 15th to the 19th century, the use of baths and seawater continued to evolve and develop. Seaside resorts were developed along the coasts, offering facilities for seawater baths. These baths were considered beneficial for health, especially for skin conditions, respiratory issues, and rheumatism (Bluteau 1931).

Giuseppe Valentini, born in 1798 in Cesena, Italy, studied medicine in Bologna and developed a particular interest in the health benefits of seawater. He played a crucial role in establishing thalassotherapy in Cesenatico, a coastal town in Italy.

In 1828, Valentini began using marine mud and seawater to treat various conditions and promote well-being. He developed specific techniques for applying marine mud to the body and devised treatment protocols for health problems such as arthritis, skin diseases, and respiratory disorders.

Thanks to his groundbreaking efforts and research, Giuseppe Valentini is considered one of the pioneers of thalassotherapy and the use of marine mud for therapeutic purposes. His work laid the foundation for the further development of thalassotherapy in Cesenatico, which became a renowned destination for treatments based on the benefits of seawater and marine mud (Bluteau 1931).

In France, the resurgence temporarily came from Ambroise Paré (which work was described by Gabriel Buon in 1579 in the book "*Les Œuvres d'Ambroise Paré, Conseiller et Premier chirurgien du Roy Paris*") the chief surgeon of King Henri III's court, who prescribed sea baths for him: "On Tuesday, June 3, 1578, King Henri III and the Queens, after having supper, went to sleep in Ecouen and Dieppe, where the King, on the advice of his doctors, went to bathe in the sea to cure certain rashes he was suffering from" (Bluteau 1931).

Bathing prescriptions mostly came from English physicians, with a preference for treatment in pediatric infections. The British enthusiasm crossed the Channel, and in 1778, the first marine thermal health center opened in Dieppe.

Dr. John Floyer (1649–1734) conducted the first therapeutic trials with seawater. His work on the benefits of hydrotherapy resulted in a thesis published in 1697 titled "Digression on the Proper Use of Hot and Cold Seawater Baths in England". And in 1753, Dr. Richard Russell (1687–1759) published the first modern treatise on the therapeutic benefits, particularly for glandular purposes, of seawater, whether ingested or used for immersion ("*Detabe glandularium*") (Obel 1984, Sakula 1995). During these times, "Marine Sanatorium" were little by little developed, first in Dieppe, France in 1778, and later on in Margate, England, in 1791. Heiligendamm in Germany, in1793 was dedicated to the cure of infectious diseases and childhood rickets, in which sea water were use through oral intake and baths (San Martin 1995).

In Italy, seawater cure started at the beginning of the 19th century, being the first regulation "*Regolamento per il buon servizio e il buon ordine dei bagni di mare*" published in 1822, and in Germany there are some references related to Dr Chenitz cures for goiter in 1823 (Hoareau 2011).

The use of baths and seawater continued to evolve, influenced by scientific progress, social changes, and well-being trends.

Thalassotherapy: a concept

Thalassotherapy only reached the status of a full-fledged medical concept in France on December 31, 1865 when Joseph La Bonnardière (1829–1887), a medical student from the Alps, presented his doctoral thesis on seawater titled "Introduction to Thalassotherapy" at the University of Medicine in Montpellier and coined the concept (La Bonnardière 1865). This work is considered the official birth of the discipline. Later on, professor Verneuil confirm this new term in his speech in the opening ceremony of the Ist International Congress of Sea Bathing and Hydrology (Tréguer 2002).

Thérèse-Marie Clotilde Pichard, a French physiotherapist born in 1856, founded the first thalassotherapy center in Roscoff, France, known as the "Institut Marin de Roscoff" in 1899.

This pioneering establishment was created to harness the therapeutic benefits of seawater for health and well-being. The Institute specialized in treatments using seawater, such as baths, algae wraps, showers, and inhalations of sea air. Pichard's belief in the healing properties of seawater helped popularize thalassotherapy as a method of treatment and relaxation, and her work influenced the marine wellness industry worldwide (Appendix 2).

Dr. Louis Bagot, born in 1885, was a French physician known for founding the Centre Hélio-Marin (CHM) of Sète in 1931. The CHM initially provided care to children with tuberculosis, a common respiratory disease at that time. Dr. Bagot believed in the healing properties of nature, particularly the sea and the sun, and developed the idea of using these natural elements to treat respiratory illnesses. The concept of the CHM relied on seawater, sunshine, and sea air to stimulate the immune system, improve respiratory function, and promote healing (Hoareau 2011). Thanks to his innovative vision, the CHM of Sète became one of the first centers for thalassotherapy and heliotherapy in France, and his approach influenced the development of marine medicine and well-being treatments based on the sea and the sun (retrieved from https://france-thalasso.com/la-thalasso/thalasso-histoire/).

René Quinton, a French biologist, physiologist, and hygienist born in 1866, is primarily known for his work on the therapeutic use of seawater. He developed the concept of marine therapy known as the "Quinton Method". Quinton was convinced of the healing properties of seawater and its similarity to human blood plasma. He conducted extensive research on the composition of seawater and developed a method to extract and filter seawater while preserving its biological properties. He used this filtered seawater solution, known as "Quinton Plasma", to treat various conditions and physiological imbalances. The Quinton Method was based on the idea that Quinton Plasma, by providing the body with minerals, trace elements, and essential nutrients present in seawater, could help restore balance and health. His work sparked considerable interest at the time and led to the establishment of many treatment centers using the Quinton Method, both in France and abroad. Quinton's legacy influenced the further development of thalassotherapy and maritime medicine, and today, seawater-derived products like Quinton Plasma are still used in some well-being and health treatments (San José 2004, Alberola 2013, Florez 2015).

Modern thalassotherapy centers around the world

In 1894, Boulogne-sur-Mer hosted 150 doctors at the first International Congress of Seaside Resorts and Hydrology, being a crucial moment in the scientific development of thalassotherapy.

Thalassotherapy has been developed in the seashores of the Atlantic Ocean and Mediterranean Sea mainly in France and Spain, and latterly in Italy, Greece, Morocco and Tunisia.

France: France is considered the birthplace of modern thalassotherapy and has numerous centers along its coasts. The country has a National Professional Syndicate, France Thalasso (https://france-thalasso.com/), that oversees thalassotherapy centers. The first International Association of Thalassotherapy was founded in 1913 under the presidency of Pr. Gilbert, a member of the Academy of Medicine. The first congress was held in Cannes in April 1914, with attendees from France, Italy, Spain, Portugal, Belgium, and the Netherlands, where thalassotherapy was no longer a question of sea bathing, but of seawater health treatment in the strictly scientific sense of the term. France remains a key player in the thalassotherapy industry.

France reclaimed its position as the conductor in 1925 in Arcachon, with the thalassotherapy congress placed under the high patronage of the presidency and directed by Dr. Fernand Lalesque of Arcachon.

In 1928, Fleming discovered penicillin, which competed with marine treatments, as respiratory diseases became less common. The Second World War contributed to the weakening of thalassotherapy with the investigation becoming more focused on drug development.

In 1952, the Bioclimatological Institute of Cannes opened under the direction of the National Institute of Hydrology and Climatology and the Faculty of Medicine of Marseille. The 10th congress will took place again in Cannes in April 1957.

The medical indications, therefore, evolved, and the new institutes turned towards functional rehabilitation. One such example is the HélioMarin Center in Collioure, Pyrénées-Orientales, which opened its doors in 1959, following the model of Roscoff, with the addition of heliotherapy.

In 1961, professional cyclist Louison Bobet (Bobet 1999) had a car accident and suffered fractures. He discovered thalassotherapy during his recovery and subsequently created an "Institute of Thalassotherapy" in 1964, inaugurated in the presence of the Minister of Health of the time.

The Sofitel group took over the institute, and Louison and Jean Bobet then managed the Port-Crouesty in Brittany and Mijas-Golf in Andalusia.

From 1976 onwards, a clear distinction appeared between the recreational "thalasso" on the one hand and medical thalassotherapy, with its traditional indications, on the other.

The definition adopted to describe thalassotherapy was: "the combined use, in a privileged marine site, under the constant medical supervision of an establishment's physician, for preventive or curative purposes, of elements from the marine environment, such as the marine climate, seawater, algae, marine mud, sand, or any other substances extracted from the sea".

Spain: The rise of modern thalassotherapy in Spain took place in the 19th century. During this time, doctors and researchers began to scientifically study the benefits of seawater and developed specific treatments based on these discoveries.

The first thalassotherapy resort in Spain was established in Santander, Cantabria, in the 1890s. The resort attracted visitors seeking well-being and seawater-based treatments.

In the 1970s, Spain experienced a true thalassotherapy boom. Modern and luxurious centers were built along the Costa Del Sol, the Costa Brava, and the Canary Islands. These centers started offering a wide range of well-being treatments, including seaweed wraps, massages, and beauty care.

Today, Spain is one of the leading destinations for thalassotherapy in Europe. Hotels and specialized centers offer comprehensive well-being programs, combining the benefits of seawater with other therapeutic elements such as marine mud, algae, essential oils, and massages. Visitors can enjoy relaxing stays and treatments in magnificent environments along the Spanish coast.

Portugal: Thalassotherapy centers in Portugal offer a range of services, including sea baths, seaweed wraps, marine oil massages, sea water jet showers, and much more. They aim to provide relaxing, revitalizing, and therapeutic care using the beneficial properties of the ocean.

The Algarve region: This regionis home to several thalassotherapy centers (Vilamoura, Albufeira, and Lagos) that offer wellness facilities providing treatments using seawater, algae, and marine minerals. Other notable locations include Cascais and Estoril, Nazaré, and Porto Santo in the Madeira archipelago.

Italy: Thalassotherapy is quite developed in Italy: some examples are Grand Hotel Alassio, Thalassio Medical Spa (Savona, Liguria), Terme di Cervia (Cervia, Emilia Romagna), where marine peloid from Saline di Cervia are applied, Thalasso Spa Kalidria (Castellaneta, Puglia) or Rimini Terme (Rimini, Emilia Romagna).

Morocco: Thalassotherapy centers are mainly in the area of Agadir and Essaouira. Traditional Moroccan treatments, such as clay baths and essential oil-based massages, are often combined with elements of thalassotherapy.

Tunisia: Tunisia also offers numerous thalassotherapy centers along its Mediterranean coast. Places like Hammamet, Sousse, and Djerba are popular among wellness enthusiasts.

Greece: Greek islands such as Crete, Rhodes, and Corfu provide popular thalassotherapy centers. The crystal-clear waters of the Aegean Sea and the Ionian Sea are used in thalassotherapy treatments because of their healing properties.

Central and Northern countries: Central and Northern Europe also offer some thermal baths and spa options in Iceland, Norway, Sweden (Varberg and Gothembourg), and Denmark (Skagen) and often use products based on marine minerals and algae.

While the countries bordering the Black Sea may not be as well-known for their thalassotherapy centers as other maritime regions, there are still a few centers in these countries: Turkey (Trabzon, Samsun, and Sinop), Bulgaria (Varna, Albena), and Romania (Mamaia and Ephorie).

These establishments offer treatments based on the benefits of seawater and well-being therapies.

The elements that make thalassotherapy a therapy

Over the centuries, thalassotherapy has been developed and has spread to different parts of the world, particularly France, where it has become a thriving industry. This is due to the fact that the benefits of thalassotherapy are numerous and varied, and contribute to overall health.

Thalassotherapy, as defined by La Bonnardière, involves the combined use, under medical supervision, of the beneficial elements of the marine environment, including seawater, climate, mud, algae, and sand, for preventive or curative purposes.

Seawater and its properties

The composition of seawater is relatively constant across all seas and oceans. In 1884, the German chemist William Dittmar (1833–1892) studied the composition of seawater and drew conclusions from the chemical analysis of 77 seawater samples collected from different parts of the world. The relative proportions of its major constituents (with concentrations exceeding 1 mg/l) are consistent throughout all seas and oceans. The various properties of seawater and its effects on the body are briefly explained below (Ditmmar 1884).

Physical properties

Archimedes principle: According to Archimedes' principle, the buoyant force on an object immersed in a fluid is equal to the weight of the fluid displaced by the object. One practical application of this principle in thalassotherapy is the explanation of the human body's buoyancy in water, which is utilized in rehabilitation or relaxation therapies. As seawater has a higher density than freshwater, the buoyant force is stronger in a pool of seawater (especially if it is saltier) compared to a pool of freshwater (Menauhé 1967).

Pressure gradient: There is a difference in pressure between the surface of a pool and its bottom. The venous network of the human body belongs to the low-pressure system of the circulatory system. The linear pressure gradient in the veins is weak, and several physiological mechanisms assist venous return. A driving force is generated by the pressure difference between the venules, created by the left ventricle, and the central venous pressure, kept low by the right ventricular pump. This mechanism is significant in a thalassotherapy pool, where resistance and hydrodynamic thrust promote muscle relaxation, increased skin tone, and most importantly, aid in venous return (Menauhé 1967).

Chemical and mechanical properties

Seawater is composed of 96.5% pure water and 3.5% dissolved substances, including salts (chlorine and sodium being the most common components) and gases (the dissolved gases mainly comprise nitrogen at 64%, oxygen at 38%, and carbon dioxide at less than 2%—about 60 times its proportion in the Earth's atmosphere).

The gases in seawater are absorbed through contact with air during dissolution, at a rate of 15 to 30 cm^3 of gas per liter. The type of gas depends on the depth; for example, oxygen decreases with depth while carbon dioxide increases. Nitrogen, on the other hand, is more influenced by temperature than depth and is combined in the form of nitrates.

The gases present in seawater circulate very little through the skin's hair follicles (thanks to the pressure from the jets used in thalassotherapy treatments). However, note that gas exchange primarily occurs through the lungs during respiration.

The biocatalysts present in seawater, such as trace elements and minerals, are essential for the chemical reactions occurring in all living tissues, including our bodies:

- Trace elements: Iron, zinc, chromium, copper, etc., although present in the body in small proportions, are indispensable for various functions, including respiration and fighting infections. They aid in wound healing, the structure consolidation of cells and tissues, contribute to collagen production, and regulate metabolism through their role in numerous enzymatic systems.
- Minerals: These macroelements are present in large quantities in our bodies and perform critical functions. They are responsible for regulating the heart, muscles, bones, blood, and more. Macroelements like calcium and phosphorus (for skeletal mineralization and density), magnesium (for muscle relaxation and nerve transmission), sodium (for proper vascular function and nerve transmission while maintaining blood pressure), potassium (to prevent heart disorders and regulate nerve impulse transmission), among others, constitute more than 50 mg/kg of our body mass.

These biocatalysts are not directly utilized in the chemical reactions of living tissues. Their action occurs through contact with the skin or inhalation during seawater-based treatments, stimulating various physiological processes in the body (Menauhé 1967).

Electrical properties

There is a natural electric current known as the trans-epidermal current that flows between the skin's surface and its interior. This current is generated by the electrical activity of skin cells and plays a role in various physiological processes, such as regulating skin hydration and wound healing.

The skin's surface is covered by a thin layer of lipids called the hydrolipidic film. This film, primarily composed of lipids such as ceramides, fatty acids, and cholesterol, plays a crucial role in protecting the skin by forming a barrier against external aggressions, excessive water loss, and the infiltration of harmful substances. Ion channels, present in skin cells, are responsible for the selective transport of ions such as sodium, potassium, calcium, and magnesium across the cell membrane (Menauhé 1967).

Biological properties

Seawater remains "alive" for about 24 hours after being pumped.

The lifespan of phytoplankton and zooplankton outside their natural environment depends on several factors, such as specific species, environmental conditions, and preservation methods.

Phytoplankton, composed of photosynthetic microorganisms, can be relatively fragile outside their natural habitat, with a lifespan ranging from a few hours to a few days. Zooplankton, which includes a variety of animal organisms, is generally more resilient than phytoplankton. Some zooplankton, such as copepods and fish larvae, can survive longer outside their original environment.

Professor Dubarry (Dubarry 1966) presented his results at the 13th International Thalassotherapy Congress in Westerland in 1966. The trans-mineralization of chemical elements in seawater has been demonstrated to occur through the trans-follicular and trans-epidermal routes, rather than through the sweat glands.

These results were confirmed five years later in an article in "Bordeaux Médical" by Jean Bobet. This penetration through the skin is experimentally demonstrated using isotopic compounds of the same chemical nature as those contained in mineral water or seawater. The isotope follows the same path as the unmarked element, with the same fate within the organism.

The proof of trans-mineralization is established. However, it depends on the temperature conditions (at least 32°C) and the duration (at least 12 minutes per day) of bathing.

Rich in minerals: Seawater and its derivatives provide minerals and trace elements (magnesium, calcium, potassium, sodium, zinc) are essential for the proper functioning of the body, contributing to a remineralizing effect.

Purifying effect

Seawater can have purifying properties due to its salt content, especially marine salt. These salts can help cleanse the skin, eliminate impurities, and promote healing.

Anti-inflammatory effect: Magnesium and trace elements have anti-inflammatory properties. Sea water baths can be beneficial for people with inflammatory conditions such as arthritis or skin conditions.

Soothing and relaxing effect: Salty water can promote muscle relaxation, reduce stress, and improve sleep quality.

Effects on the skin: Due to its hydrating, exfoliating, and healing properties, seawater improves the appearance of the skin, promotes cellular regeneration, and helps alleviate certain skin conditions such as eczema or psoriasis.

Thalassotherapy treatments are often combined with physical activities such as swimming, aquagym, or walking by the sea. Physical exercise combined with seawater provides natural resistance, strengthening muscles, improving blood circulation, and aiding in recovery after an injury.

Another undeniable contribution of the sea is highlighted by neurosciences. In 2014, Wallace Nichols, a marine biologist from the California Academy of Sciences, examined the psychological and emotional benefits that water can offer. His book "Blue Mind" supports the idea that merely being near the sea (or lakes, rivers, or even a small fountain) can have positive effects on our mental and emotional health (Wallace 2018).

The human brain reacts positively to the presence of water, stimulating feelings of calm, relaxation, and happiness. These scientific studies emphasize the therapeutic advantages of water for individuals suffering from stress, anxiety, and depression, showcasing the deep connection between humans and water.

Marine aquatic activities are thus beneficial for both physical and mental health.

Mud and algae as poultices

The demonstration of the therapeutic qualities of algae and marine mud is now abundant, since in 1983 Bikfalvi's documented the uses of algae in the PhD Thesis untitled "Aspects biochemiques et pharmacologiques des algues fucacae" (University of Brittany 1983) (Bikfalvi 1983).

Marine mud:

In thalassotherapy, mud is used for the strong analgesic properties it exhibits when applied to joints. Throughout history, humans have used wet and malleable clay to heal their wounds, as evidenced by mutilated handprints found on the walls of certain prehistoric caves. As an example, the Kahun papyrus from 19th century BCE reveals that mud was part of the Egyptian pharmaceutical arsenal.

In France, this therapy, called pelotherapy, was instituted in Collioure in 1958.

Marine mud is a true ionic concentrate, composed of disintegrated marine fauna and flora remains. It acts through its plasticity and its ability to retain water and heat.

After mud application, a decrease in its calcium and magnesium content is observed, along with an increase in its ammonium content (a waste product of the human body) (François et al. 2005).

Algae and microalgae

In thalassotherapy, algae are applied to the skin for their mineralizing and detoxifying effects.

They are most commonly used in micronized form, mixed with seawater.

There are around 2500 species of algae around the world, but only about 20 species have been subject to in-depth scientific studies, being the most commonly used different species of the genera *Laminaria* sp. and *Fucus* sp.

France produces a little over 70000 tons of algae annually, ranking 10th worldwide. China is the leader, producing and consuming 12 million tons. 90% of the French production comes mostly from Brittany, from the coasts of North Finistère and the archipelago of Molène where 35 boats harvesting goemon bring in 65000 tons of laminaria per year.

Algae and skin tissue have the same intercellular exchange mechanisms with the surrounding environment.

Algae are a concentrate of seawater and contain minerals and trace elements in a balanced form (all are present!) as sodium, chloride, sulfate, magnesium, calcium, potassium, copper, gold, silver, etc. They also contain vitamins, carbohydrates, enzymes, and proteins in the form of amino acids.

We regularly consume algae, sometimes without knowing it. For example, alginates and agars, known as E401 to E407, are widely used in many processed foods to replace butter and cream without adding calories.

Algae provide 10 to 20 times more minerals and vitamins than land-based foods.

Researchers from the INRA, in partnership with the Olmix group (https://biotechinfo.fr/article/des-algues-marines-pour-reduire-lutilisation-des-antibiotiques-en-elevage/), showed in March 2016 that a compound extracted from green algae inhibits the growth of pathogenic bacteria in vitro and stimulates the production of immune mediators by intestinal epithelial cells.

More recent studies by IFREMER (France) demonstrated that the coronavirus does not survive in seawater.

Sand

Sand has interesting thermal properties. It can absorb and retain heat, making it an ideal material for heat therapy treatments. The use of warm sand can help relax muscles, improve circulation, and relieve pain. Walking on sand can also boost mood (Chevalier et al. 2012).

Sand Baths: Warm sand baths are typically used for their relaxing and soothing properties. Patients lie on preheated sand at an appropriate temperature. The weight and warmth of the sand can promote muscle relaxation, relieve tension, and improve blood circulation.

Exfoliation: Sand can also be used as a natural exfoliator for the skin. Gently rubbing sand on the skin can remove dead cells and promote cellular renewal, leaving the skin softer and smoother.

Climate and negative ions

The marine climate that covers the sea and a thin coastal strip possesses the physicochemical properties of seawater itself. Its physical characteristics (pressure, temperature, humidity) are more stable than those of continental air. It is slightly richer in oxygen and poorer in CO_2 than continental air due to photosynthesis by algae, and it has a higher water vapor content.

Marine air contains fewer pollutants and suspended particles than continental air because it is constantly stirred by land breezes at night and sea breezes during the day, as well as winds.

Winds in the coastal strip are stronger than on land, enriching the air with volatile chemical elements from the sea: sodium chloride, iodine, silica, and even microscopic algae.

Solar radiation is intense, direct, and facilitated by the atmosphere's purity, while indirect radiation is due to reflection from the sea and sand surfaces.

The average air temperature and wind strength are likely play a role in thalassotherapy. It is generally accepted that the Mediterranean water has a sedative effect, the Atlantic water is tonic, and the water of the English Channel is invigorating.

Marine air is rich in negative ions. Regardless of the coastline, negative ionization is a process where negative ions are present in the ambient air by the seaside. These negative ions are created by the movement of the sea and waves, as the water spray breaks molecules, charging them electrically. They consist mainly of negative oxygen and iodine ions.

Breathing air rich in negative ions can have a relaxing and beneficial effect on the body and mind by:

- Stimulating the release of serotonin in the brain; serotonin is a neurotransmitter that plays a key role in mood regulation, sleep, and well-being. Increased serotonin levels can promote a sense of calm and relaxation.
- Reducing stress levels: studies suggest that exposure to negative ions can decrease cortisol levels (the stress hormone) and increase endorphin levels (the feel-good hormones), leading to a sense of relaxation and tranquility (Jiang et al. 2018).

Negative ionization is also associated with natural environments such as waterfalls or forests. The relaxing effect is perceived differently by individuals and is probably due to the unconscious association of these ions with soothing natural environments. Negative ions can also help purify the air by attracting suspended particles such as dust, allergens, or pollutants. Breathing cleaner air can create a feeling of freshness and relaxation.

In conclusion, thalassotherapy benefits from the numerous therapeutic properties of seawater, mud, algae, and the marine climate. The combined effect of these elements contributes to the physical and mental well-being of individuals, making thalassotherapy an effective and popular therapy for various health conditions.

Thalassotherapy represents an indisputable health benefit

Health and well-being are often mentioned in relation to thalassotherapy and there is no doubt that thalassotherapy it can be beneficial. However, in medicine, health is defined as the absence of disease, but in reality it not only just that. The World Health Organization, in the preamble to its 1946 constitution, defined health as a state of complete physical, mental, and social well-being, and not merely the absence of disease or infirmity.

Health can be evaluated by medical criteria, such as biological indicators and specific examinations. Well-being is a broader concept that includes not only the absence of disease but also a state of physical, mental, and emotional satisfaction and fulfillment. Well-being focuses on the quality of life and encompasses aspects such as happiness, satisfaction, emotional balance, mindfulness, positive interpersonal relationships, personal growth, and self-realization.

It is quite challenging to scientifically measure well-being, but the effects of a thalassotherapy treatment can be measured.

Let's cite some recent studies. A study conducted at the Thermes Marins de Saint-Malo with the participation of the Biomedical Research Institute of the Armed Forces (IRBA) in June 2017 focused on various items. 11 voluntary participants aged 24 to 63, stayed 4 nights at the Grand Hôtel des Thermes and had 12 thalassotherapy treatments over 3 days. Equipment: EEG monitoring, saliva sample collection; Tests: KSS, Pichot's fatigue scale, VAS, HAD, PVT-B, POMS.

All tested parameters were improved up to 106% (Table 1) ("Thermes Marins de Saint-Malo Study/Biomedical Research Institute of the Armed Forces (IRBA)." Study published on line June 2017: www.France-thalasso.com):

The study conducted by Côté Thalasso between September 2013 and May 2018, involving 598 spa-goers with disabling knee joint pain requiring the use of pain relievers, demonstrates significant improvements in pain, functionality, and a significant decrease in medication consumption. The participants attended the spa for 6 days and received 5 days of treatments, with 4 sessions per day, including relaxing massages, marine mud wraps, aquatic gymnastics, and hydro massage in

Table 1. Results.

WELL-BEING	+31%
PAIN	–55%
RELAXATION	+88%
SLEEP	+20% of sleep time from the 1st night +27% of sleep time after 3 days of treatment +51 minutes of NREM sleep (recovery) from the 1st night +23 minutes of NREM sleep (restorative) after 3 days of treatment
STRESS AND ANXIETY	–31% score on the HAD scale
MOOD AND FATIGUE	–25% score on the KSS scale –70% score on the Pichot's fatigue scale –97% total score of mood disturbance (POMS scale)
COGNITIVE ABILITIES	–49% reduction in errors in the PVT-B test
FLEXIBILITY AND VERTICAL RELAXATION	+106% improvement in flexibility; +13% improvement in vertical relaxation

bathtubs. Transcutaneous marine Mg^{2+} was added to the treatments starting from December 2015, and the results indicate that Mg^{2+} enhances the effects of other therapies.

The evaluations were based on "self-assessment":

- FUNCTIONALITY (using the WOMAC score: an assessment of the functional impact of knee osteoarthritis on the quality of life).
- PAIN (using the VAS score: a self-assessment scale that is sensitive, reproducible, reliable, and validated for both acute and chronic pain situations).
- INCIDENCE OF MEDICATION USAGE to alleviate pain.

RESULTS AT DAY 3 OF TREATMENT Overall improvement 60%; functionality 45%; pain reduction 50%; medication consumption reduction 33%.

RESULTS AT DAY 5 OF TREATMENT Overall improvement 92%; functionality 72%; pain reduction 57%; medication consumption reduction 65%.

RESULTS AT DAY 30 OF TREATMENT Overall improvement 99%; functionality 77%; pain reduction 62%; medication consumption reduction 62%.

(Study published on line may 2018: www.France-thalasso.com)

Other recent scientific studies also reach compelling conclusions: thalassotherapy treatments improve health. For instance:

- "The therapeutic efficacy of Dead Sea mud treatment for knee osteoarthritis" (2019): This study evaluated the effectiveness of Dead Sea mud treatment for knee osteoarthritis. The results showed significant improvements in pain and joint function among the treated participants compared to the control group. (Chary-Valckenaere et al. 2018-2019) (Study published on line, https://www.grandnancy.eu/fileadmin/fichiers/_PRESSE/DOSSIERS/2018_07_DP_Grand_nancy_thermal.pdf)
- "Thalassotherapy with marine mud: A review" (2018): The authors concluded that marine mud could have positive effects on the skin, blood circulation, and muscle relaxation, among others (Dai et al. 2023).
- "Biological and therapeutic effects of marine mud: A review" (2017): The authors found that marine mud could possess anti-inflammatory, antibacterial, antioxidant, and healing properties (Demay et al. 2020).
- "Effects of peloid therapy on pain, functional status, and quality of life in patients with knee osteoarthritis" (2016): The results showed a significant reduction in pain and improvement in joint function and quality of life after the treatment (Varzaityte et al. 2020).

Several other scientific studies have reached unequivocal conclusions: thalassotherapy treatments improve health conditions, linked to its composition, as sodium and chloride can penetrate the skin (Carbajo and Maraver 2018).

Thalassotherapy is not limited to adults. Although adults make up the majority of spa-goers, many centers also offer programs adapted to children and adolescents, offering playful and educational activities focused on discovering the marine environment and learning healthy lifestyles.

The rise of the so-called diseases of civilization (cancer, obesity, diabetes) and the emergence of more recent health issues related to technology usage and pollution are leading to an increased demand for targeted treatments.

Since the health crisis, thalassotherapy centers have seen an influx of requests for stays to treat and soothe various health issues. Sedentary lifestyles, lack of activities, limited social contacts, disruption of daily routines, and the use of technology have affected our physical and emotional balance. Many consequences on health have been observed, such as weight gain, physical pain, depressive moods, disrupted biological rhythms, and various sleep disorders.

As a result of an increasing fast-paced life, of feeling under pressure in personal, family, or professional lives, there is a growing need for recovery time. Practices such as slowing down, savoring the moment, temporarily disconnecting from technological devices, and improving sleep have become more prevalent and require guidance and specific environments. The ways of seeking treatment are evolving towards more autonomy and personalization, as evidenced by the emergence of numerous electronic applications allowing continuous health monitoring and even self-medication. In this context, thalassotherapy is opening its doors to a world of gentle medicines often associated with sea water and its derivatives, such as relaxation, meditation, and yoga.

The various thalassotherapy treatments

Thalassotherapy centers offer a wide range of health and well-being benefits: these benefits are both physical and psychological.

Treatments are typically combined in packages, in groups of three or four treatments, thereby creating programs to meet medical expectations and increasingly to allow a more holistic approach, addressing well-being and the pleasure of self-care.

The programs are varied: Health and Prevention, Fitness, Relaxation and Stress Relief, Beauty and Anti-Aging, Weight Loss and Nutrition, Sports Enhancement, Special for Moms, Special for Men, Discovery for Adolescents…

Water treatments

Water treatments are the very essence of thalassotherapy, and they were the first treatments to be implemented in all thalassotherapy centers, always using natural seawater (more information can be found in Chapter 5).

The development of increasingly specific water treatments in thalassotherapy has been gradual over time, concurrent with the development of treatments in thermal spas.

Sea baths "in-door": these treatments are carried out in specially designed tubs. Seawater jet baths, air bubble baths, foot or hand baths.

Aquatic circuits: Simple or against the current, more recently pebble walking.

Jet showers: Whether gentle underwater (in a tub or pool) or invigorating non-immersed showers.

Affusion showers: Also known as Vichy showers. Interestingly, we owe them to Monsieur Berthe in 1893, the originator of the first hydrotherapy service at the Vichy Thermal Establishment, who introduced this innovation.

Other specific water treatments are inhalation of sea water aerosols, and nasal rinsing with sea water.

Wraps: The first therapeutic wraps were made of marine mud and clay, dating back to ancient times.

Algae wraps have also been recognized and traditionally used for centuries in Asia. Their use in the Western world is a more recent practice that has developed in recent centuries, where the exploitation and utilization of this resource is not yet as advanced. Algae are also used in form of pastes, poultices or cataplasm, which can be commercialized as a powder or ready to use (mixed with water and other ingredients), promotes detoxification and remineralization.

Marine algae wraps are rich in minerals, trace elements, and vitamins. They are used to stimulate blood circulation, promote skin detoxification, and deeply nourish the skin. Various types of algae can be used, including bladderwrack, kelp, and lithothamnium. Red algae, rich in antioxidants, are sometimes used for revitalizing and moisturizing wraps.

Marine clay is also used in thalasso centers. Marine clay is mixed with seawater to form a paste applied to the body. It is appreciated for its purifying and detoxifying properties, as well as its soothing effect on the skin.

Marine mud contains minerals and beneficial salts for the skin. Mud wraps can be either warming or refreshing, depending on the individual's needs. They are often used to relieve muscle and joint pain.

Other wraps are green clay and marine salt wraps. Although not specifically marine mud, green clay is sometimes used in thalassotherapy because of its purifying properties. It can be mixed with seawater or other ingredients to create a beneficial skin wrap.

Salt wraps are exfoliating and revitalizing, stimulating blood circulation and removing dead skin cells, leaving the skin softer and brighter. Although Epsom salts are not of marine origin, they are sometimes included in thalassotherapy treatments for their soothing and relaxing properties.

Mechanical and other technological treatments

In recent decades, particularly since the 2010s, the use of modern technologies and advanced medical equipment in thalassotherapy has become more common. Technological and medical advancements have allowed the integration of these techniques with traditional seawater-based treatments, thereby enhancing treatment capabilities and providing a more comprehensive well-being experience for clients.

The equipment found in thalassotherapy centers is diverse and includes techniques such as pressotherapy, ultrasound therapy, oxygen therapy, and cryotherapy.

Some centers maintain a focus on traditional treatments, while others actively incorporate these more modern technologies to meet the needs and preferences of contemporary clients.

Massages

Massages have a long history as a method of therapy, and their formal incorporation into modern thalassotherapy gained momentum in the 19th century, evolving over time to become a standard practice in thalassotherapy centers.

These massages were exclusively performed by specialized physiotherapists trained in rehabilitation and manual medicine. Thalassotherapy was considered a distinct form of medical treatment, and physiotherapists played a crucial role in the care process, especially for rehabilitation, reconditioning, and massage treatments.

Over time, and particularly since the 2000s, in several countries, including France, therapeutic massages performed by physiotherapists have been complemented in many centers by a variety of massages designed to meet clients' well-being and relaxation needs.

Here are some of the types of massages commonly offered in thalassotherapy, which are not always performed by physiotherapists: Swedish massages, Californian massages, Thai massages, lymphatic drainage, slimming massages, and massages using tools like bamboo, hot stones, or herbal compresses.

The new clients or spa-goers seek relaxation "à la carte" and not just treatment for potential medical conditions. This is also why beauty treatments have been introduced in thalassotherapy, significantly expanding the range of services offered over time.

Complementary medicine

Thalassotherapy centers increasingly complement their offerings with naturopathy, aromatherapy, reflexology, sophrology, yoga, meditation, acupuncture, and Ayurveda. Thalassotherapy treatments are combined with relaxation sessions, specific nutrition, and other therapies to provide a holistic well-being and relaxation experience in a maritime environment.

The new concepts of thalassotherapy: pleasure and protection of marine ecosystem

A newcomer to the realms of health and well-being is the notion of PLEASURE, which has become an essential lever in the thalassotherapy market, closely tied to the ideas of preventive and curative health.

Thalassotherapy centers are often situated in picturesque coastal environments, offering panoramic views of the ocean and a setting that promotes the therapeutic qualities of the sea (d'Erm 2021):

- It provides a comforting respite, with the sea perceived as a remedy for anxiety, offering relaxation and a true source of happiness.
- It brings back the pleasure of childhood, a return to the source, to the essential, and to purity. Reconnecting with oneself by rekindling childhood sensations. Visiting the sea is a need, and all seasons are conducive to it.
- The sea is considered a natural and effective medication against stress, anxiety, burnout, and depression.
- It rejuvenates and energizes: The sea has invigorating virtues. Hyperactive individuals find the environment conducive to peak performance. Those experiencing diminishing physical abilities perceive the promise of renewed vitality.
  The sea is a source of happiness; it does good and allows people to disconnect.

Respect for nature

Respect for nature and particularly for seawater and the coastline, is gaining importance among thalassotherapy stakeholders. It is essential to preserve our environment and ensure the sustainability of marine resources, the protection of the marine ecosystem, and its biodiversity.

By integrating environmental preservation and social commitment into their practices, thalassotherapy centers contribute to nature preservation while offering quality care to clients. Thalasso therapies are mostly in line with nature and adhere to Corporate Social Responsibility (CSR) contracts:

- Thalassotherapy centers ensure the responsible use of seawater and its derivatives (algae, mud, salts), avoiding overexploitation or ecological damage. Measures for conserving water and energy are also put in place.
- They protect the marine ecosystem by ensuring that their thalassotherapy activities do not negatively affect it. This involves respecting protected areas, avoiding the release of harmful substances into port waters, and promoting marine biodiversity.
- The use of natural and organic products in thalassotherapy treatments is preferred, contributing to reducing environmental impact and minimizing clients' exposure to chemicals.
- Thalassotherapy centers can engage socially by favoring local employment, supporting community initiatives, and contributing to the region's sustainable economic development. Some centers, particularly in France, implement training and employment programs for local populations.

- Thalassotherapy centers play an important role in raising public awareness about environmental issues and the preservation of marine resources. They often organize educational activities, conferences, and workshops to inform clients about the importance of sustainability and environmental protection.

To demonstrate their commitment to sustainability and social responsibility, thalassotherapy centers sometimes seek recognized environmental certifications or labels. These certifications assure clients that the establishment follows environmentally friendly practices.

Conclusion

Thalassotherapy offers a natural and holistic approach to health and well-being, utilizing the benefits of seawater and marine elements to promote regeneration, relaxation, and healing. Its physical, chemical, and environmental properties make it an attractive option for those seeking to improve their quality of life and take care of their health. Whether to alleviate muscle pain, improve skin health, or simply rejuvenate, thalassotherapy offers a unique experience of well-being and reconnection with nature.

However, the main challenge lies in recognizing the properties of seawater used as thermal-mineral water, as it encompasses them all. This abundance leads to the multiplicity of thalassotherapy indications and, consequently, to confusion, unlike thermalism, which focuses on specific indications assigned to each station based on the thermo-mineral properties of its waters.

The sea remains a great unknown that worries modern humans, much like Neptune concerned his predecessor thousands of years ago.

Nevertheless, the scientific community largely agrees that the medicines of tomorrow are likely hidden in seaweeds and marine bacteria. This is the field of investigation for some teams in Biotechnologies and Bioresource Chemistry for Health. Their objective is to biosource, i.e., identify in the marine natural environment the molecules that could be the basis for future drugs, combining the expertise of biochemists, organic chemists, molecular biologists, and enzymologists.

Let us mention the CNRS Mixed Unit 7266 LIENS, directed by Dr. Ingrid Arnaudin in La Rochelle, France.

Let us also acknowledge the work of the teams at the Oceanological Observatory (Boeuf and Kornprobst Jean-Michel 2009) of Banyuls-sur-Mer, a component of the Faculty of Sciences and Engineering of Sorbonne University, studying marine biodiversity's specificities, adaptation to the environment, the role of water in living organisms, and marine living resources.

References

Alberola, J. and F. Coll. 2013. Marine therapy and its healing properties. Curr. Aging Sci. 6(1): 63–75.

Bikfalvi, A. 1983. Aspects biochemiques et pharmacologiques des algues fucacae. These de Doctoral de Medicine. (in French).

Bluteau, A. 1931. Emploi thérapeutique des eaux potables et des eaux minérales dans l'antiquité Gréco-latine. PhD Th. Médecine Bordeaux, n° 61 - 1931-32.

Bobet, J. 1999. Une histoire de la Thalassothérapie Presse Thermale et Climatique, 136, (n° 4), pp. 188–190 (XXIe Congrès International de Thalassothérapie à Hammamet, Tunisie).

Boeuf, G. and J.M. Kornprobst. 2009. Marine Biodiversity and Chemiodiversity. Biofutur 28(301) .

Carbajo, J.M. and F. Maraver. 2018. Salt water and skin interactions: New lines of evidence. Int. J. Biometeorol. 62: 1345–1360. https://doi.org/10.1007/s00484-018-1545-z.

Chary-Valckenaere, I. "Etude Clinique, gonarthose" Grand Nancy Thermal. 2018–2019 (https://www.grandnancy.eu/fileadmin/fichiers/_PRESSE/DOSSIERS/2018_07_DP_Grand_nancy_thermal.pdf) (in French).

Chary-Valckenaere, I., D. Loeuille, N. Jay, F. Kohler, J.N. Tamisier, C.F. Roques et al. 2018. Spa therapy together with supervised self-mobilisation improves pain, function and quality of life in patients with chronic shoulder pain: a single-blind randomised controlled trial. Int. J. Biometeorol. 62(6): 1003–1014. doi: 10.1007/s00484-018-1502-x.

Chevalier, G., S.T. Sinatra, J.L. Oschman, K. Sokal and P. Sokal. 2012. Earthing: health implications of reconnecting the human body to the earth's surface electrons. J. Environ. Public Health 291541. doi: 10.1155/2012/291541.
d'Erm, P. La santé bleue, la nouvelle médecine de l'eau. Ed. Massot, 2021.
Dai, D., X. Ma, X. Yan and X. Bao. 2023. The biological role of dead sea water in skin health: A review. Cosmetics, 10, 21. https://doi.org/10.3390/cosmetics10010021.
Demay, J., S. Halary, A. Knittel-Obrecht, P. Villa, C. Duval, S. Hamlaoui et al. 2020. Anti-inflammatory, antioxidant, and wound-healing properties of cyanobacteria from thermal mud of Balaruc-Les-Bains, France: A Multi-Approach Study. Biomolecules 11(1): 28. doi: 10.3390/biom11010028.
Des algues marines pour réduire l'utilisation des antibiotiques en élevage. Retrieved from: BIOTECH.INFO: groupe Olmix, INRA MARS 2016 (in French).
Dittmar, W. 1884. Report on researches into the composition of ocean water, collected by. H. M. S. Challenger, during the years 1873–1876. Voy. Challenger, Physics and Chemistry 1(1): 1–251.
Dubarry, J.C. 1966. Les diverses zones climatiques de la côte atlantique du Sud-Ouest, leur intérêt climatothérapique. Société d'Hydrologie. Séance du 21 novembre 1966. (in French)
Flórez, D.A., Bernabé, B.V. El agua de mar en la alimentación y en la terapéutica. Bol. Soc. Esp. Hidrol. Med. 2015; 30(1): 37–55. (in Spanish)
François, G., A. Micollier and I. Rouvie. 2005. Les Boues Thermales. Atelier Santé Environnemental. ENSP, 2005 (https://www.academie-medecine.fr/wp-content/uploads/2017/03/6-BOUES-ENSP-2005.pdf) (in French).
https://france-thalasso.com/la-thalasso/thalasso-histoire/
Hoareau, D. 2011. Thalassotherapie: vitalité, forme, prevention, santé. Paris, Ed. Chiron.
Jiang, S.Y., A. Ma and S. Ramachandran. 2018. Negative air ions and their effects on human health and air quality improvement. Int. J. Mol. Sci. 28; 19(10): 2966. doi: 10.3390/ijms19102966.
Klarić, I.V, S. Jurdana and I.G. Klarić. 2007. Hippocrates's Thalassotherapy from ancient times to a moder thalassotherapeutic in Crikvenica (Croatia). Acta med-hist Adriat 5(1): 125–132.
La Bonnardière, 1885. Introduction à la thalassothérapie (https://france-thalasso.com/la-thalasso/thalasso-grands-hommes/#docteur-la-bonnardiere (accessed, 15th May, 2023). (in French).
Menauhé, M. 1967. Du problème de la masse volumique de l'eau. Metrologia 3: 58. DOI 10.1088/0026-1394/3/3/002. (in French).
Obel, P. 1984. La santé par la thalassothérapie. París: MA Editions (in French).
Sakula, A. 1995. Doctor Brighton: Richard Russell and the sea water cure. J. Med. Biogr. 3(1): 30–3.
San José, J.C. 2004. René Quinton: centenario de "L'eau de mer, milieu organique". Bol. Soc. Esp. Hidrol. Med. 19(1): 17–22. (in Spanish).
San Martin, J. 1995. Talasoterapia: proyección terapéutica actual. An. R. Acad. Nac. Med. 112(2): 347–79 (in Spanish).
Thermes Marins de Saint-Malo Study/Biomedical Research Institute of the Armed Forces (IRBA). Study published on line June 2017: www.France-thalasso.com
Tréguer, Y. 2002. La Thalassothérapie, idées reçues. Paris, Ed. Le Cavalier Bleu (in French).
Varzaityte, L., R. Kubilius, L. Rapoliene, R. Bartuseviciute, A. Balcius, K. Ramanauskas et al. 2020. The effect of balneotherapy and peloid therapy on changes in the functional state of patients with knee joint osteoarthritis: a randomized, controlled, single-blind pilot study. Int. J. Biometeorol. 64(6): 955–964. doi: 10.1007/s00484-019-01785-z.
Wallace, J.N. 2018. Blue Mind: How Water Makes You Happier, More Connected and Better at What You Do. Ed. Abacus.

APPENDICES

Appendix 1

Figure 1. The Ebers Papyrus.

Appendix 2

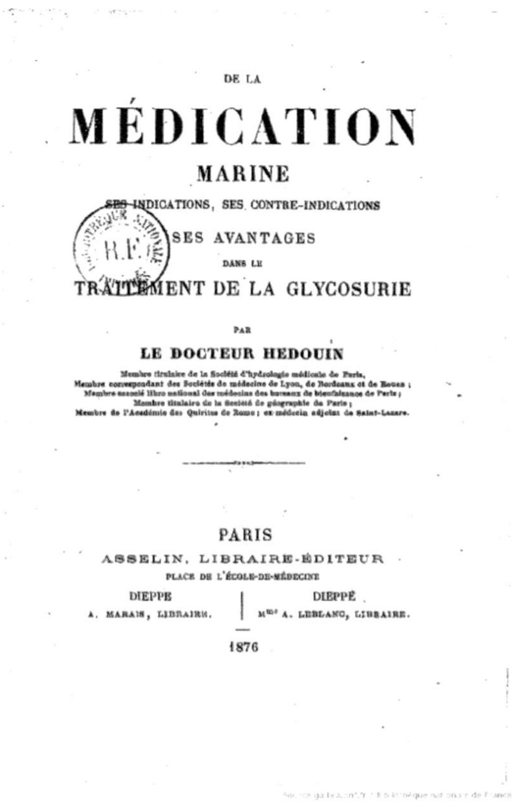

DE LA

MÉDICATION

MARINE

SES INDICATIONS, SES CONTRE-INDICATIONS

SES AVANTAGES

DANS LE

TRAITEMENT DE LA GLYCOSURIE

PAR

LE DOCTEUR HEDOUIN

Membre titulaire de la Société d'hydrologie médicale de Paris,
Membre correspondant des Sociétés de médecine de Lyon, de Bordeaux et de Rouen ;
Membre associé libre national des médecins des bureaux de bienfaisance de Paris ;
Membre titulaire de la Société de géographie de Paris ;
Membre de l'Académie des Quirites de Rome ; ex-médecin adjoint de Saint-Lazare.

PARIS

ASSELIN, LIBRAIRE-ÉDITEUR

PLACE DE L'ÉCOLE-DE-MÉDECINE

DIEPPE — A. MARAIS, LIBRAIRE.

DIEPPE — Mlle A. LEBLANC, LIBRAIRE.

1876

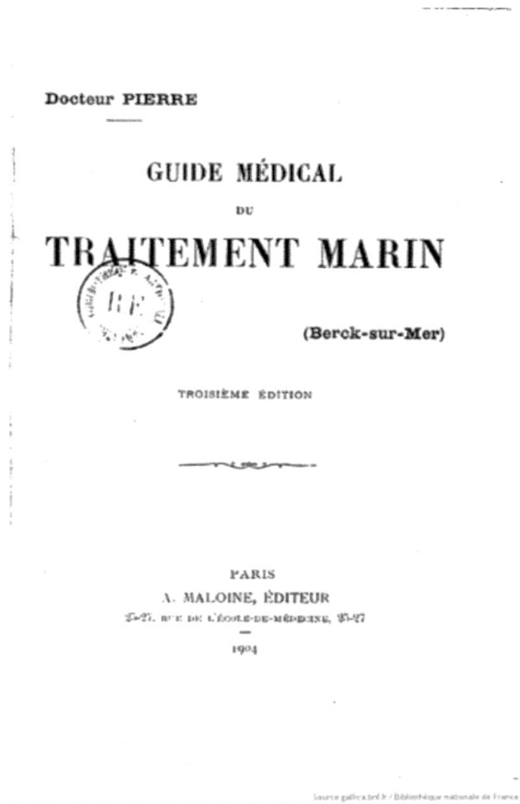

Docteur PIERRE

GUIDE MÉDICAL

DU

TRAITEMENT MARIN

(Berck-sur-Mer)

TROISIÈME ÉDITION

PARIS

A. MALOINE, ÉDITEUR

25-27, RUE DE L'ÉCOLE-DE-MÉDECINE, 25-27

1904

Figures 2 and 3. Cover Page of Books : Guides to Marine Medicine, 1876 and 1904.

Chapter 2
Seawater
Composition, Physical and Chemical Properties

*José Luis Legido Soto** and *Carmen P. Gómez Pérez*

Introduction

Seawater is found in oceans, seas, and salt lakes, which cover more than 70% of the surface of planet Earth. Seawater is a complex mixture with a high percentage of water, salts and smaller amounts of other types of substances, including dissolved organic and inorganic materials, particles, and atmospheric gases. Around eighty-eight chemical elements have been found in seawater.

The salt content in seawater is indicated by salinity (S) and it varies from around 35 g/kg in the oceans to approximately 330 g/kg in the Dead Sea. The most abundant components of seawater are chloride and sodium, potassium, magnesium, sulphate, and calcium can also be mentioned; other components are also found in smaller proportions. These are of great importance in applications in the field of thalassotherapy and cosmetics.

From the chemical point of view, the behavior of the pH is a subject of study; it is slightly alkaline and can oscillate between 7.5 and 8.4 depending on the temperature and salinity. This last characteristic of chemical types (salinity) is of great importance in cosmetic and thalassotherapy applications, for which reason it will be the subject of a special analysis.

When talking about the compounds of oceans and other masses of salt water, it is necessary to mention an important part such as the microbiological part. Among these, it is worth highlighting phytoplankton, which are a set of organisms that float in bodies of water and are mainly composed of unicellular algae and microscopic aquatic plants. Phytoplankton is a basic link in the food chain in the oceans and in marine ecosystems.

The aquatic ecosystem obtains much of its energy through phytoplankton. And this is regulated by different factors such as nutrients, temperature, light, and other environmental factors, which are of great importance for its adaptation to the environment. Algal productivity accounts for approximately 45% of global gross primary productivity (Neeley et al. 2018). Phytoplankton is essential for life in the oceans and for the balance of global climate. Its role in producing oxygen,

FA2 Group. Department of Applied Physics. University of Vigo.
* Corresponding author: xllegido@uvigo.gal

sequestering carbon, and feeding marine life makes it a critical component of marine ecosystems and a topic of interest in conservation and scientific research.

Most of the characteristics of seawater correspond to those of water in general, due to their common chemical and physical structures. Some of the different properties of seawater, for example, electromagnetic properties, viscosity, density, and melting and boiling points are due to its salinity. It is important to analyze all the properties, fundamentally and consider their behavior with salinity and temperature during the intervals in which sea water is most frequently used. Among the physical properties, thermal properties will be studied (specific heat, thermal conductivity, thermal diffusivity, boiling point, freezing point, latent heat...); other properties that will be studied are density, viscosity, surface tension, optical properties (behavior of electromagnetic waves, reflection index, refractive index, absorption...), sound transmission speed, electrical conductivity, dielectric permittivity, among others.

Seawater has a recognized action in the treatment of eczema, dermatosis, psoriasis, nasopharyngeal inflammations, conjunctivitis, vaginitis, and other infections of the external genital organs. In cosmetology, it has also been used with good results to hydrate the skin and improve its elasticity. In addition, it helps regulate sebaceous production at the level of hair follicles and scalp flaking (dandruff). It also has an important role in the absorption of saline and metallic ions, favoring the excretion of toxic residues and certain oxygenation of the tissues (Pereira 2017).

Composition

Evolution of the composition in geological time

The chemical composition of seawater has evolved through the geological times of planet Earth (Komiya et al. 2008). In recent years, and from centuries and millennia, the composition has not shown substantial changes. Figure 1 shows the evolution of salt fluxes into the ocean in the last 150 million years (Hansen and Wallmann 2003). From 1950 to the present, the surface salinity of the

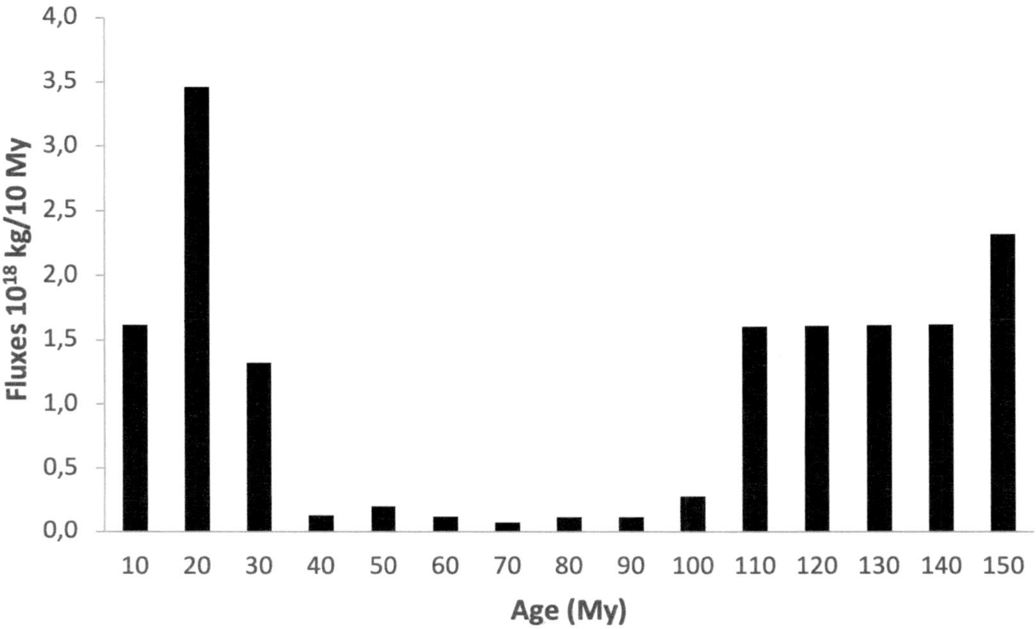

Figure 1. Evolution of salt fluxes into the ocean in the last 150 million years (Hansen and Wallmann 2003).

oceans has intensified (Gould and Wallmann 2021); in general, areas where salinity has increased have become are larger (Kennedy 2014). This change might be due to the increase in the global mean sea surface temperature.

Variation of the composition in oceans, seas, and salt lakes

Composition also changes in the different parts of the oceans, seas and salt lakes. Salinity varies in the different oceans (Figure 2), so on average the salinity of the Atlantic Ocean is 35.5 g/kg, while in the Pacific Ocean it is 34 g/kg. The salinities of the seas (Figure 3) are also different, such as the Mediterranean Sea, has an average salinity of 39 g/kg. On the contrary, salinity in the Baltic Sea is around 6 g/kg. In the inland seas and salty lagoons (Table 1) the salinities can reach very high values as in the Dead Sea which has a alinity average of 330 g/kg.

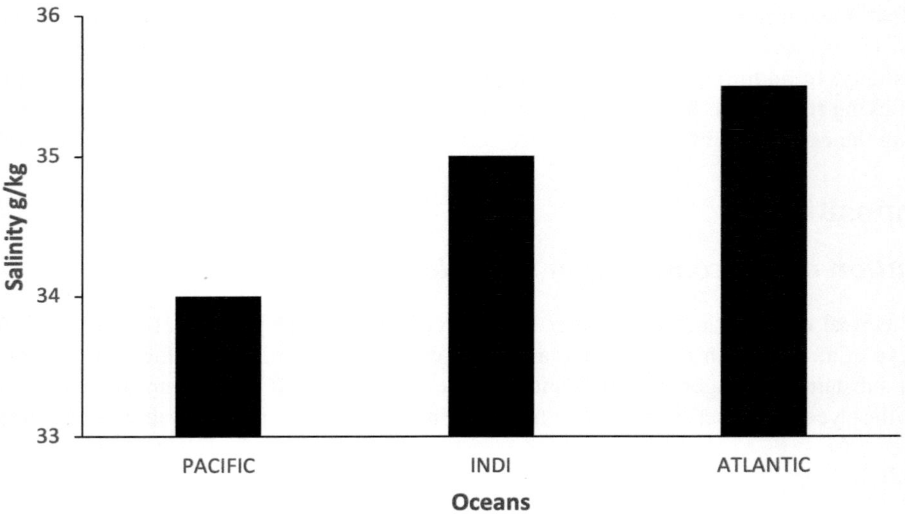

Figure 2. Salinity in the different oceans.

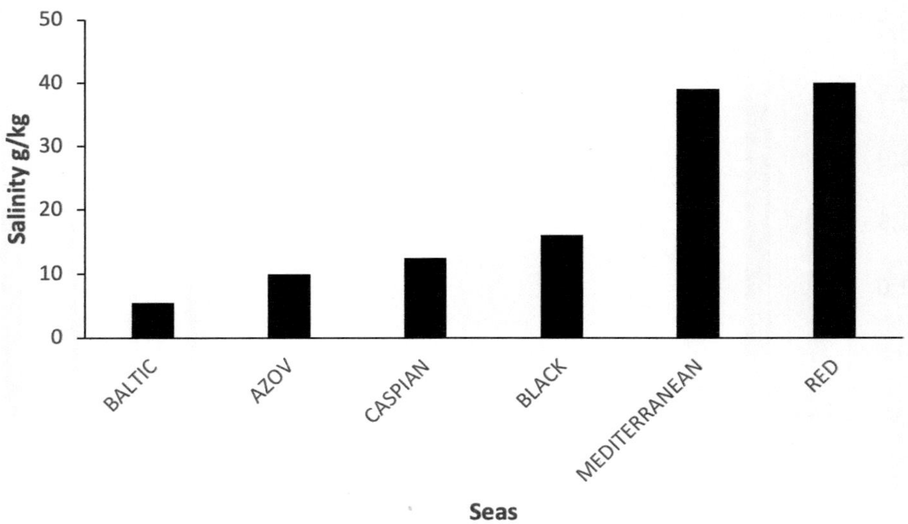

Figure 3. Salinity of the different seas.

Table 1. Salinity of inland seas and salty lagoons.

List of water bodies by salinity	g/kg
Retba (Senegal)	400
Garabogazköl (Turkmenistan)	350
Dead Sea (Israel, Jordan)	330
Tuz (Turkey)	324
Great Salt Lake (USA)	317
Baskunchak (Rusia)	300
Sarat (Rumania)	300
Urmia (Iran)	180
Litle Manitou (Canada)	180
Pikrolimni (Greece)	153
Albert (USA)	120
Vanda (Antarctica)	75
Hamelin Pool (Australia)	65
Great Bitter (Egipt)	43
Natron (Tanzania)	35
Van (Turkey)	23
Quinghai (China)	14
Mar Chiquita (Argentina)	60

From the therapeutic and cosmetic point of view, the different salinities can explain the different flotation and minority and trace elements should be taken into consideration, as in the case of the Dead Sea, which is the salt sea most used to prepare muds and cosmetics (Kamitsou et al. 2018).

Variation of the composition with depth

Another important aspect is the chemical composition and its variation related to the depth of the ocean. In general, the composition varies very little with depth, except for dissolved atmospheric gases, which decrease with increasing depth, as well as some elements such as Ca, which is more abundant at greater depths. Another ion that varies is bicarbonate, due to changes in the presence of CO_2. What does vary with the depth in the ocean is the salinity depending on the geographical situation and the surface temperature (Talley 2002). Figure 4 shows the behaviour of salinity with depth in the South Atlantic Ocean, which fluctuates up to 2000 m and then stabilizes.

At present there are numerous publications on the composition of seawater, Millero et al. 2008 carried out a detailed study on the composition of seawater and the so-called standard seawater. The first studies were performed by the prestigious scientist Edmond Halley who published a book on salt deposits in the oceans in 1715 (Deming 2021), but we can say that the first chemical composition data was published in 1864 by Dittmar (Dittmar 1984), in which the proportions of the different ions were presented. Lyman and Fleming (1940) carried out a meticulous quantitative analysis of the components of seawater. Today we know that seawater contains, in addition to ions, gases, and particles, other inorganic and organic substances not dissolved in a suspended state, as well as microorganisms (microalgae, cyanobacteria, bacteria, etc.).

There are different ways to quantitatively show the composition of seawater since they can be considered chemical elements, ions, or compounds. We are not going to consider the contribution

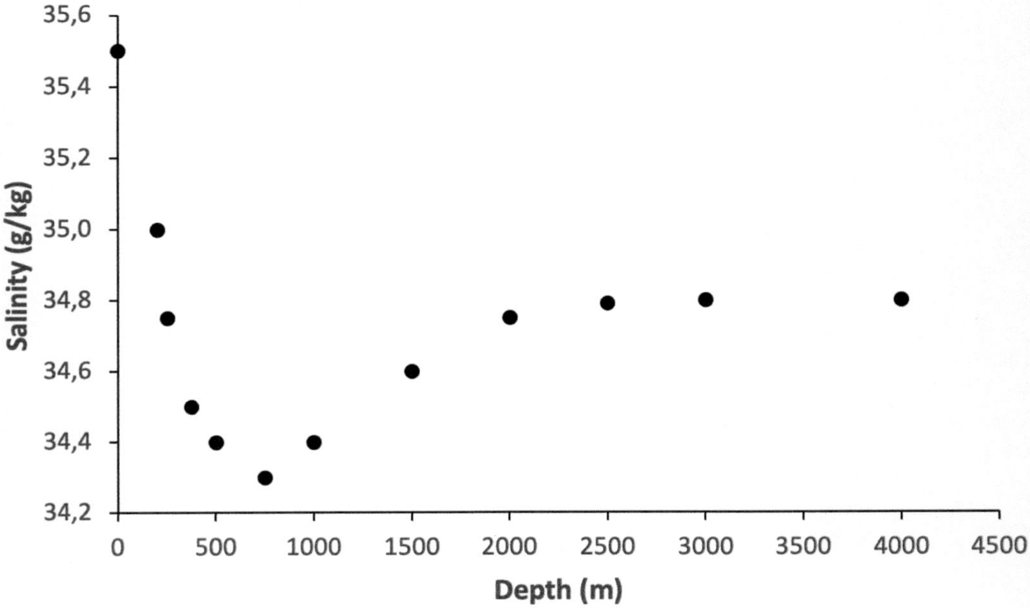

Figure 4. Salinity with depth for the South Atlantic Ocean.

of suspended particles and neither the contribution of compounds nor of the elements present in microorganisms. In this chapter, we will focus on the components in a solution considering a standard composition of seawater, since the proportions may vary slightly depending on the temperature, geographical location, and depth.

Elemental composition

The chemical elements present in seawater exceed 80 and range from g/L to values below pg/L. There are two elements with the greatest presence because they are part of the water molecule that represents more than 70% in percentage terms; thus 857 g/L corresponds to the oxygen element and 108 g/L to the hydrogen element. Figure 5 shows those most abundant elements after oxygen and hydrogen that are found in greater amounts than 1 g/L, chlorine and sodium being the most abundant, and magnesium is also found in smaller amounts. Figure 6 shows the most abundant elements in amounts of mg/L above 10 mg/L, with sulphur being the most abundant in this interval followed by calcium and potassium; below 30 mg/L are bromine and carbon. Figure 7 shows the elements in amounts of mg/L above 0.1 mg/L, with strontium being the most abundant element, followed by boron, silicon, and fluorine, below 1 mg/L and in that range, argon, lithium, and rubidium are also found.

In Figure 8 the elements are shown in amounts between 100 mg/L and higher than 10 mg/L; in this range are phosphorus and iodine in amounts around 60 mg/l; below 50 mg/l are indium, barium, and molybdenum. Figure 9 shows the abundance of elements between 1 and 5 mg/L where zinc, arsenic, uranium, vanadium, aluminium, and iron are also present.

In the interval between 1 and 600 ng/L, there are 30 chemical elements. Figure 10 shows the elements between 100 and 600 ng/L where the most abundant is nickel with 560 ng/L, with 300 ng/L are cerium and chromium, followed by copper, antimony, krypton, selenium, and manganese; lastly below 200 ng/L are neodymium, cadmium, and tungsten. Figure 11 shows the interval between 10 and 100 ng/L where elements such as germanium, xenon, silver, zirconium, gallium, mercury,

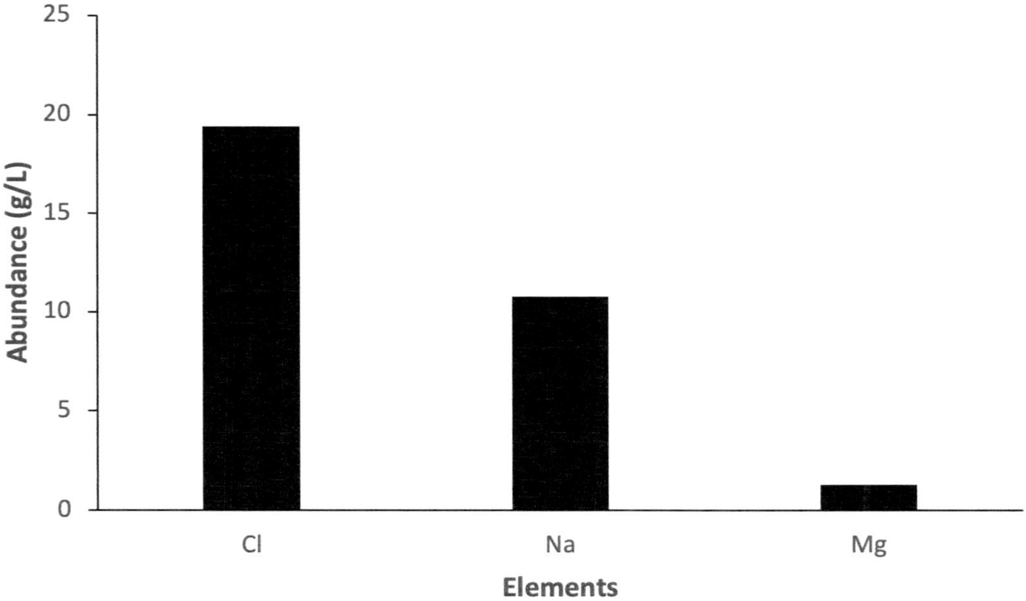

Figure 5. Abundance of chemical elements in the interval 1 – 25 g/L (Handbook of Chemistry and Physics, 89th Edition).

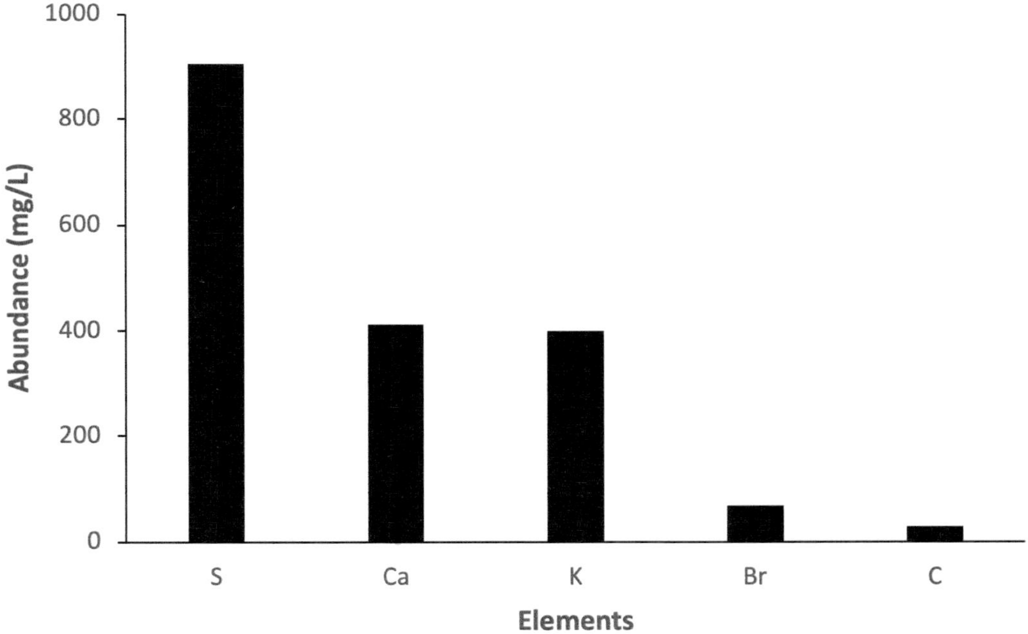

Figure 6. Abundance of chemical elements in the interval 10 – 1000 mg/L (Handbook of Chemistry and Physics, 89th Edition).

cobalt, bismuth, thallium, yttrium, and niobium are found. In the last interval of the ng/L area and between 1 and 10 ng/L (Figure 12) we find beryllium, gold, tin, lanthanum, neodymium, tantalum, cesium, and thorium. Below 1 ng/L there are 16 more elements, ending in radon with an abundance of around 0.6 fn/L.

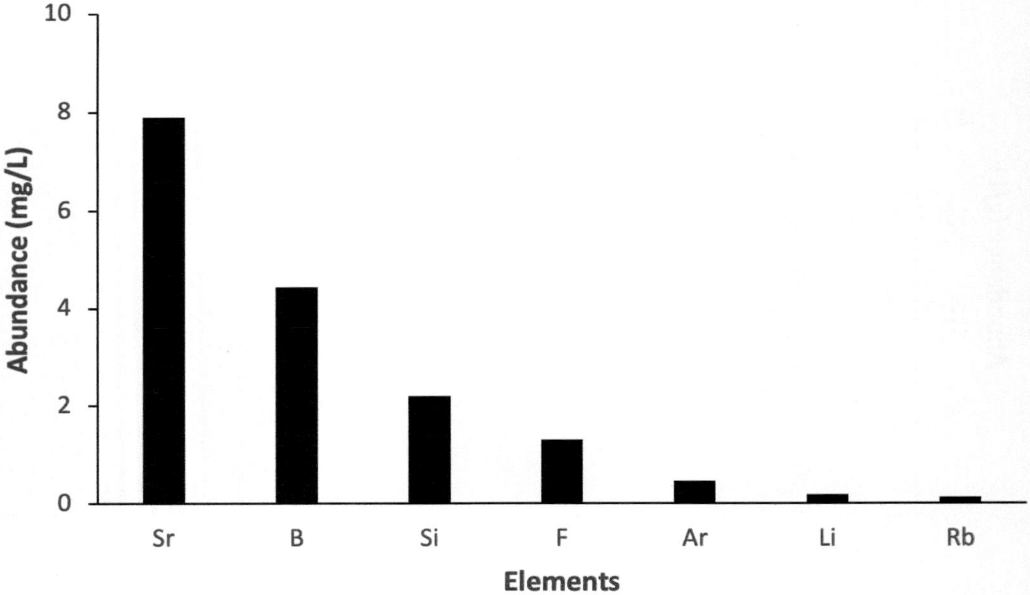

Figure 7. Abundance of chemical elements in the interval 0.1 – 10 mg/L (Handbook of Chemistry and Physics, 89th Edition).

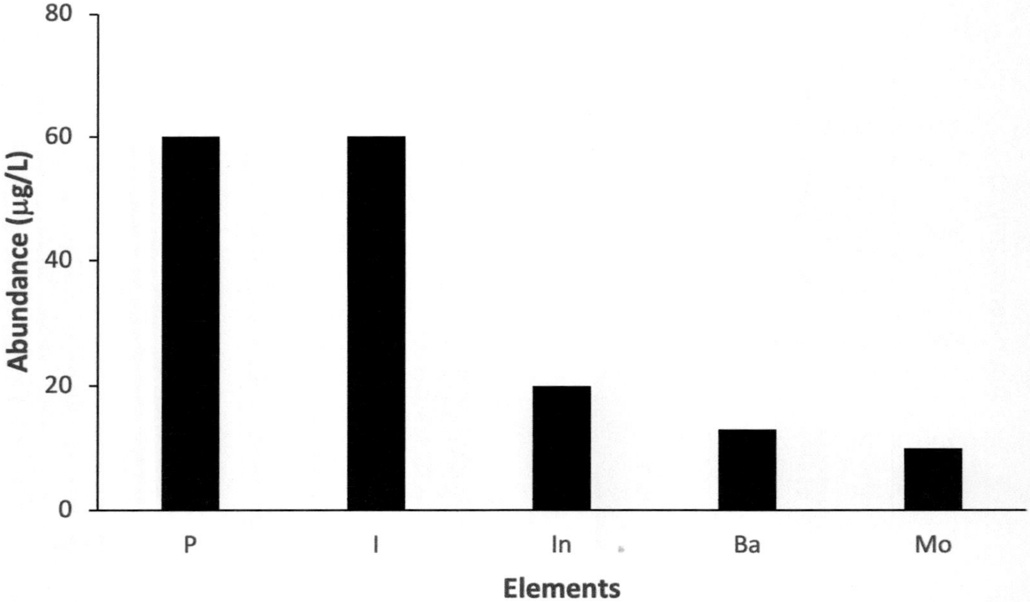

Figure 8. Abundance of chemical elements in the interval 10 – 100 mg/L (Handbook of Chemistry and Physics, 89th Edition).

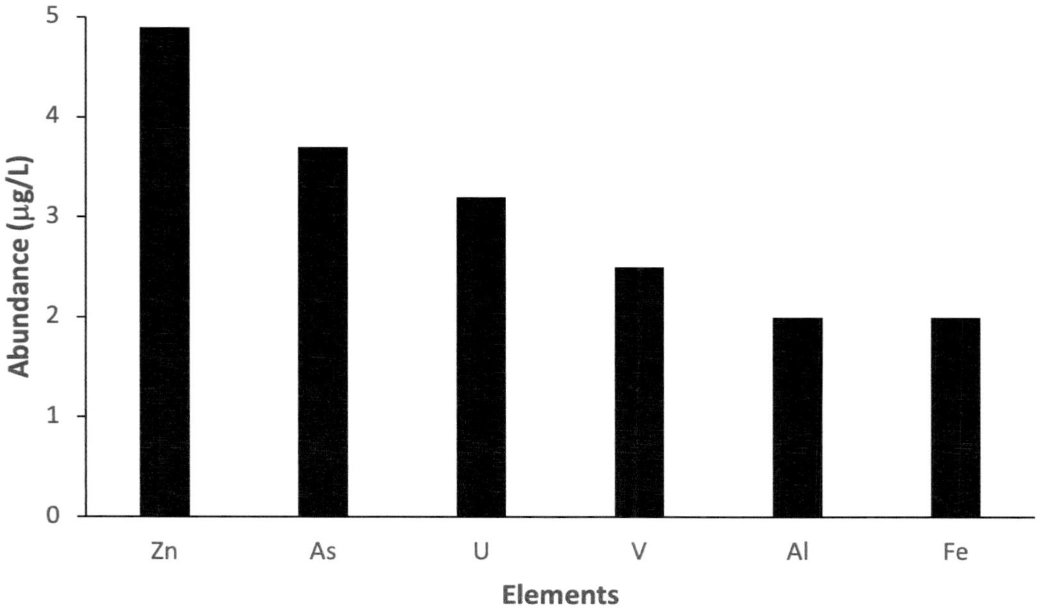

Figure 9. Abundance of chemical elements in the interval 1 – 5 mg/L (Handbook of Chemistry and Physics, 89th Edition).

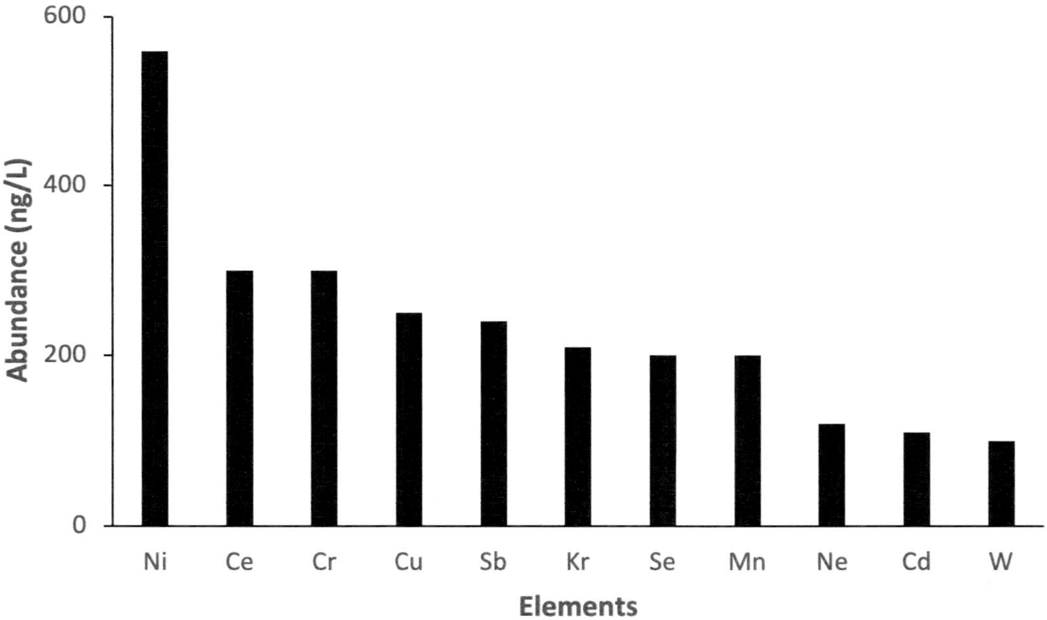

Figure 10. Abundance of chemical elements in the interval 100 – 600 ng/L (Handbook of Chemistry and Physics, 89th Edition).

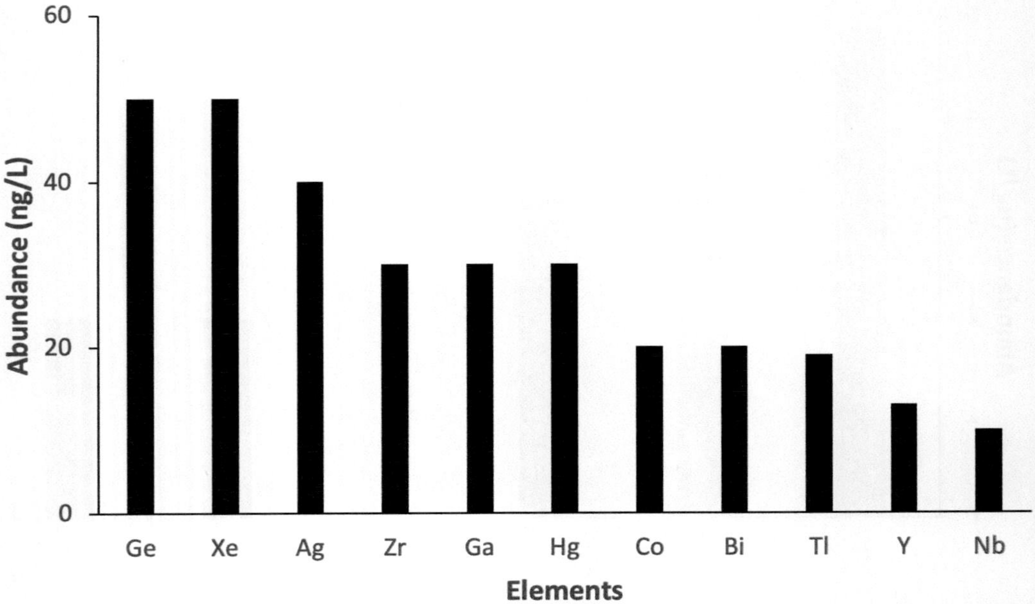

Figure 11. Abundance of chemical elements in the interval 10 – 100 ng/L (Handbook of Chemistry and Physics, 89th Edition).

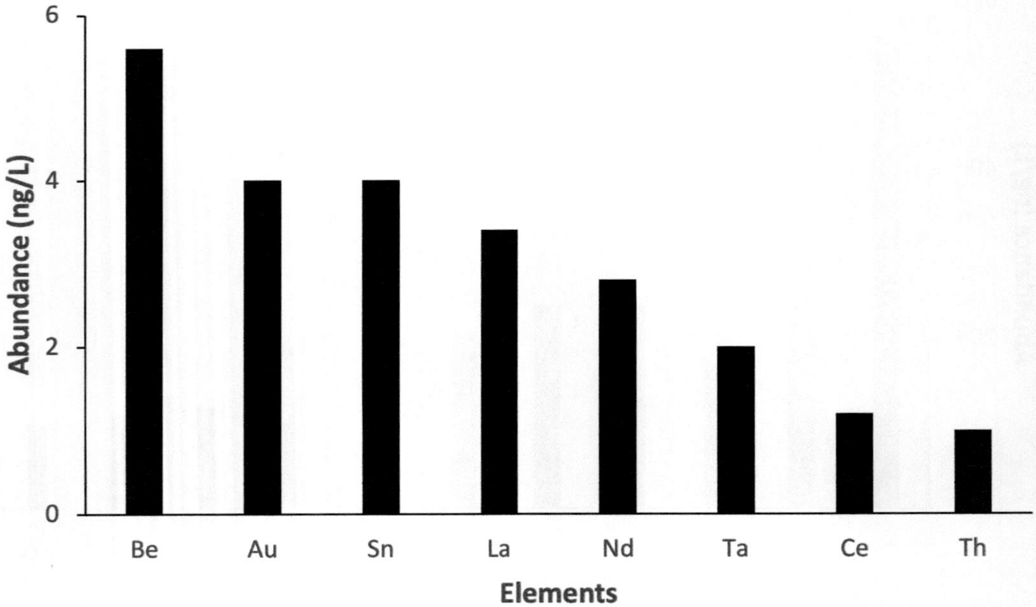

Figure 12. Abundance of chemical elements in the interval 1 – 10 ng/L (Handbook of Chemistry and Physics, 89th Edition).

Seawater ionic composition

In general, when referring to seawater composition, it is understood by the number of dissolved ions. There are numerous works related to ionic composition (Millero 1974, Millero et al. 2008) and as was already mentioned, the ionic composition in superficial seawater presents slight variations between oceans, seas, and salt lakes. Figure 13 shows the ionic amount between 1 and 600 mmol/kg in which it can be seen that chloride and sodium ions are the most abundant with concentrations between 450 and 550 mmol/kg; followed by magnesium ions with a concentration around 50 mmol/kg and, in this range, sulphate can be found with a concentration close to 28 mmol/L. Figure 14 shows the ionic content in the range between 1 and 12 mmol/kg, where the calcium ion and potassium ion are included; finally, Figure 15 shows the ionic content between 10 and 100 nmol/kg, where the bromide, the strontium^{+2} ion, and fluoride are found.

Other ions present in seawater are carbonate and bicarbonate. These are formed by the presence of dissolved carbon dioxide. The percentage of carbon dioxide present in seawater is changing. In recent years there has been an increase in the amount of carbon dioxide present in seawater, due to the increase in the presence of carbonate and bicarbonate ions. Carbonate content, due to the presence of calcium (with which it forms deposits), is continuously changing, and decreases, while bicarbonate increases and acidifies seawater. Bicarbonate is currently in a quantity of around 1.75 nmol/kg. There is another ion in seawater—that is $B(OH)_4^-$ in an amount of around 5 nmol/kg.

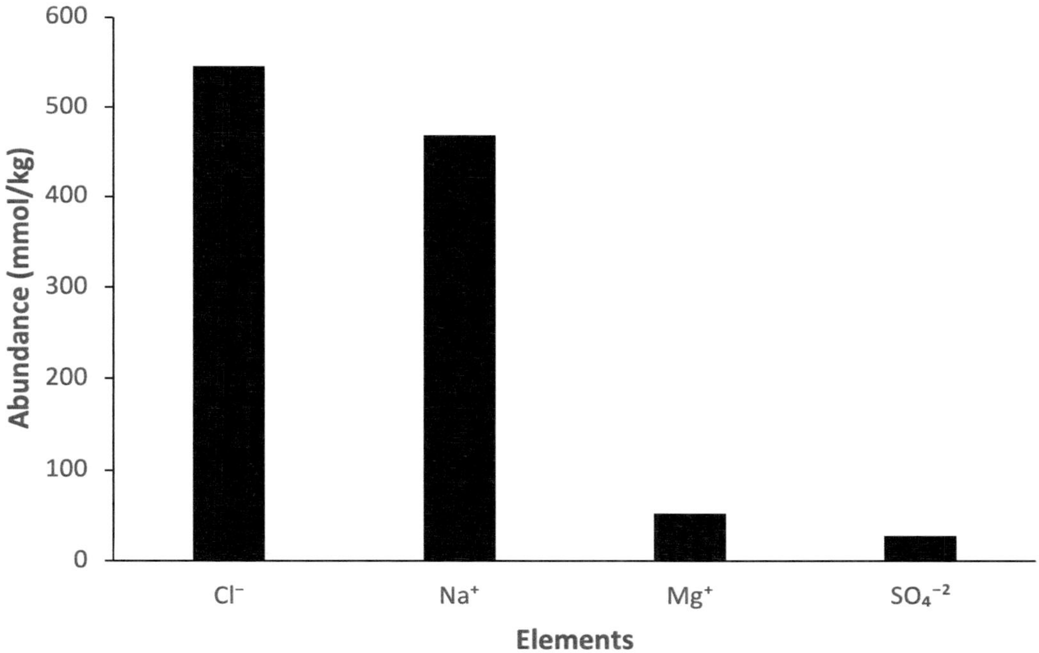

Figure 13. Ion concentration in seawater in the range 1 – 600 mmol/kg (Millero et al. 2008).

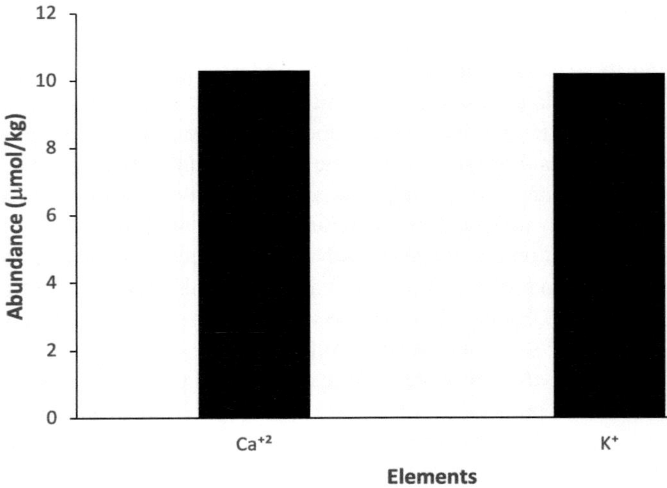

Figure 14. Ion concentration in seawater in the range 1 – 12 mmol/kg (Millero et al. 2008).

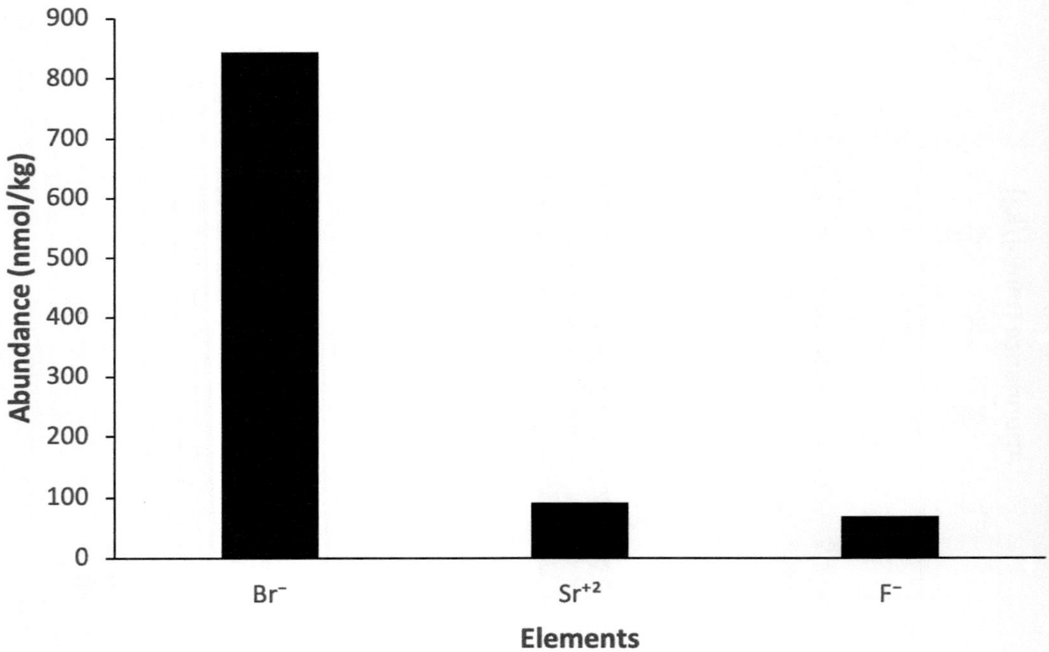

Figure 15. Ion concentration in seawater in the range 1 – 1000 nmol/kg (Millero et al. 2008).

Gases present in seawater

Table 2 shows the dissolved gases that are found in seawater. Nitrogen (N_2) at 62% and O_2 at 35% are the most abundant—like they are as in the atmosphere. A part of the nitrogen is converted into nitrates and ammonia. The oxygen content in seawater is higher at the surface as it is dissolved from the atmosphere and is produced by phytoplankton photosynthesis; surface water contains around 8 mg/L of oxygen. The noble gases (argon, neon, helium, krypton, and xenon) are also present in seawater (Bieri 1974).

Table 2. Dissolved gases in seawater.

Gases	N_2	O_2	Ar	CO_2	Ne	He	CH_4	Kr	CO	N_2O	Xe	SO_2	SH_2	H_2

Carbon dioxide is a very important gas that is dissolved in the ocean. Most of it reacts with water to form carbonic acid or with carbonates to form bicarbonates, which reduces the presence of carbon dioxide presence in seawater. The presence of carbon dioxide in seawater comes mainly from the atmosphere, from the respiratory processes of marine living beings and from submarine volcanic emissions. Nowadays, the amount of carbon dioxide in seawater is increasing.

Methane, despite its low solubility, is present in seawater thanks to sediments that have a large amount of organic matter. At very low temperatures, methane, carbon dioxide, and water form a kind of ice called methane hydrate.

Sulphur dioxide is highly soluble in seawater, being released in submarine volcanic eruptions. When sulphur dioxide is combined with water it forms bisulphates. Other gases present are nitrous oxide, carbon monoxide, and hydrogen (Seiler and Schmid 1974).

Marine pollution

Currently, another important aspect of seawater composition is the pollutants caused by the effects of human development. Table 3 summarises other aspects that are linked to the oceans, seas, and salty lakes contamination: placing urban waste in first place. In this wastewater discharges are included as are oil spills due to oil extraction. Further, another source of contamination is marine accidents that occur in the hydrocarbon's maritime transportation, as well as during ship cleaning.

Likewise, since the beginning of the industrial revolution, the residue that come from all chemical industries has increased; no less important are the pollutants that come from agricultural waste, such as fertilizers and pesticides that end up in most cases in oceans. Finally, it deserves to be mentioned, that deep-sea mining and the multitude of dredging that occurs in all coastal areas are also responsible for pollution.

Pollution in seawater gives rise to the presence of elements and compounds that play an increasingly important role. Table 4 shows a list of elements and compounds from marine pollution. Many of these elements or compounds are fixed in living organisms, which is a serious food health problem, among them, heavy metals should be highlighted.

Finally, in the last decades, the amount of plastic from urban waste that is later transformed into microplastics has caught the eye and as they are not soluble in seawater, it has been considered a serious new health problem.

Table 3. Marine Pollutants.

Seawater Pollution
Urban waste
Oil spills
Industrial waste
Agricultural waste
Deep-sea mining and dredging

Table 4. Polluting elements of seawater.

Polluting elements	Sb	As	S	Br	Cd	Cl	Cr	P	Mn	Hg	Pb

Seawater physical and chemical properties

The physical and chemical properties of seawater are important for application techniques in thalassotherapy and cosmetics. This section will focus on those properties that are related to mentioned uses: temperature, pressure, density, specific heat, thermal conductivity, thermal diffusivity, dynamic viscosity, surface tension, and pH. Salinity has already been mentioned in the composition section.

Temperature

The temperature of seawater in the oceans depends on the geographical position, the time of the season, and depth. Temperatures are higher towards the equator and on the water surface; It also varies during the year and besides, depends on the geographical position.

Figure 16 shows the average temperature evolution in the oceans related to a reference temperature (adjusted to a polynomial function).

In recent years, the increasing estimated temperature on the planet is 0.89 Kelvin.

Figure 17 shows the temperature variation with depth in the oceans. The surface temperature of the ocean varies greatly with geographical position, so it can range between 35 °C to –1 °C; however, at great depths, the temperatures do not experience such considerable variations.

The seawater coast's surface temperature is of great importance as millions of people bathe in the oceans and seas around the world. The therapeutic properties of bathing in saline water are typified among the uses of thalassotherapy in most cultures. The surface temperature of the coasts varies on the seas and oceans, as well as the time of the year and the geographical location. Nowadays, there are numerous websites that report coastal surface temperatures around the planet (https://fr.tutiempo.net/temperature-mer). If the coastal temperatures are analyzed in different seas and oceans, it is possible to highlight the following:

Mediterranean Sea: The average temperature in summer is about 27 °C on the Spanish, French, and Italian coasts, being higher in North Africa and the eastern Mediterranean which can reach 30 °C; in winter temperatures can drop to 14 °C.

North Atlantic: The temperature in summer is about 17 °C –19 °C on the northern coasts of Portugal, western Spain, and France, and 25 °C on the eastern coast of the USA; these temperatures drop to 13 °C –14 °C in winter.

South Atlantic: The temperature in summer in the North of Argentina can go up to 21 °C, in Brazil 28 °C, and in South Africa 16 °C; while in winter they can drop to 10 °C, 25 °C, and 14 °C respectively.

North Pacific: The summer temperature on the West coast of the USA is 15 °C and in the South of Japan is 30 °C; in winter they can drop to 12 °C and 18 °C respectively.

South Pacific: The temperature in summer is between 21 °C in Peru and 15 °C in Chile; in winter the temperatures can drop to 16 °C and 12 °C respectively.

Indian: The temperature on most coasts is very stable on the northern and central coasts, and on Central Africa and South India coasts it is about 27 °C in both, summer, and winter.

Baltic Sea: Average temperatures range between 3 °C in winter and 20 °C in summer.

The temperatures in the equatorial zone oceans have very few variations and are at an average of 27 °C.

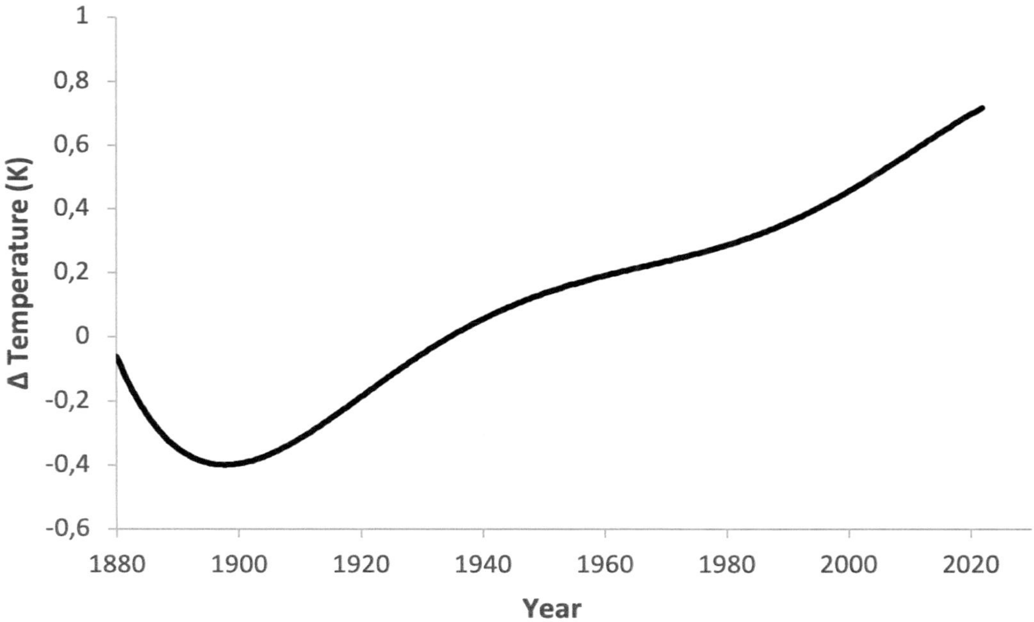

Figure 16. Evolution in the average temperature in the oceans relative to a reference temperature, in recent years (https://www.climate.gov/news-features/understanding-climate/climate-change-global-temperature).

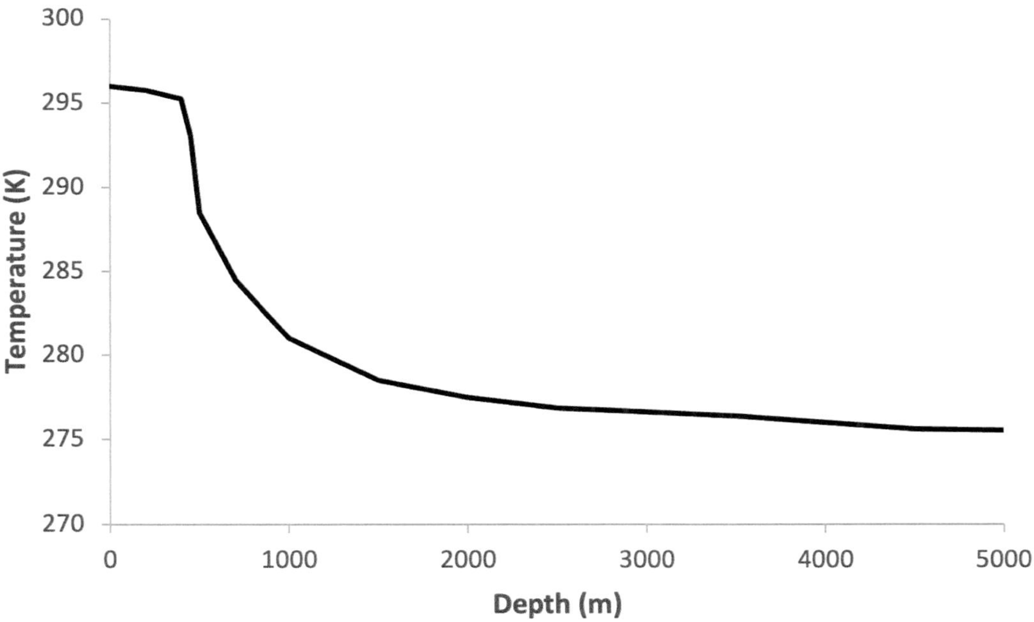

Figure 17. Variation of ocean temperature with depth (Alexander and Scott 2000).

Pressure

The pressure in the seas and oceans depends on depth and salinity. It can be estimated based on the fundamental hydrostatics principle:

$$\Delta p = dg\Delta h$$

where p is the pressure, d is density, g the acceleration of gravity and h the depth.

At the ocean surface, the pressure is the same as that of the atmospheric pressure, which presents small changes depending on the geographical location and salinity (which influences density).

The effect of hydrostatic pressure is one of the applications of thalasso, as density increases with salinity: therefore, an increase in hydrostatic pressure will favor some therapeutic applications.

Density

Water density measures the amount of mass per unit of volume. In liquid water, the density has an intermediate value between high-density substances and low-density substances. Moreover, salinity produces an increase in density.

Figure 18 shows the behavior of density as a function of salinity and temperature. As can be seen, density increases with salinity and has a variable behavior with temperature. Density increases with increasing pressure.

In thalassotherapy, high salinity pools will favor flotation since Archimedes' principle depends on the density of the fluid (the higher the density, the greater the flotation). On the other hand, as said before in the pressure section, thalassotherapy increases hydrostatic pressure in parts of the body that are more submerged.

The density of ice is lower than that of liquid water. This anomaly (solids have a higher density than liquids) is of great importance for the planet. If this were not so, the oceans would be completely frozen and ice would be deposited on the bottom of the ocean and seas and thus freeze them, leaving only the surface layer in a liquid phase. This would cause a profound change in our planet.

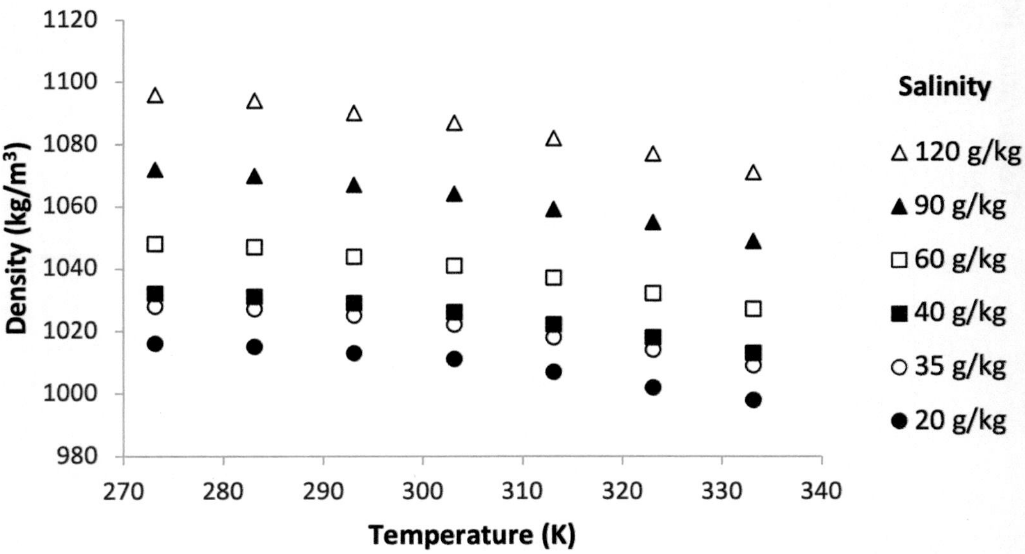

Figure 18. Behavior of density with temperature and salinity at atmospheric pressure (Nayar et al. 2016).

Specific heat

Specific heat is an important element in the ocean, as it corresponds to the amount of energy necessary to raise the temperature per unit of mass.

Water has a high specific heat when it is in the liquid phase, this means that the oceans are a store of energy that is recharged by the sun's infrared radiation and is key to the planet's climate.

The specific heat of seawater is slightly lower than that of pure water. Figure 19 shows the behavior of specific heat as a function of salinity and temperature: specific heat decreases with increased salinity. It increases lightly with increased temperature. Additionally, specific heat decreases with increased pressure.

The specific heat of water in the solid phase is around half of the liquid phase.

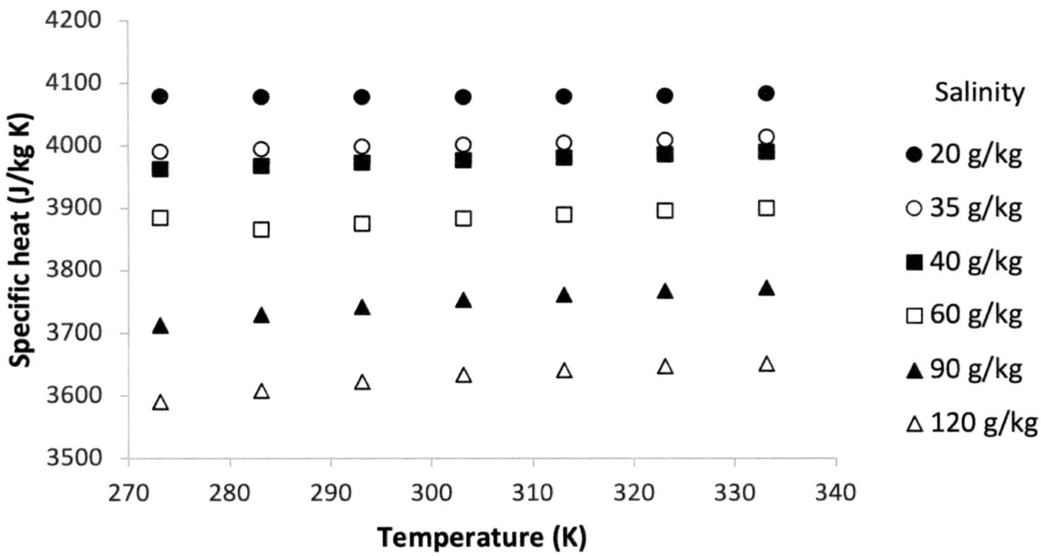

Figure 19. Specific heat behavior with temperature and salinity at atmospheric pressure (Nayar et al. 2016).

Thermal conductivity

The thermal conductivity of water is related to the rate of heat transfer. In water, thermal conductivity has intermediate values compared to high heat-conducting substances and thermal insulators; also, salinity produces a small decrease in thermal conductivity.

Figure 20 shows the behavior of thermal conductivity as a function of salinity and temperature. As it can be seen, thermal conductivity increases with temperature and decreases slightly with salinity. Thermal conductivity increases with increasing pressure.

The thermal conductivity of ice is greater than that of liquid water, increasing approximately four times.

Thermal diffusivity

The thermal diffusivity of water measures heat loss over time when the system is in contact with an environment at a lower temperature. Mathematically, it is the ratio between thermal conductivity and specific heat and density. In water, thermal diffusivity has a low value, so it can be considered that it retains thermal energy, which represents an important ocean effect on the planet, also, salinity produces an increase in thermal diffusivity.

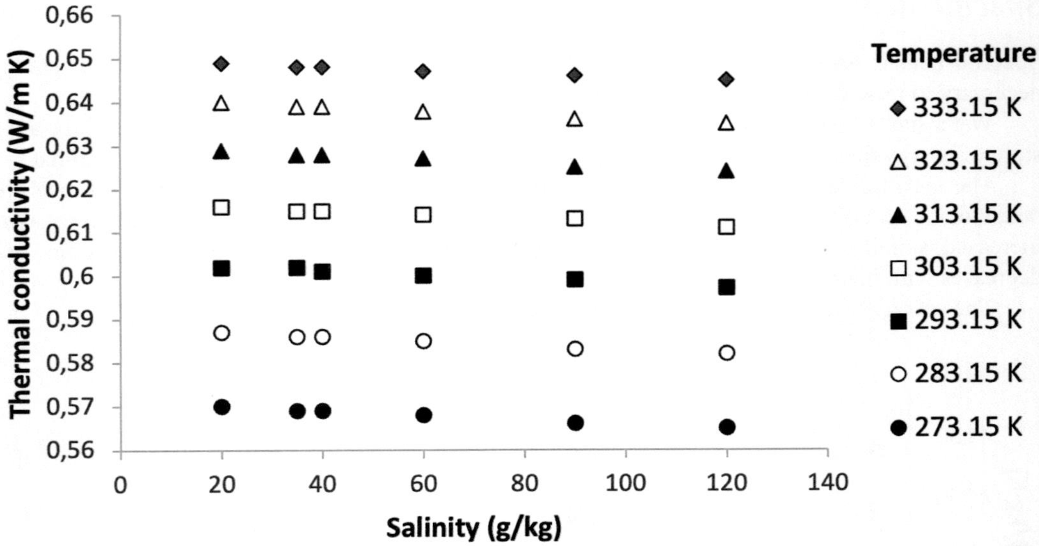

Figure 20. Thermal conductivity behavior with salinity and temperature at atmospheric pressure (Nayar et al. 2016).

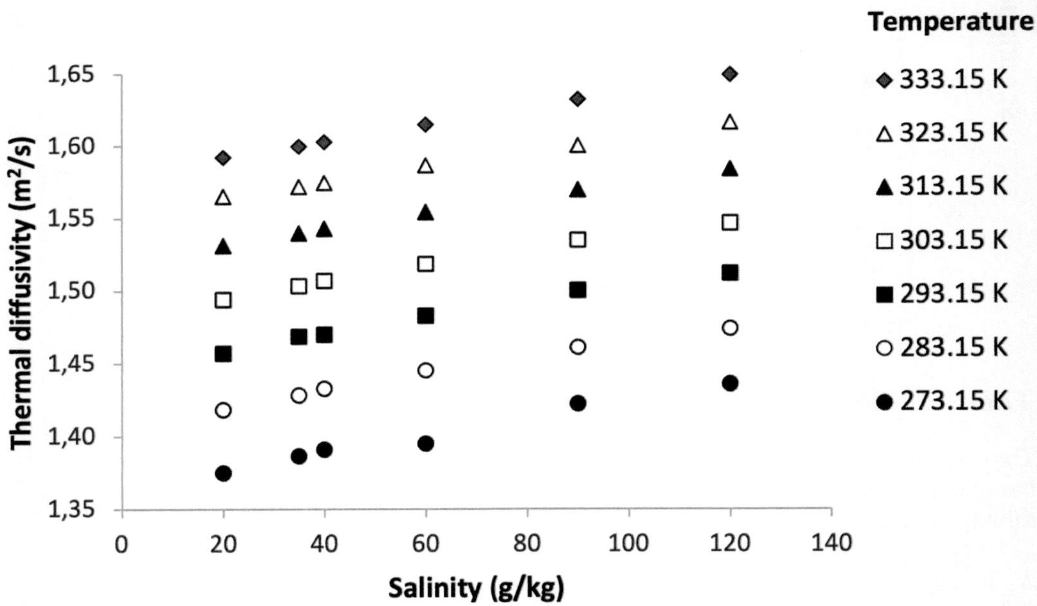

Figure 21. Thermal diffusivity behavior with salinity and temperature at atmospheric pressure (Nayar et al. 2016).

Besides, thermal diffusivity is responsible for thermal sensations; water produces much greater thermal sensation than dry air and this is the basis of most thermotherapeutic applications.

Figure 21 shows thermal diffusivity behavior as a function of salinity and temperature. As can be seen, thermal diffusivity increases with temperature and salinity. Thermal diffusivity increases with increasing pressure.

The thermal diffusivity of ice is greater than liquid water, increasing approximately eight times.

Dynamic viscosity

Liquid water dynamic viscosity indicates the resistance to fluid mobility when it is subjected to external forces. This property in thalassotherapy is of great importance, especially in rehabilitation because movements in water require more effort than they do in air. In liquid water, dynamic viscosity has an intermediate value between liquids. Salinity produces an increase in dynamic viscosity; also, temperature decreases to produce a significant increase in it.

Figure 22 shows the behavior of dynamic viscosity as a function of salinity and temperature.

As can be seen, dynamic viscosity increases with salinity and temperature, and also, with the increase of pressure.

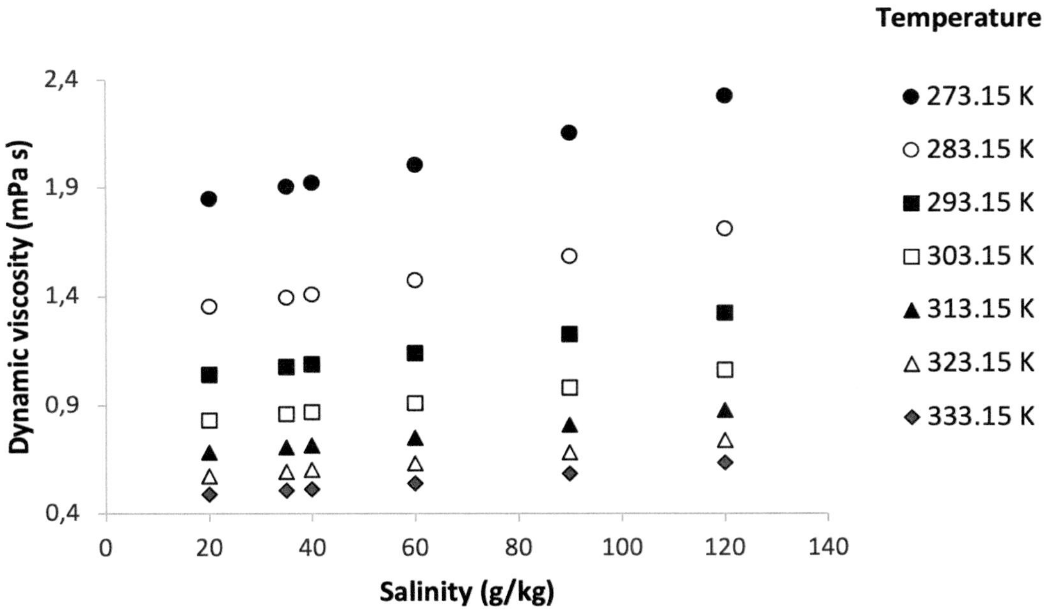

Figure 22. Dynamic viscosity behavior with salinity and temperature at atmospheric pressure (Nayar et al. 2016).

Surface tension

The surface tension of liquid water is related to the intermolecular forces on the surface of the fluid. In physics, it is a measure that relates the potential energy per unit surface. It is responsible for the formation of drops and the rise and fall of capillaries.

In thalassotherapy use, it influences the layer of water that remains on the body after a bath.

The surface tension on liquid water has a high value among other liquids. Salinity produces a slight increase in surface tension, and the increasing temperature produces a significant decrease in it.

Figure 23 shows the behavior of the surface tension as a function of salinity and temperature. As can be seen, surface tension increases with salinity and decreases with temperature.

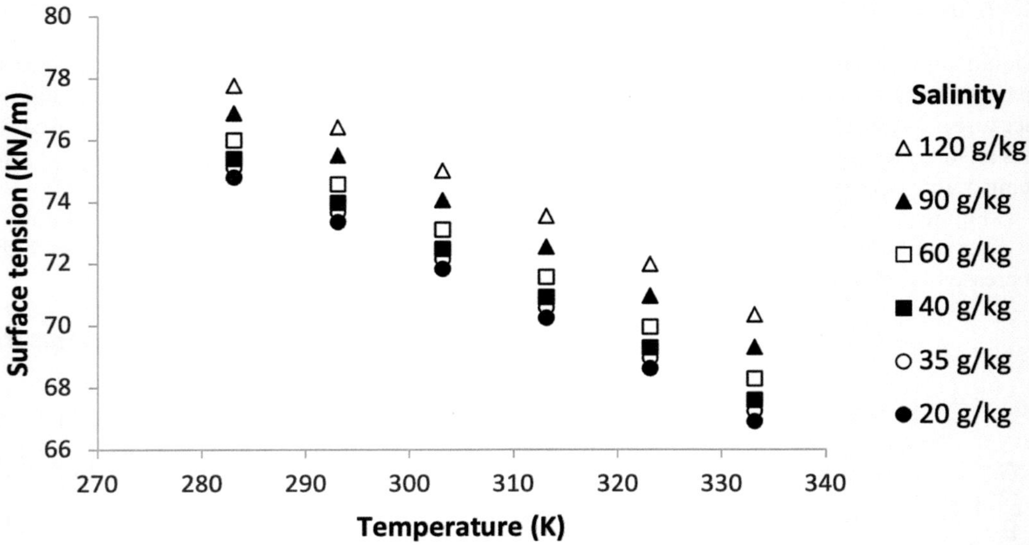

Figure 23. Surface tension behavior with salinity and temperature at atmospheric pressure (Nayar et al. 2016).

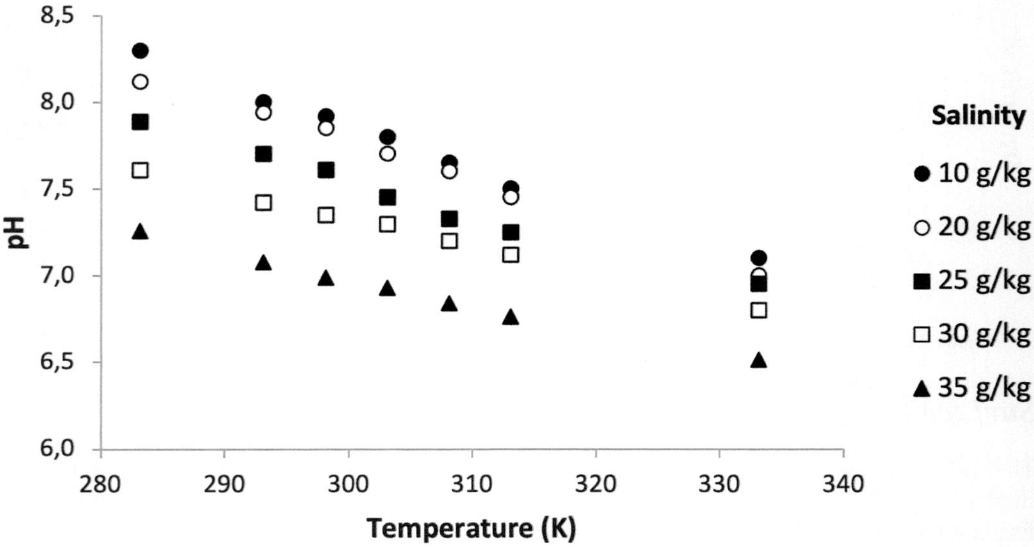

Figure 24. Behavior of pH with salinity and temperature (Marion et al. 2011).

pH

The pH of seawater depends on salinity and temperature; many chemical reactions within aquatic organisms (cellular metabolism) are necessary for survival and growth. Therefore, for thalassotherapy, it is important to preserve phytoplankton and microalgae that are part of the thalasso-wellness treatments. Furthermore, since its pH is close to neutrality, it allows seawater to be used in oral therapies (see Chapter 5) and/or in cosmetics.

Figure 24 shows the pH behavior with salinity and temperature as it can be seen the pH increases with salinity and decreases with temperature.

References

Alexander, M.A. and J.D. Schott. 2000. Processes that influence sea surface temperature and ocean mixed layer depth variability in a coupled model. Geophysical Research 105(C7): 16,823–16,842.

Bieri, R.H. 1974. Dissolved conservative gases in seawater. In Global Coastal Ocean Ed Golberg, page 199.

CRC Handbook of Chemistry and Physics, 89th Edition. Ed. David R. Lide. Taylor & Francis, (2008) ISBN 142006679X, 9781420066791.

Deming, D. 2021. Edmond Halley's Contributions to Hydrogeology. Ground Water, Jan–Feb. 59(1): 146–152.

Dittmar, W. 1884. Report on researches into the composition of ocean water, collected by. H. M. S. Challenger, during the years 1873–1876. Voy. Challenger, Physics and Chemistry 1(1): 1–251.

Gould, W.J. and S.A. Cunningham. 2021. Global-scale patterns of observed sea Surface salinity intensified since the 1870s. Communications Earth & Environment, 2, 76, https://doi.org/10.1038/s43247-021-00161-3.

Hansen, K. and K. Wallmann. 2003. Cretaceous and Cenozoic evolution of seawater composition, atmospheric O_2 and CO_2: a model perspective. American Journal of Science 303: 94–148. DOI 10.2475/ajs.303.2.94.

Kamitsou, M.D., V. Sygouni, D.G. Kanellopoulou, K. Gardikis and P.G. Koutsoukos. 2018. Physicochemical characterization of sterilized muds for pharmaceutics/cosmetics applications. Environ Geochem Health. 40(4): 1449–1464. doi: 10.1007/s10653-017-0066-8.

Kennedy, C. 2014. https://www.climate.gov/news-features/featured-images/shifting-ocean-surface-saltiness-2004-2013.

Komiya Tsuyoshi, Takafumi Hirata, Kouki Kitajima, Shinji Yamamoto, Takazo Shibuya, Yusuke Sawaki Tomoko Ishikawa et al. 2008. Evolution of the composition of seawater through geologic time, and its influence on the evolution of ife. Gondwana Research 14: 159–174.

Lyman, J. and R.H. Fleming. 1940. Composition of sea water. Journal of Marine Research 3, (2). https://elischolar.library.yale.edu/journal_of_marine_research/566.

Marion, G.M., F.J. Millero, M.F. Camões, P. Spitzer, R. Feistel and C.T. Chen. 2011. pH of seawater. Marine Chemistry 126(1–4): 89–96.

Millero, F.J., R. Feistel, D.G. Wright and T.J. McDougall. 2008. The composition of Standard Seawater and the definition of the Reference-Composition Salinity Scale. Deep Sea Research Part I: Oceanographic Research 55: 50–72. doi:10.1016/j.dsr.2007.10.001.

Millero, F.J. 1974. The physical chemistry of seawater. Annu. Rev. Earth Planet. Sci. 2: 101–150.

Nayar, K.G., M.H. Sharqawy and J.H. Lienhard. 2016. Seawater thermophysical properties library. https://web.mit.edu/seawater/2017_MIT_Seawater_Property_Tables_r2b_2023c.pdf.

Neeley, A.R., L.A. Harris and K.E. Frey. 2018. Unraveling phytoplankton community dynamics in the northern Chukchi Sea under sea-ice-covered and sea-ice-free conditions. Geophysical Research Letters 45: 7663–7671. https://doi.org/10.1029/2018GL077684.

Pereira, L. 2017. Thalassotherapy and Marine Cosmeceuticals. pp. 503–522. *In*: Therapeutic and Nutritional Uses of Algae, 1st ed.; CRC Press/Taylor & Francis Group: Abingdon, UK, 2017; Chapter 12, https://doi.org/10.1201/9781315152844-12.

Seiler, W. and U. Schmid. 1974. Dissolved nonconservative gases in seawater. *In*: Global Coastal Ocean Ed Golberg, page. 219.

CHAPTER 3

Seawater

Mechanisms of Action

Carmen P. Gómez Pérez and *María Lourdes Mourelle Mosqueira**

Introduction

The word Thalassotherapy is a Greek term originating from the combination of the Greek word, θάλασσα, "Thalassa" (sea), and "Therapy". Thalassotherapy or seawater or marine cure is defined as the combination of marine elements (seawater, climate, algae, and mud, in a marine environment), for healing and well-being purposes (Gutenbrunner et al. 2010, Maraver et al. 2019).

Since Greek and Roman times, seawater cure has been considered for the treatment and prevention of different diseases, mainly related to glands. Then, the therapy was executed directedly in the sea but at the beginning of the 19th century warm baths were introduced in France and many marine centers opened first on the Atlantic coasts and later in the Mediterranean area (Charlier and Chaineaux 2009).

Along with the use of seawater, marine climatotherapy as well as heliotherapy, has shown to be useful in non-allergic bronchial asthma and psoriasis treatments (Munteanu et al. 2019).

This chapter aims to provide an overview of the mechanisms of the action of seawater, focusing on the main factors that combined may explain the effects and actions of seawater on the human body.

Sea water: effects on the human body

Thalassotherapy has been extensively applied for the treatment of different illnesses, health improvement, mental and psychological stability, human physiological vitality improvement, and for the skin. This marine cure has a specific action with biological effects linked to its mineral content. For several centuries, it has been prescribed as recovery therapy for many patients and customers who suffer from health problems such as dropsical swelling or inflammation. It is known as an effective therapy working for skin disorders, neuralgia, osteomyelitis, and rheumatism and is used instead of an antibiotic agent (Park 2007).

FA2 Group. Department of Applied Physics. University of Vigo.
Email: carmengomez@uvigo.gal
* Corresponding author: lmourelle@uvigo.es

The main medical indications of thalassotherapy are for respiratory illnesses, such as asthma, rheumatological ailments, circulatory problems, recovery from traumatic injuries, and dermatological disorders. However, it is also indicated for general well-being and the treatment of age-related symptoms.

Karagülle (2008) summarized this new indication as follows: hydrotherapy, balneotherapy, and thalassotherapy in the elderly can be considered from three different aspects. First, because of their anti-aging effects: that is their preventive effects on the biological and physiological processes that develop in the elderly or therapeutic effects on the pathological consequences. Second, they might be effective in the prevention, treatment, and rehabilitation of diseases that occur frequently in the elderly. Third, they may complement and support other treatment modalities for the elderly. In the 19th century, the French biologist René Quinton could prove experimentally the remarkable similarity between seawater and human plasm "Our organism is a kind of aquarium whose internal environment is constituted by seawater" (Mahé 1999).

Quinton published a book "L'eau de mer, milieu organique" which became the vademecum (a handbook or guide) of thalassotherapy and is regarded as the lighthouse that has guided medical hydrologists to create the first thalassotherapy centers (Mahé 1999). Later on, Dr. Louis Bagot created a center at Roscoff (Brittany), and subsequently thalassotherapy regained publicity when the champion bicycle racer Louison Bobet was cured after a car accident by Dr. Bagot. In 1964 he opened the first modern thalassotherapy center at Quiberon (Brittany) (Pereira 2017).

The effects of seawater on the human body are linked to its composition as well as other factors such as temperature and application technique, combined with other thalasso-hydric and marine elements such as climate, seaweeds, sand, and marine peloids. Thus, the factors implied in the effects and actions can be summarized as follows (Mourelle 2013):

- Hydrothermal factors: linked to the temperature of application (natural cold baths or heated baths).
- Hydromechanical factors: Hydrostatic and Hydrodynamic factors, related to the buoyancy and hydrostatic pressure, and movement in the water. Hydrokinetic factors should also be taken into consideration, linked to the technique (with or without pressure).
- Hydrochemical factors: related to the mineralogical composition.

Hydrothermal factors

For centuries, thalassotherapy was performed directly at sea, taking cold baths and making the most of the climate and heliotherapy. The seawater cure periods, therefore, were limited to the summer season. Later, balneotherapy was introduced indoors, and both cold and warm/hot baths began to be carried out in facilities designed for this purpose, in addition to incorporating other techniques, such as jets, showers, etc.

Hydrothermal factors are related to high specific heat, high heat capacity, and considerable thermal conductivity of water, as well as the high heat of vaporization—properties which allow water to be stored or lost, slowly raise or withdraw the heat of the body to which it is exposed, mainly by conduction and convection.

The thermal effects of water depend not only on the temperature of the water but also on the extent of the body surface in contact, the duration of the application, and individual sensitivity. In addition, local and general effects will occur, early (defensive) and later (restorative) (Teixeira 2022).

Physiological effect of water as a thermotherapy method

Water is used as a superficial thermotherapy method; the depth of penetration is small and, therefore, its effects are directly observed in the blood vessels and cutaneous nerve receptors. These effects will depend on the surface of the treated area (less stimulus in local applications), the application time (the longer the time, the greater the stimulus), and individual sensitivity.

The thermal stimulus provided by the water will be greater, the further its temperature is from the zone of indifference. The temperature of thermal neutrality or indifference in water is between 34 °C and 36 °C.

Generally, the temperature of the water used as a thermotherapy agent, ranges between 36.5 °C and 40.5 °C, although in the presence of cardiovascular or pulmonary diseases and in patients with sensory deficits it is necessary to adapt the water temperature to their particular characteristics. The most important physiological effects produced by hydrotherapy, as a thermotherapy method, are (San Martin 1994):

Analgesia: Heat raises the sensitivity threshold of nociceptors—pain receptors—and decreases the speed of nerve conduction and muscle contracture. In addition, it acts as a counterirritant, modifying painful sensations. It has also been implicated in the production of analgesia, after hyperthermal baths, the release of endorphins and enkephalins. When agitation is added to hot water, as occurs in whirlpool baths or hydromassage, the analgesic effect is greater.

Increased temperature and tissue vasodilation: After the application of heat, an increase in local temperature is observed in the treated area. If the application is sufficiently wide, the body temperature rises between 0.5 °C and 3 °C and produces an increase in all organic functions due to overheating. This increase in temperature, through the activation of the efferent thermoregulatory mechanisms, causes a progressive decrease in the tone of the vascular muscles and the consequent increase in blood flow or hyperemia, of an intensity proportional to the temperature. Hyperemia produces an analgesic and anti-inflammatory effect, increases cellular metabolism, and improves nutrition, increasing the reabsorption and elimination of catabolites and accelerating the tissue repair processes. If a hydrotherapy method with agitation is used, a lower water temperature is necessary to achieve the same level of increased blood flow.

Sedative effect: In balneation, when the temperature is close to thermoneutral and is applied for a long period of time, body relaxation and a feeling of general well-being occur.

If the application is very hot and of short duration it has the opposite effect: it produces an exciting effect on the free nerve endings and causes insomnia. If the hyperthermal applications are long, the initial unpleasant sensation, produced by the high temperature, is replaced by a pleasant sensation of well-being.

Antispasmodic and muscle relaxing effect: Short applications, especially if they are very hot, increase muscle tone and improve performance, while prolonged applications, between 36 °C and 38 °C, exert a relaxing effect on the tone of skeletal muscles, combat contracture, and muscle fatigue. This antispasmodic effect is also observed, acting in a reflex manner, on the smooth muscles of the internal organs and viscera, which is why these applications are used to relieve abdominal cramps.

Effect on connective tissue: Heat produces an increase in the elasticity and a decrease in the viscosity of the connective tissue. Therefore, it is very useful to reduce joint and periarticular stiffness produced in certain rheumatic processes.

Effects of exercise in hot water: The use of exercise in water and different hydrotherapy techniques is recommended in many different neurological and musculoskeletal conditions.

The main therapeutic effects of exercise in water are the following:

- relaxation of the patient
- decreased pain and muscle spasm
- increased circulation
- maintenance or increase of joint mobility
- muscle re-education, developing power and resistance
- early training for walking, with a decrease in joint and muscle load
- improvement of psychological and emotional state

Physiological effect of water as a cryotherapy method

Cold water applications produce more or less intense effects on the body depending on the temperature, the exposed surface, and the time of application. In hydrotherapy, water as a cryotherapeutic method is used primarily through local techniques and is frequently used alternating with hot water, using contrast baths.

The main effects of hydrotherapy, as a cryotherapeutic method, are:

Decrease in temperature: Tissue vasoconstriction. Thermal chill.

The primary effect of an application with cold water (up to 15–20ºC), and also with ice, consists of a decrease in the local temperature of the skin and underlying tissues. Vasoconstriction then occurs to reduce heat loss and is manifested by the paleness of the skin.

In this first phase, the sensation of cold can become painful. If the application is very prolonged or the water temperature is very low, bluish discoloration of the skin may occur, due to decreased venous drainage.

When cold applications are of short duration, cutaneous vasoconstriction usually gives way to great reactional vasodilation, with a sensation of heat and well-being.

When the applications are of low intensity, in terms of temperature or duration, vasoconstriction occurs, but if they are of greater intensity, a compensatory mechanism is launched that tries to maintain homeothermy: increased muscle tone and, subsequently, contractions. involuntary (shivering and chills). They are used in the acute phase of trauma injuries, especially sports injuries.

Application time varies between 5 and 45 minutes of direct contact; As a general rule, at least 15 minutes of application is needed to produce an analgesic effect.

Immersion in cold water does not allow comfort. However, the succession of hot and cold-water baths (contrast baths), due to alternating vasodilation/vasoconstriction, is used due to its marked vascular action and is better tolerated.

Analgesia and muscle relaxation: Cold in hydrotherapy is used mainly for its analgesic and muscle antispasmodic effect. The analgesic effects are due to the fact that cold reduces the excitability of free nerve endings, increases the pain threshold, and reduces muscle spasms; chills should be avoided (Mourelle and Sabater 2013).

General reactions to hydrothermal stimuli: Exposure of the body to hydrotherapeutic stimuli at different temperatures not only has local effects, but also systemic ones. The greater the temperature difference between the water and the skin, and the longer the application time, the more intense the reaction will be. These general effects will be all the more effective, if the submerged body surface area is high.

The use of hydrotherapy as a thermotherapy and cryotherapy method causes significant changes in blood volumes, which require a healthy cardiocirculatory system in perfect conditions of adaptation.

General hot hydrotherapeutic applications produce an initial increase in blood pressure, followed by a decrease, as a result of peripheral vasodilation. There will be an increase in the heart rate and an increase in minute volume. After immersion, there is also a slight increase in the respiratory rate. What is described will vary in intensity depending on whether the bath is partial or complete.

It is necessary to be extremely vigilant in the response produced by balneation in patients with cardiac or peripheral vascular disease and in patients with hypertension and hypotension.

General cold hydrotherapeutic applications decrease heart rate and lengthen the duration of diastole; increase cardiac muscle tone and blood pressure as a result of peripheral vasoconstriction. They decrease the amount and speed of circulating blood, as well as decrease the stroke volume and heart rate, and decrease minute volume.

Sudden cold applications, especially on the neck and shoulders, act reflexively, producing a deep inspiration. Long-term cold-water baths cause rapid, shallow breathing.

Exposure to cold causes a release of adrenaline that modifies energy metabolism (release of fatty acids, glycogenolysis...) and alters the distribution of blood flow, directing blood from the superficial to the deep parts of the muscles (Mourelle and Sabater 2013).

The physiological effects and applications are summarized in Table 1.

As has been mentioned before, water (and especially seawater) is an excellent heat vehicle because of its abnormally high specific heat. The transmission of heat in topical applications determine changes in vascularization peripheral. When the application temperature is superior to indifference immediate vasoconstriction occurs, followed by prolonged peripheral vasodilation, activation of circulation, the opening of new capillaries and arterioles of superficial tissues, with

Table 1. Effects of hydrothermal factors and applications.

Physiological effects	**Applications**
Water as thermotherapy method	
Analgesia: raises the sensitivity threshold of the nociceptors; release of endorphins and cephalins	Rehabilitation in degenerative pathologies, neurological injuries and, in general, those pathologies that cause pain
Increase in temperature and tissue vasodilation: analgesic and anti-inflammatory effect, increase in cellular metabolism and better nutrition at the muscle level	Rehabilitation in all types of pathologies and muscle injuries
Sedative effect: muscle relaxation and feeling of well-being	Treatment of injuries and pathologies that cause pain
Antispasmodic and muscle relaxing effect: on skeletal muscles and on internal organs and viscera	Rehabilitation; sports recovery
Increase in elasticity and decrease in viscosity of connective tissue: decrease in joint and periarticular stiffness	Rehabilitation and treatment of rheumatic processes
Water as cryotherapy method	
Decreased temperature, tissue vasoconstriction, and thermal chills: - Short-term applications: cutaneous vasoconstriction and then great reactional vasodilation, with a sensation of heat and well-being - Of greater intensity: compensatory mechanism, with increased muscle tone	Short-term applications: acute phase of trauma/sports injury Contrast: vascular training
Analgesia and muscle relaxation: reduces the excitability of free nerve endings, increases the pain threshold and reduces muscle spasm	In acute conditions or traumas, as an anti-inflammatory and analgesic

consequent improvement in tissue trophism. Heat applications also produce analgesia and a general sedative action by influencing the nervous system, elevating the pain threshold, and driving difficult painful sensitivity to attenuate or abolish the sensitivity of peripheral receptors, with repercussions on the modulating center's pain relief and releasers of endorphin substances, all of which produce analgesia (San Martin 1994, San Martin 2006).

When the water temperature is higher than the indifferent, between 36 °C –38 °C, and the application is prolonged, relaxation also occurs in muscle tone, reducing contracture and muscle fatigue.

Hydromechanical factors

When baths are taken directly in the sea, an intense action of pressure and thrust occurs, favored by the passive massage that stimulates the coming and going of the waves, which acts by favoring return circulation. At the same time, staying upright against the push of the waves is an important muscle-toning exercise. When taking baths in a hot tub, this mechanic factor has less influence than the thermal factors, but pressure can be added, and hydro-kinetic factors exerted by hydromassage jets, and the bubbles are of great importance.

Thus, in immersion techniques, general baths, or partial in tank or pool, for therapeutic purposes, thermal factors are added to the other physical factors such as hydrostatic, and hydrodynamic: pressure hydrostatic, the principle of flotation or thrust, cohesion, viscosity, surface tension, turbulence, etc., which are factors that intervene in the facilitation or difficulty of the exercise in water, especially in balneo-kinesitherapy.

Floatation and hydrostatic factors: Buoyancy achieves great relevance in the rehabilitation performed in a pool. The flotation effect, derived from Archimedes' principle, facilitates the reduction of apparent weight so that the submerged body appears lighter. Buoyancy in seawater is considered two or three times higher than in normal water (San Martin and Armijo 1994). According to Lecrenier's scheme (Figure 1), when a body is submerged up to its neck, it will weigh about 10%

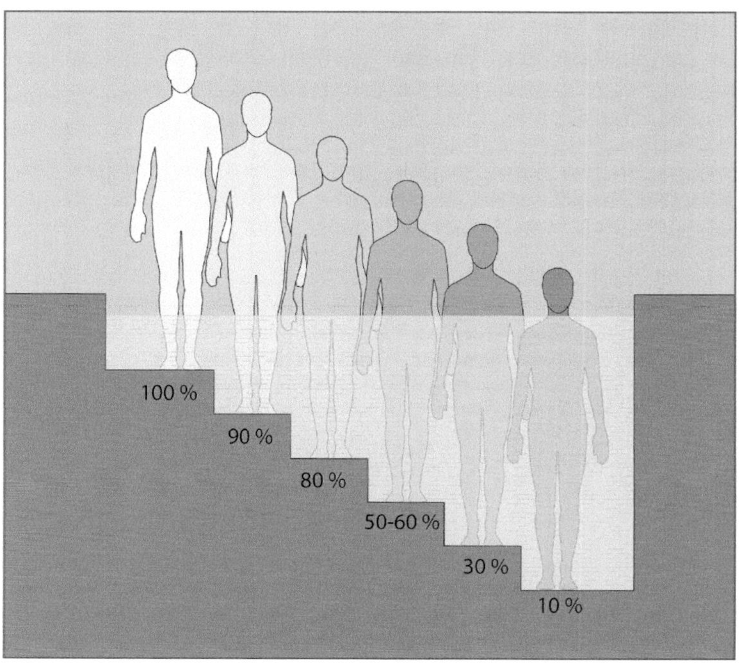

Figure 1. Lecrenier's principle.

of its actual body weight; this is why submerged bodies tend to float (Teixeira 2022): this being more evident in seawater due to greater density when compared with ordinary water.

Thus, seawater supports a large part of body weight, facilitating the execution of movements with correct muscle work even in conditions in which muscle tone is decreased. This apparent reduction in body weight achieved in the saline water environment also makes it possible to eliminate pain due to the load and helps initiate early rehabilitation (Mourelle and Sabater 2013).

The other aspect is hydrostatic pressure, which causes compression in the body structure and facilitates return circulation. The greater the hydrostatic pressure, the greater is the depth, and of the same value in all points on each horizontal plane. This manifests a direct compression action on the venous system and the large body cavities and facilitates return circulation, which is more pronounced if the individual is standing in the swimming pool. This treatment can be very beneficial in those patients with small varicose veins, and slight fluid retention in lower extremities (Figure 2). This circumstance forces the heart to adapt to greater cardiac activity, as well as a significant increases diuresis due to the stimulation of the release of atrial natriuretic factor (San Martin 2006).

The effects of hydrostatic pressure are also evident in respiratory functions. Due to the compression exerted on the rib cage and the diaphragm, expiration is facilitated and inspiration

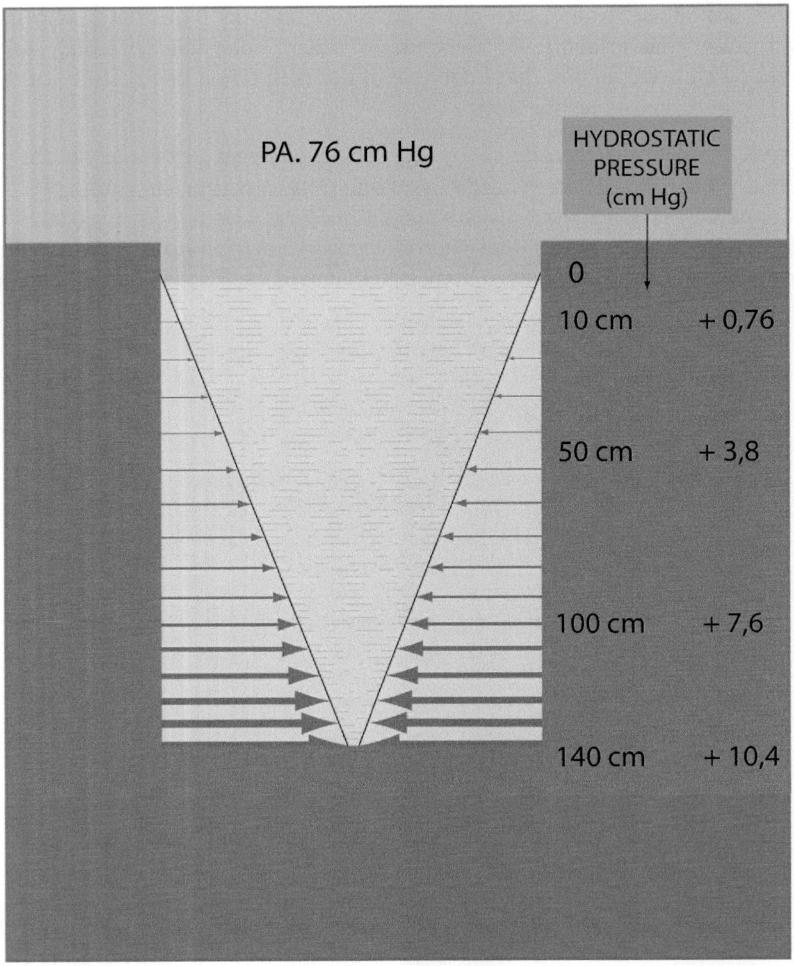

Figure 2. Hydrostatic pressure.

hindered: a circumstance that is favorable in some cases of respiratory pathology (for example, in emphysematous patients). This should be taken into account for people who are going to carry out exercises in the water, giving them instructions to take advantage more effectively of such a situation (San Martin 2006).

Furthermore, the hydrostatic pressure and forces intrinsic to water, such as cohesion, viscosity, and surface tension, can be the origin of stereoceptive stimuli, detected by specific receptors, and result in a better perception of the body scheme, balance, and sense of movement, which is very useful in the treatment of people with post-traumatic or neurological processes.

Hydrokinetic factors: In applications with pressure, showers, or jets, applied directly to the skin or under water, the effect of percussion or massage is added to the temperature, and is a source of receptor stimulation on the skin. Acting reflexively or by direct action, it facilitates muscle relaxation, the release of adhesions, increased blood flow, sedation, and analgesia (San Martin 2006).

Hydrodynamic factors: Other forces intrinsic to water, such as intermolecular cohesion, surface tension, viscosity, and density, which oppose the movement of a body in water, can offer resistance to it of up to 600 times greater than in air. This resistance hydrodynamics can be modified by circumstances or factors extrinsic to the water, such as turbulence, water agitation, direction and speed of movement, surface to mobilize, etc. This makes it possible to program a wide range of exercises—from the most facilitated to the most resisted, always according to the need, individual convenience, and tolerance—to be included in balneo-kinesitherapy.

It can be concluded that movement therapy in seawater, is of great importance for its usefulness in functional recovery, alterations of the musculoskeletal system, rheumatic, post-traumatic, or of a neurological type. It is the technique par excellence to recover or improve function, flexibility, and power muscle, without forgetting the beneficial action that can be obtained by large numbers of people who, without any pathology, wish to recover and maintain physical health and thus avoid the harmful effects of a sedentary and stressed life.

It is an irreplaceable therapy, useful in many processes, mainly in the recovery of functional disorders of the musculoskeletal system, especially in osteoarthritis as adjuvant therapy, with local and general effects, with immediate and late effects, such as vascular, immunological, mediator intervention chemicals, release of endorphins, etc., and which must be specifically prescribed attending to the needs of each individual. It is one of the most sophisticated techniques and the only one with which you can obtain simultaneously the effects derived from exercise, heat application, and massage (Pemberton triad) (San Martin 2006).

The physiological effects of floatation and hydrostatic pressure are summarized in Table 2.

Table 2. Physiological effects of floatation and hydrostatic pressure.*

Effects on the cardiovascular and renal systems	Totable increase in cardiac output, approximately 50% An increase in peripheral blood flow, due to a decrease in peripheral resistance Decrease in the concentration of hemoglobin, the number of cells and the concentration of albumin Decrease in blood viscosity Mobilization of the edema and improvement of this symptom in certain diseases Significant increase in diuresis and natriuresis
Effects on the respiratory system	Overload of the inspiratory muscles, which makes inspiration difficult and expiration easier Increase of intrathoracic pressure, modifying respiratory function, and decrease of residual volume

* Thermoneutral temperature

Therapeutic applications of the immersion: Hydromechanical factors are of great importance in several health problems, which can be summarized as:

- rehabilitation and treatment of degenerative pathologies
- general health and well-being
- care, prevention, and recovery of vascular disorders and, in general, alterations in venous return
- toning and relaxation by practicing exercises in the water

The therapeutic applications of the immersion are the following (Sinclair 2020: Therapeutic applications derived from flotation and hydrostatic pressure:

- exercise assistance: water facilitates active movement when the movement is carried out in the direction of the water surface, very important when there is muscle weakness.
- resistance means to improve muscle strength: when the movement is carried out in the opposite direction to the surface of the water.
- reduction of compressive force on joints by providing support or resistance improving joint mobility.

To vary the thrust force, flotation or ballast devices or devices can be added to the extremities, using hydrodynamic resistance factors (blades, fins, creating turbulences, ...) (2013).

Effects and applications of immersion on proprioception, balance, and coordination:

- hydrokinesitherapy allows kinesthetic memory to be safeguarded so that exercises can be performed with less effort and load on the joints
- hydrostatic pressure, together with viscosity, gives rise to sensory stimuli that allows a better perception of the position of the limbs and the body scheme
- improved balance and coordination

Other effects of immersion:

- helps improve the psychological and emotional state of patients with pool therapy (a sense of security, which is followed by a feeling of euphoria, and physical and mental relaxation)
- facilitates the return circulation of the lower limbs, thanks to the action of hydrostatic pressure on the veins and lymphatic system of the lower extremities
- muscle relaxation and antispasmodic action (in prolonged immersion at a thermoindifferent temperature). It produces generalized relaxation but, if prolonged for too long, fatigue and tiredness
- respiratory re-education: pressure on the abdomen and thorax offers resistance to inspiration, which can be used to strengthen the respiratory muscles
- added to the hydrokinetic factors due to the mechanical factor of pressure acting on the skin, the massage effects in baths with water agitation are added

Chemical factors

Seawater is a specific type of hypertonic sodium chlorine water, with a high content of calcium and magnesium which are of interest to human health. The salt contained in seawater consists mainly of six saline components, namely, the anions chloride and sulfate, and the cations sodium, magnesium, calcium, and potassium. These saline components account for over 99% of the total salt content of seawater. The remaining around 1% of salt content consists mainly of cations strontium and barium and the anions bicarbonate, carbonate, bromide, borate, fluoride, and nitrate (Ludwig 2022). (For more information about seawater composition, see Chapter 2).

Seawater contains more than 104 minerals and trace elements, some of them essential in human blood (sodium, potassium, calcium, magnesium, chloride, sulfate, etc.), and some others, such as fluoride, iodine, phosphorus, iron, silicon, rubidium, and lithium (Millero 2002).

Seawater also contains undissolved organic and inorganic substances in a suspended, finely distributed, or very finely distributed state. Organic matter is present both in inanimate form and as microbes and microorganisms (algae, plankton, bacteria, etc.). Suspended matter is usually a mixture of organic and inorganic substances and living organic substrates (Ludwig 2022).

Thus, its main actions are (San Martin and Armijo 1994):

- General stimulants of the body.
- Increase hemopoietic functions.
- Response capacity enhancer.
- General toning.

Bathing in seawater is a stimulant of the organic functions and especially metabolic, endocrine, and nervous functions, especially when it is cold since hot seawater baths attenuate the effect. Hot baths are peripheral vasodilators, blood pressure modifiers, and respiratory and nutritional stimulants, but the most interesting thing is their sedative and analgesic power.

They are generally combined with climatotherapy, so that the heliomarine cure is a general stimulant of metabolic turnover, hematopoiesis, and digestive functions. It also has a euphoric action characteristic of, at least in part, the better and more harmonious functioning of the various organs and systems (San Martin and Armijo 1994).

The crenotherapeutic effects of seawater can be also considered in terms of local effects and systemic effects.

Local effects: as seawater is a hypertonic sodium chloride water, an osmotic effect is produced, from which effects are derived at the nasopharyngeal level, with anti-congestive effect and recovery of ciliary motility, mainly with hot sea water, in nasal irrigations, inhalations, or sprays.

Seawater also improves skin hydration, and several studies suggest that can be useful in the treatment of psoriasis and atopic dermatitis (Lee et al. 2016, Chun et al. 2017).

Systemic effects: This refers to the possibility of percutaneous absorption of the multiple components of seawater: a topic under constant discussion by the scientific community. Although more research is needed, it appears that percutaneous absorption of certain chemical elements, such as sulfur, iodine, potassium, magnesium, and calcium, may occur, especially in hot baths, since transmineralization is considered to occur above 32 °C (Teixeira 2022).

Thalassotherapy and marine cure: main indications

The effects of marine cures on different alterations and diseases have been widely studied in the 60–80s of the last century, especially in Eastern countries and France, producing abundant bibliography, although generally not written in English. Some examples are: dermatoses (Uhrik et al. 1964), psoriasis (Zlatkov et al. 1973), eczema (Pürschel 1973), asthma (Etspüler et al. 1975), venolymphatic pathology (Mulot 1988), gynecology (de la Farge 1961), cardiovascular diseases (Bogutskiĭ et al. 1977), and ophthalmology (Sedan 1962), among others. Despite that, the study of the benefits of thalassotherapy has been neglected for decades, and recently, more interest in this natural therapy has been raised.

As has been before mentioned, the health benefits of marine cures are derived from the combination of thalasso-hydric elements, such as seawater, climate, sun, sand, seaweed, and marine mud. This marine cure should be performed at least for a week to make the most of the benefits.

Many different treatments can be found in the different seaside resorts and thalassotherapy centers, both for recovery and wellness purposes.

Antonelli and Donelli (2021) revised evidence-based clinical uses of thalassotherapy, finding 560 articles but only fourteen of them were included in the revision, showing that the clinical effectiveness of thalassotherapy was mainly tested in skin problems (psoriasis, atopic dermatitis, vitiligo) and rheumatic disorders (fibromyalgia and ankylosing spondylitis). Studies showed that the severity of the disease and the patient's quality of life improved significantly after thalassotherapy intervention, with the majority of evidence in psoriasis and fibromyalgia.

The effectiveness of aerobic exercise in a water pool with aerobic exercise performed in the sea by women with fibromyalgia was compared by de Andrade et al. (2008). Patients were evaluated baseline using the usual test in fibromyalgia: Visual Analogue Scale (VAS), number of tender points, Fibromyalgia Impact Questionary (FIQ), Short Form 36 Health Survey (SF-36), Pittsburgh Sleep Quality Index (PSQI), and FMS Beck's Depression Index (BDI). After 12 weeks, results showed that both groups improved significantly in post-treatment for all the evaluated variables. There were no significant differences between the two groups, except for BDI ($F = 2.418$, $P < 0.0001$). An aerobic exercise program performed in water (pool or sea) was effective for patients with fibromyalgia. However, seawater exercises have shown to bring advantages related to emotional aspects. The authors concluded that exercise performed in sea water (thalassotherapy) is an option for effective treatment at a low cost for patients with fibromyalgia.

In a quasi-experimental prospective study Morer et al. (2020) evaluated the feasibility of a two-weeks course of aquatic therapy and thalassotherapy in patients with a mild post-stroke disability. Patients were evaluated pre- and post-treatment consisting of a thalassotherapy regimen including aquatic therapy in a seawater pool at 32–34 °C for 45 minutes, daily, five times a week. The outcomes measured were the Berg Balance Scale, the Timed Up and Go test, the 10-meter walking test, the 6-min walking test, the Pain Visual Analogue Scale, the WHO Well-being index, EuroQoL VAS, and EuroQoL 5D. A significant improvement was observed in all outcomes measured ($p < 0.001$, except mobility EuroQoL $p < 0.05$), and in the other four dimensions of the EuroQoL 5D and 10-metre walking test (NS). The authors concluded that a two-week intensive course of aquatic therapy and thalassotherapy may be beneficial in the short term by reducing pain and improving the functional status and overall well-being of post-stroke patients.

Vanitha et al. (2023) revised the impact of thalassotherapy on the prevalent condition of asthma and the findings suggest that activities such as seashore walks and residing at sea level are associated with improvements in lung function test results. Coasts rich in halogen elements, iodine, bromide, fluorine, and mineral salts, as the Black Sea coast, effectively warm and humidify the air, facilitating the softening and drainage of bronchial secretions. Results also showed a reduction in the use of prescribed medications due to the mucolytic effects of thalassotherapy.

Recently, Mratskova (2023) revised scientific literature related to the therapeutic benefits of conducting rehabilitation of COVID-19 patients in a resort setting, with the aim to investigate the potential benefits of the application of climate-resort treatment in patients in a period of recovery with symptoms of Long COVID-19 or post-COVID-19 syndrome. The application of natural and preformed physical factors resulted in overcoming residual symptoms in the broncho-pulmonary system, improving the functional capacity of the respiratory system and skeletal musculature, increasing the tolerance to exercises and physical activity, and influencing the psycho-emotional consequences of COVID-19. The author concluded that resort and climatotherapy can be an important factor in the recovery process after COVID-19, for prevention and improvement of the body's resistance to the adverse effects of environmental factors, infectious agents, and the negative effects of modern lifestyle.

Deep Sea Water (DSW) was also investigated, comparing the control group with the water exercise group and DSW exercise group, finding that DSW can have a positive effect on muscle fatigue and muscle damage recovery and can contribute to improving national health and quality of life (Kim et al. 2020).

Chennaoui et al. (2018) investigated the effects of three days of thalassotherapy care combined with sleep management on psychological and physical capacities using subjective and objective tests in healthy workers, finding that three days of thalassotherapy combined with sleep management have immediate beneficial effects on self-reported mood states, anxiety, fatigue, well-being, and daytime sleepiness, as well as on the objective evaluation of sustained attention, in healthy workers. Beneficial effects on lower limb flexibility were observed after 3 days of care, and the positive effects were still observed after the weekend back home and after 3 days of return to work.

Finally, it is worth mentioning that many thalassotherapy centers have been oriented towards wellness treatments, in which thalassotherapy techniques are combined (see Chapter 5) with nutritional advice (see Chapter 9), exerting a synergy that improves the health conditions of the patients.

The combination factors in the marine cure are summarized in Figure 3.

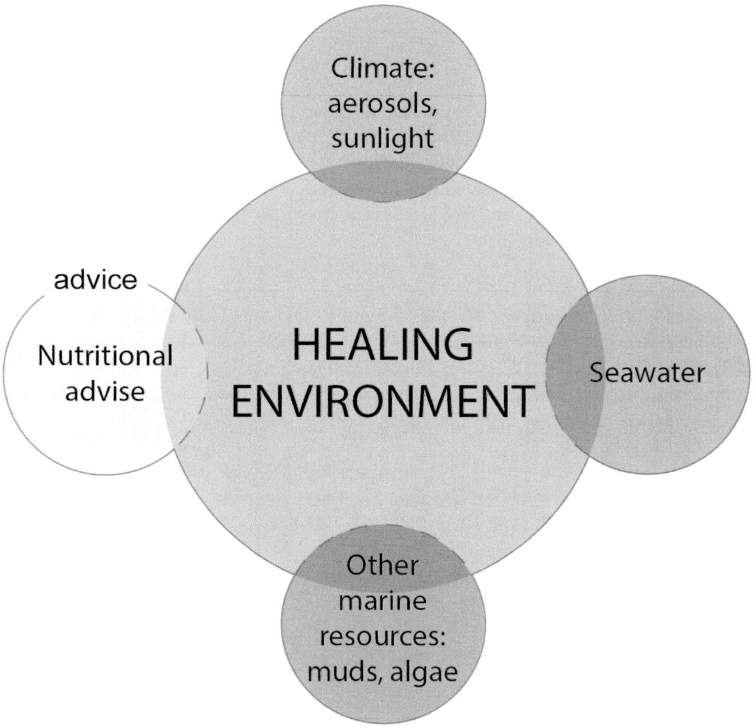

Figure 3. Sinergy in the marine cure.

Indications

After revising scientific literature, the main indications for marine cure can be summarized as follows:

- Respiratory ailments and upper airway processes: Asthma, rhinitis, etc.
- Rheumatic disorders: Fibromyalgia and ankylosing spondylitis.
- Skin diseases: Psoriasis, atopic dermatitis.
- Rehabilitation: Post-stroke disability, rehabilitation post-surgery.
- Others: stress, anxiety, muscle fatigue, etc.

Contraindications

There are no absolute contraindications in the thalassotherapy cure, although, as in crenotherapy, marine cures are contraindicated in acute processes in general, and renal failure.

Conclusions

Although cures in the marine environment have been carried out since time immemorial, thalassotherapy has been forgotten in recent years, being used only as a wellness resource. Experience and some clinical studies show that it is useful in rehabilitation processes and the recovery and improvement of health in certain diseases, such as fibromyalgia or asthma, as well as in situations of stress or chronic fatigue. In any case, improvements in health seem to come from the set of synergistic actions of an extraordinary marine environment, seawater, climate, and other marine resources, such as algae and marine mud, together with nutritional advice.

Therefore, more studies are needed to understand the mechanisms of action involved.

References

Antonelli, M. and D. Donelli. 2021. Thalassotherapy, Health Benefits of Sea Water, Climate and Marine Environment: A Narrative Review. Proceedings of the 6th International Electronic Conference on Water Sciences (retrieved from: https://sciforum.net/manuscripts/11606/manuscript.pdf).

Bogutskiĭ, B.V., V.G. Boksha, M.Iu. Akhmedzhanov, L.A. Kunitsyna and A.N. Gol'dman. 1977. Talassoterapiia bol'nykh s zabolevaniiami serdechno-sosudistoĭ sistemy [Thalassotherapy of cardiovascular diseases]. Vopr Kurortol Fizioter Lech Fiz Kult. (1): 36–9. (in Russian).

Charlier, R.H. and M. Chaineux. 2009. The Healing Sea: A Sustainable Coastal Ocean Resource: Thalassotherapy. Journal of Coastal Research 254: 838–856, DOI: 10.2112/08A-0008.1.

Chennaoui, M., D. Gomez-Merino, P.V. Beers, M. Guillard and C. Drogou. 2018. Benefits of thalassotherapy with sleep management on mood states and well-being, and cognitive and physical capacities in healthy workers. J. Sleep Disord. Ther. 7(297): 2167–0277.

Chun, S.Y., K.S. Lee and K.S. Nam. 2017. Refined deep-sea water suppresses inflammatory responses via the MAPK/AP-1 and NF-κB signaling pathway in LPS-Treated RAW 264.7 Macrophage Cells. Int. J. Mol. Sci. 18(11): 2282. doi: 10.3390/ijms18112282.

de Andrade, S.C., R.F. de Carvalho, A.S. Soares, R.P. de Abreu Freitas, L.M. de Medeiros Guerra and M.J. Vilar. 2008. Thalassotherapy for fibromyalgia: a randomized controlled trial comparing aquatic exercises in sea water and water pool. Rheumatol. Int. 29(2): 147–152. doi: 10.1007/s00296-008-0644-2.

de La Farge, G. 1961. Thalassotherapy in gynecology] Gynecol Prat. 12: 319–34. (in French).

Etspüler, R., K.P. Voget, M. Siegmund, H. Eigelsreiter, P. Blanck and A. Fenner. 1975. Pulmonary function in asthmatic children during thalassotherapy. Padiatr Padol. 10(4): 330–335.

Gutenbrunner, C., T. Bender, P. Cantista and Z. Karagülle. 2010. A proposal for a worldwide definition of health resort medicine, balneology, medical hydrology and climatology. Int. J. Biometeorol. 54: 495–507. https://doi.org/10.1007/s00484-010-0321-5.

Karagülle, Z. 2008. Hydrotherapy, SPA, Balneotherapy, Thalassotherapy [Hidroterapi, SPA, Balneoterapi, Talassoterapi]. Turkiye Klinikleri J. Med. Sci. 28: 224-229. (in Turkish).

Kim, N.I., S.J. Kim, J.H. Jang, W.S. Shin, H.J. Eum, B. Kim et al. 2020. Changes in fatigue recovery and muscle damage enzymes after deep-sea water thalassotherapy. Applied Sciences 10(23): 8383.

Lee, S.-H., I.-H. Bae, D.-J. Min, N.-H. Park, J.-H. Choi, J.-S. Shin et al. 2016. Skin hydration effect of jeju lava sea water. J Soc Cosmet Sci Korea 42(4): 343–349. doi:10.15230/SCSK.2016.42.4.343. (in Korean).

Ludwig, H. 2022. Chapter 3, seawater: composition and properties. pp. 72–203. *In*: Ludwig, H. [ed.]. Reverse Osmosis Seawater Desalination Volume 1. Springer, Cham. https://doi.org/10.1007/978-3-030-81931-6_3.

Mahé, A. El Plasma de Quinton: El agua de mar, nuestro medio interno. 1999. Icaria Milenrama: Barcelona. ISBN 9788474264494 (in Spanish).

Maraver, F., C. Morer and A. Muela. Evolution of Thalassotherapy in Spain. Conference: 72° General Assembly and International Scientific Congress FEMTEC 2019. At: Greece: Halkidiki. 16–20 October 2019.

Mratskova G. 2023. Medical resort treatment in post COVID-19 patients with persistent musculoskeletal symptoms KNOWLEDGE -International Journal 59(4): 309–315. (Retrieved from https://ikm.mk/ojs/index.php/kij/article/view/6219).

Millero, F.J. 2002. Sea water as an electrolyte. pp. 3–34. *In*: Gianguzza, A., E. Pelizzetti and S. Sammartano [eds.]. Chemistry of Marine Water and Sediments. Environmental Science. Springer, Berlin, Heidelberg. https://doi.org/10.1007/978-3-662-04935-8_1.

Morer, C., A. Michan-Doña, A. Alvarez-Badillo, P. Zuluaga and F. Maraver. 2020. Evaluation of the feasibility of a two-week course of aquatic therapy and thalassotherapy in a mild post-stroke population. Int. J. Environ. Res. Public Health. 5; 17(21): 8163. doi: 10.3390/ijerph17218163.

Mourelle, M.L. 2013. La esencia de la cura marina. Estetica & Wellness ed., Madrid, ES. (in Spanish).

Mourelle, M.L. and I. Sabater. 2013. Spa and Wellness. Manual for therapists. Estetica & Wellness ed., Madrid, ES. (in Spanish).

Mulot, J.P. 1988. Possibilités thérapeutiques des troubles veinolymphatiques par les cures marines [Therapeutic possibilities of venolymphatic disorders using thalassotherapy]. Phlebologie. 41(2): 413–420. (in French).

Munteanu, C., D. Munteanu, M. Hoteteu and G. Dogaru. 2019. Balneotherapy—Medical, scientific, educational and economic relevance reflected by more than 250 articles published in Balneo Research Journal. Balneo PRM Res. J. 10: 174–203. https://doi.org/10.12680/balneo.2019.257.

Park, Y.E. 2007. A Theoretical Study on Thalassotherapy. Kor. J. Aesthet. Cosmetol. 5(2): 229–238. (in Korean).

Pereira, L. 2017, Chapter 12, Thalassotherapy and marine cosmeceuticals. pp. 503–522. *In*: Pereira, L. [ed.]. Therapeutic and Nutritional Uses of Algae, 1st ed.; CRC Press/Taylor & Francis Group: Abingdon, UK. https://doi.org/10.1201/9781315152844

Pürschel, W. 1973. Dermatologische Klimatherapie an der Nordsee. Klinisch-analytische Untersuchungen am konstitutionelle Ekzematoid mit-ohne Asthma bronchiale und-oder Rhinitis atopica [Dermatological climatotherapy on the North Sea. Clinical-analytical studies of constitutional eczematoid with-without bronchial asthma and-or atopic rhinitis]. Dermatologica. 146(1): 1–98. (in German).

San Martín, J. 1994. Crenotecnia. Chapter 15, pp. 195–215. *In*: Armijo, M. and J. San Martín. [eds.]. Curas balnearias y climáticas. Talasoterapia y Helioterapia. Ed. Complutense, Madrid, ES. (in Spanish).

San Martín, J. and M. Armijo. 1994. Talasoterapia. Chapter 42, pp. 611–652. *In*: Armijo, M. and J. San martín. [eds.]. Curas balnearias y climáticas. Talasoterapia y Helioterapia. Ed. Complutense, Madrid, ES. (in Spanish).

San Martín, J. 2006. Chapter 8, Balneocinesiterapia. Tratamientos rehabilitadores en piscina. pp. 73–77. *In*: Hernández-Torres, A. [ed.]. Técnicas y tecnologías en hidrología médica e hidroterapia. Informe de Evaluación de Tecnologías Sanitarias N°50, Madrid, ES. (in Spanish).

Sedan, J. 1962. Thalassotherapy in ophthalmology] Ann Ocul (Paris). 195: 929–36. (in French).

Sinclair, M.B. 2020. Hydrotherapy for Bodyworkers: Improving outcomes with water therapies. Jessica Kingsley Publishers, London, UK.

Teixeira, F. Hidrologia Médica. Princípios gerais. Lidl, Ed. Lisbon, PT. (in Portuguese).

Uhrik, J., B. Kopecka, P. Bartak and S. Sk´usensti. 1964. Thalassoteraoiou dermat´oz [Experiences with the thalassotherapy of dermatoses]. Cesk Dermatol. 39: 42–5. (in Czech).

Vanitha, R., B. Rajendran, M. González-Subirà, M. Schneider and L. Panayotova-Ovcharova. 2023. Conference Paper Varna Medical Forum, vol. 12, supplement 1.

Zlatkov, N.B., B. Bozhkov and D. Genov. 1973. Serum copper and ceruloplasmin in patients with psoriasis after helio- and thalassotherapy. Arch. Derm. Forsch. 247: 289–294. https://doi.org/10.1007/BF00596249.

CHAPTER 4

Thalassotherapy and Marine Resources

Health Benefits

Franco Mauro Lamacchia

Historical notes

Thalassotherapy constitutes a fundamental branch of Medical Hydrology. The History of Medicine dates back millennia to the first therapeutic uses of seawater, marine elements such as sand and silt, and the marine climate.

The beneficial virtues of the sea were already known to the Egyptians, Phoenicians, Greeks and Romans, who exploited the therapeutic properties of the various components of the marine environment. Both the Greeks and the Romans recognized seawater's many therapeutic activities not only for the treatment of illnesses but also for the maintenance of the body's health and beauty. No less important among the Romans was exposure to the sun. The solarium, in fact, in Roman baths, was used for sunbathing and Cicero called sunbathing "*sol arsus*" if performed on dry skin and 'sol unctus' if exposure occurred on greasy skin. In various parts of Europe, the roots of thalassotherapy run very deep and we find an early trace of it in a description by Michele Savonarola (1384–1468), a professor in Padua and Ferrara, in his '*Le malattie che hanno dalla cura marina beneficio e quelle che ne traggono danno*' (Diseases that benefit from marine treatment and those that harm it), while in Venice in 1553 a masterpiece of medical hydrology and perhaps the first treatise on balneotherapy was printed: "*De Balneis omnia quae extant apud Graecos, Latinos et Arabas scriptores*" (Bonsignori 2011).

Modern' thalassotherapy originated in England in 1700 with the studies of Richard Russel (1687–1759) whose most important works were "*De tabe glandulari, sive de usu aquae marinae in morbis glandularum*" and "*On the use of sea water*". Both published in 1750, they earned him the appellation "L'inventeur de la mer" by Michelet. Russel used the marine environment for therapeutic purposes in his Brighton establishment and described the therapeutic benefits of natural agents from the sea: sun, sand, heated seawater. In 1769, Lazzaro Spallanzani sensed the bactericidal action of the sun, while the development of medical science, physics, chemistry and, above all, the spirit of research in the 18th century led to interesting observations on the effectiveness of heliotherapy in certain dermatitis, ulcers, wounds, fractures, ascites, bone TB and rickets. In that same period,

Riminiterme - Via Principe di Piemonte, 56 47924 Rimini Italy.
Email: direzionesanitaria@riminiterme.com

thalassotherapy also had a strong development in France, where Maret's works were published, describing the mechanism of action of freshwater and seawater baths and their use, Blot's "*Manual des bains de mer*", and Roccas's "*Hygienic and therapeutic manual of seawater baths*". And also in France, in Dieppe, the first 'House of Marine Thermal Health' was founded in 1778, while the term 'thalassotherapy' appeared for the first time, in 1867, in the work of French physician La Bonardiére. In the 19th century, thalassotherapy became a true discipline following research by Claude Bernard and Rene Quinton, who demonstrated the affinity of the pH and chemical composition of seawater with the extracellular fluid (Bonsignori 2011).

We cannot fail to mention the masterful work of Rene Quinton (1867–1925) who produced numerous important scientific papers including "*Sur l'action therapeutique de l'eau de mer*" and "*L'eau de mer, milieu organique*" presented in 1904 at important international congresses. Thalassotherapy, as we know it today, is thanks to Louis Bagot who established the first Thalassotherapy Centre in Roscoff, France, in 1899. It was in France that thalassotherapy institutes underwent their greatest development, and it was the French School that drew up the currently recognized and codified criteria on the application of thalassotherapy methods. In fact, the International Federation of Thalassotherapy, founded in France in 1986, defined thalassotherapy as "the simultaneous utilization under medical supervision with preventive and curative purposes, of the benefits of the marine environment including: the marine climate, seawater, sea mud, algae, sands and all other substances of the sea, in a privileged marine site" (Luccheta 2008, Monti 1948, Padrini 2010).

Introduction

Thalassotherapy is a particularly complete integrated climatotherapy method as it offers the possibility of exploiting the complex therapeutic synergy between seawater and the factors of the marine climate (Napi 1996).

Seawater is an important and "complete" mineral solution, possessing a large number of fundamental elements for carrying out biological and therapeutic actions.

The main characteristic of seawater is its salinity, which expresses, with the prevalence of chlorides, the total amount of salts expressed in units of weight contained in one liter of water and which corresponds to the salt residue obtained by evaporating the water and drying the sediment at different temperatures. In addition to saline compounds, seawater also contains gases present in the atmosphere: oxygen, carbon dioxide, nitrogen, etc. (Bonsignori 2011).

Another characteristic of seawater is its alkalinity. The pH of seawater varies between 7.5 and 8.4.

Thalassotherapy, which uses marine elements for therapeutic purposes, can be applied with baths and pools, poultices, sandblasting, irrigations, inhalations, aerosols, and nebulizations (Gualtierotti 1974).

Biological actions of seawater

The main action of seawater is anti-inflammatory. This action can be observed at the level of different apparatuses. Seawater stimulates the immune system, has an anti-oedemigenous and resolving action, stimulates mucous secretion, the endothelial reticulum system, and the production of secretory and circulating immunoglobulins, and has an antiseptic action. These actions are due to the osmotic and hypertonic properties of water and the halogens such as chlorine, iodine, and bromine to enhance the action of certain lysosomal enzymes (Bonsignori 2011).

Actions of seawater

The actions of seawater depend on the physical properties of the medium and the method of application. A hydro-mineral medium can be used in a liquid state (bathing, irrigation), gaseous state (aerosol, nebulization), semi-solid state (poultices and silts) and the resulting actions are related to heat, buoyancy, external and internal mechanical pressure, humidification, washing, osmotic pressure. Heat represents the most important effect.

The therapeutic actions for which heat is most frequently used are reduction of inflammatory infiltrates and anti-oedemigenous action, analgesic effect, improvement of joint motility, alleviation of muscle spasm, local hyperemia, and improvement of trophism. There are also numerous actions on metabolism and cell physiology. Tissue metabolism increases and many enzymatic reactions only occur in the presence of increased temperature. Heating can immediately provoke an inflammatory response, but thermal heat is also used for its anti-inflammatory properties on chronic and subacute inflammation and to prevent acute episodes of relapsing disease processes.

There are several reasons for the analgesic effect; the best known is the ability of thermal stimuli (hot and cold) to raise the excitability threshold of pain receptors. However, other mechanisms may be involved, including interference with prostaglandin synthesis, central and hormonal events such as ACTH synthesis, endorphin production, and opioid peptides.

Heat is capable of modifying the physical properties of both fibrous tissue and tendon and capsular ligament structures. As far as the myorelaxant action is concerned, it is thought to occur through two levels of intervention: the first on the neuromuscular spindles that heat makes less sensitive to stretching, the second, reflexively, through the stimulation of skin thermoreceptors, which would imply the intervention of a central control on the same mechanism. Blood overflow is linked to arteriolar and capillary dilation causing active hyperemia generated by both direct and reflex mechanisms.

Cataplasm is a moist-hot medium that adheres to the skin via a thin layer of water. The solid component has physical and chemical characteristics that make it peculiar for adherence, heat storage, heat loss, and heat transfer to the body.

Due to the porosity of sand, psammotherapy allows very high temperatures to be reached without causing side effects. This therapy produces thermal stress on the patient from which benefits arise. The organism responds by setting in motion a series of mechanisms in order to adapt to the situation: in short, it is stimulated to react.

Regarding the role of buoyancy, it should be remembered that the human body immersed in water loses a large percentage of its weight in proportion as the specific weight of the water increases.

The property of buoyancy, together with the external mechanical (hydrostatic) pressure that the liquid exerts on all submerged parts, is used in hydro-kinesiotherapy.

In fact, the discharge of body weight and the density of the medium allow controlled movements with minimal algic components, aspects that make the recovery of motor function particularly effective, even in early phases and in less time.

Humidification and washing, possessing a modest soothing and anti-inflammatory action, are used both on the skin and mucous membranes. The osmotic pressure of a particularly concentrated water such as seawater is exploited on the mucous membranes as it creates fluid currents towards the outside of the mucous membrane and exerts a drawing action on particles and microorganisms (Bonsignori 2011).

Application methods and indications

The main application methods of thalassotherapy are sunbathing, immersion, application of sand and silt, inhalation, and irrigation. The indications are related to the technique; general indications and contraindications are summarized in Table 1 (San Martin and Armijo 1994).

Table 1. Indications and contraindications of thalassotherapy.

Indications
Rheumatisms: osteoarthrosis, chronic inflammatory rheumatism, extraarticular rheumatism.
Circulatory problems: constitutional hypotonic phlebopathy, varicose vein syndrome, saphenectomy and post-phlebitis syndrome, oedema, venous insufficiency of the lower limbs, peripheral vascular diseases, outcomes of phlebitis, thrombophlebitis surgery sequelae, varicose syndrome, atherosclerotic arteriopathies, Buerger's disease, lymphatic stasis, post-traumatic oedemas.
Trauma and surgical outcomes, surgery sequelae, dislocations, sprains, muscular contractures and atrophies, myopathies and polyneuritis.
Respiratory illness: chronic catarrhal, sinusitis, otitis, rhinitis, pharyngitis, tracheitis and bronchitis.
Dermatological diseases: psoriasis, atopic dermatitis, and ichthyosis
Others: convalescence, gerontology and chronic fatigue
Contraindications
Acute illnesses in general Heart failure Renal insufficiency Serious liver disease

Heliotherapy

Heliotherapy is performed by exposing the body without clothing to the sun. The resulting effects are a feeling of relaxation, a tonic-stimulating action, especially on vitamin D, and an increase in metabolism. These are in turn linked to a resolving action on exudates and serous infiltrates, an analgesic action with an increase in the pain threshold, an immunostimulating and pigmenting action. The main indications for heliotherapy are: convalescence following serious illness, anemia, arthralgia, osteoarthritis, fibromyalgia, acne and seborrhoeic dermatitis (Hentschel 1965, Antonelli and Donelli 2019).

Balneotherapy

It consists of full-body immersion in seawater.

Balneotherapy with seawater heated to a temperature of about 38 °C, applied in individual tubs with complete immersion of the body for 15 to 30 minutes for 12 days, exerts a hydrodynamic action, a thermal action, and a chemical action. After the bath, the patient is wrapped with a blanket and held in the dressing room for the reaction, so that the transfer of heat from the body to the environment occurs gradually. The hydrodynamic action is a consequence of the high specific weight of seawater, which allows for more correct execution of movements and a reduction in muscle fatigue, the thermal action facilitates circulation and tissue nutrition, and the chemical action is related to the mineralization of the water itself (Bonsignori 2011).

The main biological actions produced by bathing with heated seawater are active vasodilation, an increase in oxygen consumption, and an increase in body temperature (Teixeira 2022).

The elective indication of balneotherapy is degenerative rheumatism: the baths can soothe pain, increase the mobility of the affected joints, and resolve muscle spasms and contractures. In arthrosis in its various localizations, marine balneotherapy produces a hyperemic action with improved vascularization of the joint complex (Bonsignori 2011).

Hydromassage is a special balneotherapy method that exploits the pressure exerted by jets of seawater. Baths equipped with multiple nozzles are used for this purpose, from which water mixed with air, also enriched with ozone, comes out under pressure. The jets are oriented in a centripetal direction, from the sole to the thorax, to promote return blood circulation.

Figure 1. Hydromassage bath.

The therapeutic actions of whirlpool bathing derive (in addition to those of balneotherapy) from the exploitation of the physical massaging action of the water jet.

Indications for hydromassage are constitutional hypotonic phlebopathy, varicose vein syndrome, saphenectomy and post-phlebitis syndrome, osteoarthrosis, chronic inflammatory rheumatism, extraarticular rheumatism, trauma, and surgical outcomes, and oedema.

Hydro-kinesitherapy

It is carried out in a rehabilitation pool with seawater at a temperature of around 34 °C. The depth varies from 110 to 140 cm.

It represents the method of maximum and best integration between the chemical-physical properties of water and the therapeutic properties of movement.

Due to its physical-biological properties, seawater decisively favors the performance of active and passive mobilization exercises.

The main characteristic is the manifestation of the antigravity action according to Archimedes' law, which determines the lightness effect of the submerged body.

Seawater supports a large part of the body's weight, facilitating the execution of movements with correct muscle work even in conditions of reduced tone-trophism and difficult loading. For this reason, a muscle that has reduced strength and does not allow proper work can perform various movements in water.

The reduction in body weight that is achieved in the saline water medium also makes it possible to eliminate pain due to loading and to start motor rehabilitation at an early stage.

Equally important is the analgesic effect of the heat, which is obtained by using seawater heated to a temperature of 34 °C. The heat in fact raises the threshold of pain sensitivity, also leading to general sedation and muscle relaxation (Talenti 1970).

Another effect is the manifestation of sensory phenomena such as a better appreciation of one's body position and sense of movement due to the perception of the pressure exerted by the water on the entire body, which translates into a feeling of greater safety when performing movements.

The marine water medium, which is denser than air, also allows controlled movements performed against slight resistance. This reduces the risks of uncontrolled movements that can trigger symptoms, worsen the pathological situation, give fear and insecurity to the patient, and slow down their re-education (Bonsignori 2011).

The indications for its use are arthrosis, periarthritis, fracture outcomes, surgery sequelae, dislocations, sprains, muscular contractures and atrophies, myopathies and polyneuritis, and all

Figure 2. Hydro-kinesitherapy

pathological forms that restrict movement. Another recent indication is fibromyalgia (Zijlstra et al. 2005).

Vascular path

The vascular path consists of two corridors complete with whirlpools with seawater at different temperatures (33 °C and 23 °C) and the depth of the water is of about 80–90 cm.

It allows for walking exercises and vascular gymnastics; the bottom equipped with blowers produces a massage that promotes venous outflow where there is stagnation. The salt in the seawater also has a stimulating action on the microcirculation and an osmotic outward action.

Lastly, the temperature difference between the hot and cold corridor allows for vascular gymnastics with repeated vasoconstriction and vasodilatation. The rapid temperature change stimulates the peripheral circulation by alternating dilation and constriction of vessels and capillaries; the rapid succession of hot and cold stimuli has a spasmolytic effect on the vascular smooth muscles and an activating and favorable action on the collateral circles.

Figure 3. Vascular path

The indications for the use of the vascular pathway are all forms of venous insufficiency of the lower limbs, peripheral vascular diseases, outcomes of phlebitis, thrombophlebitis surgery sequelae, varicose syndrome, atherosclerotic arteriopathies, Buerger's disease, lymphatic stasis, post-traumatic oedemas (Bonsignori 2011).

Psammotherapy

Psammotherapy consists of using sand as a therapeutic medium. It is an effective treatment method, especially in the treatment of arthrosis in its various localizations.

The dryness and physical discontinuity of the sand facilitate sweating and allow the body to maintain thermal equilibrium even at temperatures not achievable with other therapies. It is an intense dry thermotherapy to which is added the stimulation of sand granules and the various substances, organic and inorganic, contained in seawater.

This complex physical-chemical stimulus brought to the patient's skin exerts local and general effects. These are mechanisms in which the environmental climatic factors and the more properly thalassotherapeutic ones interact, mutually influencing each other.

Sand is the product of the fragmentation of minerals derived from the disintegration of siliceous and calcareous rocks by the action of water and weathering.

Sand, frequently removed and renewed, must receive periodic spraying with seawater to enrich itself with salt components.

The ideal temperature for application varies from 45 °C to 55 °C; the hole must be 25–30 cm deep, 2 meters long, and 1 m wide; on the bottom layer 15 cm of dry sand, while on the edges sand is piled in such a quantity as to cover the patient with a layer of 2–3 cm. In fact, given the poor conductivity of heat, a thicker layer would cause the sand in contact with the skin to cool.

The patient, who has previously undergone a thorough medical examination, lies naked in the pit and is covered with sand. The applications can be total or partial. If the patient is free of contraindications, it is clearly preferable for the sandblasting to be total, i.e., covering the entire body surface. In this case, local and general effects can be exploited. In other cases, sandblasting is only performed on the affected area (knee, spine, etc.), leaving the remaining part of the body uncovered.

The duration of a single sandblasting is 5–10 minutes on the first day, increasing thereafter by 5 min for each day up to a maximum of 20 min for general sandblasting and 30 min for partial sandblasting. A maximum of 12 total and 15 partial applications are generally applied. At the end of the sandblasting, the patient is taken out of the pit and wrapped in a towel. He/she is then led into the Calidarium where he/she is wrapped in a woolen cloth and laid on a couch where he/she remains for about 20 min during which the patient retains heat and produces an abundant diaphoresis. When the reaction is over, the patient goes to the shower room where he or she will have a copious full-body wash with heated seawater. After the shower, wrapped in a bathrobe, he goes to another closed room (Tepidarium) where he can rest and gradually recover his basic condition.

Psammotherapy has broad indications in the field of rheumato-arthropathy and the main indications are as follows: arthrosis in its various localizations, outcomes of fractures, after-effects of osteo-articular traumas, dysmetabolic arthropathies, outcomes of myositis, bursitis, tendinitis.

Psammotherapy produces an antalgic and muscular decontracting effect that favors the functional re-education of joints. It is also able to act on pain, muscular contracture, and the stiffness of capsular ligament and tendon structures (Federici 1980).

Psammotherapy constitutes a complex physical-chemical stimulus in which the actions of the marine climate—in its particular hydroclimatic characteristics of the area in which it is carried out—the size of the sand grains, and the inorganic chemical composition of the sand are invoked along with heat.

The main activity of psammotherapy is local and the general effects are caused by heat. Sand, due to its composition of fine granules between which air is interposed, is a porous system with low thermal conductivity and limited specific surface area, which explains why burns do not occur despite the high sand-skin thermal gradient. In the contact of the skin with sand at 50 °C–60 °C heat is transferred by conduction and convention.

Skin thermoreceptors transform the energy they receive into modulated impulses that are sent to the brain. Our cutaneous apparatus constitutes a vast reflex assembly: heat is also able to stimulate other non-specific receptors such as those of tactile and pain sensitivity, thus explaining the algosedative action of psammotherapy.

The application of psammotherapy induces an impressive and complex response of the organism by activating the regulatory centers of thermoregulation and stimulating all the mechanisms deputed to the maintenance of thermal homeostasis (Antonelli and Donelli 2019).

The general activity is expressed through a stimulus on the pituitary-adrenal axis and the release of corticosteroid hormones, whose activity is of fundamental importance in the short- and medium-term effects produced by this treatment method, which can be interpreted as an intense stress-inducing increase in ACTH.

Like other thermoergic applications, psammotherapy would also produce the release of tissue hormones at skin level, almost histamine and bradykinin, which locally induce vasodilation with increased capillary permeability and profuse sweating. In the case of application to the whole body, they pass into the circulation and induce general reactions such as an increase in cardiac output and increased oxygen consumption. The heat improves tissue circulation through hyperemia (Antonelli and Donelli 2019).

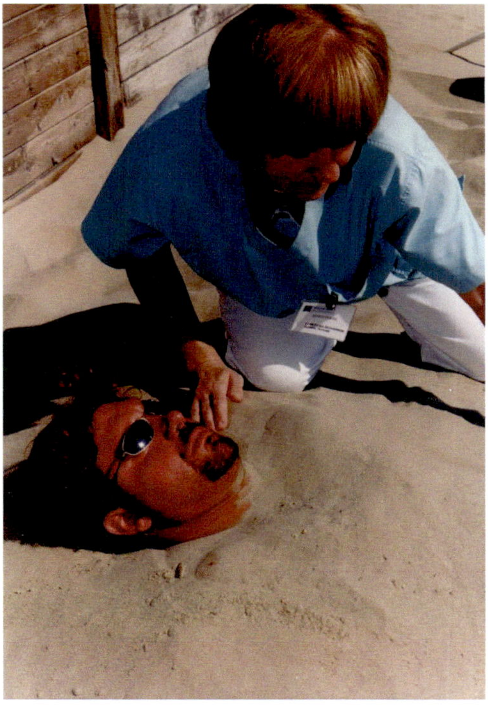

Figure 4. Psammotherapy.

Cataplasms

Another application of thalassotherapy is with silts of marine sedimentary origin whose solid component is sand and clay soaked in seawater. The applications, called poultices and often associated with balneotherapy, are carried out on all or parts of the body.

At 'room' temperature, the action is analgesic, sedative, myorelaxant, keratoplastic, and eutrophic.

At higher temperatures (42 °C) the stimulating, vasodilating, dehydrating action prevails. Topical applications can also be made on the face not only for cosmetic purposes but also in the treatment of acne rosacea and eczema.

Typical components of cataplasm are Fucus, Laminaria, Litotamnion as well as Sea Water of course. Marine muds are also used in the form of poultices or cataplasms. The mechanism of action of cataplasms is above all linked to a physical action brought about by thermal stress, which leads to neuroendocrine reactions and local hyperemia with consequent elimination of phlogogenic substances produced by the cellular component of chronic phlogosis, such as proflogistic enzymes of lysosomal origin released by neutrophils and monocytes. In addition, a reduction in muscle hypertonus occurs with the interruption of the vicious circle:

pain>muscle contracture>altered joint dynamics>pain, which characterizes many chronic arthropathies (Sukenik et al. 1990, 1992).

Furthermore, there is an improvement in the trophism of the articular connective tissue and the metabolism of cartilaginous tissues through intervention in the synthesis of various mediators such as the reduction of interleukin-1β and tumor necrosis factor-α, which are mediators of cartilaginous damage and inflammation, and through an increase in insulin growth factor-1 (IGF1), which stimulates cartilaginous metabolism, and transforming growth factor-β (TGF-β), which is a powerful cytokine with antiphlogistic activity.

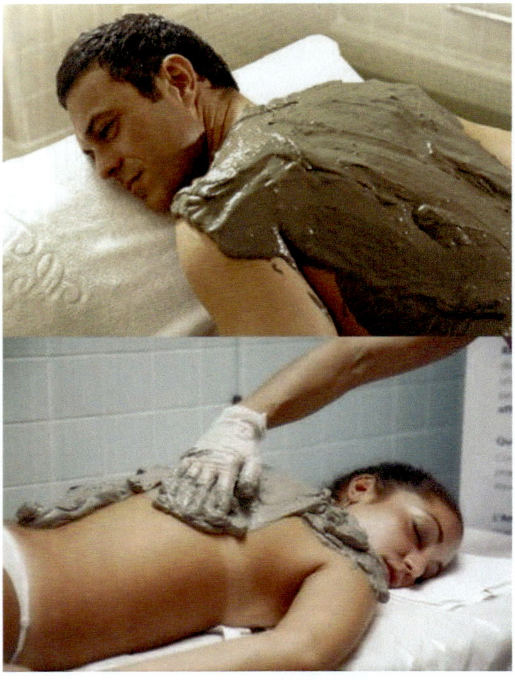

Figure 5. Application of mud cataplasm.

There is also a pain-relieving action through the release of endogenous opiates such as beta-endorphins with a powerful morphine-like action with increased production of cytokines and killer lymphocytes, and an anti-inflammatory action through non-specific stimulation of the hypothalamus-hypophysis-adrenal axis with increased ACTH synthesis and decreased pro-inflammatory mediators, prostaglandin E2 and leukotriene B4 (Sukenik et al. 1990, 1994).

The indications for this type of treatment are rheumatism and arthrosis, fracture outcomes, and diseases of the musculoskeletal system (Federici 1980).

Inhalation treatments

They constitute an important field of application of thalassotherapy through inhalations, aerosols, nebulizations and nasal irrigations. In the respiratory field, local and general biological effects can be recognized.

The local activity is due to hypertonia, sympathicotonic action, and colloid-osmotic phenomena.

At the same time the saline molecules, due to the high concentration of the various solutions, carry out their antiseptic and resolving action by contact or absorption at the mucosal level and by stimulation of mucociliary activity.

The general activity is secondary to an energetic stimulation of the saline compounds on the peripheral nerve endings of the mucous membranes of the respiratory system, a sort of mild "irritation" which, by favoring cellular activity, improves tissue trophism and activates turnover.

Hyperreactivity goes hand in hand with hypersensitivity to seawater, which acts as a modulator of microcirculatory reactivity (with a synthesis of prostaglandin E). In fact, the advantages of marine inhalation treatment are greater in pathologies linked to a lack of reactivity to climatic factors (perfrigeration) than to seasonal viral phlogosis (de Bernardi et al. 1983).

The therapeutic effects of respiratory treatment with seawater are many and varied.

The anti-inflammatory effect in chronic non-specific phlogosis treated with seawater never occurs through fibrotic processes. Seawater induces the appearance of an exudative imprint in the repair process of the phlogosis itself. In fact, during the first few days of thermal therapy, a proflogistic stimulation occurs, which then turns into an anti-inflammatory effect that lasts for a long time.

There is also an effect on the immune system through the enhancement of the activity of endogenous pulmonary proteases: enzymes that have the task of lysing the thick and viscous mucus. This action has also been highlighted in exudative-catarrhal diathesis where much of the pathogenesis seems to lie in the lack of response from organic mechanisms.

The increased defensive power of the respiratory districts also makes use of the mechanical influences of the circulatory and muco-ciliary matrix with greater possibilities of bacterial cleansing and dilution.

The stimulating action exerted on the blood circulation and submucosal nerve endings is also very important, causing vascular congestion with submucosal oedema, exfoliation of the most superficial layers of the epithelium and its eutrophication, which is particularly important when there is a state of atrophy or subatrophy. The anti-oedemigenic and resolving effect, particularly evident in the mucous membranes and due to its osmotic power, should also be noted. The hypertonia of the mucous membrane not only generates a current of fluids from the deep layers of the mucous membrane towards the outside, capable of conveying and removing corpuscular elements (microorganisms, pollutants, etc.) but also products of phlogosis, catabolites, and proflogistic enzymes. Finally, there is the trophic effect linked to iodine, which induces epithelial proliferation of the respiratory mucosa and fluidification of secretions, properties that make seawater treatment particularly suitable in atrophic forms. Seawater thus improves mucociliary clearance (de Bernardi et al. 1983, 1998).

Seawater crenotherapy makes use of the inhalation methods used for all thermal waters, except for humage.

They are used as follows (de Bernardi et al. 1983, Zaoli 1990):

- warm-moist inhalation which is a delivery of a vaporized jet with water particles around 100 microns. It lasts about 10 minutes;
- nasal or oral aerosol (with particle fragmentation by compressed air, sonic) with a flow of air with water particles of 1 to 5 microns. It lasts about 10 minutes;
- nebulization, which involves spending an increasing amount of time, from 10 to 20 minutes, in a chamber filled with a mist of micelles ranging in size from a few microns (dry) to more voluminous micelles up to 60 microns that aggregate (wet).

In addition to inhalation methods, irrigation methods are extremely useful. These consist of injecting marine water solution into the nasal cavities through a probe. They last approximately 10 to 15 minutes.

There are different ways of classifying the inhalation methods, but certainly, from a therapeutic point of view, the main one is the one that takes into account the physical characteristics of the inhaled substances, and in particular the particle size of the mineral water and the heat.

As is well known, particles with a diameter of around 100 microns stop at the level of the upper airways (nose, larynx, pharynx); those with a diameter of between 10 and 5 microns can reach the tracheo-bronchial mucosa, while only those of around 3 microns can reach the finest bronchial branches, down to the level of the terminal bronchioles and the alveolar wall.

Inhalations and aerosols use devices for single applications, while nebulizations are mainly a collective method.

The indications for marine inhalation treatments are therefore all chronic catarrhal and atrophic forms of the upper and lower respiratory tract, from sinusitis to otitis, from rhinitis to pharyngitis, from tracheitis to bronchitis.

The respiratory mucosa responds as a whole from a physiopathological point of view (rhinosinuso-bronchial syndrome) and therefore the indications are linked more to the type of reactivity that the mucosa as a whole demonstrates than to the site affected (Canzi 1984).

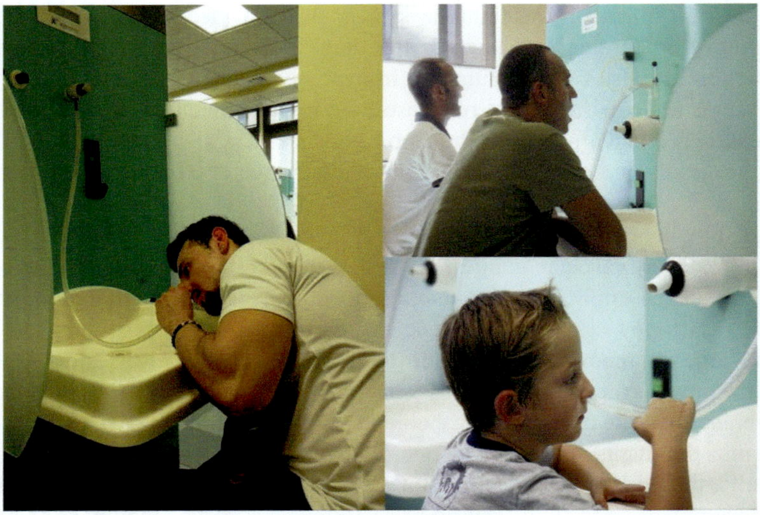

Figure 6. Inhalation therapy.

The reaction to inflammatory stimuli is evidenced by (de Bernardi et al. 1998):

- serous, mucous, mucopurulent or purulent hypersecretion
- lymphatic hyperplasia of the airways within Waldeyer's ring;
- vascular reactions.

Conclusions

Thalassotherapy, through the use of seawater and associated thalasso-hydric resources (seaweed, mud, climate, sun, and sand), is a highly effective therapy in the prevention and treatment of respiratory, rheumatological, and dermatological diseases, when used alone or in combination with pharmacological therapies, always under medical supervision.

However, more research is needed on its effects on the body and greater awareness of the population is required regarding the use of these natural therapies, for preventing diseases and improving quality of life.

References

Antonelli, M. and D. Donelli. 2019. Hot sand baths (psammotherapy): A Systematic Review. Complement. Ther. Med. 42: 1–6, doi:10.1016/j.ctim.2018.10.020.

Bonsignori, F. 2011. La Talassoterapia. Cure e benessere alle terme marine e al mare. ETS ed., Pisa. (in Italian).

Canzi, P. 1984. Tecniche ed apparecchiature. Atti del Convegno A.O.O.I., Terapia medica ORL, (in Italian).

de Bernardi, M., A. Re, G.M. Pedrinazzi, L. Passaroni, L. Ariati, F. Berté et al. 1983. L'azione diretta sulle mucose delle acque minerali - Clinica Termale 1983. (in Italian).

de Bernardi, M., A. Re, G.M. Pedrinazzi and A. Zisiadou. 1998. Influenza della crenoterapia inalatoria sul trasporto muco-ciliare. Clinica Termale 1988 (in Italian).

Federici, P.C. 1980. Sabbie, limi e alghe nella talassoterapia. World Thermalism Congress on Progress in Medical Hydrology, Climatology and Thalassotherapy. Lido di lesolo, Venezia, 1980.

Gualtierotti, R. 1974. Elementi di idrologia medica. Ed. Lucidano, Milano (in Italian).

Hentschel, G. 1965. Report on thalassotherapy in 1965. 1. Bioclimatic principles of the thalassotherapeutic voyage.

Lucchetta, M.C., G. Monaco, V. Valenzi, M.V. Russo, J. Campanella, S. Nocchi et al. 2008. The historical-scientific foundations of thalassotherapy: State of the art. Clin. Ter. 158: 533–541. (In Italian).

Monti, G. 1948. Idrotalassoclimatologia, ricchezza e salute d'Italia. Ed. Nistri Lischi, Firenze, Italy. (in Italian).

Nappi, G. Medicina e Clinica Termale. Selecta Medica, Pavia, 1996 (in Italian). https://www.amazon.it/Medicina-clinica-termale-Giuseppe-Nappi/dp/8808180034.

Padrini, F., M.T. Lucheroni and U. Solimene. 2010. La Talassoterapia. Il Mare Che Cura. Xenia ed., Pavia, Italy (in Italian).

San Martin & Armijo, 1994. Curas Balnearias y Climáticas. Talasoterapia y Helioterapia. Editorial Complutense, Madrid. (in Spanish).

Sukenik, S., D. Buskila, L. Neumann and A. Kleiner-Baumgarten. 1992. Mud pack therapy in rheumatoid arthritis. Clin. Rheumatol. 11(2): 243-7. doi: 10.1007/BF02207966.

Sukenik, S., L. Neumann, D. Buskila, A. Kleiner-Baumgarten, S. Zimlichman and J. Horowitz. 1990. Dead Sea bath salts for the treatment of rheumatoid arthritis. Clin. Exp. Rheumatol. 8(4): 353–7.

Talenti, M. 1970. Idrologia Generale e crenologia. Bulzoni ed., Roma. (in Italian).

Teixeira, F. 2022. Hidrologia médica. Princípios gerais. Lidel ed., Lisboa. (in Portugese).

Zaoli, G. 1990. La talassoterapia nelle affezioni delle prime vie aeree. Comunicazione a convegno. (in Italian).

Zijlstra, T.R., M.A. van de Laar, H.J. Bernelot Moens, E. Taal, L. Zakraoui and J.J. Rasker. 2005. Spa treatment for primary fibromyalgia syndrome: a combination of thalassotherapy, exercise and patient education improves symptoms and quality of life. Rheumatology 44(4): 539-546. doi: 10.1093/rheumatology/keh537.

CHAPTER 5
Seawater Cure Techniques

María Lourdes Mourelle Mosqueira

Introduction

As it has been mentioned in the previous chapter: "Thalassotherapy is the combined use of marine elements (seawater, algae, mud and climate), in a marine environment, for healing and well-being purposes" (Lucchetta et al. 2007). In 2010, Gutenbrunner et al. (2010) offered a proposal for a worldwide definition of health resort medicine, balneology, medical hydrology, and climatology. Though the authors do not mention the word "thalassotherapy", the word Climatotherapy (the use of climatic factors) suggests thalassotherapy is included as a part of Health Resort Medicine. Later, Maraver et al. (2010) suggested adding to that definition the use of seawater and its peloids (thalassotherapy), including its modalities, full-body or local baths, showers, inhalations, irrigations, and peloid packs, and its agents, seawater, marine peloids and sand, among others. After that, for years thalassotherapy was barely mentioned in scientific articles. However, recently, the term has become relevant again and has been mentioned in treatises that, although not specific to thalassotherapy, are related to it, focusing on algae and mineral salts (Pereira 2017, Gomes et al. 2021).

Balneotherapy ("Taking the waters") at a spa was a long-established practice of proven popularity, but sea-bathing and salt-water therapy was a recent arrival on the scene and was the subject of considerable contemporary debate as to its virtues and dangers (Durie 2003). At the very beginning, seawater cure was performed directedly in the sea, so the cold bath was the most common practice (using different contraptions such as tubs on wheels placed under tents). Later around 1820 warm baths were introduced in France and actual thalassotherapy centers were opened in the Atlantic coasts (Charlier and Chaineaux 2009).

Nowadays, seawater is applied to the body using different respiratory (aerosol) or topical techniques, such as bathing and pressure techniques, whether partial (legs and feet, arms and hands) or full body, as well as, sauna and steam baths, and applications of algae and marine mud. The so-called complementary techniques are also used, which include mechanical techniques, electrotherapy, or different types of massages (Mourelle et al. 2023). Added to this, marine

FA2 Group. Department of Applied Physics. University of Vigo.
Email: lmourelle@uvigo.es

climatotherapy has been used in non-allergic bronchial asthma and psoriasis treatments as well (Schuh and Nowak 2011, Ezhov 2021), as heliotherapy in specific cases (Munteanu et al. 2019). Furthermore, a recent study by Shim et al. (2023) with the aim of exploring the relationship between oceanic environmental factors—particularly water, sunlight, and mud—and human health, with a specific emphasis on their therapeutic applications, concluded that marine therapy shows promise as a viable therapeutic approach for improving chronic skin conditions (highlighting atopic dermatitis) and alleviating musculoskeletal discomfort, and improving QoL (Shim et al. 2023).

This chapter provides an overview of the main techniques used in thalassotherapy centers, the actions and effects they exert, and treatment procedures. The complementary techniques are also briefly described, which are combined with the previous ones to organize customized treatment programs.

Thalassotherapy facilities and equipment

According to ISO 17680/2015, thalassotherapy centers should include a minimum of facilities and equipment (Table 1), and also include general specifications (ISO 17680/2015, reviewed and confirmed in 2020, https://www.iso.org/standard/60244.html).

Regarding the general specifications, the standard appoints:

- The design of the facilities shall be such that ensures service provision according to the quality levels established in this International Standard at peak attendance, maintaining the diversity of the services offered.
- The choice of materials shall be specifically adapted to the aggressiveness—especially on metals—of an environment using seawater as well as cleaning and disinfection agents.
- The thalassotherapy center facilities shall be located at a maximum of 1000 m from the coastline in a healthy natural environment and in an area of non-polluted water.
- All facilities open to the public shall be prepared to meet the needs of people with disabilities.

The standard also includes requirements for personnel who are in charge of applying the techniques and/or providing health care treatments, as hydro-therapist/thalasso-therapist qualifications, medical and paramedical staff (medical doctor, physiotherapist, nurse, dietitian, etc.), as well as technical staff, administration staff, and others (cleaning and laundry personnel, etc.).

Procedures and techniques classification

Thalassotherapy requires the combination of sea-derived elements such as seawater, algae, mud, and thermal plankton, obtained from a certain depth of the sea -and at a specific distance from the seashore-, and heated to the body temperature of the human being, to be used in appropriate dosages and durations (Yuncu et al. 2019).

Thalassotherapy techniques can be classified in different ways: according to the pressure (high, medium, and low pressure); according to the body part where it is applied (full body, back, shoulders, etc.); or according to the temperature (cold, warm, hot), but the most usual classification is those based on the type of application: immersion or pressure techniques, drinking cure, respiratory applications, muds, and seaweed wraps and poultices. Climatotherapy, heliotherapy, and sand therapy are also included (Table 2). The so-called complementary therapies can be also found in many centers; some examples are shown in Table 3.

Table 1. Thalassotherapy facilities and equipment (adapted from ISO 17680/2015).

Facilities	**Specifications**
Reception	Independent of all other areas In accordance with the capacity of the center
Offices (administrative)	Doctor´s offices Management office
Dressing room	Possibility for men and women to change separately Equipped by showers Adapted to the hosting capacity of the center
Sanitary facilities	Number according to the hosting capacity
Technical area	Including pumping station, boiler and calories recuperation tanks Recommended to have automatic control system
Health care areas	Structured in "Care Units" Separation of the wet area from the dry zone Care units: - Medical unit - First aid unit - Dry areas: individual treatments (treatment rooms); common treatments (treatment rooms, rest areas and day beds) - Humid areas: individual treatments (treatment rooms); common treatments (treatment rooms, pools, rest areas and day beds). Should include: treatment showers, bath, wraps and pools - Pools: should operate according to the principle of reverse hydraulicity; maximum water height shall be 1,40 m - Optional areas: tea room, vitamin bar, gym
Facilities for the personal	Sanitary facilities Dressing rooms
Equipment	**Specifications**
General	-
Light and ventilation	Natural lighting preferred Indirect lighting Air circuits composed of installations enabling air renewal, circulation, and treatment in a specific unit Ventilation system shall allow for the exclusive circulation of air from clean areas to less clean areas
Heating and air conditioning systems	Temperature should be kept between 18 °C and 25 °C for healthcare areas Recommended individual temperature control in each cabin
Equipment to treat waste sea water	Specific equipment to treat the waste sea water before it is sent back to the sea Reuse of sea water is not allowed
Drinking (not sea) water supply	Supply drinking water (tap water) should comply with national legislation in force

Seawater drinking cure

Seawater has been used as a drinking cure since ancient times, but it was in the last century that this therapy was promoted by René Quinton. He was recognized for the leaps he achieved in ocean water analysis and applications and for his book "*L'eau de mer, milieu organique*", published in Paris in 1904 (Mahé 1999). As it has been described in previous chapters, seawater contains more than 104 minerals and trace elements, some of them essential to human blood (sodium, potassium, calcium, magnesium, chloride, sulfate, etc.), and some more, like fluoride, iodine, phosphorus, iron, silicon, rubidium, and lithium (Millero 2002), and this is the reason why seawater drinking cure might be recommended in some illness as René Quinton proposed.

Table 2. Thalassotherapy techniques and procedures.

Technique	**Types**	**Subtypes / Specifications**
Hydropinic cure	Drinking cure	-
Oral / Dental therapy	Mouthwash / Rinse	-
Respiratory therapy	Nasal irrigation Aerosol Nebulization	-
Bathing techniques	Cold bath Warm/hot bath	In the sea Bath tube Swimming pool
	Bath with additives* or whirlpool	Hydromassage bath Bubbling bath Underwater massage Algae or salt bath
	Phlebologic /vascular path	With jets With bubbles
	Hand and Foot Baths	Warm or contrast
Pressure techniques	Jet showers	High, medium pressure
	Affusion showers	With or without massage
	Scottish showers	Warm/Cold
	Vichy showers	With massage
Saunas and Steam baths	Sauna	Different temperatures
	Steam baths	Different temperatures and humidity
Seaweed applications	Wraps Poultices	Different types of seaweeds
Marine mud	Wraps Poultices Cataplasms	Different mineralization and composition
Seawater Pool therapies	Watsu Ai Chi Halliwick Bad Ragaz Jahara Aquatic PNF	Individual Groups
Seawater Pool activities	Aquagym Aquaerobic AquaRelax Cardio-aquagym Aquabuilding Aquastreching Gymswin	Groups
Marine climatotherapy	Sea breeze	Different temperature and humidity
Heliotherapy	Sun bathing**	Specific timetable and time
Psammotherapy	Sand therapy	Specific timetable and time
Other techniques	Kneipp therapy Halotherapy (Salt chamber) Floatation therapy	Individual Groups Individual

* Adding air, hydro-jets or marine derivatives.

** Medical prescription for a specific illness.

Table 3. Complementary techniques.

Technique	Types/Specifications
Pressotherapy	For circulatory problems
Electrotherapy	High frequency Medium Frequency
Thermo and Criotherapy	Hot packs Cold packs
Massages	Therapeutic, anti-stress Tonic, etc.
Oriental massages	Shiatsu, Thai, etc.
Lymphatic drainage	Vodder Leduc, etc.
Relaxing and strengthening activities	Yoga, Tai-chi and others
Re-educational therapies	Adapted: back-ache, after injuries, etc.
Nutritional therapies	Nutritional counselling Specific diets

After decades of neglect of the seawater drinking cure, it was at the beginning of this century that it was reassumed in Japan: thus developing deep seawater therapy once again.

Deep seawater between about 500 m and 1000 m of depth circulates through the world. Deep seawater (DSW) is known to have such characteristics as being clean and rich in mineral components compared to surface seawater and its application has been attempted in many fields (Miyamura et al. 2004). There are some differences between surface and deep seawater; the abundance of inorganic material becomes higher as the depth of the seawater increases. These characteristics have derived attention for research regarding DSW, especially for its numerous beneficial minerals, which include magnesium (Mg), calcium (Ca), potassium(K), chromium (Cr), selenium (Se), zinc (Zn), and vanadium (V). DSW is claimed to be high in minerals compared to other sources of water (Katsuda et al. 2008). Several studies focused on metabolic disorders (obesity, diabetes), and on cardiovascular disease, cancer, and atopic dermatitis, can be found in the scientific literature.

Due to the increasing of beverage products manufactured using deep seawater (DSW), desalted deep seawater, and concentrated deep seawater, Tsuchiya et al. (2004) investigated the safety of desalted seawater, compared with surface seawater and purified water, studying the effects on hematologic and blood chemical values in mice, concluding that, after 12 weeks of treatment, both seawaters are safe.

Yoshioka et al. (2003), in an *in vivo* study, investigated the pharmacological activity of DSW on hyperlipemia prevention, concluding that DSW controlled the increase of serum lipid values (T-Cholesterol and LDL-Cholesterol) of cholesterol-fed rabbits, and promoted the reduction of serum lipid hyperlipemia. Later, the same group studied the effects of DSW in atherosclerosis prevention, comparing with surface seawater, showing that deep seawater could be useful for the prevention of hyperlipidemia and arteriosclerosis compared to the surface seawater. They also found that reduction of the LDL-Cholesterol level and enhancement of GPx activity were involved in its effects (Miyamura et al. 2004).

Hsu et al. (2011) studied the effects of different concentrations of DSW (hardness 300, 900, and 1500 ppm) showing that all of them decreased liver size, as well as serum total cholesterol, triacylglycerol, atherogenic index, and serum malondialdehyde levels in high-fat/cholesterol-fed hamsters. They also observed that DSW drinking water was also able to increase the serum Trolox equivalent antioxidant capacity, as well as fecal total cholesterol, triacylglycerol, and fecal bile acid contents. And Chen et al. (2013), using the same DSW dilutions, demonstrated the preventive effect of DSW on the development of hepatosteatosis induced by a high-fat diet.

Other health problems could benefit from DSW therapy. Bernardo et al. (2014) investigated the use of DSW in induced anemia in mice. The results showed that DSW isotonic seawater was useful as a complementary therapy in the experimental models, in cases of anemia caused by diets poor in protein, vitamins, and minerals and triggered by medications.

Chun et al. (2017) studied the effects of DSW on atopic dermatitis, finding that DSW suppressed the inflammatory responses via the MAPK/AP-1 and NF-B Signaling Pathway in lipopolysaccharide-treated RAW264.7 macrophage cells.

In a review performed by Mohd-Nani et al. (2016), the authors described the effects on improvement of cholesterol profiles, protection from cardiovascular problems, prevention from atherogenesis, reduction of blood pressure, protection from obesity, treatment for diabetes, prevention from hepatic problems, treatment for fatigue, prevention of cancer, treatment of stomach ulcer and treatment of skin problems as atopic eczema. They also found some studies related to recovery from osteoporosis, treatment for cataracts, and antibacterial activity.

And recently, Takeuchi et al. (2020) demonstrated the beneficial effects of DSW on gut health, showing that DSW was able to improve the intestinal environment, increasing fecal short-chain fatty acids and urinary isoflavones in humans. The same research group, in a randomized double-blind controlled trial, showed that DSW could improve human health by increasing the production of three major fecal short-chain-fatty-acids (acetic, propionic, and butyric acids) and fecal biomarker secretory immunoglobulin A, and accelerated daidzein-to-equol metabolic conversion (Takeuchi et al. 2022).

Finally, other reviews focused on DSW intake in long-duration sports should be taken into account.

This review aims to evaluate the efficacy of seawater consumption in both performance and post-exercise recovery in long-duration sports events. After revising more than five hundred studies, eight studies matched the criteria (six were randomized clinical trials, and two were observational studies), showing that DSW consumption accelerated the recovery of aerobic capacity and leg muscle capacity on running performance. In addition, lactate production after the running exercise in seawater was significantly lower than in pure water. The main conclusion was that seawater consumption could significantly improve the capacity of recovery after exercise (Aragón-Vela et al. 2022).

Oral and dental therapy

There are several studies related to sea salt-based mouthwashes and rinses for oral hygiene and treatment after oral surgery, and also against dental plaque, gingivitis, and salivary *Streptococcus mutans* load (Cantore et al. 2018, Ballini et al. 2019, Signorini et al. 2020, Cantore et al. 2020, Ballini et al. 2020). The studies showed that mouth rinse containing sea salt, xylitol, and lysozyme was able to reduce the bacterial levels of *S. mutans* (*in vitro* study). This type of salt-based mouthwash also improves the wound-healing process after oral surgery (Cantore et al. 2020). Furthermore, in a randomized double-blind study, sea salt mouth rinse (applied twice a day) for four weeks improved the wound healing process in diabetic patients after tooth extraction (Signorini et al. 2020).

Respiratory therapy

Seawater, alone or in combination with other preparations, has its place in the treatment of numerous conditions of the upper respiratory tract (URT), primarily chronic (rhino)sinusitis, allergic rhinitis, acute URT infections and postoperative recovery (Štanfel et al. 2022). In respiratory therapy, the main techniques are nasal irrigation, aerosol, and nebulization with isotonic and hypertonic seawater.

According to Štanfel (2022), the mechanism of nasal irrigation solutions action is based on two principles: physical and biological/physiological. The first principle is based on the physical

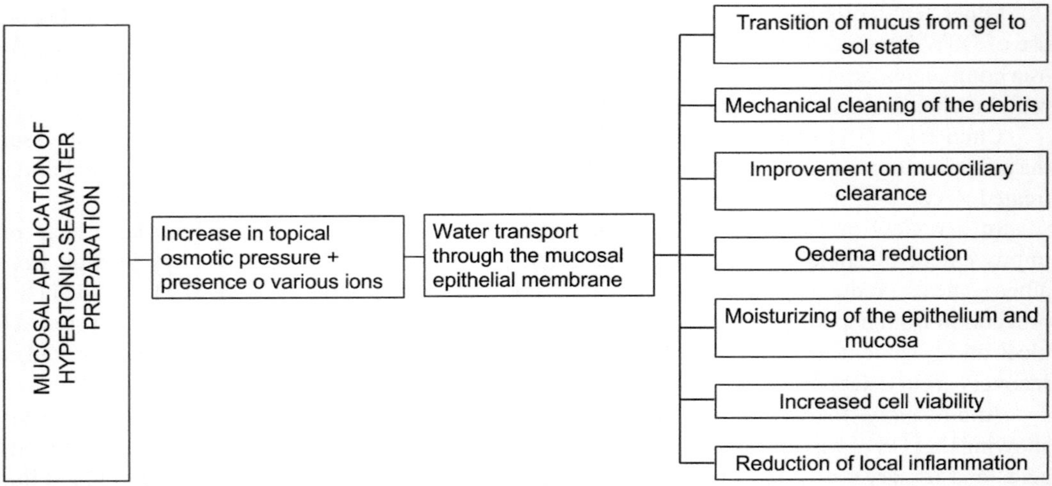

Figure 1. Proposed mechanism of action of hypertonic seawater preparations locally applied to the upper respiratory tract mucosa (taken from Štanfel et al. 2022).

(mechanical) effect of cleansing the nasal mucosa of the accumulated secretion and pathogens. The second principle depends on the effects of the ions on the physiology of the mucosal cells. In Figure 1, the mechanisms proposed by Štanfel et al. (2022) are summarized, based on the transport of water through the mucosal epithelial membrane, provoked by the local application of hypertonic solution.

In this extensive review, authors examined the clinical trials and *in vitro* studies performed over more than 20 years, concluding that nasal irrigations with seawater preparations are safe and the main benefits range from promoting cell repair and reducing inflammation to reducing viscosity of the mucus and increasing ciliary beat frequency. In addition, side effects were rare, and consisted mostly of burning feelings and nasal drainage, stating that serious adverse events were practically non-existent.

When applied as aerosol or nebulization, seawater is used in aerosol/nebulization equipment and projected directly or through a mask into the respiratory tract. After 10–15 minutes of treatment, the thick phlegm becomes liquid, less adhesive, and starts melting down. Aerosols exert the deepest action, at the level of the pulmonary alveoli, due to their small particle size (< 3 μm). Moreover, aerosols can also be applied collectively in rooms set up for this purpose, or naturally on the seashore.

Additionally, Atar et al. (2022) compared the effect of isotonic seawater nasal spray containing chamomile liquid extract with isotonic seawater nasal washing solutions in allergic rhinitis patients and showed that isotonic seawater spray containing chamomile liquid extract might be a better alternative treatment option for these kinds of patients.

Other uses of seawater

Dry eye drops

Despite not having been studied in depth, seawater is used to improve dry eye syndrome and other related disorders; Díaz-Llopis et al. (2019), in a randomized multicenter study, compared seawater washes and carmellose artificial tears eyedrops in the treatment of dry eye syndrome (DES), finding that the application of seawater is more effective than treatment with carmellose artificial tears in reducing symptoms and pro-inflammatory molecules (IL-1 beta and IL-6) in tears of patients with DES.

Seawater in cosmetics

Seawater is well-known as a source of minerals that can have benefits in quite a few skin disorders, particularly inflammatory dermatitis such as atopic eczema (Couteau and Coiffard 2020), but also as an excipient or active ingredient in cosmetics which is listed as "*Aqua maris*" (more information about seawater and skin can be found in the following section).

Bathing techniques

According to ISO 17680/2015 definitions, a bath consists of "body immersion in seawater that provides pain release and relaxation effects. Baths result in peripheral vasodilatation or vasoconstriction depending on the person having hydrotherapy status of health and medical prescription".

Bathing in seawater could be carried out directly in the sea, in swimming pools, or in bath tubes; the so-called partial techniques as hand and foot baths, walking corridors, and phlebologic paths can also be found in many thalassotherapy facilities.

Sea bathing

The use of immersion in cold water is very well-known in sport recovery and it is also regarded as a readily available and inexpensive method to gain resistance to adverse conditions, a natural type of athletic recovery, and an adaptive stimulus against respiratory tract infections and musculoskeletal pains, especially in northern countries (Lubkowska 2013). Also, swimming in cold water is not exempt from risks; for example, susceptibility to pulmonary and other infections has been attributed to the negative effects of swimming pool chlorine but swimming in open-sea-water venues avoids all these potential risks, while acclimation appears to lead to body hardening with increased immune defense, instead of the weakening of the immune system because of the overtraining of professional swimmers (Manolis et al. 2019).

Kim et al. (2020) investigated the effect of deep-sea water thalassotherapy) on recovery from fatigue and muscle damage; the results showed a significant decrease in the parameters studied (lactate, ammonia, lactate dehydrogenase, and creatinine kinase) compared with the control group. And Sutkowy et al. (2015) studied the effects of ice-cold bathing after exercise, showing that short-term ice-cold water bath decreases postexercise lipid peroxidation.

Additionally, Harper et al. (2012) found a relationship between the surgical stress response and the response to cold exposure and hypothesized how a program of sea bathing may be used to reduce preoperative complications and to enhance postoperative recovery.

Furthermore, a recent study showed the positive influence of sea swimming on depression and anxiety (Burlingham et al. 2022).

Seawater and skin

Seawater is very well-known for its curative actions in the treatment of skin diseases such as eczemas, dermatoses, and psoriasis. For skincare, it has also been used as a moisturizing agent, and regularizes the sebaceous production, avoiding the consequent formation of scaling of the scalp (dandruff). It is worth mentioning the important role of seawater in the absorption of saline and metallic ions, favoring the excretion of toxic residues and oxygenation of certain tissues (Pereira 2017). Seawater has been shown to have important effects on skin hydration and regeneration, although there are few scientific studies since they are mostly testimonials. Most of the studies focus on the hypersaline water of the Dead Sea; a few of them study the seawater of the Atlantic coasts and others the deep seawater.

Yoshizawa et al. (2001) studied the effects of seawater on experimental irritant contact dermatitis induced by sodium lauryl sulfate, concluding that seawater inhibited the increase of TEWL as compared with deionized water, and this effect may be attributed to skin barrier preservation by NaCl and KCl, and an emollient effect by NaCl. Later , the same research group compared different concentrations of salt water, concluding that these solutions could be useful, as adjunctive therapy, in the treatment of some skin diseases as atopic dermatitis and other chronic dermatitis, being able to prevent skin barrier disruption and stratum corneum water content, and improving skin dryness and pruritus (Yoshizawa et al. 2003).

In recent years, several studies related to the beneficial effects of deep seawater on skin diseases have been carried out. Bak et al. (2012), in a mice model, investigated the preventive effect of mineral water from deep-sea bedrock (rich in minerals such as Ca, Mg, Na, K, Fe, and others) on developing Atopic dermatitis, concluding that this water inhibits the development of atopic dermatitis-like skin lesions. In addition to this, Lee et al. (2016) studied, in a skin equivalent model, Jeju lava seawater, showing that this water (rich in sodium, magnesium, and calcium) increased the CD44 (hyaluronic acid receptor) which is related to skin hydration; the authors concluded that the Jeju lava seawater may help in skin hydration. Other studies focused on dermatitis; Lee et al. (2018) evaluated the effects of mineral-balanced deep-sea water on Atopic dermatitis-like skin damage in both *in vitro* and *in vivo*. The results showed that DSW is effective in the treatment and prevention of atopic-type skin lesions. The mechanism of action involved seems to be the suppression of the expressions of proinflammatory chemokines and cytokines, as well as the inhibition of histamine and IgE production. In addition, other mechanisms involved are the positive regulation of filaggrin, the recovery of involucrin expressions, and the inhibition of IL-4 production, which help restore skin health. Moreover, Chun et al. (2017), investigated the effect of deep-sea water (high levels of Mg and Ca) on inflammation showing that this water suppressed inflammatory responses via the MAPK/AP-1 and NF-κB Signaling Pathway.

Dead Sea is a special case, considering its high mineral content, with an average concentration of 280 g/K. Bathing in the Dead Sea waters has been extensively studied: Magnesium salts are known to bind water, influence epidermal proliferation and differentiation, and enhance permeability barrier repair (Schempp et al. 2000). In a study carried out by Proksch et al. (2005), immersion in magnesium-rich salt solutions from Dead Sea was investigated, showing that the high content of magnesium ions enhanced stratum corneum hydration, improved skin barrier function and reduced skin roughness and inflammation.

Portugal-Cohen et al. (2019) studied the topical effects of Dead Sea minerals and concluded that topical application of Dead Sea water encouraged the expression of the barrier-related proteins: filaggrin, involucrin, and transglutaminase, while transdermal penetration of calcium ions was not detected. Additionally, Dead Sea water application had increased skin secretion of β-endorphin and attenuated the expression of inflammatory and irritation-related cytokines.

On the other hand, Carbajo et al. (2018) revised the mechanism of saline water action; when topically administered, this water rich in sodium and chloride penetrated the skin where it was able to modify cellular osmotic pressure and stimulate nerve receptors in the skin via cell membrane ion channels known as "Piezo" proteins. They postulated that the actions of salt mineral waters are mediated by mechanisms linked to the concentration and type of salts involving cellular osmosis-mediated activation/inhibition of cell apoptotic or necrotic processes, and in turn, this osmotic mechanism modulates the mechanosensitive piezoelectric channels.

Full bath

The most common in most of the thalasso centers is not only a warm/hot bath, but also the so-called contrast bath, which consists of alternating hot baths with short cold plunges. Total body immersion can produce different effects depending on the water temperature(s) used (either singly or in succession). Cold immersion dilates blood vessels of the skin and increases metabolic rate and circulation of body fluids while reducing pulse, respiratory rate, and blood pressure. A hot bath acts as a sedative on the sensory and motor nerves, relieving pains, cramps, and spasms. It can also raise oxygen consumption, increasing blood volume and pulse rate (Moss 2020).

The so-called bath with additives consists of adding water jets (hydromassage bath, underwater massage), air (bubbling bath), or marine-derived substances such as marine salts, and algae (macro and microalgae).

Hydromassage bath

Hydromassage/multi-jet and whirlpool baths consist of bathing in a tub with several jets of water automated to target specific parts of the body. This treatment provides relaxation to the muscles and joints and improves circulation, among other benefits (Anup and Christopher 2020).

Some studies have shown the benefits of bathing in hydromassage tubs. Patients with peripheral artery disease who experience intermittent claudication were treated with hydrotherapy, consisting of 10 whirlpool treatments of the lower limbs and an individual training program (compared with a control group which only followed the training program); authors have found that the combination of exercise with hydrotherapy gives more beneficial effects than exercise alone, and these could be a promising form of conservative treatment in patients with atherosclerotic ischemia of the lower limbs (Kapusta 2022).

Figure 2. Hydromassage bath.

Bubbling bath

It consists of adding air to create bubbles that can be of different sizes (the smallest ones promote microcirculation, and the largest ones have a massage effect) (Mourelle 2013).

Underwater massage

It consists of applying a massage or drainage through the body with the help of a jet, while the patient is submerged in a bathtub. The temperature of the jet is about 4–5 °C higher than that of the bath water, which is generally 36 °C–38 °C.

Underwater massage is used to treat specific areas such as the shoulders, and dorsal/lumbar back, or to carry out a circulatory and lymphatic drainage (Mourelle 2013).

Baths with marine derivatives

Algae extract, microalgae powder, salts, and other marine derivatives can be added to the bathtub with different aims, depending on the additive type (e.g., seaweed for relieving joint pain, cellulitis, etc.) (Mourelle 2013).

Pool bathing

When people are immersed up to neck-deep, water supports 90% of the body's weight and, as a result, stress is relieved on joints (mainly knee and hip joints) encouraging freedom of movement, which is one of the great benefits for those with chronic problems. This general effect of hydrotherapy is more remarkable in seawater due to its greater density and therefore better body buoyancy. Also, water provides resistance in multiple directions since hydrostatic pressure is exerted equally on all surfaces of the body; ensuring that movement in any side-to-side or forward-backward direction meets equal resistance, and opposing muscle groups work equally (Moss 2020).

Phlebologic path and waking corridor

Phlebologic path, also called "vascular path", is a walkway or walking corridor in which the water reaches the middle of the thighs. The treatment consists of walking alternately in warm and cold seawater (temperature range 32 °C–28 °C/14 °C–20 °C) in two parallel pools with jets or air bubbles that massage the lower limbs. It could also have an irregular floor that is covered by large pebbles or tiles. The alternation between hot and cold stimulates a sort of vascular gymnastics that boosts the tone of the vein walls, alleviates circulatory disorders in the lower limbs, and helps to increase joint mobility and balance. At the same time, the tiny air bubbles stimulate proprioception and microcirculation. It also increases venous return by the calf muscle pump which accelerates when walking against the current. Balance therapy using an inclined surface (uphill and downhill) of pebbles or tiles enables stimulation of the plantar arch and venous pump, relaxation of the ankles as well as an improved physical perception of the ground (Blain et al. 2016).

The main indications are: chronic venous insufficiency, lymphedema, and the results and after venous peripheral vascular surgery. The scientific literature regarding the use of the phlebologic path in thalassotherapy is scarce, but there are some studies related to the use of thermal water and those related to hydrotherapy (tap water). For example, Forestier et al. combined Kneipp therapy (alternating warm and cold showers), walking the pool, underwater massage, and a bath in a tub in the treatment of chronic venous insufficiency showing the improvement of quality of life after three months of treatment (Chronic Venous Insufficiency Questionnaire, CIVIQ 2, was applied). The authors concluded that the clinical status and quality of life of patients with CVI treated with balneohydrotherapy were improved when compared with usual care and treatment and are well tolerated (Forestier et al. 2014).

Figure 3. Phlebologic path.

Hand and foot baths

Consists of immersing the hand, arms, and/or feet in a shallow tub of water; it has been recommended for varicose veins, headaches, low blood pressure, and a sprained ankle. A warm foot bath has been recommended for sleeplessness while alternating hot/cold improves circulation in the feet and legs.

In scientific literature, there are no studies on the use of hand and foot baths with seawater, but there are some experiences with tap water. Kudo et al. (2019) conducted a study in order to clarify the effects of warm hand bath at 40 °C for 10 min on the blood flow in the shoulder, skin and deep body temperature, autonomic nervous activity, and subjective comfort in healthy women, concluding that hand and foot baths can improve the level of subjective comfort and increase the heart rate and might affect autonomic nervous activity. And Maheshkumar et al. (2020) studied the effects of warm hand and foot baths on bronchial asthma concluding that, this economic therapy intervention might stabilize or reduce mucus accumulation or alleviate bronchospasm, mimicking the effects of bronchodilators, nevertheless, more studies are needed to evaluate this therapy. And Armat et al. (2021) investigated the effect of the warm foot bath on the sleep quality of older adults (foot bath group with water at 40 °C, footbath group with water at 37 °C, and a control group), showing that both foot baths caused significant improvement in the participants' sleep quality.

Finally, it is worth mentioning a multicenter, double-blind, randomized controlled trial conducted by Fan et al. (2018), in which they assessed the efficacy and safety of foot bath treatment with a Chinese herbal decoction (mixture of different plants as cayenne, red Sichuan peppercorns, frankincense, myrrh, honeysuckle vine, borneol from camphor tree, and safflower) in diabetic patients suffering diabetic peripheral neuropathy (DPN), concluding that this formula used to promote circulation and eliminate stasis to improve symptoms of DPN, and when applied as a foot bath, was safe.

Pressure techniques

These group of spa techniques include jet showers, affusion showers, Scottish showers, and the so-called Vichy shower. Their commonality is water is projected onto the body with a greater or lesser degree of pressure through tubular structures. The effects are linked to the mechanical action of water on the different body structures and skin receptors. Likewise, as in balneotherapy techniques, the water temperature will exert one effect or another (Mourelle 2013).

Jet shower

A jet system is used to project water on the body, making a route in which pressure is exerted through the water on the different muscles. In addition to the stimulation or relaxation that is exerted (depending on the pressure, which can be strong or moderate), the circulatory system is activated, favoring venous return.

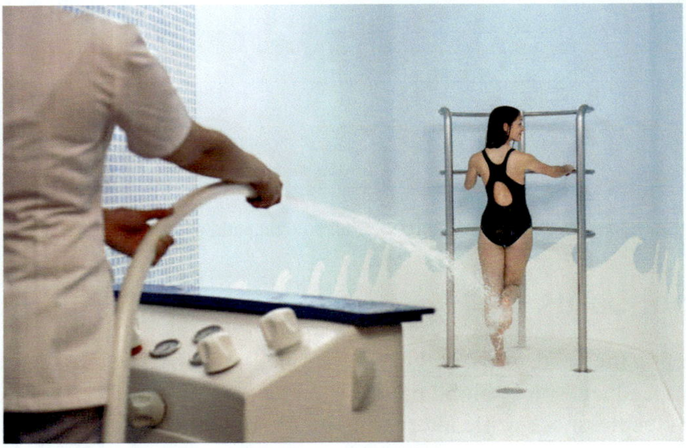

Figure 4. Jet shower.

Affusion shower

A fine spray of warm sea water is poured over the body, with the main objective of relaxation.

Vichy shower

It consists of a type of affusion shower to which is added a massage applied by a hydro-therapist, which can also be performed with four hands (two therapists at the same time).

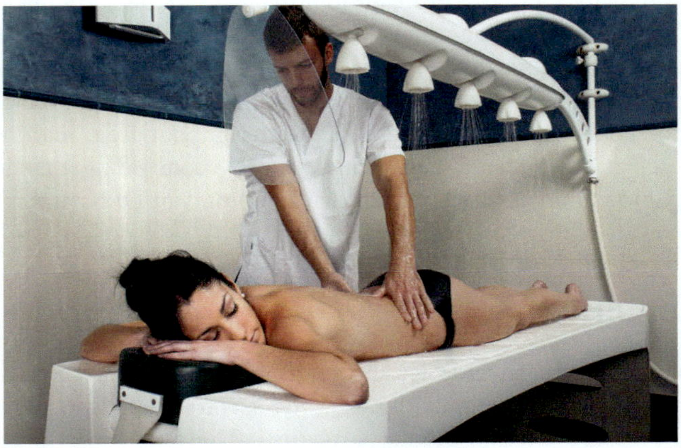

Figure 5. Vichy shower

Scottish shower

It is a jet or multi-jet shower in which hot and cold water is combined to stimulate and improve the circulatory system, and general condition.

Sauna and steam baths

Traditional Finish sauna bathing involves exposure to high environmental temperatures (80 °C–100 °C, humidity less than 25%) for a brief period (10–15 minutes), combined with a refreshing technique that could be an immersion or a shower in cold water, ice application or a short outdoor walk. Sauna bathing, which has been a tradition in Finland for thousands of years and is mainly used for pleasure and relaxation, is becoming increasingly popular in many other populations. Emerging evidence suggests that beyond its use for pleasure, sauna bathing may be linked to several health benefits, which include a reduction in the risk of vascular diseases such as high blood pressure, cardiovascular disease, and neurocognitive diseases; nonvascular conditions such as pulmonary diseases; mortality; as well as amelioration of conditions such as arthritis, headache, and flu. The beneficial effects of sauna bathing on these outcomes have been linked to its effect on circulatory, cardiovascular, and immune functions. It has been postulated that regular sauna bathing may improve cardiovascular function via improvement of endothelium-dependent dilatation; reduction of arterial stiffness; modulation of the autonomic nervous system; beneficial changes in circulating lipid profiles; and lowering systemic blood pressure (Laukkanen et al. 2018). There are also studies about the improvement of exercise-heat tolerance following post-exercise sauna bathing (Kirby et al. 2021), and in body composition, showing improvements in bone and muscle mass (Toro et al. 2021). And in a review Patrick and Johnson (2021) postulated that sauna can be used as a lifestyle practice to extend the health span. It also deserves to be mentioned as beneficial for the skin: a more stable epidermal barrier function, an increase in stratum corneum hydration, and a faster recovery of both elevated water loss and skin pH after exposure to 2 x 15 min sauna at 80 °C could be demonstrated in volunteers who took regular saunas (Kowatzki et al. 2008).

Steam baths (also known as Hammam and Turkish baths) consist of exposure to a hot and humid environment (up to 50 °C, humidity between 50-100 %) to promote sweating and to eliminate toxins (Ansari 2019). Vapors of steam baths can penetrate the lungs and respiratory tract to facilitate the release of phlegm. Furthermore, Pandiaraja et al. (2021), in a randomized control trial including 80 healthy volunteers, demonstrated that steam baths improve cardiovascular function in normal healthy subjects.

Seaweed applications

Algae produce both primary metabolites (polysaccharides, proteins, amino acids, and fatty acids) involved in normal growth, and secondary metabolites (pigments, phenolic compounds, sterols, vitamins, and other bioactive agents) performed under different stress conditions such as UV radiation exposure, salinity, temperature changes, or environmental pollutants (Pereira 2018). These compounds are of great interest not only in the functional food and nutraceutical, pharmacological, and cosmetic industries (Mourelle et al. 2021), but also in thalassotherapy applications and the so-called thalasso-wellness treatments. In Chapter 6 of this book, the algae and microalgae used in thalassotherapy are exhaustively reviewed, which is why, in this section, only the most widely used techniques for their application in health and wellness treatments will be cited.

Seaweed wraps

Seaweeds are applied as a thin layer on the entire body with the aim of skin remineralization and/or hydration. It could also be applied partially, for example, on the thighs in anti-cellulite treatments. Seaweeds are usually applied warm, and the treatment could last between 20–30 min (Mourelle et al. 2021).

Seaweed poultices

Consists of a soft, moist mass of seaweed, applied on the skin in the form of a thick layer, which is wrapped around a part of the body (knee, elbow, etc.) to reduce pain or swelling. They are also called seaweed cataplasms and seaweed packs.

In a study comparing the effects of seaweed packs and mud packs in osteoarthritis (OA), the authors found that seaweed packs and mudpacks had similar positive effects on knee OA, improving pain, function, and quality of life throughout the four-week follow-up period after both treatments. Additionally, the seaweed pack group showed decreased levels of TNF-α at two weeks post-treatment, which may explain the reduced inflammatory reaction. They concluded that, for rehabilitation therapy, the use of seaweed packs may serve as an alternative modality for the treatment of knee OA (Lee et al. 2019).

Marine peloids applications

Marine mud packs, either natural or artificially prepared, are used in many thalasso centers. Theses packs consist of the application of 1–2-cm-thick layers of heated peloid to 38 °C–46 °C on a limited region or the entire body surface for 20–40 min.

Peloids as defined by Gomes et al. (2013) are used for therapeutic treatments, rheumatic pathologies, arthritis, muscle and bone traumas, and for dermatological pathologies or aesthetic purposes (Gomes et al. 2013). Marine peloids are enriched with seawater minerals and trace elements, as well as the seaweed or marine microalgae that can be included in their composition.

The most well-known marine peloids are those from Dead Sea. They have shown to be effective in the treatment of dermatological diseases such as psoriasis and rheumatology, its main therapeutic indications (Sukenik et al. 1994, Katz et al. 2012, Abu-Shakra 2014, Melandri 2020), and also for other skin diseases and cosmetic uses (Ma'or et al. 2006, 2015, Portugal-Cohen et al. 2009, Hamed et al. 2021). Other marine peloids of interest are from Techirghiol Lake and other Romania saline lakes (Calin et al. 2020, Baricz et al. 2021), and Crimean saline lakes (Kotova et al. 2016).

The mineralogical, chemical composition and thermophysical properties of peloids from the Adriatic coast (Mihelčić et al. 2012, Komar et al. 2015, Glavaš et al. 2017) and those from Adjara region lakes (Masiukovichi et al. 2017) have also been studied, with the aim of determining their possible therapeutic applications, for both therapeutics and cosmetic use (Masiukovichi et al. 2018).

As marine peloids can be collected in the natural environment or elaborated from natural sediments, the raw materials used for this purpose must be well characterized to guarantee safe use, and this is why toxic elements are evaluated (Calin et al. 2020).

More information about marine peloids can be found in Chapter 7.

Figure 6. Microalgae mud application.

Combining techniques for specific treatments

In most thalassotherapy centers the usual way of performing treatments is combining techniques.

Zijlstra et al. (2005) studied the effect of a combination of thalassotherapy, exercise, and patient education in people with fibromyalgia, showing that this combination therapy improved fibromyalgia symptoms and health-related quality of life.

Grozeva and Stoicheva (2015) investigated the efficacy of the combined application of seawater, Pomorie therapeutic mud, and lye electrophoresis in diabetic polyneuropathy rehabilitation, showing that this combination of techniques improved the clinical symptoms and quality of life of patients with diabetic polyneuropathy because of the trophic and pain-reducing properties of the applied techniques.

And finally, another study by Chennaoui et al. (2018) deserves to be mentioned. It is related to the improvement of sleep management, in which the authors showed that three days of thalassotherapy combined with sleep management have immediate beneficial effects on self-reported mood states, anxiety, fatigue, well-being, and daytime sleepiness, as well as on the objective evaluation of sustained attention, in healthy workers. Beneficial effects on lower limb flexibility were also observed after three days of care, and the positive effects were still observed after the weekend back home and after three days of return to work.

Seawater pool therapies

Aquatic physiotherapy has increased among professionals, patients, and researchers due to the benefits provided by water. As the body is immersed in water, it is unloaded, which enables the joints to be mobilized by the patient with reduced effort. Additionally, aquatic rehabilitation provides motor and sensory stimuli that may induce neuronal plasticity, improving motor function and the static and dynamic balance of people who have suffered acquired brain injury (Pérez-de la Cruz 2021).

Seawater pool therapies can be carried out both in small groups and individually. The most common are listed in Table 2.

Watsu

Watsu (portmanteau word: water and shiatsu) is a form of passive hydrotherapy in chest-deep thermoneutral water (35°C) that combines elements of myofascial stretching, joint mobilization, massage, and shiatsu and is reported to be used to address physical and mental issues (Schitter et al. 2020). This technique was first described by Dull in the 1980s. The therapist stands in thermoneutral water (35°C), supporting the patient in a supine position with hands, forearms, or shoulders and softly moving her/him in slow and spacious circular motion sequences following elaborated movement patterns related to the receiver's and therapist's level of experience. Gentle tractions are applied to the body of the receiver to mobilize joints and stretch myofascial structures, as well as meridians, channels through which the life-energy flows. The thermoneutral temperature of 35°C allows passive immersion of about 60 minutes without causing temperature-induced stress, and hydrostatic pressure influences fluid distribution, metabolism, and respiration. The impact of gravity is greatly reduced, thus decreasing joint loads and allowing maximal flexibility in the positioning of the treated individual (Dull 2004, Becker 2009).

In a systematic review and meta-analyses, Schitter et al. (2020) concluded that Watsu has been applied to individuals of all ages. Indications covered acute (e.g., pregnancy-related low back pain) and chronic conditions (for example, cerebral palsy) with beneficial effects of Watsu regarding relaxation or sleep quality. Meta-analyses suggested the beneficial effect of Watsu on pain, physical function, and mental issues.

Ai Chi

The Ai Chi is a physiotherapy method developed by Jun Konno in 1990s, from the principles of Tai Chi and Qigong methods associated with the techniques of Shiatsu and Watsu methods. Ai chi is performed standing in shoulder-depth water using a combination of deep breathing and slow, broad movements of the arms, legs, and torso. The benefits are associated with the stabilization of the trunk, pain control, and improvement of the metabolism and blood flow (Brody and Geigle 2009). Several studies have shown the efficacy of Ai Chi method in chronic low back pain (Camilotti et al. 2015).

Halliwick

The Halliwick method was developed at the turn of the 1940s and 1950s by James McMillan to teach physically disabled children independence in water. This method is based on the application of a ten-point-program (mental adjustment, disengagement, transversal, sagittal, longitudinal and combined rotation control, mental inversion, balance in stillness, turbulent gliding, simple progression, and basic swimming movement) to develop balance control, swimming skills, and independence (Brody and Geigle 2009). Aquatic exercises based on the Halliwick concept have shown to be better than conventional exercises to improve sitting, standing and walking, running and jumping activities in children aged 3 to 5 years with spastic cerebral palsy (Hamed et al. 2022). And Korelc et al. (2019), used Halliwick techniques in a program for children with impaired mobility helping them to learn to swim. The authors concluded that the Cognitive Orientation to daily Occupational Performance as part of the Halliwick-concept-based swimming program could be effective so that learning is faster and focused on a clear goal.

Bad Ragaz Ring

The Bad Ragaz Ring methods, developed in the 1950s in the Bad Ragaz thermal waters in Switzerland, are based on proprioceptive neuromuscular facilitation. Therapist-assisted strengthening and mobilization are applied to a patient who lies in the water in a horizontal position supported by rings or floats around the neck, arms, pelvis, and legs (Brody and Geigle 2009). Cha et al. (2017) evaluated the effects of the Bad Ragaz Ring method on functional recovery in chronic stroke patients, finding differences in post-training gains in the activations of tibialis anterior and gastrocnemius muscles and balance index were observed between the experimental and control groups (Cha et al. 2017).

Jahara

Jahara is an aquatic therapy method founded in body awareness, healthy body mechanics and the gentle power of water. It consists of micro-adjustments of the structural alignment of the body. A flexible floatation device called a 'third arm' is used to support the participant's body. This allows the therapist to move the participant without effort which generates a sense of lightness. Gentle, gradual, and continuous traction of the spine is applied, which results in 'expansion' and release of the muscular-skeletal system. The creators of the method assert that there is no stated objective but suggest that the technique promotes relaxation and feelings of well-being for the client (Gaab and Steinhorn 2015).

Aquatic PNF

PNF (performing proprioceptive neuromuscular facilitation) utilizes a typical helical or diagonal pattern to stimulate proprioceptive sensation and promote a nerve root response, enhancing functional movement (Klein 2002). It improves muscle strength, flexibility, and balance; by applying it on the non-paretic side, force is transferred to the paretic side, effectively promoting muscle activity. When applied in stroke patients, performing aquatic PNF patterns in the lower extremity enhances balance and activities of daily living (Kim 2015).

Other aquatic therapies

Other aquatic therapies are aquatic treadmill walking, deep water running, and water Pilates, among others.

Veldema and Jansen (2021) performed a systematic review and meta-analysis with the aim of collecting the available evidence of aquatic therapy in stroke rehabilitation; the authors found that water-based therapy (such as the Halliwick, Ai Chi, Watsu, or Bad Ragaz Ring methods) are the most effective and aquatic treadmill walking is the least effective. In comparison with land-based interventions, aquatic therapy shows superior effectiveness on balance, walking, muscular strength, proprioception, health-related quality of life, physiological indicators, and cardiorespiratory fitness. Only in independence in daily life activities does exercise on land and in water induce similar effects.

Deep water running is a form of aerobic aquatic exercise that is used in the treatment of chronic musculoskeletal pain conditions such as fibromyalgia and in non-specific chronic low back pain (Cuesta-Vargas et al 2012).

A pilot observational cohort study, Skinner et al. (2016) showed that Aquabalance, an aquatic therapy based on Ai Chi principles, improved balance and reduced falls in community-dwelling adults.

Water Pilates is a water-based extension of Pilates, termed Aqua-Pilates, that has been shown to improve balance, muscle strength, and flexibility. Bagherzadeh-Rahmani et al. (2023) performed a preliminary study about the use of Aqua-Pilates to improve lung function in patients with a history of COVID-19, showing that this water therapy can improve pulmonary function by as much as 34%.

Seawater pool activities

Pool activities are very popular; they have been developed with the aim of making the most of the mechanical properties of the water, mainly the buoyancy, which enables movements that could be difficult in land, depending on the state of health, mobility, physical condition, and even age.

Among them, Aquagym, Aquaerobic, AquaRelax, Cardio-aquagym, Aquabuilding, Aquastreching, Gymswin, and others can be found on specialized web pages and training centers.

A study performed by de Andrade et al. (2008) deserves to be mentioned. It compared groups of patients suffering from fibromyalgia who followed an exercise program, both in natural seawater and pool. There was significant improvement in both groups in post-treatment (pain intensity, fatigue, number of tender points, physical functional capacity, general health status, sleep quality, and depression), but a greater improvement in the sea group for all the evaluated parameters was observed. The authors postulated that these results are probably due to the effects of thalassotherapy, which can offer advantages due to the effects of the biophysical properties of immersion (buoyancy, hydrostatic pressure gradient, water viscosity, specific heat of water), salts, and trace elements of seawater, and wellbeing benefits arising from contact with nature.

Marine climatotherapy

As has been mentioned before (and will be extensively developed in Chapter 8), marine climatotherapy is the exposure to coastal climatic conditions, that can be in Mediterranean climates (with mild temperatures, higher atmospheric pressure, a variable degree of humidity, and relatively high wind speed); or in the Atlantic coast (small or moderate diurnal and annual thermal oscillations, high degree of humidity, abundant rainfall, and high wind speed). Special cases are the Dead Sea (dry arid climate, with dry air, high temperatures, weaker ultraviolet radiation, and the air has a slightly higher oxygen content –3.3% in summer to 4.8% in winter– as compared to oxygen concentration at sea level), and Black Sea (humid subtropical climate, with abundant precipitation, warm winters, and humid summer). In all these areas, marine climatotherapy has been developed as a part of thalassotherapy, adapting the marine cures to the climate of each one of them (https://scijinks.gov/climate-zones/). For example, in France, when the marine cure is carried out, the climate of the Atlantic coast is considered invigorating, while that on the Mediterranean coast is considered to have a sedative effect.

Marine climatotherapy is combined with heliomarine cure (heliotherapy), sand therapy, and soft exercises in the natural marine environment.

On the other hand, nowadays more and more importance has been given to healing environments -understanding that the thalassotherapy center environment can be part of this tendency, highlighting cures in natural environments, may be idoneous for healing. For example, Liamputtong and Suwankhong (2015) described the importance of healing landscapes for “emotional healing”, and Zhang et al. (2021) highlight the effects of waterscapes for promoting mental health in the general population.

Heliotherapy

Heliotherapy is controlled sunlight exposure (under medical supervision) that is used for improving different health conditions such as endocrine or depression, but mainly for chronic skin diseases such as psoriasis, eczema, atopic dermatitis, and vitiligo (Antonelli and Donelli 2021) (further explanation can be found in Chapter 8).

Psammotherapy

Gomes et al. (2021) described "special sands" as those of biogenic carbonate minerals, those of radioactive minerals yielding radon gas evolution, and those of naturally heated volcanoclasts which are used in the form of "sand baths" either outdoors in the natural environment or indoors inside Spas facilities. In thalassotherapy they are used for therapeutical purposes and are called Psammotherapy or Arenotherapy.

In most thalassotherapy centers, psammotherapy is performed in natural environments, usually on the beach, to make the most of the sand therapy and the climate and heliotherapy. It includes physical exercise (with contracts between warm and cold sand) and relaxing or stretching activities (yoga, tai-chi, etc.).

The main benefits of sand baths are not only those derived from thermotherapy but also through the transfer of the main ions to the skin. Natural or artificial heating is essential to produce sweat. The pH acid and the interface with the human body and sand dissolves and turns free Ca, Mg, Sr, and other elements that interact within the body.

In some places, sand baths can have an added value; for example, in southern Japan (Beppu and Ibusuki, Kyushu Island), where the sand of volcanic nature is naturally heated by exposure to volcanic heat, and the case of certain beaches of Guarapari (in Victória do Espirito Santo district, Brazil) where health benefits are assumed to be related to the radon emitted from the radioactive monazite-rich sands (Gomes et al. 2021). Another famous place is Porto Santo (in the archipelago of Madeira, Portugal) where the sand is exposed to solar radiation, taking care that the sand temperature is not over 41 °C. The climatotherapy in Safaga (Red Sea, Egypt) includes psammotherapy which is known for its content on natural radiation levels, and this is why studies about the radiation hazard indices were carried out, concluding that the levels are below the internationally accepted values (El-Arabi 2005). The "Enseki sandbath" tries to imitate the effects of a sand bath using a ceramic which can radiate far-infrared (FIR) rays when heated adequately (Aragane and Higashino 2019).

Antonelli and Donelli (2019) revised the scientific studies about psammotherapy, noting that, based on available data, no conclusions could be driven on the clinical efficacy of psammotherapy; however, very limited evidence suggests that hot sand baths might be useful in improving symptoms and functionality of patients with some rheumatic and respiratory chronic illnesses.

Other therapies

Kneipp therapy

Kneipp therapy comprises the five naturopathic elements of water, exercise, nutrition, medicinal herbs and order of life. As a traditional method, it has been proven not only in self-treatment at home, but also as a holistic therapy concept. It is also used in many medical practices both for the prevention and therapy of functional disorders and in the treatment of organic diseases and ailments.

The implementation of evidence-based medicine (EbM) in internal medicine practice represents a major challenge for Kneipp therapy; however, existing studies point to positive or significant effects. Ehnert and Geiser (2022) reviewed several studies on the application of Kneipp hydrotherapy. This therapy, alone or in combination, improves the quality of life and significantly helps in chronic venous insufficiency, hypertension, mild heart failure, discomfort caused by menopause, and sleep disturbances. Kneipp cold infusions help patients with chronic obstructive pulmonary disease by reducing respiratory infections and improving subjective well-being. In addition, it alleviates respiratory tract infections in children between the ages of 3 and 6. The authors concluded that a classical Kneipp cure in a spa can significantly affect and improve the state of health over a year.

Halotherapy

Inhalation therapy in natural caves or salt chambers has regained popularity in the last decades. Halotherapy consists of inhalation of natural pure sodium chloride (NaCl) in a controlled environment (air temperature 18 °C to 24 °C and relative humidity 40% to 60%). When the inhalation is performed in natural caves, it is also called speleotherapy. The main active ingredient in halo- and speleotherapy is NaCl aerosol particles, which penetrate all layers of the respiratory tract. In addition to their antibacterial and anti-inflammatory properties, salt particles also facilitate mucociliary transport and reduce immunogloblin E (IgE) levels (Wasik and Tuuminen 2021). Halotherapy has been worldwide used to treat chronic obstructive pulmonary disease (Rashleigh et al. 2014).

Floatation therapy

The floatation therapy consists of floating supine in a salt-saturated pool or tank, at about 35 °C. In addition, a quiet and dark environment is provided and earplugs are used to further minimize auditory stimuli. The high salt content in the water gives high buoyancy, which makes it possible to float on top of the water. By keeping the water at skin temperature, sensory input to the somatosensory system is reduced. The so-called Floatation-REST (Reduced Environmental Stimulation Therapy) is used mainly for reducing anxiety, stress, and pain. Some studies reported significant reductions in stress, muscle tension, pain, depression, and negative affects, accompanied by a significant improvement in mood characterized by increases in serenity, relaxation, happiness, and overall well-being, in patients suffering from different kinds of illnesses such as anxiety and stress-related disorders (posttraumatic stress, generalized anxiety, panic, agoraphobia, and social anxiety) (Feinstein et al. 2018), in which water saturated with Epsom salt was used. And this technique was also used in patients with insomnia, showing that 12 sessions of 45 min of t flotation-REST improved the insomnia symptoms (Norell-Clarke et al. 2022).

Complementary techniques

Complementary techniques are those that come from the field of physiotherapy, aesthetics and / or well-being that are used, combined with thalassotherapy techniques, to organize treatment programs or to complement them. They are generally included in wellness programs. The following table summarizes their main characteristics and uses (Table 4).

Conclusions and future perspectives

Thalassotherapy techniques have evolved over recent years, parallel to the development and improvement of hydrotherapy and balneotherapy equipment. These improvements are both from a technological and material point of view, since in the case of the use of seawater, materials resistant to saline corrosion are required.

In addition, other well-being techniques have been incorporated, such as water therapies, along with other techniques such as yoga, tai-chi, etc. Multiple types of massages complete thalasso-wellness programs, so that thalassotherapy centers evolve towards being considered emotional healing and recovery centers due to their privileged environment and the combination of all the aforementioned techniques. Likewise, they are centers where diseases of civilization such as stress, anxiety, and eating disorders are treated.

Table 4. Complementary techniques: description and uses.

Technique	**Description**	**Effects/Uses**
Pressotherapy	A device that covers the body and blows air through a suit, putting pressure on different parts of the body as arms, limbs and abdomen	Improves lymphatic drainage Stimulates the reabsorption of interstitial fluids Promotes the lymphatic and venous return circulation
Electrotherapy		
- Low frequency	Currents whose frequency is in the range of 0 to 100 Hz	Muscles and nerves stimulation Alleviates chronic and acute pain Strengthen muscles
- High frequency	Currents whose frequency is in the range of 10 Mhz	Produces deep heat inside the tissues Anti-inflammatory and analgesic effects Recovery of damaged tissues Promotes wound healing
Thermotherapy	Application of heat	Decreases pain in joints and muscles Improves metabolism, blood flow, inflammation, oedema and connective tissue extensibility
Criotherapy	Application of cold	Decreases the neuronal excitability Decreases inflammation
Massages	Manipulation of the body's soft tissues	Helps to reduce stress Lessens pain and muscle tightness Increases relaxation Boosts the immune system
Oriental massages	Different types: Shiatsu, Thai, Lomi-Lomi, Tui-Na, etc.	Improvement of blood circulation Increases cellular activity, boosts immunity Regulates the autonomic nervous system and improves sleep Strengthens bones and muscles Helps in recovery from sports injuries
Manual lymphatic drainage	Stimulation of the lymph vessels under the skin applying soft pressure massage	Relieves painful swelling in arms and legs caused by lymphedema
Relaxing and strengthening activities	Many types: Yoga, Tai-chi, Pilates, etc.	Flexibility amelioration Muscle strength and tone improvement Respiration, energy and vitality improvement
Re-educational therapies	Many types and adapted to many conditions: back-ache, after injuries, etc.	Improvement of body conditions
Nutritional therapies	Nutritional counselling Specific diets	Improvement of metabolic diseases Obesity control Improvement of body conditions

References

Abu-Shakra, M., A. Mayer, M. Friger and M. Harari. 2014. Dead Sea mud packs for chronic low back pain. Isr. Med. Assoc. J. 16(9): 574–577.

Ansari, M.A., H. Noor and A. Quadeer. 2019. Application of Ḥammām (The Steam Bath), in the Past and Present; An Overview. IJRAR 6(1): 1038–1315.

Antonelli, M. and D. Donelli. 2021. Thalassotherapy, health benefits of sea water, climate and marine environment: a narrative review. Proceedings of the 6th International Electronic Conference on Water Sciences (retrieved from: https://sciforum.net/manuscripts/11606/manuscript.pdf).

Antonelli, M. and D. Donelli. 2019. Hot Sand Baths (psammotherapy): A Systematic Review. Complement. Ther. Med. 42: 1–6, doi: 10.1016/j.ctim.2018.10.020.

Anup, J.R. and J.G. Christopher. 2020. Marine products and methods of thalassotherapy: A review. IJAEM 2(3): 126–130.

Aragane, Y. and H. Higashino. 2019. The "Enseki" sandbath: A novel, safe and effective far-infrared bathing procedure for health. Photodermatol Photoimmunol Photomed. 35(1): 31–39. doi: 10.1111/phpp.12418.

Aragón-Vela, J., O. González-Acevedo, J. Plaza-Diaz, R.A. Casuso and J.R. Huertas. 2022. Physiological benefits and performance of sea water ingestion for athletes in endurance events: A systematic review. Nutrients. 14(21): 4609. doi: 10.3390/nu14214609.

Armat, M.R., H. Mortazavi, H. Akbari and Z. Ebrahimzadeh. 2021. The effect of footbath on the quality of sleep in older adults: A pilot study. Geriatr. Nurs. 42(5): 1178–1183. doi: 10.1016/j.gerinurse.2021.07.008.

Atar, Y., S. Karaketir, I. Aydogdu, H. Sari, H.S. Bircan, Y. Uyar et al. 2022. Comparison of isotonic seawater nasal spray containing chamomile liquid extract and other isotonic seawater nasal washing solutions for allergic rhinitis. Ann. Otol. Rhinol. Laryngol. 131(4): 427–434. doi: 10.1177/00034894211025411.

Bagherzadeh-Rahmani, B., N. Kordim, A.H. Haghighi, C.C.T. Clark, L. Brazzi, E. Marzetti et. al. 2023. Eight weeks of pilates training improves respiratory measures in people with a history of COVID-19: A preliminary study. Sports Health. 15(5): 710–717. doi: 10.1177/19417381221124601.

Bak, J.P., Y.M. Kim, J. Son, C.J. Kim and E.H. Kim. 2012. Application of concentrated deep sea water inhibits the development of atopic dermatitis-like skin lesions in NC/Nga mice. BMC Complement Altern. Med. 12: 108. doi: 10.1186/1472-6882-12-108.

Ballini, A., S. Cantore, E.A. Fotopoulou, I.P. Georgakopoulos, E. Athanasiou, D. Bellos et al. 2019. Combined Sea salt-based oral rinse with xylitol in orthodontic patients: clinical and microbiological study. J. Biol. Regul. Homeost. Agents. 33(1): 263–268.

Ballini, A., S. Cantore, L. Signorini, R. Saini, S. Scacco, A. Gnoni et al. 2020. Efficacy of sea salt-based mouthwash and xylitol in improving oral hygiene among adolescent population: A pilot study. Int. J. Environ. Res. Public Health. 18(1): 44. doi: 10.3390/ijerph18010044.

Baricz, A., E.A. Levei, M. Şenilă, S.C. Pînzaru, M. Aluaş, A. Vulpoi et al. 2021. Comprehensive mineralogical and physicochemical characterization of recent sapropels from Romanian saline lakes for potential use in pelotherapy. Sci. Rep. 11(1): 18633. doi: 10.1038/s41598-021-97904-1.

Becker, B.E. 2009. Aquatic therapy: Scientific foundations and clinical rehabilitation applications. PM R. 1(9): 859–872. https://doi.org/10.1016/j.pmrj.2009.05.017.

Bernardo, M.L., A.M. Castro, Y. Morales, S. Boueiri, S. Brito, C. Rondón et al. 2014. Valoración terapéutica del agua de mar en modelos experimentales como terapia complementaria en anemia. MÉD.UIS 27(3): 9–18 (in Spanish).

Blain, H., P.L. Bernard, G. Canovas, N. Raffort, H. Desfour, L. Soriteau et al. 2016. Combining balneotherapy and health promotion to promote active and healthy ageing: the Balaruc-MACVIA-LR® approach. Aging Clin. Exp. Res. 28(6): 1061–1065. doi: 10.1007/s40520-016-0596-4.

Brody, L.T. and P.R. Geigle. 2009. Aquatic exercise for rehabilitation and training. Human Kinetics ed. Champaign (EEUU).

Burlingham, A., H. Denton, H. Massey, N. Vides and C.M. Harper. 2022. Sea swimming as a novel intervention for depression and anxiety-A feasibility study exploring engagement and acceptability. Ment. Health Phys. Act. 23: 100472.

Calin, M.R., I. Radulescu, A.C. Ion, L. Capra and E.R. Almasan. 2020. Investigations on chemical composition and natural radioactivity levels from salt water and peloid used in pelotherapy from the Techirghiol Lake, Romania. Environ. Geochem. Health. 42(2): 513–529. doi: 10.1007/s10653-019-00382-8.

Camilotti, B.M., N. Leite, D. Alberti, I.A. Francisco and V.L. Israel. 2015. Effects of Ai Chi and Yamamoto new scalp acupuncture on chronic low back pain. Fisioterapia em Movimento 28: 723–730.

Cantore, S., A. Ballini, R. Saini, V. Altini, D. De Vito, F. Pettini et al. 2020. Effects of sea salt rinses on subjects undergone to oral surgery: a single blinded randomized controlled trial. Clin. Ter. 170(1): e46–e52. doi: 10.7417/CT.2020.2188.

Cantore, S., A. Ballini, R. Saini, D. De Vito, V. Altini, S.R. Saini et al. 2018. Efficacy of a combined sea salt-based oral rinse with xylitol against dental plaque, gingivitis, and salivary *Streptococcus mutans* load. J. Biol. Regul. Homeost. Agents. 32(6): 1593–1597.

Carbajo, J.M. and F. Maraver. 2018. Salt water and skin interactions: New lines of evidence. Int. J. Biometeorol. 62: 1345–1360. https://doi.org/10.1007/s00484-018-1545-z.

Cha, H.G., Y.J. Shin and M.K. Kim. 2017. Effects of the Bad Ragaz Ring Method on muscle activation of the lower limbs and balance ability in chronic stroke: A randomised controlled trial. Hong Kong Physiother J. 37: 39–45. doi: 10.1016/j.hkpj.2017.02.001.

Charlier, R.H. and M. Chaineux. 2009. The Healing Sea: A Sustainable Coastal Ocean Resource: Thalassotherapy. Journal of Coastal Research, 254: 838–856, DOI: 10.2112/08A-0008.1.

Chen, I.S., Y.Y. Chang, C.L. Hsu, H.W. Lin, M.H. Chang, J.W. Chen et al. 2013. Alleviative effects of deep-seawater drinking water on hepatic lipid accumulation and oxidation induced by a high-fat diet. J. Chin. Med. Assoc. 76(2): 95–101. doi: 10.1016/j.jcma.2012.10.008.

Chennaoui, M., D. Gomez-Merino, P.V. Beers, M. Guillard and C. Drogou. 2018. Benefits of thalassotherapy with sleep management on mood states and well-being, and cognitive and physical capacities in healthy workers. J. Sleep Disord. Ther. 7(297): 2167–0277.

Chun, SY., K.S. Lee and K.S. Nam. 2017. Refined deep-sea water suppresses inflammatory responses via the MAPK/AP-1 and NF-κB signaling pathway in LPS-Treated RAW 264.7 macrophage cells. Int. J. Mol. Sci. 18(11): 2282. doi: 10.3390/ijms18112282.

Couteau, C. and L. Coiffard. 2020. Phycocosmetics and other marine cosmetics, specific cosmetics formulated using marine resources. Mar Drugs 18(6): 322. doi: 10.3390/md18060322.

Cuesta-Vargas, A.I., N. Adams, J.A. Salazar, A. Belles, S. Hazañas and M. Arroyo-Morales. 2012. Deep water running and general practice in primary care for non-specific low back pain versus general practice alone: randomized controlled trial. Clin. Rheumatol. 31(7): 1073–1078. doi: 10.1007/s10067-012-1977-5.

de Andrade, S.C., R.F. de Carvalho, A.S. Soares, R.P. de Abreu Freitas, L.M. de Medeiros Guerra and M.J. Vilar. 2008. Thalassotherapy for fibromyalgia: a randomized controlled trial comparing aquatic exercises in sea water and water pool. Rheumatol. Int. 29(2): 147–152. doi: 10.1007/s00296-008-0644-2.

Diaz-Llopis, M., M.D. Pinazo-Duran and L.A. Diaz-Guiñon. 2019. Randomized multicenter study comparing seawater washes and carmellose artificial tears eyedrops in the treatment of dry eye syndrome. Clin. Ophthalmol. 13: 483–490. doi: 10.2147/OPTH.S185409.

Dull, H. 2004. Watsu: Freeing the body in water. Trafford Publishing, Victoria, BC, Canada.

Durie, A. 2003. Medicine, health and economic development: promoting spa and seaside resorts in Scotland C. 1750–1830. Med Hist. 47(2): 195–216.

Ehnert, L. and C. Geiser. 2022. What is secured in the Kneipp Therapy? The internal perspective. Inn. Med. (Heidelb). 63(12): 1229–1236. doi: 10.1007/s00108-022-01423-8. (in German).

El-Arabi, A.M. 2005. Natural radioactivity in sand used in thermal therapy at the Red Sea Coast. J. Environ. Radioact. 81(1): 11–19. doi: 10.1016/j.jenvrad.2004.11.002.

Ezhov, V. 2021. Climate-therapy at seaside resorts in modern medical and wellness practice. Vopr. Kurortol. Fizioter. Lech. Fiz. Kult. 98: 60–66. https://doi.org/10.17116/kurort20219803160. (In Russian).

Fan, G., H. Huang, Y. Lin, G. Zheng, X. Tang, Y. Fu et al. 2018. Herbal medicine foot bath for the treatment of diabetic peripheral neuropathy: protocol for a randomized, double-blind and controlled trial. Trials. 19(1): 483. doi: 10.1186/s13063-018-2856-4.

Feinstein, J.S., S.S. Khalsa, H.W. Yeh, C. Wohlrab, W.K. Simmons, M.B. Stein et al. 2018. Examining the short-term anxiolytic and antidepressant effect of Floatation-REST. PLoS One. Feb 2; 13(2): e0190292. doi: 10.1371/journal.pone.0190292.

Forestier, R.J., G. Briancon, A. Francon, F.B. Erol and J.M. Mollard. 2014. Balneohydrotherapy in the treatment of chronic venous insufficiency. Vasa. 43(5): 365–371. doi: 10.1024/0301-1526/a000374.

Gaab, E. and D.M. Steinhorn. 2015. Families' views of pediatric palliative aquatics: a qualitative study. Pain Manag. Nurs. 16(4): 526–533. doi: 10.1016/j.pmn.2014.09.012.

Glavaš, N., M.L. Mourelle, C.P. Gómez, J.L. Legido, N.R. Smuc, M. Dolenec et al. 2016. The mineralogical, geochemical, and thermophysical characterization of healing saline mud for use in pelotherapy. Appl. Clay Sci. 135: 119–128. https://doi.org/10.1016/j.clay.2016.09.013.

Gomes, C.S.F., J.V. Fernandes, F.V. Fernandes and J.B.P. Silva. 2021. Salt mineral water and thalassotherapy. Chapter 16, pp. 631–656. *In*: Gomes, C. and M. Rautureau, M. [eds.]. Minerals Latu Sensu and Human Health, 1st ed.; Springer: Cham, Switzerland, https://doi.org/10.1007/978-3-030-65706-2_16.

Gomes, C., M.I. Carretero, M. Pozo, F. Maraver, P. Cantista, F. Armijo et al. 2013. Peloids and pelotherapy: Historical evolution, classification and glossary. Appl. Clay Sci. 75: 28–38.

Grozeva, A. and M. Stoicheva. 2015. Thalassotherapy in diabetic polyneuropathy: a study in Pomorie, Bulgaria. J. Jpn. Assoc. Phys. Med. Balneol. Climatol. 78(3): 271–275.

Gutenbrunner, C., T. Bender, P. Cantista and Z. Karagülle. 2010. A proposal for a worldwide definition of health resort medicine, balneology, medical hydrology and climatology. Int. J. Biometeorol. 54: 495–507. https://doi.org/10.1007/s00484-010-0321-5.

Hamed, S., A.M. Almalty and H.S. Alkhatib. 2021. The cutaneous effects of long-term use of Dead Sea mud on healthy skin: a 4-week study. Int. J. Dermatol. 60(3): 332–339. doi: 10.1111/ijd.15304.

Hamed, S.A., M.M. ElMeligie and E. Kentiba. 2022. The effects of Halliwick aquatic exercises on gross motor function of children aged from 3 to 5 years with spastic cerebral palsy. Pedagogy Phys. Cult. Sports 27(1): 24–31. https://doi.org/10.15561/26649837.2023.0103.

Harper, C.M. 2012. Extreme preconditioning: cold adaptation through sea swimming as a means to improving surgical outcomes. Med Hypotheses 78(4): 516–519. doi: 10.1016/j.mehy.2012.01.022.

Hsu, C.L., Y.Y. Chang, C.H. Chiu, K.T. Yang, Y. Wang, S.G. Fu et al. 2011. Cardiovascular protection of deep-seawater drinking water in high-fat/cholesterol fed hamsters. Food Chem. 1, 127(3): 1146–1152. doi: 10.1016/j.foodchem.2011.01.116.

ISO 17680/2015, reviewed and confirmed in 2020, https://www.iso.org/standard/60244.html.

Kapusta, J. and R. Irzmański. 2022. The impact of controlled physical training with hydrotherapy on changes in swelling and claudication distance in patients with atherosclerotic ischemia of the lower limbs. Int. J. Environ. Res. Public Health. 19(23): 15715. doi: 10.3390/ijerph192315715.

Katsuda, S., T. Yasukawa, K. Nakagawa, M. Miyake, M. Yamasaki, K.A. Katahira et al. 2008. Deep-sea water improves cardiovascular hemodynamics in Kurosawa and Kusanagi-Hypercholesterolemic (KHC) rabbits. Biol. Pharm. Bull. 31(1): 38–44. doi: 10.1248/bpb.31.38.

Katz, U., Y. Shoenfeld, V. Zakin, Y. Sherer and S. Sukenik. 2012. Scientific evidence of the therapeutic effects of dead sea treatments: a systematic review. Semin Arthritis Rheum. 42(2): 186–200. doi: 10.1016/j.semarthrit.2012.02.006.

Kim, E.K., D.K. Lee and Y.M. Kim. 2015. Effects of aquatic PNF lower extremity patterns on balance and ADL of stroke patients. J. Phys. Ther. Sci. 27(1): 213–215. doi: 10.1589/jpts.27.213.

Kim, N.I., S.J. Kim, J.H. Jang, W.S. Shin, H.J. Eum, B. Kim et al. 2020. Changes in fatigue recovery and muscle damage enzymes after deep-sea water thalassotherapy. Applied Sciences 10(23): 8383.

Kirby, N.V., S.J.E. Lucas, O.J. Armstrong, S.R. Weaver and R.A.I. Lucas. 2021. Intermittent post-exercise sauna bathing improves markers of exercise capacity in hot and temperate conditions in trained middle-distance runners. Eur. J. Appl. Physiol. 121(2): 621–635. doi: 10.1007/s00421-020-04541-z.

Klein, D.A., J.S. William and T.P. Wayne. 2002. PNF training and physical function in assisted-living older adults. J. Aging Phys Act, 41: 476–488.

Komar, D., T. Dolenec, M. Dolenec, P. Vrhovnik, S. Lojen, Ž.L. Belak et al. 2015. Physico-chemical and geochemical characterization of Makirina Bay peloid mud and its evaluation for potential use in balneotherapy (N Dalmatia, Republic of Croatia). IJTK, 1(1): 5–12.

Korelc, S., A. Snedic, J. Brodnik and K. Groleger Sršen. 2019. Effectiveness of the cognitive approach in learning swimming skills in a program based on the Halliwick concept. Rehabilitacija. 18(2): 37–43. (In Slovenian).

Kotova, I., E. Kayukova and S. Kotov. 2016. Peloids of Crimean salt lakes and the Dead Sea: Controls on composition and formation. Environ. Earth Sci. 75: 1207.

Kowatzki, D., C. Macholdt, K. Krull, D. Schmidt, T. Deufel, P. Elsner et al. 2008. Effect of regular sauna on epidermal barrier function and stratum corneum water-holding capacity *in vivo* in humans: a controlled study. Dermatology 217(2): 173–80. doi: 10.1159/000137283.

Kudo, Y., M. Sasaki, Y. Kikuchi, R. Sugiyama, M. Hasebe and N. Ishii. 2019. Effects of a warm hand bath on the blood flow in the shoulder, skin and deep body temperature, autonomic nervous activity, and subjective comfort in healthy women: An experimental cross-over trial. Jpn. J. Nurs. Sci. 16(1): 88–100. doi: 10.1111/jjns.12216.

Laukkanen, J., T. Laukkanen and S. Kunutsor. 2018. Cardiovascular and other health benefits of sauna bathing: A review of the evidence. Mayo Clin. Proc. 93: 1111–1121.

Lee, K.S., S.Y. Chun, M.G. Lee, S. Kim, T.J. Jang and K.S. Nam. 2018. The prevention of TNF-α/IFN-γ mixture-induced inflammation in human keratinocyte and atopic dermatitis-like skin lesions in Nc/Nga mice by mineral-balanced deep sea water. Biomed Pharmacother 97: 1331–1340. doi: 10.1016/j.biopha.2017.11.056.

Lee, S-H., I-H. Bae, D.-J. Min, N.-H. Park, J.-H. Choi, J.-S. Shin et al. 2016. Skin hydration effect of Jeju Lava sea water. J. Soc. Cosmet. Sci. Korea 42(4): 343–349. doi:10.15230/SCSK.2016.42.4.343. (in Korean).

Lee, S.H., J.H. Han, S.J. Lee, H.Y. Choi, J.H. Baek and J.G. Kim. 2019. Comparison of seaweed pack and mudpack as treatment for knee osteoarthritis: a prospective randomized controlled study. Phys. Ther. Rehabil. Sci. 8(1): 22–31. https://doi.org/10.14474/ptrs.2019.8.1.22.

Liamputtong, P. and D. Suwankhong. 2015. Therapeutic landscapes and living with breast cancer: The lived experiences of Thai women. Soc. Sci. Med. 128: 263–271. https://doi.org/10.1016/j.socscimed.2015.01.031.

Lubkowska, A., B. Dołęgowska, Z. Szyguła, I. Bryczkowska, M. Stańczyk-Dunaj, D. Sałata et al. 2013. Winter-swimming as a building-up body resistance factor inducing adaptive changes in the oxidant/antioxidant status. Scand. J. Clin. Lab. Invest. 73(4): 315–325. doi: 10.3109/00365513.2013.773594.

Lucchetta, M.C., G. Monaco, V.I. Valenzi, M.V. Russo, J. Campanella, S. Nocchi et al. 2007 The historical-scientific foundations of thalassotherapy: State of the art. Clin. Ter. 158: 533–541. (In Italian).

Ma'or, Z., L. Halicz, M. Portugal-Cohen, M.Z. Russo, F. Robino, T. Vanhaecke et al. 2015. Safety evaluation of traces of nickel and chrome in cosmetics: The case of Dead Sea mud. Regul. Toxicol. Pharmacol. 73(3): 797–801. doi: 10.1016/j.yrtph.2015.10.016.

Ma'or, Z., Y. Henis, Y. Alon, E. Orlov, K.B. Sørensen and A. Oren. 2006. Antimicrobial properties of Dead Sea black mineral mud. Int. J. Dermatol. 45(5): 504–511. doi: 10.1111/j.1365-4632.2005.02621.x.

Mahé, A. 1999. El Plasma de Quinton: El agua de mar, nuestro medio interno. Icaria Milenrama: Barcelona. ISBN 9788474264494 (in Spanish).

Maheshkumar, K., M. Pandiaraja, V. Venugopal, S. Poonguzhali and L. Sundareswaran. 2020. Effects of hot foot and arm bath in bronchial asthma: A single case report. Foot (Edinb). 42: 101651. doi: 10.1016/j.foot.2019.10.006.

Manolis, A.S., S.A. Manolis, A.A. Manolis, T.A. Manolis, N. Apostolaki and H. Melita. 2019. Winter Swimming: Body Hardening and Cardiorespiratory Protection Via Sustainable Acclimation. Curr. Sports Med. Rep. 18(11): 401–415. doi: 10.1249/JSR.0000000000000653.

Maraver, F., A. Michan-Doña, C. Morer and L. Aguilera. 2010. Is thalassotherapy simply a type of climatotherapy? Int. J. Biometeorol. 55: 107–108. https://doi.org/10.1007/s00484-010-0382-5.

Masiukovichi, T., M. Jokhadze, T. Murtazashvili and A. Bakuridze. 2017. Characterization of sulfide silt peloids ins some lakes within Adjara region and determination of biologically active compounds content. Georgian Med News. 270: 115–120.

Masiukovichi, T., T. Murtazashvili and A. Bakuridze. 2018. Development if the formulation and technology hydrogel, containing Adjara sulfide silt peloid. Georgian Med. News. Feb; (Issue): 157–162.

Melandri, D., V.M. Albano, M. Venturi, A. Flamigni and M. Vairetti. 2020. Efficacy of combined liman peloid baths and heliotherapy in the treatment of psoriasis at Cervia Spa, Emilia, Italy. International Journal of Biometeorology 64, 7: 1145–1152. http://dx.doi.org/10.1007/s00484-020-01888-y.

Mihelčić, G., G. Kniewald, G. Ivanišević, R. Čepelak, V. Mihelčić and N. Vdović. 2012. Physico-chemical characteristics of the peloid mud from Morinje Bay (eastern Adriatic coast, Croatia): suitability for use in balneotherapy. Environ. Geochem. Health. 34(2): 191–198. doi: 10.1007/s10653-011-9434-y.

Millero, F.J. 2002. Sea water as an electrolyte. pp 3–34. *In*: Gianguzza, A., E. Pelizzetti and S. Sammartano. [eds.]. Chemistry of Marine Water and Sediments. Environmental Science. Springer, Berlin, Heidelberg. https://doi.org/10.1007/978-3-662-04935-8_1.

Miyamura, M., S. Yoshioka, A. Hamada, D. Takuma, J. Yokota, M. Kusunose et al. 2004. Difference between deep seawater and surface seawater in the preventive effect of atherosclerosis. Biol. Pharm. Bull. 27(11): 1784–1787. doi: 10.1248/bpb.27.1784.

Mohd-Nani, S.Z., F.A. Majid, A.B. Jaafar, A. Mahdzir and M.N. Musa. 2016. Potential health benefits of deep sea water: A review. Evid Based Complement Alternat Med. 6520475. doi: 10.1155/2016/6520475.

Moss, G.A. 2020. Water and health: a forgotten connection? Perspect. Public Health. 130(5): 227–232. doi: 10.1177/1757913910379192.

Mourelle, M.L., C.P. Gómez and J.L. Legido. 2021. Microalgal peloids for cosmetic and wellness uses. Mar. Drugs 19: 666. https://doi.org/10.3390/md19120666.

Mourelle, M.L. and I. Sabater. 2013. Spa and Wellness. Manual for therapists. Estetica & Wellness ed., Madrid (ES) (in Spanish).

Mourelle, M.L., C.P. Gómez and J.L. Legido. 2023. Cosmeceuticals and thalassotherapy: Recovering the skin and well-being after cancer therapies. Appl. Sci. 13: 850. https://doi.org/10.3390/ app13020850.

Munteanu, C., D. Munteanu, M. Hoteteu and G. Dogaru. 2019. Balneotherapy—Medical, scientific, educational and economic relevance reflected by more than 250 articles published in Balneo Research Journal. Balneo. PRM Res. J. 10: 174–203. https://doi.org/10.12680/balneo.2019.257.

Norell-Clarke, A., K. Jonsson, A. Blomquist, R. Ahlzén and A. Kjellgren. 2022. A study of flotation-REST (restricted environmental stimulation therapy) as an insomnia treatment. Sleep Sci. 5(Spec 2): 361–368. doi: 10.5935/1984-0063.20210012.

Pandiaraja, M., A. Vanitha, K. Maheshkumar, V. Venugopal, S. Poonguzhali, L. Radhika et al. 2021. Effect of the steam bath on resting cardiovascular parameters in healthy volunteers. Adv. Integr. Med. 8(3): 199–202. https://doi.org/10.1016/j.aimed.2020.06.001.

Patrick, R.P. and T.L. Johnson. 2021. Sauna use as a lifestyle practice to extend healthspan. Exp. Gerontol. 154: 111509. doi: 10.1016/j.exger.2021.111509.

Pereira, L. 2017, Chapter 12, pp. 503–522. Thalassotherapy and Marine Cosmeceuticals. *In*: Pereira, L. [ed.]. Therapeutic and Nutritional Uses of Algae, 1st ed.; CRC Press/Taylor & Francis Group: Abingdon, UK. https://doi.org/10.1201/9781315152844.

Pereira, L. 2018. Seaweeds as Source of Bioactive Substances and Skin Care Therapy—Cosmeceuticals, Algotheraphy, and Thalassotherapy. Cosmetics, 5: 68. https://doi.org/10.3390/cosmetics5040068.

Pérez-de la Cruz, S. 2021. Comparison between three therapeutic options for the treatment of balance and gait in stroke: A randomized controlled trial. Int. J. Environ. Res. Public Health. 18(2): 426. doi: 10.3390/ijerph18020426.

Portugal-Cohen, M., Y. Soroka, Z. Ma'or, M. Oron, T. Zioni, F.M. Brégégère et al. 2009. Protective effects of a cream containing Dead Sea minerals against UVB-induced stress in human skin. Exp. Dermatol. 18(9): 781–788. doi: 10.1111/j.1600-0625.2009.00865.x.

Portugal-Cohen, M., D. Cohen, E. Ish-Shalom, Y. Laor-Costa and Z. Ma'or. 2019. Dead Sea minerals: New findings on skin and the biology beyond. Exp. Dermatol. 28(5): 585–592. doi: 10.1111/exd.13918.

Proksch, E., H.-P. Nissen, M. Bremgartner and C. Urquhart. 2005. Bathing in a magnesium-rich Dead Sea salt solution improves skin barrier function, enhances skin hydration, and reduces inflammation in atopic dry skin. Int. J. Dermatol. 44: 151–157. https://doi.org/10.1111/j.1365-4632.2005.02079.x.

Rashleigh, R., S.M. Smith and N.J. Roberts. 2014. A review of halotherapy for chronic obstructive pulmonary disease. Int. J. Chron. Obstruct. Pulmon. Dis. 21(9): 239–246. doi: 10.2147/COPD.S57511.

Schempp, C.M., H.C. Dittmar, D. Hummler, B. Simon-Haarhaus, E. Schöpf, J.C. Simon et al. 2000. Magnesium ions inhibit the antigen-presenting function of human epidermal Langerhans cells *in vivo* and *in vitro*. Involvement of

ATPase, HLA-DR, B7 molecules, and cytokines. J. Investig. Dermatol. 115: 680–686. https://doi.org/10.1046/j.1523-1747.2000.00090.x.

Schitter, A.M., J. Fleckenstein, P. Frei, J. Taeymans, N. Kurpiers and L. Radlinger. 2020. Applications, indications, and effects of passive hydrotherapy WATSU (WaterShiatsu)-A systematic review and meta-analysis. PLoS One 15(3): e0229705. doi: 10.1371/journal.pone.0229705.

Schuh, A. and D. Nowak. 2011. Evidence-based acute and long-lasting effects of climatotherapy in moderate altitudes and on the seaside. DMW—Dtsch. Med. Wochenschr. 136: 135–139. https://doi.org/10.1055/s-0031-1272496. (In German).

Signorini, L., A.D. Inchingolo, L. Santacroce, E. Xhajanka, V. Altini, I.R. Bordea et al. 2020. Efficacy of combined sea salt based oral rinse with xylitol in improving healing process and oral hygiene among diabetic population after oral surgery. J. Biol. Regul. Homeost. Agents. 34(4): 1617–1622. doi: 10.23812/20-418-L.

Shim, S.R., D. Shin, S.-J. Kim, Y.K. Kim and K.J. Lee. 2023. Unveiling the therapeutic potential and healthcare applications of marine therapy: A systematic review with meta-analysis and meta-regression. Mar. Drugs 2023, 21, 604. https://doi.org/10.3390/md21120604.

Skinner, E.H., T. Dinh, M. Hewitt, R. Piper and C. Thwaites. 2016. An Ai Chi-based aquatic group improves balance and reduces falls in community-dwelling adults: A pilot observational cohort study. Physiother. Theory Pract. 32: 581–590.

Štanfel, D., L. Kalogjera, S.V. Ryazantsev, K. Hlača, E.Y. Radtsig, R. Teimuraz et al. 2022. The role of seawater and saline solutions in treatment of upper respiratory conditions. Mar Drugs 20(5): 330. doi: 10.3390/md20050330.

Sukenik, S., H. Giryes, S. Halevy, L. Neumann, D. Flusser and D. Buskila. 1994. Treatment of psoriatic arthritis at the Dead Sea. J. Rheumatol. 21(7): 1305–1309.

Sutkowy, P., A. Woźniak, T. Boraczyński, C. Mila-Kierzenkowska and M. Boraczyński. 2015. Postexercise impact of ice-cold water bath on the oxidant-antioxidant balance in healthy men. Biomed. Res. Int. 2015: 706141. doi: 10.1155/2015/706141.

Takeuchi, H., K. Higuchi, Y. Yoshikane, R. Takagi, S. Tokuhiro, K. Takenaka et al. 2020. Drinking refined deep-sea water improves the gut ecosystem with beneficial effects on intestinal health in humans: A randomized double-blind controlled trial. Nutrients 12(9): 2646. doi: 10.3390/nu12092646.

Takeuchi, H., Y. Yoshikane, H. Takenaka, A. Kimura, J.M. Islam, R. Matsuda et al. 2022. Health effects of drinking water produced from deep sea water: A randomized double-blind controlled trial. Nutrients 14(3): 581. doi: 10.3390/nu14030581.

Toro, V., J. Siquier-Coll, I. Bartolomé, M. Pérez-Quintero, A. Raimundo, D. Muñoz et al. 2021. Effects of twelve sessions of high-temperature sauna baths on body composition in healthy young men. Int. J. Environ. Res. Public Health. 18(9): 4458. doi: 10.3390/ijerph18094458.

Tsuchiya, Y., A. Watanabe, N. Fujisawa, T. Kaneko, T. Ishizu, T. Fujimoto et al. 2004. Effects of desalted deep seawater on hematologic and blood chemical values in mice. Tohoku J. Exp. Med. 203(3): 175–82. doi: 10.1620/tjem.203.175.

Veldema, J. and P. Jansen. 2021. Aquatic therapy in stroke rehabilitation: systematic review and meta-analysis. Acta Neurol. Scand. 143(3): 221–241. doi: 10.1111/ane.13371.

Wasik, A.A. and T. Tuuminen. 2021. Salt therapy as a complementary method for the treatment of respiratory tract diseases, with a focus on mold-related illness. Altern. Ther. Health Med. 27(S1): 223–239.

What Are the Different Climate Types? https://scijinks.gov/climate-zones/ (accessed, June 2023).

Yoshioka, S., A. Hamada, T. Cui, J. Yokota, S. Yamamoto, M. Kusunose et al. 2003. Pharmacological activity of deep-sea water: examination of hyperlipemia prevention and medical treatment effect. Biol. Pharm. Bull. 26(11): 1552–1559. doi: 10.1248/bpb.26.1552.

Yoshizawa, Y., K. Kitamura, S. Kawana and H.I. Maibach. 2003. Water, salts and skin barrier of normal skin. Skin Res. Technol. 9(1): 31–3. doi: 10.1034/j.1600-0846.2003.00346.x.

Yoshizawa, Y., H. Tanojo, S.J. Kim and H.I. Maibach. 2001. Sea water or its components alter experimental irritant dermatitis in man. Skin Res. Technol. 7(1): 36–9. doi: 10.1034/j.1600-0846.2001.007001036.x.

Yuncu, E.N., G. Yurcu and Z. Akıncı. 2019. Thalassotherapy (Swot Analysis). pp. 58–72. *In*: Tüfekci, Ö.K. [ed.]. New Horizons in Social Sciences and Humanities. Sra Academic Publishing Press, Klaipeda.

Zhang, X., Y. Zhang, J. Zhai, Y. Wu and A. Mao. 2021. Waterscapes for promoting mental health in the general population. Int. J. Environ. Res. Public Health 18(22): 11792. doi: 10.3390/ijerph182211792.

Zijlstra, T.R., M.A. van de Laar, H.J. Bernelot Moens, E. Taal, L. Zakraoui and J.J. Rasker. 2005. Spa treatment for primary fibromyalgia syndrome: a combination of thalassotherapy, exercise and patient education improves symptoms and quality of life. Rheumatology 44(4): 539–546. doi: 10.1093/rheumatology/keh537.

CHAPTER 6
Marine Resources
Macro- and Microalgae

Leonel Pereira

Introduction

Overview of the importance of macro- and microalgae in the fields of thalassotherapy and cosmeceuticals

Macro- and microalgae hold significant importance in the fields of thalassotherapy and cosmeceuticals due to their diverse bioactive compounds and therapeutic properties. In thalassotherapy, macroalgae, or seaweeds, are widely utilized in treatments such as seaweed baths, wraps, and poultices, offering relaxation, detoxification, hydration, improved circulation, and skin nourishment. Microalgae, though less commonly known in thalassotherapy, provide high nutritional value and detoxifying properties. They aid in replenishing essential nutrients, supporting cellular regeneration, and promoting overall well-being (Ampofo and Abbey 2022).

In cosmeceuticals, macroalgae extracts are incorporated into skincare products like facial masks, creams, serums, and body care products. They provide moisturizing, anti-aging, and skin-conditioning properties, improving skin elasticity, reducing inflammation, and protecting against environmental damage. Microalgae, with their microscopic nature and high nutritional content, offer significant potential in skincare. They are rich in vitamins, minerals, amino acids, and antioxidants, making them ideal for enhancing skin health, combating aging signs, and improving texture and tone.

The future prospects in these fields include exploring untapped species of macro- and microalgae, optimizing cultivation methods, and identifying novel bioactive compounds (Gosh et al. 2022). However, sustainable harvesting practices and conservation efforts are crucial to ensure the long-term availability and ecological balance of these marine resources (Cotas et al. 2023). Challenges include standardizing extraction processes, ensuring product quality and stability, and overcoming regulatory barriers. Ongoing research aims to address these challenges and develop innovative formulations that maximize the benefits of algae in thalassotherapy and cosmeceuticals (Rumin et al. 2021).

Marine Resources, Conservation and Technology, Marine Algae Lab, CFE—Centre for Functional Ecology: Science for People & Planet, Department of Life Sciences, University of Coimbra, 3000-456 Coimbra, Portugal.
Email: leonel.pereira@uc.pt

Brief explanation of the diversity and abundance of these marine resources

Macroalgae and microalgae exhibit remarkable diversity and abundance in marine environments. Macroalgae, commonly known as seaweeds, encompass a wide range of species belonging to three major groups: brown algae (phylum Ochrophyta, class Phaeophyceae), red algae (phylum Rhodophyta), and green algae (phylum Chlorophyta). Each group comprises numerous species with distinct characteristics, including varying sizes, shapes, and pigmentation (Pereira 2021). Brown algae are prevalent in temperate and cold coastal regions, while red algae dominate tropical and subtropical waters. Green algae can be found in various marine habitats, including intertidal zones and shallow waters (Duarte et al. 2022).

The diversity of microalgae is even more extensive, as they encompass a vast array of microscopic organisms such as diatoms, dinoflagellates, and cyanobacteria. Diatoms (class Bacillariophyceae) are single-celled algae encased in intricate silica cell walls and are known for their distinct shapes and patterns. Dinoflagellates (phylum Dinoflagellata) are flagellated algae that are responsible for bioluminescence and harmful algal blooms. Cyanobacteria, often referred to as blue-green algae, are photosynthetic bacteria that can form colonies or filaments and contribute to primary production in aquatic ecosystems (Gonçalves and Figueredo 2020).

The abundance of macroalgae and microalgae is influenced by factors such as nutrient availability, light intensity, water temperature, and ecological interactions (Menaa et al. 2020). Macroalgae are commonly found attached to substrates such as rocks, reefs, and seafloor sediments, forming extensive underwater forests and beds. They can thrive in various marine environments, ranging from intertidal zones to deep-sea habitats (Wiktor et al. 2022). In contrast, microalgae are present throughout the water column, ranging from surface waters to deeper layers. They often form blooms under favorable conditions, leading to visible discoloration of water bodies (Mobin et al. 2022).

Both macroalgae and microalgae play crucial roles in marine ecosystems. Macroalgae provide habitat and food for numerous organisms, contribute to primary production, and help maintain coastal biodiversity. Microalgae are fundamental components of the marine food web, serving as primary producers and supporting the growth of higher trophic levels. Their photosynthetic activity is responsible for a significant portion of global oxygen production (Cotas et al. 2023, Lenzo et al. 2023).

The diversity and abundance of macroalgae and microalgae underline their significance as valuable marine resources in various industries, including thalassotherapy, cosmeceuticals, food production, and biofuel development. However, proper management and conservation efforts are necessary to ensure the sustainable utilization of these resources and protect their ecological roles within marine ecosystems (Cotas et al. 2020, da Rosa et al. 2023).

Chapter objectives

The mains objectives of this chapter are:

- Provide a comprehensive overview of the importance of macro- and microalgae in the fields of thalassotherapy and cosmeceuticals;
- Explore the diverse applications and benefits of macroalgae and microalgae in thalassotherapy treatments, including their therapeutic properties and contributions to relaxation, detoxification, hydration, improved circulation, and skin nourishment;Examine the role of macroalgae and microalgae in cosmeceuticals, focusing on their skincare benefits such as moisturizing, anti-aging, skin-conditioning, and protection against environmental damage; highlight the abundant availability and ecological significance of macroalgae and microalgae in marine environments, emphasizing their role in nutrient cycling, primary production, and support of marine ecosystems;

- Discuss the future prospects and potential challenges in the utilization of macroalgae and microalgae in thalassotherapy and cosmeceuticals, including the need for sustainable harvesting practices, conservation efforts, and ongoing research and development to optimize extraction methods and product formulations;
- Foster awareness of the diverse species and bioactive compounds found in macroalgae and microalgae, promoting their exploration as potential sources of innovative thalassotherapy and cosmeceutical products; interdisciplinary collaboration between researchers, industry professionals, and environmental stakeholders in order to maximize the benefits of macroalgae and microalgae while ensuring the preservation of marine ecosystems (Mourelle et al. 2017, Pereira 2018, Kalasariya and Pereira 2022).

Classification and characteristics of macro- and microalgae

Macroalgae

Seaweed, also known as macroalgae, is a diverse group of marine plants that play a vital role in marine ecosystems and have numerous practical applications. Seaweeds are distinct from microalgae, which are microscopic unicellular organisms. They are commonly found in coastal areas and are classified into three main groups: brown algae (Phaeophyceae), red algae (Rhodophyta), and green algae (Chlorophyta).

As primary producers, macroalgae are at the base of the marine food chain supporting several communities of herbivorous animals (invertebrates, such as some sea urchins and/or gastropods, and vertebrates, such as herbivorous fish) where they end up finding refuge from their predators, the carnivores. In order to escape this herbivory, which is sometimes intense (in natural reefs or rocky walls of the continental shelf), many macroalgae have been improving defense strategies, with calcification (common in red algae, such as the Corallinaceae family) being one of the most common, although they can also produce secondary metabolites (terpenes, aromatic substances, polyphenols, etc.) that act as demotivators of their ingestion (García-Poza et al. 2022, Pereira 2022).

Seaweeds are multicellular organisms that have evolved to adapt to marine environments. They lack true roots, stems, and leaves like terrestrial plants but possess specialized structures that help them anchor to substrates and absorb nutrients from the surrounding water (Salehi et al. 2019). Some seaweeds have air bladders or gas-filled structures to provide buoyancy and keep them suspended in the water column (Salehi et al. 2019).

The main characteristics of seaweed include their large size, which can range from a few cm to several m in length, depending on the species. They have a complex morphology with thallus, the plant body, consisting of blades, fronds, stipes, holdfasts, or a combination of these structures. The coloration of seaweeds varies among groups and species, with brown, red, and green hues being the most common (Pereira 2015).

Seaweeds are photosynthetic organisms, utilizing sunlight, carbon dioxide, and nutrients from the water to produce energy through photosynthesis. They contain chlorophyll, as well as accessory pigments, which give them their characteristic colors. Seaweeds can grow in a wide range of marine habitats, including intertidal zones, rocky shores, kelp forests, coral reefs, and deep-sea environments (Pereira 2021, Freitas et al. 2022).

One of the defining features of seaweeds is their ability to extract and concentrate various minerals and trace elements from seawater. They are considered a rich source of essential nutrients, including iodine, calcium, iron, and vitamins such as vitamin C and vitamin K. Seaweeds also contain a diverse array of bioactive compounds such as polysaccharides, polyphenols, phlorotannins, and antioxidants, which contribute to their numerous beneficial properties (Wells et al. 2017, Michalak et al. 2022).

Due to their ecological importance and valuable bioactive compounds, seaweeds have been utilized by humans for centuries. They are harvested for various purposes, including food, fertilizer, animal feed, and applications in industries such as thalassotherapy, cosmeceuticals, and biofuel production. Seaweed farming, also known as seaweed aquaculture, has gained traction as a sustainable method for cultivating these marine resources (Morais et al. 2020).

In conclusion, seaweed represents a fascinating group of marine plants with unique characteristics. Their diverse forms, adaptability to marine environments, nutritional value, and bioactive compounds make them an important resource with a wide range of applications in various industries and contribute significantly to the health and balance of marine ecosystems (Choudhary et al. 2021, Quitério et al. 2021).

Microalgae

Microalgae are microscopic algae that encompass a diverse group of photosynthetic organisms, including diatoms, dinoflagellates, and cyanobacteria (Hachicha et al. 2022). Despite their small size, microalgae hold immense importance in the fields of thalassotherapy and cosmeceuticals due to their unique properties and bioactive compounds (Mourelle et al. 2017).

Microalgae exhibit remarkable diversity and are classified into various groups based on their characteristics and evolutionary relationships (Borics et al. 2023). The classification of microalgae is constantly evolving as new research uncovers their genetic and physiological complexities. Here, we provide a brief discussion on the diversity and classification of microalgae (Khan et al. 2018).

Microalgae can be broadly categorized into three main groups: diatoms (Bacillariophyceae), dinoflagellates (Dinoflagellata), and blue-green algae (Cyanobacteria) (Guiry and Guiry 2023).

Diatoms: Diatoms are unicellular algae that are enclosed in intricate silica cell walls called frustules. They are known for their unique shapes and patterns, which contribute to their taxonomic identification (Reid et al. 2021). Diatoms are found in both freshwater and marine environments, and their abundance and ecological importance are significant. They play a crucial role in primary production, nutrient cycling, and carbon sequestration. Additionally, diatoms are a valuable source of lipids, silica, and bioactive compounds with various applications in industries such as biofuels, bioremediation, and biotechnology (Serôdio and Lavaud 2020).

Dinoflagellates are unicellular algae with two flagella that enable their movement in the water column. They are known for their diverse forms and ecological roles. Some dinoflagellates are photosynthetic, while others are heterotrophic or mixotrophic. Dinoflagellates can form blooms under favorable conditions, leading to the phenomenon known as "red tides" or harmful algal blooms (Guo et al. 2013, Riding et al. 2023). While some dinoflagellates produce toxins that can harm marine life and humans, others play vital roles as primary producers and symbionts in coral reefs. Their bioluminescence ability is another fascinating feature exhibited by certain species (Verma et al. 2019).

Cyanobacteria, also known as blue-green algae, are a group of prokaryotic microorganisms that are capable of photosynthesis. They are widespread in various aquatic environments, including freshwater and marine habitats. Cyanobacteria have a diverse range of forms, including single cells, filaments, and colonies. They play a crucial role in primary production, nitrogen fixation, and oxygen production (Mehdizadeh Allaf and Peerhossaini 2022). Cyanobacteria are known for their ability to form blooms, which can have ecological and health implications. Some cyanobacteria produce toxins, called cyanotoxins, which can be harmful to other organisms and pose risks to human health (Rastogi et al. 2015).

It is important to note that microalgae diversity extends beyond these three groups, and other smaller groups exist, such as coccolithophores (phylum Haptophyta, class Coccolithophyceae), cryptophytes (phylum Cryptista, class Cryptophyceae), Chrysophyceae, Pavlovaceae,

Synurophyceae, Xanthophyceae, and Euglenophyceae, among others (see Table 1) (Sehgal et al. 2019). Each group has unique characteristics, including their morphology, pigmentation, ecological roles, and potential applications (Khavari et al. 2021).

Advancements in molecular techniques, such as DNA sequencing and phylogenetic analysis, have revolutionized our understanding of microalgae diversity. These techniques have allowed for the discovery and classification of previously unknown or misunderstood microalgal species (Sproles et al. 2021). Ongoing research continues to expand our knowledge of microalgae diversity and their ecological functions, as well as their potential applications in various industries, including biotechnology, pharmaceuticals, and environmental management (Araújo et al. 2021).

Thus, microalgae display a vast array of forms, ecological roles, and potential applications. Their diversity, ranging from diatoms to dinoflagellates and cyanobacteria, contributes to the functioning of aquatic ecosystems and presents opportunities for scientific exploration and innovation. Understanding and conserving microalgae diversity is crucial for maintaining the health of aquatic environments and harnessing their potential benefits for human well-being (Thoré et al. 2023).

In the thalassotherapy industry, microalgae are gaining recognition for their high nutritional value and detoxifying properties (Kiran and Mohan 2021). They are rich sources of essential nutrients such as vitamins, minerals, amino acids, and omega-3 fatty acids. Microalgae extracts and supplements are utilized in thalassotherapy treatments to replenish essential nutrients, support cellular regeneration, and promote overall well-being. They also contribute to the detoxification process by helping the body eliminate toxins and supporting liver function. The therapeutic effects of microalgae in thalassotherapy contribute to relaxation, revitalization, and improved vitality (Quitério et al. 2021, Lucakova et al. 2022).

In the cosmeceutical industry, microalgae are being increasingly utilized for their beneficial effects on skin health and beauty. They are abundant in antioxidants, including carotenoids, phycocyanin's, and tocopherols, which help protect the skin from oxidative stress and damage

Table 1. Algae taxonomy.

Domain	Kingdom	Phylum	Class	Example
Prokaryota/ Bacteria	Eubacteria	Cyanobacteria	Cyanophyceae	*Rivularia bullata*
Eukaryota	Plantae	Glaucophyta	Glaucophyceae	*Glaucocystis*
Eukaryota	Plantae	Rhodophyta	Porphyridiophyceae Bangiophyceae Florideophyceae	*Porphyridium purpureum* *Porphyra umbilicalis* *Osmundea pinnatifida*
Eukaryota	Plantae	Chlorophyta	Chlorophyceae Trebouxiophyceae Ulvophyceae	*Dunaliella salina* *Chlorella vulgaris* *Ulva lactuca*
Eukaryota	Protozoa	Euglenozoa	Euglenophyceae	*Euglena viridis*
Eukaryota	Chromista	Dinoflagellata	Dinophyceae	*Kryptoperidinium foliaceum*
Eukaryota	Chromista	Haptophyta	Coccolithophyceae Pavlovophyceae	*Gephyrocapsa oceanica* *Pavlova gyrans*
Eukaryota	Chromista	Ochrophyta	Chrysophyceae	*Chrysocapsa planktonica*
Eukaryota	Chromista	Ochrophyta	Synurophyceae	*Synedra uvella*
Eukaryota	Chromista	Ochrophyta	Xathophyceae	*Tribonema bombycinum*
Eukaryota	Chromista	Bacillariophyta	Bacillariophyceae	*Phaeodactylum tricornutum*
Eukaryota	Chromista	Ochrophyta	Phaeophyceae	*Fucus vesiculosus*

caused by free radicals (Thiyagarasaiyar et al. 2020). Microalgae-based ingredients are incorporated into anti-aging creams, serums, sunscreens, and other skincare products. They promote collagen synthesis, improve skin elasticity, reduce the appearance of wrinkles, and enhance skin texture and tone (Favas et al. 2021). Additionally, microalgae extracts have shown potential in combating skin pigmentation disorders and promoting an even skin tone (Wang et al. 2015).

The unique advantage of microalgae lies in their ability to produce a wide range of bioactive compounds with diverse properties. Some microalgae species produce polysaccharides, which possess moisturizing and film-forming properties, making them valuable ingredients in skincare formulations (Martins et al. 2023). Other microalgae produce peptides with antioxidant and anti-inflammatory properties, contributing to the overall health and appearance of the skin (Choo et al. 2020).

Furthermore, microalgae have a high potential for sustainable production. They can be cultivated using various methods, including open ponds, photobioreactors, and closed systems. Microalgae require minimal land, freshwater, and nutrients to grow, making them a more environmentally friendly alternative to traditional land-based crops. Their fast growth rates and ability to convert carbon dioxide into oxygen through photosynthesis further contribute to their environmental benefits (Khor et al. 2022).

So, microalgae play a significant role in thalassotherapy and cosmeceuticals. Their high nutritional content, detoxifying properties, and bioactive compounds make them valuable ingredients in thalassotherapy treatments and skincare products. Microalgae offer numerous benefits for skin health and beauty, providing antioxidant protection, anti-aging effects, and improved skin texture. Their sustainable cultivation methods also make them an attractive option for the industry (Mourelle et al. 2017, Martínez-Ruiz et al. 2022).

Bioactive properties of macroalgae and microalgae chemical compounds

Seaweed is rich in various chemical components that contribute to its nutritional value and numerous health benefits. Some of the main chemical components found in seaweed include polysaccharides, polyphenols, and carotenoids (El-Beltagi et al. 2022, Michalak et al. 2022).

Polysaccharides: Seaweed contains a wide range of polysaccharides, which are complex carbohydrates made up of sugar molecules. These polysaccharides play important roles in seaweed's structural integrity and provide health-promoting properties (Cunha and Grenha 2016, El-Beltagi et al. 2022). Examples of polysaccharides found in seaweed include carrageenans, alginates, fucoidans, and ulvans (Hentati et al. 2020). Carrageenans, derived from red seaweeds, are used as gelling and thickening agents in food and pharmaceutical industries (Pereira 2016). Alginates, obtained from brown seaweeds, have excellent gelling and stabilizing properties, and are used in various food and medical applications (Abka-khajouei et al. 2022). Fucoidans, found in brown seaweeds, have been studied for their potential anticoagulant, antiviral, and anti-inflammatory properties (Wang et al. 2019). Ulvans, present in green seaweeds, have antioxidant and immunostimulant activities (Cindana Mo'o et al. 2020).

Polyphenols: Seaweeds are a rich source of polyphenols, which are bioactive compounds with antioxidant and anti-inflammatory properties. Polyphenols found in seaweed include phlorotannins, flavonoids, and phenolic acids (Cotas et al. 2020). Phlorotannins, specifically found in brown seaweeds, are unique polyphenols with strong antioxidant activity. They have been studied for their potential benefits in preventing oxidative stress, inflammation, and various chronic diseases (Catarino et al. 2021). Flavonoids, also present in seaweed, contribute to the antioxidant and anti-inflammatory properties (Michalak et al. 2022). Phenolic acids, including caffeic acid, ferulic acid, and protocatechuic acid, have been associated with various health benefits, including anti-cancer and cardiovascular protective effects (Kiokias et al. 2021).

Carotenoids: Seaweed contains a diverse range of carotenoids, which are pigments responsible for its characteristic colors. The most common carotenoids found in seaweed include beta-carotene, fucoxanthin, and lutein. β-carotene, a precursor of vitamin A, is known for its antioxidant properties and its role in maintaining healthy vision and immune function (Freitas et al. 2022). Fucoxanthin, found specifically in brown seaweeds, has gained attention for its potential anti-obesity, anti-diabetic, and anti-inflammatory properties (Mumu et al. 2022). Lutein, found in green seaweeds, is known for its role in eye health and has antioxidant activity (Mitra et al. 2021).

These chemical components in seaweed not only contribute to their nutritional value but also offer potential health benefits. They exhibit antioxidant, anti-inflammatory, anti-cancer, and immune-enhancing properties. Seaweed-derived polysaccharides, polyphenols, and carotenoids are being explored for their applications in functional foods, nutraceuticals, and pharmaceuticals (Lomartire and Gonçalves 2022).

It is important to note that the specific chemical composition and concentrations of these components vary among seaweed species and can be influenced by environmental factors such as light, temperature, and nutrient availability (Salehi et al. 2019). Ongoing research is focused on identifying and understanding the bioactive compounds present in seaweed and their potential health implications, paving the way for their utilization in various industries (Juárez-Portilla et al. 2019).

Microalgae are known for their diverse range of chemical compounds, which contribute to their nutritional value and bioactive properties. Here is a brief overview of the main chemical compounds found in microalgae (Silva et al. 2022, Eze et al. 2023).

Pigments: Microalgae contain a variety of pigments that give them their characteristic colors. Chlorophylls, including chlorophyll a and chlorophyll b, are responsible for capturing light energy during photosynthesis. Other pigments found in microalgae include carotenoids, such as beta-carotene, astaxanthin, and lutein, which serve as antioxidants and play a role in light harvesting and protection against oxidative stress (Patel et al. 2022, Sun et al. 2023).

Polyunsaturated fatty acids (PUFAs): Microalgae are a primary source of omega-3 polyunsaturated fatty acids (PUFAs) such as eicosapentaenoic acid (EPA) and docosahexaenoic acid (DHA). These essential fatty acids are important for human health, particularly in supporting brain function, cardiovascular health, and inflammation regulation. Microalgae are considered a sustainable alternative to fish as a source of omega-3 PUFAs (Barta et al. 2021, Liu et al. 2022).

Proteins and amino acids: Microalgae are rich in proteins, which are composed of amino acids. Some microalgae species have a high protein content, making them a potential source of plant-based protein for food and feed applications. Microalgae also contain essential amino acids, including lysine, tryptophan, and leucine, which are important for human nutrition (Koyande et al. 2019, Eilam et al. 2023).

Polysaccharides: Microalgae produce various polysaccharides, including sulfated polysaccharides. These polysaccharides contribute to the structural integrity of microalgae cells and offer potential health benefits, as they have antioxidant, anti-inflammatory, and immunomodulatory properties. They have applications in the food, pharmaceutical, and cosmetic industries (Costa et al. 2021, Moreira et al. 2022).

Bioactive peptides: Microalgae can produce bioactive peptides with potential health benefits. These peptides exhibit various bioactivities, including antioxidant, antimicrobial, anti-inflammatory, and antihypertensive properties. Microalgae-derived peptides are being explored for their potential therapeutic applications in functional foods, nutraceuticals, and pharmaceuticals (Ampofo et al. 2022, Kaur et al. 2023).

Vitamins and minerals: Microalgae are a rich source of vitamins and minerals. They contain vitamins such as vitamin C, vitamin E, vitamin K, and various B vitamins (Del Mondo et al. 2020).

Additionally, microalgae are known for their high mineral content, including calcium, iron, magnesium, and potassium, among others (Mahata et al. 2022).

The specific chemical composition of microalgae can vary among species, growth conditions, and environmental factors. As a result, different microalgae species offer unique profiles of bioactive compounds and potential applications (Silva et al. 2022). Ongoing research is focused on further exploring and harnessing the diverse chemical compounds of microalgae for various industries, including food, pharmaceuticals, nutraceuticals, and biofuels. Exploration of bioactive compounds found in microalgae, such as polyunsaturated fatty acids (PUFAs), pigments and peptides (Balasubramaniam et al. 2021, Zhuang et al. 2022).

Seaweed derived metabolites useful for cosmetics

Owing to the presence of the bioactive constituents mentioned above and their potential biological activities, seaweeds have wide-ranging applications in cosmeceutical products. Macroalgae-derived compounds have been recognized as significantly important for the formulation of cosmeceuticals (Brunt et al. 2018). Polysaccharides play a crucial role in cosmetics, serving as moisturizers, emulsifiers, wound healing agents, and thickening agents (Percival 1979).

In a study conducted by Fernando et al. (2017), the anti-inflammatory activity of Fucoidan from *Chnoospora minima* (Phaeophyceae) was reported. It was found to inhibit the production of nitric oxide, inducible nitric oxide, Cyclooxygenase-2, and Prostaglandin E2 levels induced by Lipopolysaccharides, thus targeting RAW macrophages. Similarly, Ariede et al. (2017), Wang et al. (2017), and Teas and Irhimeh (2017) reported the beneficial activities of polysaccharides derived from *Fucus vesiculosus* (Figure 1a) (brown alga). These activities included anti-aging, anti-melanogenic, anti-cancer, and antioxidant effects.

The polysaccharides stimulated collagen production, inhibited tyrosinase, reduced melanoma growth, and prevented oxidative damage. Another study focused on the anti-inflammatory activity of sulphated polysaccharides from *Padina tetrastromatica* (brown alga) in an experimental model of Paw edema in rats. The polysaccharides inhibited COX-2 and iNOS, demonstrating anti-inflammatory effects (Mohsin et al. 2011).

Khan et al. (2009) investigated the anti-inflammatory activity of polyunsaturated fatty acids derived from *Undaria pinnatifida* (Figure 1b) (brown alga) and observed their effects on mouse ear edema and erythema.

In vitro studies assessed the antioxidant activity of methanolic extracts from *Osmundaria obtusilo* and *Palisada flagellifera* (red algae) using various assays (Santos et al. 2019, Vasconcelos et al. 2019). Sargachromanol E, a phenolic compound found in *Sargassum horneri* (brown alga), exhibited anti-aging properties by inhibiting matric metalloprotein expression on UVA irradiated

Figure 1. Brown seaweed species images: a—*Fucus vesiculosus*; b—*Undaria pinnatifida*; c—*Laminaria digitata*; d—*Himantalia elongata*; Scale = 1 cm.

dermal fibroblast (Kim et al. 2013). Additionally, Sargachromanol E from *Sargassum serratifolium* (brown alga) showed anti-melanogenic activity by downregulating microphthalmia-associated transcription factors on B16F10 melanoma cells (Azam et al. 2017).

Yoon et al. (2009) demonstrated the inhibition of tyrosinase and melanin, along with anti-melanogenic activity, in the ethanolic extract of the brown macroalga *Petalonia binghamiae*. Mycosporine amino acids (MAAs) derived from *Porphyra* sp. (Figure 2b) (red alga) were found to inhibit collagenase and control MMP expression on human fibroblast cell lines *in vitro* (Wang et al. 2017). *Neopyropia yezoensis* (red alga)-derived MAAs exhibited antioxidant activity by scavenging ROS and regulating MMP expression on human skin fibroblast (Sakai et al. 2011). The anti-aging activity of *Palmaria palmata* (Figure 2a) (Rhodophyta)-derived MAAs was attributed to collagenase inhibition (Hartmann et al. 2015).

Several macroalgae species, such as *Laurencia pacifica*, *Laurencia rigida*, *Wilsonosiphonia howei*, *Rhodomela confervoides*, and *Schizymenia dubyi* (Figure 2c) (red alga), were identified as good sources of phenolic compounds (Li et al. 2012, Wang et al. 2015, Azam et al. 2017).

Polysaccharides have the greatest amount of water retention time and can act as humectant and moisturizers in cosmetic products. Many research studies suggested greater hydration capacity and moisturizing effects of polysaccharides from *Saccharina japonica* (brown alga) than hyaluronic acid (Jesumani et al. 2019). Isolation of sulphated polysaccharide was carried out from the green macroalgae *Ulva linza* (Figure 3a) which revealed higher moisture absorption and retention for 96 h in comparison to glycerol (Shao et al. 2015). *S. japonica*, *Pyropia haitanensis* (red alga), *Codium fragile*, *Ulva linza*, and *Bryopsis plumosa* (Figure 3b) (green algae) derived polysaccharides showed good amount of moisture absorption and retention (Wang et al. 2013). They also reported the role of sulphate content and molecular weight in the moisture-holding capacity.

Oligosaccharide zinc complex derived from *Laminaria digitata* (Figure 1c) (brown alga) reported anti-acne activity (i.e., reduction of acne signs) by reducing sebum production (Ruxton and Jenkins 2013). Sebaaly et al. (2014) reported bactericidal activity of *Corallina* sp. (Figure 2d) (red alga) derived sulphated galactan against *Enterococcus faecalis* and *Streptococcus epidermidis*.

Stengel et al. (2011) reported that macroalgae are a quality source of essential amino acids, such as histidine, tyrosine, tryptophan, and non-essential amino acids such as alanine, serine, and proline. Antioxidant activity was reported by *Ulva australis*-derived amino acids, histidine, and taurine (Houston et al. 2005). On the other hand, red macroalgae *Palmaria palmata* (Dulse) and brown alga *Himanthalia elongata* (Figure 1d) are plentiful in glutamic acid, alanine, and serine (Bedoux et al. 2014).

Figure 2. Red seaweed species images: a—*Palmaria palmata*; b—*Schizymenia dubyi*; c—*Laminaria digitata*; d—*Porphyra*; e—*Lithothamnion*; Scale = 1 cm.

Figure 3. Green seaweed species images: a—*Ulva linza*; b—*Bryopsis plumosa*; Scale = 1 cm.

Thalassotherapy

Description of the therapeutic benefits of seaweed in thalassotherapy treatments

Seaweed has long been recognized for its therapeutic benefits in thalassotherapy treatments, which utilize seawater, seaweed extracts, and seaweed-based products to promote health and well-being (Antonelli and Donelli 2021). Seaweed's rich composition of minerals, vitamins, polysaccharides, and other bioactive compounds contribute to its re-mineralizing, moisturizing, and rejuvenating actions on the body. Here is a brief description of these therapeutic benefits (Pereira 2018):

Re-mineralizing action: Seaweed is renowned for its high mineral content, including calcium, magnesium, potassium, iodine, and iron, among others. During thalassotherapy treatments, seaweed extracts or whole seaweed are applied to the body or incorporated into baths and wraps. The minerals present in seaweed are absorbed through the skin, replenishing mineral deficiencies, and promoting overall mineral balance in the body. This re-mineralizing action is beneficial for enhancing the functioning of various bodily systems, such as bone health, nerve transmission, and muscle function (Circuncisão et al. 2018, Leandro et al. 2020).

Moisturizing action: Seaweed extracts and polysaccharides found in seaweed have exceptional moisturizing properties. When applied topically, they help to hydrate and nourish the skin, improving its moisture retention capacity (Zhang et al. 2022). The polysaccharides form a protective film on the skin's surface, preventing moisture loss and maintaining optimal hydration levels. This moisturizing action is particularly beneficial for individuals with dry, dehydrated, or sensitive skin, as it helps to restore the skin's natural barrier function and promote a smoother, more supple complexion (Purnamawati et al. 2017, Pereira 2018).

Rejuvenating action: Seaweed is rich in antioxidants, such as polyphenols and carotenoids, which possess rejuvenating properties. These antioxidants help neutralize harmful free radicals, which can contribute to skin aging and damage (López-Hortas et al. 2021). By reducing oxidative stress and inflammation, seaweed-based treatments promote skin cell regeneration, improve skin elasticity, and reduce the appearance of fine lines and wrinkles. The bioactive compounds in seaweed also stimulate collagen synthesis, contributing to a more youthful and rejuvenated appearance (Pereira 2018, Pangestuti et al. 2021).

In addition to re-mineralizing, moisturizing, and rejuvenating actions, seaweed in thalassotherapy treatments offers other potential benefits. It can help detoxify the body by promoting lymphatic drainage and elimination of toxins, contributing to improved overall well-being. The soothing and anti-inflammatory properties of seaweed, can alleviate skin conditions such as psoriasis, eczema, and acne (Pereira 2018). Furthermore, the minerals and bioactive compounds in seaweed support the stimulation of circulation, which can enhance oxygen and nutrient delivery to tissues and promote a sense of relaxation and vitality (Park et al. 2023).

Thalassotherapy treatments harness the therapeutic potential of seaweed to provide a holistic approach to wellness and skincare. The combined effects that seaweed has of re-mineralizing, moisturizing, and rejuvenating make it a valuable ingredient in various thalassotherapy protocols, including body wraps, baths, and cosmetic formulations (Pereira 2018, López-Hortas et al. 2021).

Examples of thalassotherapy treatments using seaweed, such as body wraps and seaweed baths

Thalassotherapy treatments make use of seaweed in various forms, including body wraps and seaweed baths.

Seaweed body wraps: Seaweed body wraps are popular thalassotherapy treatments that involve applying a mixture of seaweed extracts or whole seaweed onto the body. The seaweed is typically combined with other ingredients such as clay, essential oils, and minerals to enhance the treatment's effects. The body is then wrapped in a thermal blanket or sheets to promote heat retention and facilitate the absorption of the beneficial compounds from the seaweed. This treatment can help detoxify the body, improve skin texture, and provide a sense of relaxation (Pereira 2018).

Seaweed baths: Seaweed baths are another common thalassotherapy treatment that involves immersing the body in a bath filled with seawater or seaweed-infused water. The seaweed used may be in the form of powdered seaweed, seaweed extracts, or whole seaweed. The bathwater is typically heated to a comfortable temperature and may be combined with other ingredients such as essential oils or mineral salts. During the bath, the skin absorbs the minerals, vitamins, and bioactive compounds present in the seaweed, providing a range of therapeutic benefits, including relaxation, detoxification, and skin nourishment (Turrin 2015).

Seaweed body scrubs: Seaweed body scrubs are exfoliating treatments that combine seaweed extracts with natural exfoliants, such as sea salt or sugar. The scrub is gently massaged onto the skin, helping to remove dead skin cells, promote blood circulation, and enhance skin renewal. The beneficial properties of seaweed, including its minerals and antioxidants, nourish and revitalize the skin. After the scrub, the body is typically rinsed with warm water, leaving the skin smooth, soft, and rejuvenated (Deer Lake Lodge Spa Resort 2023).

Healthy bones and joints: Seaweeds are naturally high in minerals that are essential for healthy bones and joints. They are also high in vitamins and antioxidants—important components of cosmetic preparations (Lupo 2001). Calcium can comprise up to 7% of dry weight (Pengzhan et al. 2003) and is also found in association with polysaccharides, such as alginate and carrageenans. The calcareous red seaweeds such as *Lithothamnion* spp. (Figure 2e) can contain calcium as high as 35 percent. An Irish Product Aquamin™ is a multi-mineral preparation from *Lithothamnion* spp. It has been shown to be high in bioavailable calcium, more so than the standard calcium carbonate that is used in supplements and is effective against mobilization of bone calcium through parathyroid hormonal activity (Aqumin 2017). Similar activity has been observed with a mix of oyster shell and the brown seaweed *Sargassum fusiforme* (Fujita 2005).

Seaweed facial treatments: Seaweed is also used in thalassotherapy for facial treatments. Facial masks and creams containing seaweed extracts or seaweed-based formulations are applied to the face to provide hydration, improve skin tone, and minimize the appearance of wrinkles. The rich nutritional profile of seaweed, including its vitamins, minerals, and antioxidants, helps nourish the skin, improve elasticity, and promote a youthful complexion (Pereira 2018, Jesumani et al. 2019).

These are just a few examples of thalassotherapy treatments that incorporate seaweed. The specific protocols and ingredients used may vary depending on the spa or treatment center. Thalassotherapy treatments using seaweed aim to provide a holistic approach to wellness by utilizing the therapeutic properties of seaweed to promote relaxation, rejuvenation, and overall well-being (Gálvez et al. 2018, Morer et al. 2020).

Cosmeceuticals

Cosmeceuticals are a category of cosmetic products that contain active ingredients with pharmaceutical properties. The term "cosmeceutical" is a combination of "cosmetics" and "pharmaceuticals," reflecting the dual nature of these products. Unlike traditional cosmetics, cosmeceuticals are designed to go beyond the surface level and provide specific therapeutic benefits to the skin (Ahmed et al. 2022, Liu and Ji-Kai 2022).

Some key points about cosmeceuticals:

Active ingredients: Cosmeceuticals are formulated with active ingredients that have been scientifically proven to have beneficial effects on the skin. These ingredients may include antioxidants, peptides, retinoids, hyaluronic acid, vitamins, plant extracts, and other bioactive compounds. Each active ingredient has specific functions and targets different skin concerns, such as aging, hyperpigmentation, acne, or dryness (Glass et al. 2020, Juncan et al. 2021).

Skin benefits: Cosmeceuticals offer a range of skin benefits beyond simple cosmetic enhancement. They are designed to improve skin health, address specific skin conditions, and provide long-term results. These products may help reduce wrinkles, improve skin texture and firmness, even out skin tone, enhance hydration, protect against environmental damage, and promote overall skin rejuvenation (Ganceviciene et al. 2012, Kalasariya et al. 2022).

Scientifically supported: Cosmeceuticals are backed by scientific research and studies that demonstrate their efficacy and safety. The active ingredients undergo rigorous testing to ensure their effectiveness and compatibility with the skin. The concentration and formulation of these ingredients are carefully designed to optimize their performance and deliver desired results (Barthe et al. 2021, Liu 2022).

Over-the-counter availability: Cosmeceuticals are typically available over the counter and do not require a prescription. They can be purchased from specialty skincare brands, beauty stores, pharmacies, and online retailers. However, it is important to note that some cosmeceuticals with higher concentrations of active ingredients may be recommended or prescribed by dermatologists or skincare professionals for specific skin concerns (Pandey et al. 2023).

Complementary to skincare routine: Cosmeceuticals are intended to be used as part of a comprehensive skincare routine. They are often recommended to be used alongside cleansers, toners, moisturizers, and sunscreens. Combining cosmeceuticals with other skincare products can help maximize their benefits and achieve optimal skin health (Shireesha et al. 2023).

It is essential to follow the instructions provided by the manufacturer when using cosmeceuticals. Adherence to the recommended application frequency, duration, and usage guidelines is crucial for achieving desired results and minimizing potential adverse effects (Gupta et al. 2022, Pandey et al. 2023).

Macro- and microalgae applications in the cosmetic industry

Seaweed- and microalgae-based cosmetic products, including creams, gels, and face masks, utilize the beneficial properties of seaweed and microalgae to provide a range of skincare benefits. Here is a brief description of these products (Pereira 2018, López-Hortas et al. 2021, Morais et al. 2021):

Seaweed- and microalgae-based creams: These creams are formulated with extracts or derivatives of seaweed and microalgae. They are designed to nourish and moisturize the skin while harnessing the bioactive compounds present in these marine resources. Seaweed and microalga extracts provide essential minerals, vitamins, antioxidants, and fatty acids that help improve skin texture, enhance hydration, and promote a healthy complexion. These creams can address various skin concerns, such as dryness, aging, and dullness (Stoyneva-Gärtner et al. 2020, Thiyagarasaiyar et al. 2020).

Seaweed- and microalgae-based gels: Gels containing seaweed and microalgae extracts offer lightweight and refreshing hydration to the skin. These gels are often water-based and quickly absorbed, making them suitable for oily or combination skin types. Seaweed and microalgae extract help replenish moisture, soothe irritated skin, and deliver essential nutrients. The gels may also have a cooling or soothing effect, making them ideal for sensitive or sun-exposed skin (Thiyagarasaiyar et al. 2020, Fonseca et al. 2023).

Seaweed- and microalgae-based face masks: Face masks enriched with seaweed and microalgae extracts are used to provide targeted skincare benefits. These masks are typically applied to the face for a specified period, allowing the active ingredients to penetrate and nourish the skin. Seaweed and microalgae extract help detoxify the skin, provide hydration, and promote a youthful appearance. These masks can help improve skin tone, minimize pores, and reduce the appearance of fine lines and wrinkles (Jesumani et al. 2019, López-Hortas et al. 2021).

Seaweed and microalgae are rich sources of bioactive compounds such as polysaccharides, proteins, vitamins, minerals, and antioxidants. These components work synergistically to provide multiple benefits to the skin, including improved hydration, enhanced skin elasticity, reduced inflammation, and increased collagen production (Pereira 2018, Kalasariya et al. 2021). Seaweed- and microalgae-based cosmetic products are known for their natural and sustainable formulations, making them popular choices for eco-conscious consumers seeking effective and environmentally friendly skincare options (Pagels et al. 2022).

It is important to note that the specific formulation and ingredients can vary among different brands and products. Therefore, it is advisable to read product labels and follow the instructions provided for optimal use and results (Vranić et al. 2004).

Seaweed- and microalgae-based cosmetic products

Algae-based cosmetic products offer several beneficial effects on skin health, including hydration, protection against free radicals, and reduction of signs of aging. Let us discuss each of these effects in detail (Mourelle et al. 2017).

Hydration: Seaweed and microalgae are known for their ability to provide deep hydration to the skin. These marine ingredients contain polysaccharides, such as alginate and carrageenan, which have excellent water-binding properties (López-Hortas et al. 2021, Morais et al. 2021). When applied topically, these compounds help to attract and retain moisture in the skin, resulting in improved hydration levels. Well-hydrated skin appears plump, smooth, and supple, reducing the appearance of dryness and fine lines (Purnamawati et al. 2017).

Protection against free radicals: Seaweed and microalgae are rich sources of antioxidants, including vitamins C and E, beta-carotene, and various polyphenols. These antioxidants help protect the skin against harmful free radicals, which are unstable molecules generated by environmental factors like UV radiation, pollution, and stress (Gomez-Zavaglia et al. 2019, Choudhary et al. 2021). Free radicals can cause oxidative stress, leading to skin damage, premature aging, and the breakdown of collagen and elastin fibers. By neutralizing free radicals, the antioxidants in seaweed and microalgae-based products help to maintain skin health and promote a more youthful appearance (Jesus et al. 2023).

Reduction of signs of aging: Seaweed and microalgae contain a range of bioactive compounds that can address signs of aging (Ampofo et al. 2022). For instance, they often contain peptides, which are amino acid chains that support collagen synthesis and improve skin elasticity. These ingredients can help minimize the appearance of wrinkles, fine lines, and sagging skin. Furthermore, the vitamins, minerals, and antioxidants present in seaweed and microalgae contribute to improved skin texture, tone, and brightness, enhancing the overall youthful appearance of the skin (Pereira 2018, Lourenço-Lopes et al. 2020).

The combination of hydration, protection against free radicals, and reduction of signs of aging contributes to healthier and more vibrant skin. Seaweed- and microalgae-based cosmetic products provide a natural and effective way to incorporate these benefits into a skincare routine. Regular use of these products can help maintain optimal skin health, improve the skin's barrier function, and promote a more youthful and radiant complexion (Mourelle et al. 2017, Pangestuti et al. 2021).

Challenges and future prospects

The sustainable use of marine resources, including algae and microalgae, poses several challenges that need to be addressed to ensure their responsible collection and long-term viability. Here, we will discuss some of the key challenges associated with the sustainable use of these marine resources (Merlo et al. 2021).

Overharvesting and habitat destruction: One of the primary challenges is the risk of overharvesting. Algae and microalgae play crucial ecological roles in marine ecosystems, providing habitat, food, and oxygen for marine organisms. Indiscriminate collection practices can disrupt these delicate ecosystems, leading to the destruction of habitats and the depletion of marine biodiversity. It is essential to establish sustainable harvesting practices that allow for the regeneration and maintenance of natural populations (Edwards et al. 2020).

Environmental impact of collection methods: The methods used to collect algae and microalgae can have environmental implications. For example, some collection techniques involve physically removing entire organisms or large portions of seaweed from their natural environment. This practice can disturb the balance of marine ecosystems and disrupt the life cycles of various species. Sustainable collection methods should be employed, such as selective harvesting or cultivating algae and microalgae in controlled environments to minimize the impact on natural populations (Campbell et al. 2019, Matos et al. 2021).

Monitoring and regulation: Adequate monitoring and regulation are essential to ensure the sustainable use of marine resources. It is crucial to establish clear guidelines and regulations for collection practices, including quotas, seasonal restrictions, and protected areas. Regular monitoring of population health, growth rates, and biodiversity levels can help assess the impact of harvesting activities and inform management decisions (Contarinis et al. 2023, Tremblay et al. 2023).

Conserving genetic diversity: Algae and microalgae exhibit a vast array of genetic diversity, which is critical for their adaptation to environmental changes and potential applications in various industries. However, unsustainable collection practices can reduce this genetic diversity, limiting the potential for future scientific and commercial use. Conservation efforts should focus on preserving the genetic diversity of algae and microalgae through in-situ and ex-situ conservation strategies (Brodie et al. 2017, Benedetti et al. 2018).

Collaboration and education: Collaboration among stakeholders, including scientists, policymakers, industry professionals, and local communities, is crucial for the sustainable use of marine resources. Sharing knowledge, promoting research, and raising awareness about the importance of responsible collection practices can help drive positive change. Education and training programs can also play a vital role in promoting sustainable practices among those involved in the collection and utilization of algae and microalgae (Cvitanovic et al. 2015, Fallah et al. 2022).

Thus, the sustainable use of marine resources, particularly algae and microalgae, requires addressing challenges such as overharvesting, environmental impact, monitoring, genetic diversity conservation, and collaboration. By implementing responsible collection practices, regulating harvesting activities, and fostering collaboration and education, we can ensure the long-term viability of these valuable marine resources while preserving marine ecosystems and biodiversity (Rotter et al. 2021, Lubchenco and Haugan 2023).

Exploration of future prospects in the research and development of thalassotherapy and cosmeceutical products, including the discovery of new bioactive compounds

The field of thalassotherapy and cosmeceuticals holds exciting prospects for future research and development, particularly in the discovery of new bioactive compounds. Some of the potential future prospects in these areas are:

Identification of novel bioactive compounds: There is still much to be discovered in the vast biodiversity of marine environments. Ongoing research efforts are focused on exploring unexplored regions, such as deep-sea ecosystems and remote coastal areas, to uncover new species of algae and microalgae with unique and potent bioactive compounds. Advanced techniques, such as metagenomics and metabolomics, enable researchers to identify and characterize previously unknown compounds that may have significant therapeutic and cosmetic potential (Ghosh et al. 2022, Karthikeyan et al. 2022).

Biotechnological advances: Biotechnology plays a crucial role in the development of thalassotherapy and cosmeceutical products. Advancements in biotechnology, including genetic engineering and synthetic biology, offer exciting possibilities for the production of bioactive compounds in large quantities. Researchers can modify and optimize the metabolic pathways of algae and microalgae to enhance the production of specific compounds with desired properties, potentially revolutionizing the availability and sustainability of these ingredients (Balasubramaniam et al. 2021, Hempel et al. 2023).

Sustainable cultivation methods: The development of sustainable cultivation methods for algae and microalgae is a growing area of interest. Cultivating specific species in controlled environments, such as photobioreactors or open pond systems, allows for optimized growth conditions and consistent production of bioactive compounds. Advancements in cultivation techniques, including nutrient management, light optimization, and carbon capture, contribute to efficient and environmentally friendly production processes (Veerabadhran et al. 2021, Chanana et al. 2023).

Synergistic formulations: The future of thalassotherapy and cosmeceuticals lies in the exploration of synergistic formulations. Combining bioactive compounds from different marine sources, such as algae, microalgae, and other marine organisms, may enhance their individual benefits and create powerful multi-functional products. Synergistic formulations can target multiple skin concerns, offering comprehensive solutions for improved skin health and appearance (Mourelle et al. 2017, Pereira 2018).

Personalized skincare: The advancement of technology, including "Artificial Intelligence" (AI) and machine learning, opens opportunities for personalized skincare in the field of cosmeceuticals. Analyzing individual skin characteristics, genetic predispositions, and lifestyle factors can help tailor thalassotherapy and cosmeceutical products to specific needs. Personalized formulations can deliver targeted results, optimize efficacy, and improve overall customer satisfaction (Arias-Echeverri et al. 2022, Freitas et al. 2023).

Sustainable packaging and practices: In addition to the development of innovative products, future prospects also include a focus on sustainable packaging and practices. The industry is moving towards eco-friendly packaging materials and reducing the carbon footprint associated with production and distribution. Responsible sourcing, waste reduction, and ethical practices are becoming increasingly important in the development and marketing of thalassotherapy and cosmeceutical products (Meherishi et al. 2019, Asim et al. 2022).

In conclusion, the future of thalassotherapy and cosmeceuticals is promising, with potential advancements in the discovery of novel bioactive compounds, biotechnological applications, sustainable cultivation methods, synergistic formulations, personalized skincare, and sustainable practices. These developments have the potential to revolutionize the industry, providing more effective, eco-friendly, and tailored solutions for skincare and wellness (Mourelle et al. 2017, Kalasariya et al. 2022).

Acknowledgments

The authors also would like to thank MARE—Marine and Environmental Sciences Centre—ARNET, and the University of Coimbra, within the scope of the project LA/P/0069/2020 granted to the Associate Laboratory ARNET, and UIDB/04292/2020 granted to MARE—Marine and Environmental Sciences Centre.

References

Abka-khajouei, R., L. Tounsi, N. Shahabi, A. Kumar Patel, S. Abdelkafi and P. Michaud. 2022. Structures, Properties and Applications of Alginates. Marine Drugs 20(6): 364. https://doi.org/10.3390/md20060364.

Ahmed, I.A., M.A. Mikail, N.H. Zamakshshari, M.R. Mustafa, N.M. Hashim and R. Othman. 2022. Trends and challenges in phytotherapy and phytocosmetics for skin aging. Saudi Journal of Biological Sciences 29(8): 103363. https://doi.org/10.1016/j.sjbs.2022.103363.

Ampofo, J. and L. Abbey. 2022. Microalgae: bioactive composition, health benefits, safety and prospects as potential high-value ingredients for the functional food industry. Foods 11(12): 1744. https://doi.org/10.3390/foods11121744.

Antonelli, M. and D. Donelli. 2021. Thalassotherapy, health benefits of sea water, climate and marine environment: a narrative review. In Proceedings of the 6th International Electronic Conference on Water Sciences, 15–30 November 2021, MDPI: Basel, Switzerland. https://doi.org/10.3390/ECWS-6-11606.

Aqumin. 2017. Marine minerals for health. Available online at: http://aquamin.com; searched on June 5, 2023.

Araújo, R., F.V. Calderón, J.S. López, I.C. Azevedo, A. Bruhn, S. Fluch et al. 2021. Current status of the algae production industry in Europe: An emerging sector of the blue bioeconomy. Frontiers in Marine Science 7. https://doi.org/10.3389/fmars.2020.626389.

Arias-Echeverri, J.P., P.A. Zapata-Ramírez, M. Ramírez-Carmona, L. Rendón-Castrillón and C. Ocampo-López. 2022. Present and future of seaweed cultivation and its applications in Colombia. Journal of Marine Science and Engineering 10(2): 243. https://doi.org/10.3390/jmse10020243.

Ariede, M.B., T.M. Candido, A.L.M. Jacome, M.V.R. Velasco, J.C.M. de Carvalho and A.R. Baby. 2017. Cosmetic attributes of algae—A review. Algal Research 25: 483–487. https://doi.org/10.1016/j.algal.2017.05.019.

Asim, Z., I.R. Al Shamsi, M. Wahaj, A. Raza, S. Abul Hasan, S.A. Siddiqui et al. 2022. Significance of sustainable packaging: A case-study from a supply chain perspective. Applied System Innovation 5(6): 117. https://doi.org/10.3390/asi5060117.

Azam, M.S., J. Choi, M.-S. Lee and H.-R. Kim. 2017. Hypopigmenting effects of brown algae-derived phytochemicals: A review on molecular mechanisms. Marine Drugs 15(10): 297. https://doi.org/10.3390/md15100297.

Balasubramaniam, V., R.D.-N. Gunasegavan, S. Mustar, J.C. Lee and M.F.M. Noh. 2021. Isolation of industrial important bioactive compounds from microalgae. Molecules 26(4): 943. https://doi.org/10.3390/molecules26040943.

Barta, D.G., V. Coman and D.C. Vodnar. 2021. Microalgae as sources of Omega-3 polyunsaturated fatty acids: Biotechnological aspects. Algal Research 58: 102410. https://doi.org/10.1016/j.algal.2021.102410.

Barthe, M., C. Bavoux, F. Finot, I. Mouche, C. Cuceu-Petrenci, A. Forreryd et al. 2021. Safety testing of cosmetic products: overview of established methods and New Approach Methodologies (NAMs). Cosmetics 8(2): 50. https://doi.org/10.3390/cosmetics8020050.

Bedoux, G., K. Hardouin, A.S. Burlot and N. Bourgougnon. 2014. Chapter twelve—bioactive components from seaweeds: Cosmetic applications and future development. *In*: Bourgougnon, N. [ed.]. Advances in Botanical Research: Academic Press. https://doi.org/10.1016/B978-0-12-408062-1.00012-3

Benedetti, M., V. Vecchi, S. Barera and L. Dall'Osto. 2018. Biomass from Microalgae: The potential of domestication towards sustainable biofactories. Microbial Cell Factories 17(1): 173. https://doi.org/10.1186/s12934-018-1019-3.

Borics, G., G. Várbíró, J. Falucskai, Z. Végvári, E. T-Krasznai, J. Görgényi et al. 2023. A two-dimensional morphospace for cyanobacteria and microalgae: Morphological diversity, evolutionary relatedness, and size constraints. Freshwater Biology 68(1): 115–126. https://doi.org/10.1111/fwb.14013.

Brodie, J., C.X. Chan, O. De Clerck, J.M. Cock, S.M. Coelho, C. Gachon et al. 2017. The algal revolution. Trends in Plant Science 22(8): 726–738. https://doi.org/10.1016/j.tplants.2017.05.005.

Brunt, E.G. and J.G. Burgess. 2018. The promise of marine molecules as cosmetic active ingredients. International Journal of Cosmetic Science 40(1): 1–15. https://doi.org/10.1111/ics.12435.

Campbell, I., A. Macleod, C. Sahlmann, L. Neves, J. Funderud, M. Øverland et al. 2019. The environmental risks associated with the development of seaweed farming in europe—prioritizing key knowledge gaps. Frontiers in Marine Science 6. https://doi.org/10.3389/fmars.2019.00107.

Catarino, M.D., S.J. Amarante, N. Mateus, A.M.S. Silva and S.M. Cardoso. 2021. Brown algae phlorotannins: A marine alternative to break the oxidative stress, inflammation and cancer network. Foods 10(7): 1478. https://doi.org/10.3390/foods10071478.

Chanana, I., P. Kaur, L. Kumar, P. Kumar and S. Kulshreshtha. 2023. Advancements in microalgal biorefinery technologies and their economic analysis and positioning in energy resource market. Fermentation 9(3): 202. https://doi.org/10.3390/fermentation9030202.

Choo, W.-T., M.-L. Teoh, S.-M. Phang, P. Convey, W.-H. Yap, B.-H. Goh et al. 2020. Microalgae as potential anti-inflammatory natural product against human inflammatory skin diseases. Frontiers in Pharmacology 11. https://doi.org/10.3389/fphar.2020.01086.

Choudhary, B., O.P. Chauhan and A. Mishra. 2021. Edible seaweeds: A potential novel source of bioactive metabolites and nutraceuticals with human health benefits. Frontiers in Marine Science 8. https://doi.org/10.3389/fmars.2021.740054.

Cindana, M., F. Ratu, G. Wilar, H.P. Devkota and N. Wathoni. 2020. Ulvan, a polysaccharide from *Macroalga Ulva* sp.: A review of chemistry, biological activities and potential for food and biomedical applications. Applied Sciences 10(16): 5488. https://doi.org/10.3390/app10165488.

Circuncisão, A.R., M.D. Catarino, S.M. Cardoso and A.M.S. Silva. 2018. Minerals from macroalgae origin: Health benefits and risks for consumers. Marine Drugs 16(11): 400. https://doi.org/10.3390/md16110400.

Contarinis, S., C. Kastrisios and B. Nakos. 2023. Marine protected areas and electronic navigational charts: Legal foundation, mapping methods, IHO S-122 portrayal, and advanced navigation services. Euro-Mediterranean Journal for Environmental Integration 8(1): 67–87. https://doi.org/10.1007/s41207-023-00343-9.

Cotas, J., A. Leandro, P. Monteiro, D. Pacheco, A. Figueirinha, A.M.M. Gonçalves et al. 2020. Seaweed phenolics: From extraction to applications. Marine Drugs 18(8): 384. https://doi.org/10.3390/md18080384.

Cotas, J., L. Gomes, D. Pacheco and L. Pereira. 2023. Ecosystem services provided by seaweeds. Hydrobiology 2(1): 75–96. https://doi.org/10.3390/hydrobiology2010006.

Costa, J.A.V., B.F. Lucas, A.G.P. Alvarenga, J.B. Moreira and M.G. de Morais. 2021. Microalgae polysaccharides: An overview of production, characterization, and potential applications. Polysaccharides 2(4): 759–772. https://doi.org/10.3390/polysaccharides2040046.

Cunha, L. and A. Grenha. 2016. Sulfated seaweed polysaccharides as multifunctional materials in drug delivery applications. Marine Drugs 14(3): 42. https://doi.org/10.3390/md14030042.

Cvitanovic, C., A.J. Hobday, L. van Kerkhoff, S.K. Wilson, K. Dobbs and N.A. Marshall. 2015. Improving knowledge exchange among scientists and decision-makers to facilitate the adaptive governance of marine resources: A Review of Knowledge and Research Needs. Ocean & Coastal Management 112: 25–35. https://doi.org/10.1016/j.ocecoaman.2015.05.002.

da Rosa, M.D.H., C.J. Alves, F.N. dos Santos, A.O. de Souza, E.d.R. Zavareze, E. Pinto et al. 2023. Macroalgae and microalgae biomass as feedstock for products applied to bioenergy and food industry: A brief review. Energies 16(4): 1820. https://doi.org/10.3390/en16041820.

Del Mondo, Angelo, Arianna Smerilli, Elisabet Sané, Clementina Sansone and Christophe Brunet. 2020. Challenging microalgal vitamins for human health. Microbial Cell Factories 19(1): 201. https://doi.org/10.1186/s12934-020-01459-1.

Deer Lake Lodge Spa Resort. 2023. Body Wraps & Scrubs. https://deerlakelodge.com/body-scrubs; searched on June 4, 2023.

Duarte, C.M., J.-P. Gattuso, K. Hancke, H. Gundersen, K. Filbee-Dexter, M.F. Pedersen et al. 2022. Global estimates of the extent and production of macroalgal forests. Global Ecology and Biogeography 31(7): 1422–1439. https://doi.org/10.1111/geb.13515.

Edwards, M., B. Konar, J.-H. Kim, S. Gabara, G.K.W. Sullaway, T. McHugh et al. 2020. Marine deforestation leads to widespread loss of ecosystem function. PLoS ONE 15(3): e0226173. https://doi.org/10.1371/journal.pone.0226173.

Eilam, Y., H. Khattib, N. Pintel and D. Avni. 2023. Microalgae—sustainable source for alternative proteins and functional ingredients promoting gut and liver health. Global Challenges 7(5): 2200177. https://doi.org/10.1002/gch2.202200177.

El-Beltagi, Hossam S., Amal A. Mohamed, Heba I. Mohamed, Khaled M.A. Ramadan, Aminah A. Barqawi and Abdallah Tageldein Mansour. 2022. Phytochemical and potential properties of seaweeds and their recent applications: A review. Marine Drugs 20(6): 342. https://doi.org/10.3390/md20060342.

Eze, C.N., C.K. Onyejiaka, S.A. Ihim et al. 2023. Bioactive compounds by microalgae and potentials for the management of some human disease conditions. AIMS Microbiology 9(1): 55–74. https://doi.org/10.3934/microbiol.2023004.

Fallah Shayan, N., N. Mohabbati-Kalejahi, S. Alavi and M.A. Zahed. 2022. Sustainable development goals (SDGs) as a framework for corporate social responsibility (CSR). Sustainability 14(3): 1222. https://doi.org/10.3390/su14031222.

Favas, Rita, J.M., R. Martins, V. Vasconcelos and G. Lopes. 2021. Cyanobacteria and microalgae bioactive compounds in skin-ageing: Potential to restore extracellular matrix filling and overcome hyperpigmentation. Journal of Enzyme Inhibition and Medicinal Chemistry 36(1): 1829–1838. https://doi.org/10.1080/14756366.2021.1960830.

Fernando, I.P.S., K.K.A. Sanjeewa, K.W. Samarakoon. W.W. Lee, H.-S. Kim, N. Kang et al. 2017. A fucoidan fraction purified from chnoospora minima; a potential inhibitor of LPS-induced inflammatory responses. International Journal of Biological Macromolecules 104: 1185–1193. https://doi.org/10.1016/j.ijbiomac.2017.07.031.

Fonseca, S., M.N. Amaral, C.P. Reis and L. Custódio. 2023. Marine natural products as innovative cosmetic ingredients. Marine Drugs 21(3): 170. https://doi.org/10.3390/md21030170.

Freitas, M.V., D. Pacheco, J. Cotas, T. Mouga, C. Afonso and L. Pereira. 2022. Red seaweed pigments from a biotechnological perspective. Phycology 2(1): 1–29. https://doi.org/10.3390/phycology2010001.

Freitas, M.V., L.G. Inácio, A. Ruas, I.A. Silva, T. Mouga, L. Pereira et al. 2023. Antioxidant and antimicrobial properties of selected red seaweeds from central portugal. Applied Sciences 13(1): 157. https://doi.org/10.3390/app13010157.

Fujita, T. 2005. Active Absorbable Algal Calcium (AAACa) Changes Calcium Paradigm. Clinical Calcium. 15(1): 87–93.

Gálvez, I., S. Torres-Piles and E. Ortega-Rincón. 2018. Balneotherapy, immune system, and stress response: A hormetic strategy? International Journal of Molecular Sciences 19(6): 1687. https://doi.org/10.3390/ijms19061687.

Ganceviciene, R., A.I. Liakou, A. Theodoridis, E. Makrantonaki and C.C. Zouboulis. 2012. Skin anti-aging strategies. Dermato-Endocrinology 4(3): 308–319. https://doi.org/10.4161/derm.22804.

García-Poza, S., D. Pacheco, J. Cotas, J.C. Marques, L. Pereira and A.M.M. Gonçalves. 2022. Marine macroalgae as a feasible and complete resource to address and promote sustainable development goals (SDGs). Integrated Environmental Assessment and Management 18(5): 1148–1161. https://doi.org/10.1002/ieam.4598.

Ghosh, S., T. Sarkar, S. Pati, Z.A. Kari, H.A. Edinur and R. Chakraborty. 2022. Novel bioactive compounds from marine sources as a tool for functional food development. Frontiers in Marine Science 9. https://doi.org/10.3389/fmars.2022.832957.

Glass, G. 2020. Cosmeceuticals: The principles and practice of skin rejuvenation by nonprescription topical therapy. Aesthetic Surgery Journal Open Forum 2(4). https://doi.org/10.1093/asjof/ojaa038.

Gomez-Zavaglia, A., M.A.P. Lage, C. Jimenez-Lopez, J.C. Mejuto and J. Simal-Gandara. 2019. The potential of seaweeds as a source of functional ingredients of prebiotic and antioxidant value. Antioxidants 8(9): 406. https://doi.org/10.3390/antiox8090406.

Gonçalves, C.d.A. and C.C. Figueredo. 2020. What we really know about the composition and function of microalgae cell coverings?—An overview. Acta Botanica Brasilica 34. https://doi.org/10.1590/0102-33062020abb0309.

Ghosh, S., T. Sarkar, S. Pati, Z.A. Kari, H.A. Edinur and R. Chakraborty. 2022. Novel bioactive compounds from marine sources as a tool for functional food development. Frontiers in Marine Science 9. https://doi.org/10.3389/fmars.2022.832957.

Guiry, M.D. and G.M. Guiry. 2023. AlgaeBase. World-wide electronic publication, National University of Ireland, Galway. https://www.algaebase.org; searched on June 3, 2023.

Guo, Z., H. Zhang, S. Liu and S. Lin. 2013. Biology of the marine heterotrophic dinoflagellate oxyrrhis marina: Current status and future directions. Microorganisms 1(1): 33–57. https://doi.org/10.3390/microorganisms1010033.

Gupta, V., S. Mohapatra, H. Mishra, U. Farooq, K. Kumar, M.J. Ansari et al. 2022. Nanotechnology in cosmetics and cosmeceuticals—A review of latest advancements. Gels 8(3): 173. https://doi.org/10.3390/gels8030173.

Hachicha, R., F. Elleuch, H.B. Hlima, P. Dubessay, H. de Baynast, C. Delattre et al. 2022. Biomolecules from microalgae and cyanobacteria: Applications and market survey. Applied Sciences 12(4): 1924. https://doi.org/10.3390/app12041924.

Hartmann, A., J. Gostner, J.E. Fuchs, E. Chaita, N. Aligiannis, L. Skaltsounis et al. 2015. Inhibition of collagenase by mycosporine-like amino acids from marine sources. Planta Medica 81(10): 813–820. https://doi.org/10.1055/s-0035-1546105.

Hempel, M.d.S.S., P. Colepicolo and L. Zambotti-Villela. 2023. Macroalgae biorefinery for the cosmetic industry: Basic concept, green technology, and safety guidelines. Phycology 3(1): 211–241. https://doi.org/10.3390/phycology3010014.

Hentati, F., L. Tounsi, D. Djomdi, G. Pierre, C. Delattre, A.V. Ursu et al. 2020. Bioactive polysaccharides from seaweeds. Molecules 25(14): 3152. https://doi.org/10.3390/molecules25143152.

Houston, M.C. 2005. Nutraceuticals, vitamins, antioxidants, and minerals in the prevention and treatment of hypertension. Progress in Cardiovascular Diseases 47(6): 396–449. https://doi.org/10.1016/j.pcad.2005.01.004.

Ismail, M.M., B.S. Alotaibi and M.M. EL-Sheekh. 2020. Therapeutic uses of red macroalgae. Molecules 25(19): 4411. https://doi.org/10.3390/molecules25194411.

Jesumani, V., H. Du, M. Aslam, P. Pei and N. Huang. 2019. Potential use of seaweed bioactive compounds in skincare—A review. Marine Drugs 17(12): 688. https://doi.org/10.3390/md17120688.

Jesus, A., S. Mota, A. Torres, M.T. Cruz, E. Sousa, I.F. Almeida et al. 2023. Antioxidants in sunscreens: Which and what for? Antioxidants 12(1): 138. https://doi.org/10.3390/antiox12010138.

Juárez-Portilla, C., T. Olivares-Bañuelos, T. Molina-Jiménez, J.A. Sánchez-Salcedo, D.I.D. Moral, T. Meza-Menchaca et al. 2019. Seaweeds-derived compounds modulating effects on signal transduction pathways: A systematic review. Phytomedicine 63: 153016. https://doi.org/10.1016/j.phymed.2019.153016.

Juncan, A.M., D.G. Moisă, A. Santini, C. Morgovan, L. Rus, A.L. Vonica-Țincu et al. 2021. Advantages of hyaluronic acid and its combination with other bioactive ingredients in cosmeceuticals. Molecules 26(15): 4429. https://doi.org/10.3390/molecules26154429.

Kalasariya, H.S. and L. Pereira. 2022. Dermo-cosmetic benefits of marine macroalgae-derived phenolic compounds. Applied Sciences 12(23): 11954. https://doi.org/10.3390/app122311954.

Karthikeyan, A., A. Joseph and B.G. Nair. 2022. Promising bioactive compounds from the marine environment and their potential effects on various diseases. Journal of Genetic Engineering and Biotechnology 20(1): 14. https://doi.org/10.1186/s43141-021-00290-4.

Kaur, M., S. Bhatia, U. Gupta, E. Decker, Y. Tak, M. Bali et al. 2023. Microalgal bioactive metabolites as promising implements in nutraceuticals and pharmaceuticals: Inspiring therapy for health benefits. Phytochemistry Reviews. https://doi.org/10.1007/s11101-022-09848-7.

Khan, M.N.A., S.-J. Yoon, J.-S. Choi, N.G. Park, H.-H. Lee, J.-Y. Cho et al. 2009. Anti-edema effects of brown seaweed (*Undaria pinnatifida*) extract on phorbol 12-myristate 13-acetate-induced mouse ear inflammation. The American Journal of Chinese Medicine 37(02): 373–381. https://doi.org/10.1142/S0192415X09006837.

Khan, M.I., J.H. Shin and J.D. Kim. 2018. The promising future of microalgae: current status, challenges, and optimization of a sustainable and renewable industry for biofuels, feed, and other products. Microbial Cell Factories 17(1): 36. https://doi.org/10.1186/s12934-018-0879-x.

Khavari, F., M. Saidijam, M. Taheri and F. Nouri. 2021. Microalgae: Therapeutic potentials and applications. Molecular Biology Reports 48(5): 4757–4765. https://doi.org/10.1007/s11033-021-06422-w.

Khor, Wei Han, Hooi-Siang Kang, Jun-Wei Lim, K. Iwamoto, C. H.-H. Tang, P.S. Goh et al. 2022. Microalgae cultivation in offshore floating photobioreactor: State-of-the-art, opportunities and challenges. Aquacultural Engineering 98: 102269. https://doi.org/10.1016/j.aquaeng.2022.102269.

Kim, J.-A., B.-N. Ahn, C.-S. Kong and S.-K. Kim. 2013. The chromene sargachromanol E inhibits ultraviolet a-induced ageing of skin in human dermal fibroblasts. British Journal of Dermatology 168(5): 968–976. https://doi.org/10.1111/bjd.12187.

Kiokias, S. and V. Oreopoulou. 2021. A review of the health protective effects of phenolic acids against a range of severe pathologic conditions (Including Coronavirus-Based Infections). Molecules 26(17): 5405. https://doi.org/10.3390/molecules26175405.

Kiran, B.R. and S.V. Mohan. 2021. Microalgal cell biofactory—therapeutic, nutraceutical and functional food applications. Plants 10(5): 836. https://doi.org/10.3390/plants10050836.

Koyande, A.K., K.W. Chew, K. Rambabu, Y. Tao, D.-T. Chu and P.-L. Show. 2019. Microalgae: A potential alternative to health supplementation for humans. Food Science and Human Wellness 8(1): 16–24. https://doi.org/10.1016/j.fshw.2019.03.001.

Leandro, A., L. Pereira and A.M.M. Gonçalves. 2020. Diverse applications of marine macroalgae. Marine Drugs 18(1): 17. https://doi.org/10.3390/md18010017.

Lenzo, D., M.A. Colangelo, A. Pasteris, F. Rindi, R. Pistocchi and L. Pezzolesi. 2023. Understanding the role of macroalgal complexity and allelochemicals production in invasive and non-invasive macroalgae in the North-Western Adriatic Sea: Effect on the Associated Communities. Water 15(9): 1697.

Li, K., X.-M. Li, J.B. Gloer and B.-G. Wang. 2012. New nitrogen-containing bromophenols from the marine red alga rhodomela confervoides and their radical scavenging activity. Food Chemistry 135(3): 868–872. https://doi.org/10.1016/j.foodchem.2012.05.117.

Liu, Y., X. Ren, C. Fan, W. Wu, W. Zhang and Y. Wang. 2022. Health benefits, food applications, and sustainability of microalgae-derived N-3 PUFA. Foods 11(13): 1883. https://doi.org/10.3390/foods11131883.

Liu, J.-K. 2022. Natural products in cosmetics. Natural Products and Bioprospecting 12(1): 40. https://doi.org/10.1007/s13659-022-00363-y.

Lomartire, S. and A. M.M. Gonçalves. 2022. An overview of potential seaweed-derived bioactive compounds for pharmaceutical applications. Marine Drugs 20(2): 141. https://doi.org/10.3390/md20020141.

López-Hortas, L., N. Flórez-Fernández, M.D. Torres, T. Ferreira-Anta, M.P. Casas, E.M. Balboa et al. 2021. Applying seaweed compounds in cosmetics, cosmeceuticals and nutricosmetics. Marine Drugs 19(10): 552. https://doi.org/10.3390/md19100552.

Lupo, Mary P. 2001. Antioxidants and Vitamins in Cosmetics. Clinics in Dermatology 19(4): 467–473. https://doi.org/10.1016/S0738-081X(01)00188-2.

Lubchenco, J. and P.M. Haugan. 2023. The ocean genome: conservation and the fair, equitable and sustainable use of marine genetic resources. In: Lubchenco, J. and P.M. Haugan. [eds.]. The Blue Compendium: From Knowledge to Action for a Sustainable Ocean Economy. Cham: Springer International Publishing. https://doi.org/10.1007/978-3-031-16277-0_4.

Lucakova, S., I. Branyikova and M. Hayes. 2022. Microalgal proteins and bioactives for food, feed, and other applications. Applied Sciences 12(9): 4402. https://doi.org/10.3390/app12094402.

Mahata, C., P. Das, S. Khan, M.I.A. Thaher, M. Abdul Quadir, S.N. Annamalai et al. 2022. The potential of marine microalgae for the production of food, feed, and fuel (3F). Fermentation 8(7): 316. https://doi.org/10.3390/fermentation8070316.

Martínez-Ruiz, M., C.A. Martínez-González, D.-H. Kim, B. Santiesteban-Romero, Reyes-H. Pardo, K.R. Villaseñor-Zepeda et al. 2022. Microalgae bioactive compounds to topical applications products; A review. Molecules 27(11): 3512. https://doi.org/10.3390/molecules27113512.

Martins, R., H. Sales, R. Pontes, J. Nunes and I. Gouveia. 2023. Food wastes and microalgae as sources of bioactive compounds and pigments in a modern biorefinery: A review. Antioxidants 12(2): 328. https://doi.org/10.3390/antiox12020328.

Matos, G.S., S.G. Pereira, Z.A. Genisheva, A.M. Gomes, J.A. Teixeira and C.M.R. Rocha. 2021. Advances in extraction methods to recover added-value compounds from seaweeds: Sustainability and functionality. Foods 10(3): 516. https://doi.org/10.3390/foods10030516.

Mehdizadeh Allaf, M. and H. Peerhossaini. 2022. Cyanobacteria: Model microorganisms and beyond. Microorganisms 10(4): 696. https://doi.org/10.3390/microorganisms10040696.

Meherishi, L., S.A. Narayana and K.S. Ranjani. 2019. Sustainable packaging for supply chain management in the circular economy: A review. Journal of Cleaner Production 237: 117582. https://doi.org/10.1016/j.jclepro.2019.07.057.

Menaa, F., P.A.U.I. Wijesinghe, G. Thiripuranathar, B. Uzair, H. Iqbal, B.A. Khan et al. 2020. Ecological and industrial implications of dynamic seaweed-associated microbiota interactions. Marine Drugs 18(12): 641. https://doi.org/10.3390/md18120641.

Merlo, S., X.G. Durany, A.P. Tonon and S. Rossi. 2021. Marine microalgae contribution to sustainable development. Water 13(10): 1373. https://doi.org/10.3390/w13101373.

Michalak, I., R. Tiwari, M. Dhawan, M. Alagawany, M.R. Farag, K. Sharun et al. 2022. Antioxidant effects of seaweeds and their active compounds on animal health and production—A review. Veterinary Quarterly 42(1): 48–67. https://doi.org/10.1080/01652176.2022.2061744.

Mitra, S., A. Rauf, A.M. Tareq, S. Jahan, T.B. Emran, T.G. Shahriar et al. 2021. Potential health benefits of carotenoid lutein: An updated review. Food and Chemical Toxicology 154: 112328. https://doi.org/10.1016/j.fct.2021.112328.

Mobin, S.M.A., F. Alam and H. Chowdhury. 2022. Environmental impact of algae-based biofuel production: A review. AIP Conference Proceedings 2681(1): 13. https://doi.org/10.1063/5.0117093.

Mohsin, S. and G.M. Kurup. 2011. Mechanism underlying the anti-inflammatory effect of sulphated polysaccharide from padina tetrastromatica against carrageenan induced paw edema in rats. Biomedicine & Preventive Nutrition 1(4): 294–301. https://doi.org/10.1016/j.bionut.2011.09.004.

Morais, T., A. Inácio, T. Coutinho, M. Ministro, J. Cotas, L. Pereira et al. 2020. Seaweed potential in the animal feed: A review. Journal of Marine Science and Engineering 8(8): 559. https://doi.org/10.3390/jmse8080559.

Morais, T., J. Cotas, D. Pacheco and L. Pereira. 2021. Seaweeds compounds: An ecosustainable source of cosmetic ingredients? Cosmetics 8(1): 8. https://doi.org/10.3390/cosmetics8010008.

Moreira, J.B., B.d.S. Vaz, B.B. Cardias, C.G. Cruz, A.C.A.d. Almeida, J.A.V. Costa et al. 2022. Microalgae polysaccharides: an alternative source for food production and sustainable agriculture. Polysaccharides 3(2): 441–457. https://doi.org/10.3390/polysaccharides3020027.

Morer, C., A. Michan-Doña, A. Alvarez-Badillo, P. Zuluaga and F. Maraver. 2020. Evaluation of the feasibility of a two-week course of aquatic therapy and thalassotherapy in a mild post-stroke population. International Journal of Environmental Research and Public Health 17(21): 8163. https://doi.org/10.3390/ijerph17218163.

Mourelle, M.L., C.P. Gómez and J.L. Legido. 2017. The potential use of marine microalgae and cyanobacteria in cosmetics and thalassotherapy. Cosmetics 4(4): 46. https://doi.org/10.3390/cosmetics4040046.

Mumu, M., A. Das, T.B. Emran, S. Mitra, F. Islam, A. Roy et al. 2022. Fucoxanthin: A promising phytochemical on diverse pharmacological targets. Frontiers in Pharmacology 13. https://doi.org/10.3389/fphar.2022.929442.

Pagels, F., A. Arias, A. Guerreiro, A.C. Guedes and M.T. Moreira. 2022. Seaweed cosmetics under the spotlight of sustainability. Phycology 2(4): 374–383. https://doi.org/10.3390/phycology2040021.

Pandey A., G.K. Jatana and S. Sonthalia. 2023. Cosmeceuticals. *In*: StatPearls [Internet]. Treasure Island (FL): StatPearls Publishing. Available from: https://www.ncbi.nlm.nih.gov/books/NBK544223/; searched on June 4, 2023.

Pangestuti, R., K.-H. Shin and S.-K. Kim. 2021. Anti-photoaging and potential skin health benefits of seaweeds. Marine Drugs 19(3): 172. https://doi.org/10.3390/md19030172.

Park, Eunyoung, Hahyeong Yu, Jeong-Ho Lim, Jeong Hee Choi, Kee-Jai Park and Jihyun Lee. 2023. Seaweed metabolomics: A review on its nutrients, bioactive compounds and changes in climate change. Food Research International 163: 112221. https://doi.org/10.1016/j.foodres.2022.112221.

Patel, A.K., F.P.J.B. Albarico, P.K. Perumal, A.P. Vadrale, C.T. Nian, H.T.B. Chau et al. 2022. Algae as an emerging source of bioactive pigments. Bioresource Technology 351: 126910. https://doi.org/10.1016/j.biortech.2022.126910.

Pengzhan, Y., Z. Quanbin, L. Ning, X. Zuhong, W. Yanmei and L. Zhi'en. 2003. Polysaccharides from Ulva pertusa (Chlorophyta) and preliminary studies on their antihyperlipidemia activity. Journal of Applied Phycology 15(1): 21–27. https://doi.org/10.1023/A:1022997622334.

Percival, E. 1979. The Polysaccharides of Green, Red and Brown Seaweeds: Their Basic Structure, Biosynthesis and Function. British Phycological Journal 14(2): 103–117. https://doi.org/10.1080/00071617900650121.

Pereira, L. 2015. Seaweed flora of the european north atlantic and mediterranean (Chapter 6), pp. 65–178. *In:* Se-Kwon Kim [ed.]. Springer Handbook of Marine Biotechnology, Part II. Springer: Berlin, Heidelberg. ISBN 978-3-642-53970-1 https://doi.org/10.1007/978-3-642-53971-8_6.

Pereira, L. [ed.]. 2016. Carrageenans: Sources and Extraction Methods, Molecular Structure, Bioactive Properties and Health Effects. Nova Science Publishers, Inc. 268 pp. ISBN: eBook - 978-1-63485-534-1.

Pereira, L. 2018. Seaweeds as Source of Bioactive Substances and Skin Care Therapy—Cosmeceuticals, Algotheraphy, and Thalassotherapy. Cosmetics 5(4): 68. https://doi.org/10.3390/cosmetics5040068.

Pereira, L. 2021. Macroalgae. Encyclopedia 2021 (1): 177–188. https://doi.org/10.3390/encyclopedia1010017.

Pereira, L. 2022. Therapeutic and nutritional uses of marine algae: A pharmacy in the Ocean. Traditional Medicine Research 7(3): 30. https://doi.org/10.53388/TMR20220310268.

Purnamawati S., N. Indrastuti, R. Danarti and T. Saefudin. 2017. The role of moisturizers in addressing various kinds of dermatitis: A review. Clinical Medicine Research 15(3-4): 75–87. https://doi.org/10.3121/cmr.2017.1363.

Quitério, E., C. Soares, R. Ferraz, C. Delerue-Matos and C. Grosso. 2021. Marine health-promoting compounds: Recent trends for their characterization and human applications. Foods 10(12): 3100. https://doi.org/10.3390/foods10123100.

Rastogi, R.P., D. Madamwar and A. Incharoensakdi. 2015. Bloom dynamics of cyanobacteria and their toxins: Environmental health impacts and mitigation strategies. Frontiers in Microbiology 6. https://doi.org/10.3389/fmicb.2015.01254.

Reid, A., F. Buchanan, M. Julius and P.J. Walsh. 2021. A review on diatom biosilicification and their adaptive ability to uptake other metals into their Frustules for potential application in bone repair. Journal of Materials Chemistry B 9(34): 6728–6737. https://doi.org/10.1039/D1TB00322D.

Riding, J.B., R.A. Fensome, M.-O. Soyer-Gobillard and L.K. Medlin. 2023. A review of the dinoflagellates and their evolution from fossils to modern. Journal of Marine Science and Engineering 11(1): 1. https://doi.org/10.3390/jmse11010001.

Rotter, A., M. Barbier, F. Bertoni, A.M. Bones, M.L. Cancela, J. Carlsson et al. 2021. The essentials of marine biotechnology. Frontiers in Marine Science 8. https://doi.org/10.3389/fmars.2021.629629.

Rumin, J., R.G. de Oliveira Junior, J.-B. Bérard and L. Picot. 2021. Improving microalgae research and marketing in the european atlantic area: Analysis of major gaps and barriers limiting sector development. Marine Drugs 19(6): 319. https://doi.org/10.3390/md19060319.

Ruxton, C.H.S. and G. Jenkins. 2013. A Novel Topical Ingredient Derived from Seaweed Significantly Reduces Symptoms of Acne Vulgaris: A General Literature Review 64(3): 219–226.

Salehi, B., J. Sharifi-Rad, A.M.L. Seca, D.C.G.A. Pinto, I. Michalak, A. Trincone et al. 2019. Current trends on seaweeds: Looking at chemical composition, phytopharmacology, and cosmetic applications. Molecules 24(22): 4182. https://doi.org/10.3390/molecules24224182.

Santos, J.P., P.B. Torres, D.Y.A.C. dos Santos, L.B. Motta and F. Chow. 2019. Seasonal effects on antioxidant and anti-HIV activities of Brazilian seaweeds. Journal of Applied Phycology 31(2): 1333–1341. https://doi.org/10.1007/s10811-018-1615-y.

Sakai, S., Y. Komura, Y. Nishimura, T. Sugawara and T. Hirata. 2011. Inhibition of mast cell degranulation by phycoerythrin and its pigment moiety phycoerythrobilin, prepared from Porphyra yezoensis. Food Science and Technology Research 17: 171-177. https://doi.org/10.3136/fstr.17.171.

Sebaaly, C., S. Kassem, E. Grishina, H. Kanaan, A. Sweidan, M.S. Chmit et al. 2014. Anticoagulant and antibacterial activities of polysaccharides of red algae Corallina collected from lebanese coast. Journal of Applied Pharmaceutical Science 4(04): 030–037. https://doi.org/10.7324/JAPS.2014.40406.

Sehgal, A., K. Goswami, M. Pal, C. Chikkaputtaiah, P. Chetia and H.P.D. Boruah. 2019. Morpho-taxonomic, genetic, and biochemical characterization of freshwater microalgae as potential biodiesel feedstock. 3 Biotech 9(4): 137. https://doi.org/10.1007/s13205-019-1664-1.

Serôdio, J. and J. Lavaud. 2020. Diatoms and their ecological importance. *In*: Leal Filho, W., A.M. Azul, L. Brandli, A. Lange Salvia and T. Wall [eds.]. Life Below Water. Cham: Springer International Publishing. https://doi.org/10.1007/978-3-319-71064-8_12-1.

Shao, P., J. Shao, L. Han, R. Lv and P. Sun. 2015. Separation, preliminary characterization, and moisture-preserving activity of polysaccharides from Ulva fasciata. International Journal of Biological Macromolecules 72: 924–930. https://doi.org/10.1016/j.ijbiomac.2014.09.04.

Shireesha, B., Shyamala and Sathoorimanasa. 2023. Comprehensive review on cosmeceuticals. International Journal of Science and Research Archive 8: 649–654. http://dx.doi.org/10.30574/ijsra.2023.8.1.0066.

Silva, M., F. Kamberovic, S.T. Uota, I.-M. Kovan, C.S.B. Viegas, D.C. Simes et al. 2022. Microalgae as potential sources of bioactive compounds for functional foods and pharmaceuticals. Applied Sciences 12(12): 5877. https://doi.org/10.3390/app12125877.

Sproles, A.E., F.J. Fields, T.N. Smalley, C.H. Le, A. Badary and S.P. Mayfield. 2021. Recent advancements in the genetic engineering of microalgae. Algal Research 53: 102158. https://doi.org/10.1016/j.algal.2020.102158.

Stengel, D.B., S. Connan and Z.A. Popper. 2011. Algal chemodiversity and bioactivity: Sources of natural variability and implications for commercial application. Biotechnology Advances 29(5): 483–501. https://doi.org/10.1016/j.biotechadv.2011.05.01.

Stoyneva-Gärtner, M., B. Uzunov and G. Gärtner. 2020. Enigmatic microalgae from aeroterrestrial and extreme habitats in cosmetics: The potential of the untapped natural sources. Cosmetics 7(2): 27. https://doi.org/10.3390/cosmetics7020027.

Sun, H., Y. Wang, Y. He et al. 2023. Microalgae-derived pigments for the food industry. Marine Drugs 21(2): 82. https://doi.org/10.3390/md21020082.

Teas, J. and M.R. Irhimeh. 2017. Melanoma and brown seaweed: An Integrative hypothesis. Journal of Applied Phycology 29(2): 941–948. https://doi.org/10.1007/s10811-016-0979-0.

Thiyagarasaiyar, K., B.-H. Goh, Y.-J. Jeon and Y.-Y. Yow. 2020. Algae metabolites in cosmeceutical: An overview of current applications and challenges. Marine Drugs 18(6): 323. https://doi.org/10.3390/md18060323.

Thoré E.S.J., K. Muylaert, M.G. Bertram and T. Brodin. 2023. Microalgae. Current Biology 33(3): R91–R95. https://doi.org/10.1016/j.cub.2022.12.032.

Tremblay, L.A., A.A. Chariton, M.-S. Li, Y. Zhang, T. Horiguchi and J.I. Ellis. 2023. Monitoring the health of coastal environments in the pacific region&mdash - A review. Toxics 11(3): 277. https://doi.org/10.3390/toxics11030277.

Turrin, S.C. 2015. Thalassotherapy, When the Sea is a Source of Health. Lifegate. Available on: https://www.lifegate.com/thalassotherapy-sea-health; searched on June 4, 2023.

Vasconcelos, J.B., E.R.T.P.P. de Vasconcelos, V. Urrea-Victoria, P.S. Bezerra, T.N.V. Reis, A.L.M. Cocentino et al. 2019. Antioxidant activity of three seaweeds from tropical reefs of brazil: Potential sources for bioprospecting. Journal of Applied Phycology 31(2): 835–846. https://doi.org/10.1007/s10811-018-1556-5.

Veerabadhran, M., S. Natesan, D.M. Ali, S. Xu and F. Yang. 2021. Using different cultivation strategies and methods for the production of microalgal biomass as a raw material for the generation of bioproducts. Chemosphere 285: 131436. https://doi.org/10.1016/j.chemosphere.2021.131436.

Verma, A., A. Barua, R. Ruvindy, H. Savela, P.A. Ajani and S.A. Murray. 2019. The genetic basis of toxin biosynthesis in dinoflagellates. Microorganisms 7(8): 222. https://doi.org/10.3390/microorganisms7080222.

Vranić, E., A. Lačević, A. Mehmedagić and A. Uzunović. 2004. Formulation ingredients for toothpastes and mouthwashes. Biomolecules and Biomedicine 4(4): 51–58. https://doi.org/10.17305/bjbms.2004.3362.

Wang, J, W. Jin, Y. Hou, X. Niu, H. Zhang and Q. Zhang. 2013. Chemical composition and moisture-absorption/retention ability of polysaccharides extracted from five algae. International Journal of Biological Macromolecules 57: 26–29. https://doi.org/10.1016/j.ijbiomac.2013.03.001.

Wang, H.-M.D., C.-C. Chen, P. Huynh and J.-S. Chang. 2015. Exploring the potential of using algae in cosmetics. Bioresource Technology 184: 355–362. https://doi.org/10.1016/j.biortech.2014.12.001.

Wang, H.-M.D., X.-C. Li, D.-J. Lee and J.-S. Chang. 2017. Potential biomedical applications of marine algae. Bioresource Technology 244: 1407–1415. https://doi.org/10.1016/j.biortech.2017.05.198.

Wang, Z.-J., X. Wei, J.-W. Liang, C.-S. Wang and Y. Kang. 2017. Effect of Fucoidan on B16 Murine melanoma cell melanin formation and apoptosis. African Journal of Traditional, Complementary and Alternative Medicines 14(4): 149–155. https://doi.org/10.21010/ajtcam.v14i4.18.

Wang, Y., M. Xing, Q. Cao, A. Ji, H. Liang and S. Song. 2019. Biological activities of fucoidan and the factors mediating its therapeutic effects: A review of recent studies. Marine Drugs 17(3): 183. https://doi.org/10.3390/md17030183.

Wells, M.L., P. Potin, J.S. Craigie et al. 2017. Algae as nutritional and functional food sources: revisiting our understanding. Journal of Applied Phycology 29(2): 949–982. https://doi.org/10.1007/s10811-016-0974-5.

Wiktor, J.M., A. Tatarek, A. Kruss, R.K. Singh and J.E. Søreide. 2022. Comparison of macroalgae meadows in warm atlantic versus cold arctic regimes in the high-arctic svalbard. Frontiers in Marine Science 9. https://doi.org/10.3389/fmars.2022.1021675.

Yoon, H.S., W.B. Koh, Y.-S. Oh and I.-J. Kim. 2009. The anti-melanogenic effects of petalonia binghamiae extracts in α-Melanocyte stimulating hormone-induced B16/F10 Murine Melanoma cells. Journal of the Korean Society for Applied Biological Chemistry 52(5): 564–567. https://doi.org/10.3839/jksabc.2009.095.

Zhang, T., Q. Guo, Y. Xin and Y. Liu. 2022. Comprehensive review in moisture retention mechanism of polysaccharides from algae, plants, bacteria and fungus. Arabian Journal of Chemistry 15(10): 104163. https://doi.org/10.1016/j.arabjc.2022.104163.

Zhuang, D., N. He, K.S. Khoo, E.-P. Ng, K.W. Chew and T.C. Ling. 2022. Application progress of bioactive compounds in microalgae on pharmaceutical and cosmetics. Chemosphere 291: 132932. https://doi.org/10.1016/j.chemosphere.2021.132932.

Chapter 7
Marine Resources
Marine Muds

Ze'evi Maor

Introduction

Mud, what makes this short word so unusual? How could a three-letter word bring together a whole world? One word that is full of so many meanings? Mud brings together soil and water, two of the "four elements of nature", believed by ancient philosophers, to empower perfection in the universe. Mud, semi solid, semi fluidic, that can be used by mankind in so many different ways.

The term "Pelotherapy", combining "pelos", the Greek word for mud, with the word "therapy" for the healing medical use, was officially adopted in 1949, by the General Assembly of "The International Society of Hydrology", during its conference, in the French city of Dax. "Pelotherapy", the remedy based on external application of mud, could be considered as a specific branch of "Thalasso-therapy", and related to "balneotherapy" and "hydrotherapy". All these therapy methods belong to the arena of alternative and complementary medicine, that stand aside from conventional medicine. France, Italy, and Germany are the three European countries that support the pelotherapy medical concept (Gomes et al. 2013).

Due to the huge variation in its nature, pelotherapy encompasses a wide spectrum of characteristics found in mud. The deviation in the place of origin affects the exact conditions of the mud's specific chemical combination, and physical parameters, such as grain particle size and shape, pH, and water content. Consequently, every kind of mud has its own identity and works in a specific manner on skin (Tian et al. 2022, Gálvez et al. 2018).

The list of various muds, exploited for pelotherapy practices, is long and generally divided into two major groups according to their origin: one major mud category is 'organic mud', containing more than 50 percent of plant originated matter, for example: peat. The other major category is 'mineral mud', also is classified as 'non-organic mud'. It includes volcano mud, sea mineral mud and other soil originated muds. Some muds are a mix of organic and non-organic muds in different ratios (Gomes et al. 2013, Tian et al. 2022).

In addition to the difference related to chemo-physical parameters, connected to the location of the origin of the mud, differences in culture, have generated further heterogenicity in mud therapy.

Head of Innovation Excellence Center, Dead Sea Hub, Fosun Jinmei Cosmetics, ex-AHAVA CTO, and a Skin Researcher at Dead Sea & Arava Science Center, Masada, Israel.
Email: zeevimaor1983@gmail.com

Figure 1. Crude Mud at Nature.

There are many working methodologies of mud treatments. The temperature of the mud, during its application on skin, is varied, categorizing mud treatments as "hyper thermal" (above 38°C and usually up to 42°C), "isothermal", or "homeo-thermal" (between 36 to 38°C) and "hypothermal" (below 36 Celsius degrees), (Gomes et al. 2013, Calin et al. 2020).

Different applications are commonly used as mud therapy procedures. A mud bath involves soaking in a bath of warm mud, often in a natural geothermal pool. A mudpack is the application on the body skin of various amounts of viscous, sticky mud. Both mud bath and mud pack treatments have been used for centuries to promote health and relaxation. Some of the mud therapy procedures use pure natural mud. Others uses processed and mixed muds or even artificially blended muds (Gomes et al. 2013, Fonseca et al. 2023, Tian et al. 2022).

Greeks and Romans are assumed to be pioneers in the realization of the therapeutic powers of muds. The first spa foundations were established near the source of thermal water and mud more than 2100 years ago. Some of the most luxurious Roman spas are still active today, maintaining old-fashioned mud pack and mud bath treatments as practiced by the original Romans and Greeks (Gomes et al. 2013).

The Greek and Roman philosophies, searching answers for health needs, that come together with the human desire for pleasure, is still reflected in the list of mud treatments, offered by many spa institutions today. In parallel to the conservative spa pelotherapy activities, cosmetic and cosmeceutical businesses, based on muds has also evolved.

In this chapter, the different types of muds, exploited for mud therapy, cosmeceutical and cosmetics are described. Scientifically proven benefits of mud treatment for health and beauty are elaborated. Speculated theories of the mode of activation, beyond the reported beneficial effects are discussed, and important facts, relating to the safety and regulation of mud treatments are touched upon.

Mud masks in skin cosmetics and cosmeceuticals

The evolution of mud masks

Clay was perhaps, the first natural ingredient, to be applied on human skin. Archeologists believe that prehistoric shamans used white, black, red, and green muds to cover their skin, aiming to connect with the good powers and expel evil spirits; while warriors painted their body and faces, using mud masks to frighten their enemies. Using primitive mud masks, as a part of their religious rituals, the prehistoric men were not aware of their pioneering contribution to the history of todays'

Figure 2. Facial Mud Mask at Home.

billion-dollar global cosmetic business. We can estimate, that hundreds of years later, the benefits and healing effects of specific muds were noticed, and as a result, mud masks were added to the list of common health and beauty products.

It is speculated that the business of mud masks, was initiated by ancient Greek and Romans in the local spas and thalassotherapy centers. The first cosmetic mud masks were probably prepared with the concept of 'packing the experience' of the locally practiced treatments; enabling customers to bring home the mud wrapped as souvenirs and gifts for the people they loved.

Today, in addition to the huge business executed in cosmetic markets, modern souvenir shops near health spas can be found. The Dead Sea region is a good example of the spa mud business. Hundreds of different brands are available in spa resorts and visitor centers for visiting tourists. (Ma'or et al. 1996).

Mud mask—from the perspective of markets and marketing

From the original primitive 'packed mud' souvenirs, muds have evolved into a massive business, with numerous offered products that are industrially manufactured and distributed globally. Mud masks are considered natural and have a health-oriented authentic aroma. This fact may explain the high popularity of mud masks in the current cosmetic business. Cosmetic products, formulated with a wide range of mud concentrations, are globally marketed for health and beauty purposes. Different levels of quality can be found in the available mud masks today. Usually the point of

Figure 3. Facial Dead Sea Mud Exfoliating Mask Application.

difference, distinguishing one mud mask from another, is the reputation of the origin of the mud, and its proven therapeutic skin effects. The high competition among brands is promoting innovation in the cosmetic industry (Fonseca et al. 2023).

In this chapter, business estimations regarding mud sales and profits from mud derived products are not included, however, published data on the relevant commercial topics can be gathered from the skincare business reviews.

Dermo-cosmetic/cosmeceutical mud masks

Cosmeceutical products, also named “dermo-cosmetic skincare”, are offered to clients who suffer from skin conditions and skin disorders. Cosmeceutical mud masks are cosmetic products by law and not categorized as medical drugs. Therefore, their key purpose should be limited to beautifying and taking care of the appearance of the skin. Moreover, customers should not be misled to imagine that the purchased mud product can substitute a visit to the dermatologist or delay the application of medical treatments (Regulation (EC) 2009, Ferreira et al. 2022).

Cosmeceutical mud masks are often recommended for use as complementary treatments, serving to assist a solution for a wide range of symptoms relating to various skin disorders. The cosmeceutical/dermo-cosmetic mud masks propose solutions to customers with oily and acne-prone skin on the one hand and sensitive and very dry skin on the other. Dermo-cosmetic mud masks could improve the quality of life of a big part of the population that suffers from a moderate level of skin conditions. It could help clients with symptoms of skin diseases, such as seborrheic dermatitis (Ferreira et al. 2022, Borda et al. 2019, Fonseca et al. 2023), psoriasis, atopic dermatitis, eczema, acne etc. The application of mud masks along with the application of prescribed drugs was found to be beneficial both as a preventive step to postpone the flare up of diseases and to attenuate the signs of inflammation (Gálvez et al. 2018, Ferreira et al. 2022, Ma’or et al. 2008).

Mud mask from the perspective of a cosmetic beauty advisor

For healthy customers, mud masks work to detox the skin, to alleviate muscle pain, and to promote a healthy skin. In some cases, mud masks are used as skin scrubs too.

Mud products are characterized by their targeted skin area for application. They can be facial mud masks or body mud masks, with a small sub-category for hand mud masks. The application of body mud mask on facial skin is not recommended, since the facial skin is more delicate when compared to body skin. Generally, it could be stated that pure mud is almost always too aggressive for facial application. Hence, special products with reduced dosage of mud are formulated for the delicate facial skin. Furthermore, oily skin can “take” richer mud than dry and sensitive skin. Consequently, if a mud mask is endorsed as a good product for every customer it needs to be checked.

Mud preparations are not completely absorbed by the skin and therefore, after a few minutes of ‘wearing’ the mud mask, customers are instructed to wash off the remaining mud using warm or tap water. This kind of mud application is categorized as a rinse-off treatment.

The standard daily ‘Western’ beauty regimen is conservatively based on four basic steps: cleaning, toning, moisturizing, and make-up. Despite the excellent reputation and long history of usage mud masks are not included in this regimen. The main reason for excluding mud baths from the daily recommended beauty regimen is practical—it takes five to fifteen minutes for a single treatment of a mud mask. Since this length of time does not meet the tough timeline frame of a day-to-day modern woman, the mud mask is added only to the weekly beauty regimen and normally recommended for the leisurely weekend.

In the seven to eleven day-to-day steps of the modern Eastern regimen, recognized as K-Beauty and J-Beauty, (for Korean, and Japanese), daily application of sheet masks is included. Several of the marketed sheet-masks contain muds. Inspired by the commercial success of K- Beauty and J- Beauty on the Western markets, you may now find sheet mud masks offered in the updated daily regimen of some leading global brands.

During the weekly treatment facial mud masks are applied first for the purpose of a 'detox' or to cleanse the skin deeply. This rinse-off application prepares the skin for the following steps to be applied later—such as leave-on toners, moisturizers and make-up products. Mud's adhesiveness helps the mask to remain attached to the surface of the skin for the recommended period. The pleasant mud texture, resulting from its tiny solid grain size is easily absorbed by the skin and comes in contact intimately with every part of the skin, including the deep wrinkles. The relatively long time required for mud mask application helps the customer relax, while the active ingredients of the mud are working on the skin cells in three different levels: chemical, physical, and biological (Ma'or 2008, Ferreira et al. 2022). The speculated mode of the activity of mud masks will be discussed in a further sub part of this chapter.

Mud Masks from the perspective of a cosmetic formulator

From the chemist point of view, mud masks are built as a suspension, in which the solid grains of mud float in a watery liquid ingredient. As a suspension, mud masks present physical stability challenges. Their homogeneity is under the risk of phase separation with time. The mud grains originally mixed and dispersed in volume tend to sink and accumulate at the lower part of the package due to their higher density. This natural and unavoidable sedimentation process may lead to a separation of the solid and liquid ingredients.

Experts in formulation of mud based products can attenuate the speed of natural sedimentation by adding anti-flocculants to the mud masks, or by adding water supper adsorbing agents to neutralize excess water, or reduce the particle size of the solid mud grains, but all this will not eliminate the possibility of separation. Another point is that it is difficult to preserve a liquid mud mask against risks of microbiological contamination.

The formulation of a new mud mask should take into consideration the type of the target skin by considering the different characteristics of the skin in different areas. Practically, a facial mud mask should not be as strong as a body mud mask, as pure mud is almost always too aggressive for facial applications. Therefore, together with the muds, additional ingredients are introduced in the mud masks. In many products the recipe of mud mask includes active ingredients, such as aloe

Figure 4. Dr. Ma'or with Crude Dead Sea Mud

vera, allantoin, panthenol, bisabol, and chamomile. They are specially added for their calming and soothing capabilities. In many facial mud preparations, a mixture of different muds, with different levels of activity strength is set. Metal silicate and carbonate clays are used as complementary ingredients for mud mask formulae, including among others: bentonite, China clay/kaolin, and attapulgite.

Another factor, to be considered, while creating a new mud mask, is the skin type of the targeted customer, namely, more active mud is required for oily and acne-prone skin, while reduced activity is chosen in mud mask formula, especially developed for the dry and very dry skin types. As already mentioned, mud masks are usually applied for the purposes of deep cleansing and skin 'detox'. Sometimes, in addition to the detox function, mud masks also claim skin scrubbing. In most cases the desired scrub activity is achieved by adding solid microbead ingredients to the fluidic mud to work as physical peeling agents. The scrubbing mud mask application includes a circular massaging motion of the product onto the skin surface, allowing the small solid grains, from botanical/plant origin or from mineral source to remove the cells of the outer layer of the skin. These factors, namely the treated skin area, customer's skin type and the desired beauty activity are all considered when creating the final formula of a mud mask.

Mud masks and cosmetic regulation

Usually, mud masks are marketed as cosmetic products, even when 'healing', 'dermo-cosmetics' or 'cosmeceutical' claims are added. Cosmetics are, per the legal definition, 'any substance or mixture intended to be placed in contact with the external parts of the human body, with the sole or main purpose to clean, perfume, change the appearance, protect, keep good condition, or alter body odors' (Regulation (EC) 2009, Ferreira et al. 2022).

Since mud masks are cosmetic products, it is the responsibility of the manufacturer and marketeers to make sure customers do not confuse mud masks with medical drugs. Therefore, their key focus should be on beautifying the face, enhancing the appearance of the skin. Claiming that they have healing properties, or can cure skin diseases is illegal. However, 'cosmeceutical' mud masks are often recommended for use as a complementary treatment. They do not replace a visit the dermatologist or medical treatments (Fonseca et al. 2023, Ferreira et al. 2022).

According to international cosmetic nomenclature (INCI), mud can appear in the list of ingredients printed on every cosmetic product's package, as 'silt', 'clay', 'moss', 'maris limmus extract' and 'peat'. In specific occasions, when first approved by the INCI authorities, the name of mud origin could be added, like in the case of "Moroccan Lava Clay", "Canadian Colloidal Clay", "Tanakura Clay", and "Sea Silt".

Figure 5. ISO 22716 – GMP Mud Processing.

Mud therapy: mode of action

Background

The title above is a kind of oxymoron.

Seeking to expose the role of the different elements, composing mud and mud's specific effect on each part of our body, is almost in contradiction to the concept of mud therapy. It is a denial of the holistic perception. By this perception, mud is considered a miracle treasure, coming from nature, and given to civilization, while human well-beings are thought to be a harmonic system that works as one. The wiliness to cut natural mud to pieces, and body health to the different components, namely cells, tissues, organs, systems, molecule, and symptoms, could be considered by holistic mud therapists as not respecting the perfection and harmony of divine gifts (Ma'or et al. 1996, Gálvez et al. 2018).

However, as scientists, we always try to dig as deep as we can, searching for logical foundations that are hiding beyond what we can see with our bare eyes. Mud was used for treatments many years before scientists had started to investigate its mode of activity. During the last five decades, scientific efforts were allocated to expose the truth beyond the legends of mud wonders. Controlled clinical tests were performed to demonstrate the scientific evidence of the beneficial effect, especially of mud treatments to dermatologic and rheumatic health conditions (Ma'or et al. 1996, Ma'or 2008). Muds are globally applied for skin health and beauty, yet the explanations to the resulted beneficial effects, are not fully clear (Portugal-Cohen et al. 2019, Fonseca et al. 2023). Naive descriptions had pictured a 'feeding' mode of activity, with a nourishing machinery of essential supplements, diffused from the upper skin layers, where mudpack is applied, deep into the inner skin layers. However, recognition of the true characteristics of skin, could not assess this permeation-by-diffusion speculated theory (Ma'or et al. 1996, Portugal-Cohen et al. 2019). This bad appreciation does not avoid wannabe-scientific brands to publish commercial brochures, describing in detail how vitamins, additives, hyaluronic acid, and enzymatic cofactors deeply penetrate all the way to the inner skin layers, and "nourish" skin cells, disregarding the well-established skin's competent barrier.

In general, muds are alleged to externally improve skin health. Attenuation of skin inflammation, and reduction of cortisol levels to calm stress conditions, are among the many described health beneficials results of mud application, as well as significantly improved the quality of life of patients (Antonelli et al. 2018, Gálvez et al. 2018, Mourelle et al. 2021). In which way does mud work to improve skin health? This question remains unsolved. We may say that all known curing effects are accomplished via mud activities in three levels: physical, chemical, and biological. Most likely, all three levels are inter-connected as chemo-physical-bio assets, by which positive health and beauty effects are caused (Mennuni et al. 2021).

Physical aspects of mud activity

Three physical properties are considered as primary factors in mud's therapeutic activity, namely hydropexy, hot melt, and plasticity. Hydropexy is related to the heavy 'muddy' wet feeling of mud when applied on the skin, due to the relatively high level of water in the mud. Hot melt property, with specific mud cooling index for each kind of mud indicates the capacity of the mud to absorb and retain heat. Since mud is frequently applied in high temperatures, its property of gradually transferring heat from the mass of mud to the treated body is believed to play a critical role. Indeed, hipper thermic machinery may rationalize a large part of mud's therapeutic effects. It basically assumes that the transferred heat, from mud to the body, increases blood flow and stimulates the immune system. This explains the reported anti-inflammatory effect of hot mud application. Plasticity is related to mud texture, that adopts the shape of the organ it is applied on and allows the mud to stay attached to skin surface for the required period of time. The physical parameter of mud plasticity is mainly determined by the shape and size of granule particles and the viscosity of semi-colloidal mud and is highly correlated to the concentration of clay minerals, presented in mud's solid phase.

Other physical parameters are sometimes suggested as contributing to the mode of action underlying beyond the mud effects on skin biology.

A *mechanical effect* is believed to occur when a mud pack is applied on skin due to friction stimulation, that works to relax tenses just as a massage does.

Magnetic fields, mostly weak, are found in specific muds, containing the mineral magnetite and other magnetic metals. Knowing the poor penetration of materials through skin, the prospects of magnetic fields to externally activate inner organs, have excited mud researchers for years. However, the mesmerizing ideas of magnetic therapy have failed to obtain scientific validation, and magnetic fields from muds have not received sufficient proof for claiming true therapeutic effects.

Radioactivity was measured in specific muds, due to the content of radionuclides such as ^{238}U, ^{226}Ra, ^{232}Th, and ^{40}K, in the muds, believed to positively affect health in specific conditions (Calin et al. 2020, Karpińska et al. 2016).

Chemical aspects of mud activity

When mud is applied on the surface of the skin, one way to evaluate its mode of activity is to think that we have applied a paste with a natural blend of chemically active ingredients. The chemical aspects of mud try to elucidate the involvement of the different molecules comprising the mud (Dai et al. 2023).

In general, the chemical aspect of mud activity is not via skin penetration. Mud application has a poor capacity to supply molecules to the inner body, and, the percutaneous penetration, in most cases, is negligible. This statement is stronger when the ingredients that are to be delivered from the mud to skin are "non-lipophilic" and not well dissolved in oils, and even more robust when considering a delivery of charged molecules, such as mud minerals, with high electrolyte content. Some exceptions to this penetration statement exist, mostly small molecules, such as Hydrogen sulfide (H2S) and other sulfuric derivatives, that effectively penetrate via the skin barrier, and could be detected in blood circulation, of customers who have been exposed to long mud treatments. However, in most cases, the effective preventing characteristic of skin, is attained, thanks to the epidermal barrier functioning. As a results, scientists are looking for other modes of activity basically related to signaling machineries, that may explain how chemical actives, obtained in mud composition, do affect skin health, despite its poor penetration through the skin.

Among the candidate molecules, scientists count dissolved ions of magnesium, zinc, iron, selenium, calcium, lithium, and a mixture of these active molecules. Indeed, high concentrations of magnesium, calcium, sodium, strontium, bromine, boron, and lithium are believed to participate in the beneficial health effects, reported in the case of Dead Sea mineral rich mud. *Signaling effects* on skin by calcium, magnesium and other metal cations were reported. In this aspect, it was found in the Dead Sea mineral that the proportion among the different metal ions is a key factor, especially the ratio between bivalent cations, such as calcium [2+], magnesium [2+], and zinc [2+], and the monovalent cations of sodium [Na+] and potassium [K+].

Biological aspects of mud activity

Skin is a complicated organ that maintains enormous biological activity. The biological research of mud tries to reveal how mud, externally applied on the surface of the skin deeply affects the biology of the skin. Mineral signaling is a biological concept and follows Peter Elias 'calcium gradient theory', that opened the scientific recognition of skin signaling or the communication of essential information, between skin cells, via a transformation of simple and smart 'mineral messages'. Such mineral messages could work as a signal and initiate profound biological activity (Portugal-Cohen et al. 2019, Antonelli et al. 2018).

The other two proposed mechanisms are the 'Osmotic pump' and 'Nrf2-Keap1' pathway. Both may contribute to a better understanding of the cellular biology process, beyond the beneficial effects of mud applications, that are already well established by clinical work. In this framework, the two proposed modes of activity, Osmotic pump and Nrf2-Keap1 transcription machinery are initiated as a respond to external stress. Indeed, one aspect, involved in mineral mud application is the stress on the surface of the skin when mud is topically applied.

The Osmotic pump is a biological mechanism believed to activate the skin, anticipating that mineral mud, when applied on the surface of the skin stimulates the activity of the skin's osmotic pump. It signals the cells of the skin to attract moisture and nutrition elements from within, up to the epidermis or top layer of the skin, where it is needed the most (Ma'or et al. 1996).

Nrf2-Keap1 is a surviving transcription biological pathway, activated by moderate osmotic stress. Indeed, moderate ionic osmotic stress (MIOS) is one of the proposed biological mechanisms behind the claimed skin therapeutic effects of minerals. Such MIOS may be the consequence of skin exposure to the high concentration of dissolved minerals that naturally exists in mineral muds. In previous research, we had investigated that MIOS occurs when mineral rich cosmetic products are applied. Dead Sea mud, as a rich mineral preparation, for example, is anticipated to impose a moderate osmotic stress to skin. The moderate stress in turn activates the endogenous production of

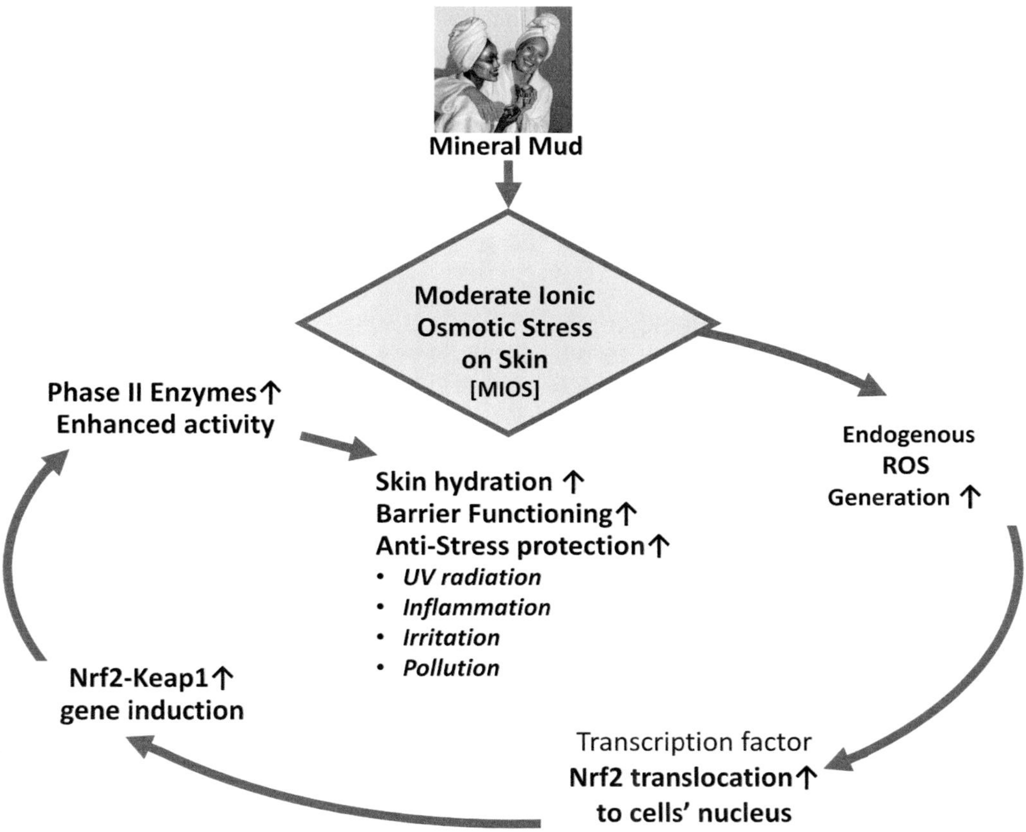

Scheme 1. Mineral Mud Biological Mode of Activity.

mitochondrial radical oxidative species (ROS). The enhanced ROS level is sufficient to stimulate the Nrf2-Keap1 pathway, that via enhanced expression of phase II protecting enzymes in skin, provides protection to skin cells against stress and insults (Cohen et al. 2022). Nrf2-Keap1 machinery, as a speculated mode of action of muds is presented in the scheme above.

Mud mode of activity—conclusions

The traditional explanation of chemical penetration of molecules as previously described as the mode of skin bioactivity was found not relevant due to the skin barrier. Therefore, other ides from physical, chemical, and biological aspects were speculated. Some of the different proposed modes of activity are described here and discussed. We assume that many of the real therapeutic or beauty proven effects, following the application of mud can be explained only by a combined mode of activity involving different levels.

Mud safety and regulation

Mud Safety

In general, mud application is safe for most of the population, and its safety perception is well assessed, based on hundreds of years of mud usage in treatments in spas and health resorts all over the world. When mud becomes an ingredient, formulated in a cosmetic product, the safety of the mud for the customers should be assessed in the same way as the safety of every cosmetic ingredient is evaluated. This safety should mainly address microbial contamination risks, as well as chemical hazardous.

Figure 6. Deep Cleansing Mineral Mud Mask

Mud safety and microbial contaminations

Muds are susceptible to microbial contaminations. The large surface area of the small mud grains and the high percentage of free water make mud an excellent medium for microorganism growth. As a natural ingredient mud is not aseptic. If contaminated mud is applied to the skin, it could interrupt the balance of the skin microbiome and lead to skin pathologies. After the crude mud is collected

Industrial processes normally add solubilizing water to it, mixing it to obtain a homogeneous fluidic mud, and napping out big particles and impurities. During handling, mud is exposed to further risks of microbial contamination, usually introduced via the added contaminated water, and sometimes by the irresponsible touch of the working staff. To ensure consumers' safety and guarantee the product's shelf of life, appropriate methods of preservation must be established. The most common way to protect mud from microbial contamination is by formulating adequate preservative systems in natural mud and selecting an effective chemical preservation system with minimal undesired side effects. Choosing preservatives for mud preparation could be accomplished via performing a set of try and error formulations with microbial challenge tests to measure the microorganism killing effect, taking into consideration the regulatory aspects and customer concern. More specifically, the formulator, when searching for the proper preservation system must evaluate negative public opinion against most of the potent preservatives, including phenoxyethanol, parabens, and formaldehyde formers. An alternative way to preserve muds is using sterilization methods, such as gamma irradiation, chlorination, and ozonation. However, since sterilization has no residual preservation effect, its efficacy is limited to the time of manufacturing. Hence, future risks of contamination should be addressed. While choosing the preservation strategy, it is expected from mud manufacturers to respect today's customers' obligatory demand, to enable full transparency, and openly share the true set of applied process and involved chemicals.

Mud safety and chemical hazards

In relation to the risks of chemical hazardous, when skin is exposed to mud, two types of chemical health threats must be considered: risks related to skin exposure to heavy and transient metals and the risk related to radioactivity.

Mud safety and toxic metal risks

Many types of mud contain naturally occurring metals. These metals may be adsorbed by the high surface area of the skin. When mud is applied on the skin surface the suspended metals in the tiny clay particles may pose a toxic threat. Muds can be contaminated with heavy metals, principally arsenic (As), cadmium (Cd), lead (Pb), and mercury (Hg), that are dangerous for human health even when they exist in low concentrations. Transition metals, mainly nickel (Ni), chrome (Cr), barium (Ba), manganese (Mn), strontium (Sr), iron (Fe), zinc (Zn), and copper (Cu), may have toxic effects in case the exposure to them is above a safe level. To avoid possible contamination of heavy and transition metals in muds, constant QC activities of mud-based products should be adopted. Analytical methods must be used to detect the level of heavy and transition metals in muds to verify that customers are not exposed to potential systemic and local chronic chemical toxicity, i.e., the actual exposed concentrations of these metals to the customer does not go beyond the international recommended safe values. In addition to ongoing quality control and mud analysis, clear labeling should facilitate customer's awareness informing them about the expected concentration of the risky metals. This data may support customers who are conscious of their specifical sensitivity to these metals. They may need to take precautions before exposing their skin to the metal via the topical mud application.

Health authorities which investigate mud safety, have concluded in some cases, that transient metals found in muds are unavoidable traces, that are present in naturally originating ingredients due to their ubiquitous characteristics. Even if impurities arise from their presence in a natural substance, it still needs to be demonstrated that the cosmetic product containing these impurities does not pose a threat to the consumer's health.

A safety evaluation of dead sea mineral mud as an example

The potential systemic and local toxicity related to the application of Dead Sea mud, taking into consideration its natural content of nickel and chrome, and the instructions on how to use the mud were assessed and described in Ma'or's work, (Ma'or et al. 2015). The detailed safety calculation presented in Ma'or's publication, could be employed as a step by step working plan to evaluate the safety of many types of mud. Based on the detected observations in Dead Sea mud, it was concluded that the reported nickel and chrome concentrations are naturally occurring metal traces. The Margin of Safety (MoS) for nickel and chrome was calculated following skin's exposure to Dead Sea mud. According to the MoS, there is no toxicological concern for systemic toxicity. Skin sensitization via chronic exposure is also not expected to occur when mud is applied on normal and healthy skin. However, Dead Sea mud topical application is not recommended for nickel or chrome-sensitive persons. Crude mud has been centrifuged and analyzed for the metal content of the interstitial water. The low levels of nickel and chrome measured demonstrate that these metals are mainly a part of the crystallographic structure or enclosed by the high surface area of solid mud particles because of which the metal release rate to the liquid phase is poor. These findings may dramatically calm the fear of the exposure of the skin to both elements via the application of Dead Sea mud.

Dead sea mud safety evaluation—conclusions

Nickel and chrome measured in Dead Sea mud are technically unavoidable as defined in Article 17 of EC Regulation 1223/2009. The nickel and chrome concentrations that were measured in Dead Sea mud are safe for human health with respect to systemic toxicity, even when the worst-case scenario is considered. They are also safe for human health with respect to local toxicity when used on non-sensitive persons. The skin exposure to nickel and chrome are actually much lower since both metals are mainly attached to the clay components of the mud and not easily released to the constituting aquatic solution. Dead Sea mud case study could be taken as an example of other muds that may contain traces of unavoidable metals (Ma'or et al. 2015).

Figure 7. Dr. Ma'or Audit at the Mud Processing Plant of Dead Sea Laboratories Ltd.

The four lessons, studied from the Dead Sea mud case:

1. The level of transient and heavy metals in muds should be monitored via ongoing sampling and chemical analysis.
2. Margin of Safety should be calculated for each suspicious material to decide if it might be toxic.
3. The effective exposure level of skin to transient metal elements following the application of muds is mainly dependent on water dissolved metals, while the metals that are retained by the clay in the solid mud particles are poorly released into the liquid and are much less dangerous to skin health.
4. As a preventive measure, metal traces, presented as impurities in the raw materials, need to be stated with their characterization and concentration, to inform customers who are sensitive to these metals (Abdel-Fattah et al. 2009, Ma'or et al. 2015).

Mud safety and radioactivity risks

In addition to metals, muds originated from some places could be contaminated with radiological hazards and since they may disrupt human health and introduce ecological risks, this possibility should be considered. More specifically, contamination of the radionuclides ^{238}U, ^{226}Ra, ^{232}Th, and ^{40}K could be presented in muds. Gamma-ray spectroscopy tests can easily monitor the existing levels of these radionuclides to confirm that each one of suspicious radiological compounds, comprise below the limit of its safe and legally approved level (Calin et al. 2020, Karpińska et al. 2016).

Innovation in mud therapy and mud skincare

Background

Innovation is a crucial issue in the struggle among competing brands. Trends are rising bringing attractive promises and encouraging customers to try 'new fresh technology'. The cosmetic customers of today believe that innovation design by science can deliver improved beauty results. The importance of newness in ingredients is growing as the awareness of the present customers to the products' composition evolves rapidly. However, the long history of thousands of years of cosmetic use of mud is in some way in contradiction to the customers expectation for a novel story. Consequently, researchers are searching for newness, to meet the market's endless demand to bring innovations to mud (Ma'or et al. 2021). Different types of novelty could be used in the exploration towards finding the new 'Super-Mud'. In the few paragraphs below, you may find selected examples', demonstrating four ways to inject innovation to mud:

A. New combination of mud plus other active ingredients
B. New modes of application of muds on skin
C. Improved definition and special process of mud preparation
D. Different ways of presenting mud to the consumers

Mud new combinations may create an innovative product. This way of innovation could be achieved via mixing different types of muds together, or mud plus botanical extracts, or mud plus any other 'surprising' active ingredient. The new desired complex is to be tested for its biological activity on skin and could be described as an innovation, if the new product can accomplish unexpected positive result, or a synergistic effect. As a demonstration sample of a 'new combination' and innovation magnetic mud is described. Knowing the poor penetration of material through skin, magnetic fields may open new prospects to externally stimulate inside organs.

Magnetic fields, mostly weak, were reported to naturally occur in some mineral muds such as Dead Sea mud. These facts inspired the design of a cosmetic product based on mud combined with dispersed magnetite or strontium hexaferrite. The new cosmetic preparation containing dispersed magnetic powder was permanently magnetized by a strong magnetic field and the product was clinically tested and proved to have an extra moisturizing effect. Surface magnetism resulted due this innovative combination and was patented and described in literature (Zioni et al. 2010).

New mode of application-leave-on mud product, and non-woven mud preparations as examples

Almost all the mud masks in the markets are designed to be applied as rinse-off skin products. As deep cleansing, detox and scrub products the mud-based items are washed off with water after a few minutes of application. A leave-on mud technology was first developed by AHAVA R&D team, and several skin products, based on the patented leave-on mud technology, were successfully launched, under the commercial trademarks of Dermud™ and Clineral™. The Leave-on mud preparations were tested, and beneficial effects of skin antioxidant and anti-inflammatory activity were significantly proven. Three per cent of Dead Sea mud containing leave-on preparations were tested and found to attenuate the biological damage to skin caused by UVB irradiation (Portugal-Cohen et al. 2009).

Another newness by changing the mode of application is not applying mud directly on the skin but first spreading it on a non-woven fabric. A body mudpack was invented for home use to be applied after being heated in a microwave. The new mud preparation was confirmed to effectively reduce body pain and rheumatic symptoms. This 'new mode of application' invention enables hyperthermal mud treatments usually performed only at spa resorts and health clinic to be applied at home. The same technology of mud attached to non-woven fabrics was also exploited for the creation of facial sheet masks. Introducing mud as a classic active ingredient to the modern sheet masks was well received by global markets.

Figure 8. Purifying Mud Sheet Mask.

Innovation via Improved definition and processes of mud preparation

Giving muds more specific definitions includes countless options always involving the introduction of a new parameter to the mud preparation as demonstrated below:

(1) 'Maturation' processes affect the chemical parameters of the mud by using new extraction and purification protocols.

For example, mud that has gone through a long soak with specific plant extracts; or sterile mud that has been irradiated with gamma rays; or washed mud to obtain a reduced level of sodium dissolved salts in mud.

(2) Choosing to obtain 'clean mud' observing precise environmentally friendly rules and social ethics.

(3) Selecting the precise solid particle size of mud and affecting the physical mud characteristics to receive more defined mud, such as mud that has been physically processed in a super grinding machine to have sub-micron grain size, or even semi-transparent nano mud, or mud that is filtered in a strong rotating nap to have extra homogenous particle size.

(4) Selecting the exact mining location, for example, mud that is collected from an extreme place, such as deep-sea mud or mud from the mouth of volcano.

Innovation via different ways of presenting mud to the consumer

1. A mud liquid extract obtained via mud centrifugation with a concentrated fluid that incorporates the dissolved actives in the mud. This extract can be applied to the skin in a new texture quite different from the semi-solid mud.
2. Encapsulated mud or mud entrapped inside the core of a vehicle, keeping it away from other compounds, enabling its formulation as an uninvolved ingredient, until its application. Encapsulating it means it will not change the color or the odor of the standard cosmetic preparation.
3. Mud globules, organized as a sticky ball of mud, thanks to the addition of adhesive elements.
4. Powder of mud, obtained via controlled water expel mud technologies for self-preparation of a mud bath at home.

Figure 9. Mud at the Innovation Lab.

These four proposed options to supply innovation to mud packs are in addition to many other options. For example, adopting new marketing concepts, such as personalized product, especially formulated and manufactured or tailor-made for a client; dermo-cosmetic natural remedy fragrance-free potions to treat skin disorders; a set of mud items for a Do It Yourself (DIY) spa at home. Any innovative idea can inject newness to the 'old-fashioned' or 'classic' mud. (Ma'or et al. 2021).

Two remarks suggested to those who need to supply innovation to their mud-based cosmetics

1. Let mud continue its position as the key active ingredient of the product. Despite its evaluation as a conservative ingredient, with the added newness, it can successfully be transformed to be considered by customers as "improved yet classic" new product.
2. Protect your innovation via patent application, or know-how. The innovation would help to differentiate your brand from all other competitors, if it is patented.

Mud and skin microbiome

From the microbiological point of view, human skin is a multifactorial setting, populated with lots of different types of microorganisms constituting what is termed today as 'skin microbiome'. The high rank of skin microbiome has been elucidated in recent decades and its involvement in skin health and diseases has been established, especially when connected to inflammatory processes, and innate skin immune functions. Therefore, the effect of mud on skin microbiome has been under investigation of late (Antonelli et al. 2018). Sea mud, containing a high concentration of salts and minerals, was found to present antibacterial activity against some pathogenic bacterial strains, including *Escherichia coli*, *Staphylococcus aureus*, *Propionic acnes*, and *Candida albicans* (Antonelli et al. 2018). Mud can affect the microbiome of the skin via its chemo-physical composition. Ma'or has demonstrated the antimicrobial properties of Dead Sea mud and its ability to attenuate the growth of skin pathogenic bacteria (Ma'or et al. 2006). Mud applications can also affect skin microbiome through a probiotic effect, achieved by introducing a huge number of non-pathogenic live bacteria onto the surface of the skin. These bacteria, such as probiotic bacillus species, may successfully compete with the pathogenic bacteria species, that already flourish on the skin, and consequently establish a more balanced population leading to a healthier skin microbiome. In general, microbiome is currently an inspiring topic in science, and many efforts are being made to better understand the relationship among the relevant involved factors, including mud and skin microflora. Whether the reported beneficial effects of mud application on skin microbiome are achieved via mud's chemo-physical features; or by presenting live probiotic bacteria on the surface of the skin; or a combination of these factors is used: we can expect to find mud applications being used in the near future as a new promising tool aimed at achieving improved skin health with a balanced microbiome.

Mudpacks and mud baths to treat non-skin pathologies, including rheumatic diseases and fibromyalgia

Mud therapy treatments are useful to relieve pain, reduce stiffness and improve the functional status of patients (Verhagen et al. 2015, Gálvez et al. 2018, Mennuni et al. 2021). The list of pathologies that gain benefits through pelotherapy, the mud-based remedy, is wide. In general, diseases that involve chronic pains and skeletal disorders are the first candidates for mud treatments (d'Angelo et al. 2021). Living with chronic pain requires an open orientation to life, hence, the patients' willingness to expose their body to non-drug complementary and alternative medical treatments is high.

Consequently, mud treatments could be experimented by people who suffer from chronic pains (Hahm et al. 2020, Kim et al 2010). Most mud treatments are used as hyper-thermal applications since the slow local heating process of the painful inflamed organ is believed to stimulate the innate immune system and attenuate the inflammation. Therefore, topical mud application could be useful to heal chronic pain diseases, such as rheumatic diseases and fibromyalgia (Maeda et al. 2018, Ma'or et al. 2008). The treatment of osteo-arthritis, for example, includes a recommendation to try mud application as a complementary approach. The efficacy of the mud-bath therapy was assessed in a clinical study with patients suffering from osteoarthritis of the knee joints. They reported significant improvement on the functional score and reduced pain (Codish et al. 2005, Verhagen et al. 2015). A review, investigating the effect of mud bathes on osteoarthritis, was concluded with a statement that mud bath therapy cannot substitute for the conventional medicinal therapy, but can be supportive when applied in addition to the drug application (Swedish Council on Health Technology Assessment 2006). In fibromyalgia, another chronic pain disease, multimodal rehabilitation, including pelotherapy, was found to improve long-term functional abilities. In one clinical work mud bath treatment had significantly reduced chronic musculoskeletal pain by 20–30% (Mennuni et al. 2021). Additional scientific reports suggest using mud treatments to reduce low back pain (Abu-Shakra et al. 2014, Hahm et al. 2020).

Conclusions: Mud therapy is a non-drug treatment modality, noninvasive, and complication-free, with scientific works to support the evaluation of mud therapy, as an effective modality, to be applied, in addition to the conservative treatments of many chronic pain and skeletal involved diseases (Gálvez et al. 2018, Verhagen et al. 2015).

Overview

Using muds for body and face application has been known since early history. The clinical therapeutic benefits of mud treatments are scientifically established, primarily proven to reduce pain, attenuate inflammation, and stimulate innate healing properties. Muds are used in cosmetics, mainly as a detox active agent to deeply cleanse the epidermis and to promote healthy skin. I believe that, only a small proportion of the full potential of mud has been exploited. Many cosmetic brands have lately turned their attention to muds as innovative, authentic, natural-based compounds for cosmetics. Further research is needed to determine and elucidate the benefits of muds. A fruitful cooperation among academics and cosmetic industries could lead to a more sustainable market answering new challenges of responsible sourcing and ecological manufacturing.

In this chapter, I had tried to deploy the full map of the different categories of mud, used for cosmeceutical therapy and as cosmetics. The benefits derived by customers from mud treatments have been described. The speculated concepts, explaining the mode of activation, beyond the beneficial effects, have been presented. The relevant safety issues, relating to mud application on skin were discussed.

Acknowledgements

This chapter could not be completed without the great support I have received through three decades from the team at Ahava R&D. This is my opportunity to thank dear friends and colleagues for their help. My personal appreciations are sent to Dr. Meital Portugal-Cohen, Dr. Miriam Oron, Dr. Dror Cohen, and Dr. Dov Effron.

Deep-felt gratitude to my excellent supervisors at The Hebrew University of Jerusalem, Professors Yoram Milner, Shlomo Magdassi and Roni Kohen.

References

Abdel-Fattah, A. and N.E. Pingitore. 2009. Low levels of toxic elements in Dead Sea black mud and mud-derived cosmetic products. Environ. Geochem. Health 31: 487–492.

Abu-Shakra, M., A. Mayer, M. Friger and M. Harari. 2014. Dead Sea mud packs for chronic low back pain. Isr. Med. Assoc. J. 2014 Sep; 16(9): 574–7.

Antonelli, M. and D. Donelli. 2018. Effects of balneotherapy and spa therapy on levels of cortisol as a stress biomarker: a systematic review. Int. J. Biometeorol. 2018 Jun; 62(6): 913–924.

Antonelli, M. and D. Donelli. 2018. Mud therapy and skin microbiome: A review. Int. J. Biometeorol. 2018 Nov; 62(11): 2037–2044.

Borda, L.J., M. Perper and J.E. Keri. 2019. Treatment of seborrheic dermatitis: A comprehensive review. J. Dermatolog Treat. 2019 Mar; 30(2): 158–169.

Calin, M.R., I. Radulescu, A.C. Ion, L. Capra and E.R. Almasan. 2020. Investigations on chemical composition and natural radioactivity levels from salt water and peloid used in pelotherapy from the Techirghiol Lake, Romania. Environ. Geochem. Health 2020 Feb; 42(2): 513–529.

Codish, S., M. Abu-Shakra, D. Flusser, M. Friger and S. Sukenik. 2005. Mud compress therapy for the hands of patients with rheumatoid arthritis. Rheumatol. Int. 2005 Jan; 25(1): 49–54.

Cohen, D., Z. Ma'or, M. Portugal Cohen, M. Oron and R. Kohen. 2022. Nrf2 pathway involvement in the beneficial skin effects of moderate ionic osmotic stress—the case of the dead sea water. Journal of Cosmetics, Dermatological Sciences and Applications 12: 109–130.

Dai, D., X. Ma, X. Yan and X. Bao. 2023. The biological role of dead sea water in skin health: A review. Cosmetics 10, 21.

d'Angelo, D., D. Coclite, A. Napoletano, A.J. Fauci, R. Latina, S. Gianola et al. 2021. The efficacy of balneotherapy, mud therapy and spa therapy in patients with osteoarthritis: an overview of reviews. Int. J. Biometeorol. 2021 Jul; 65(7): 1255–1271.

Ferreira, M., A. Matos, A. Couras, J. Marto and H. Ribeiro. 2022. Overview of cosmetic regulatory frameworks around the World. Cosmetics 9: 72.

Fonseca, S., M.N. Amaral, C.P. Reis and L. Custódio. 2023. Marine natural products as innovative cosmetic ingredients. Mar. Drugs. 2023 Mar. 8; 21(3): 170.

Gálvez, I., S. Torres-Piles and E. Ortega-Rincón. 2018. Balneotherapy, immune system, and stress response: A hormetic strategy? Int. J. Mol. Sci. 2018 Jun 6; 19(6): 1687.

Gomes, C., M.I. Carretero, M. Pozo, F. Maraver, P. Cantista, F. Armijo et al. 2013. Peloids and pelotherapy: Historical evolution, classification and glossary. Applied Clay Science 75: 28–38.

Hahm, S.C., H.J. Shin, M.G. Lee, S.J. Lee and H.Y. Cho. 2020. Mud Therapy combined with core exercise for chronic nonspecific low back pain: a pilot, single-blind, randomized controlled trial. Evid Based Complement Alternat Med. 2020 Apr 3; 2020.

Karpińska, M., K. Mnich, J. Kapała, A. Bielawska, G. Kulesza and S. Mnich. 2016. Radioactivity of peat mud used in therapy. J. Environ. Radioact. 2016 Feb; 152: 97–100.

Katz, U., Y. Shoenfeld, V. Zakin, Y. Sherer and S. Sukenik. 2012. Scientific evidence of the therapeutic effects of dead sea treatments: A systematic review. Semin. Arthritis Rheum. 42: 186–200.

Kim, J.H., J. Lee, H.B. Lee, J.H. Shin and E.K. Kim. 2010. Water-retentive and anti-inflammatory properties of organic and inorganic substances from Korean sea mud. Nat. Prod. Commun. 2010 Mar; 5(3): 395–8.

Maeda, T., Y. Kudo, T. Horiuchi and N. Makino. 2018. Clinical and anti-aging effect of mud-bathing therapy for patients with fibromyalgia. Mol. Cell Biochem. 2018 Jul; 444(1-2): 87–92.

Ma'or, Z., S. Magdassi, D. Efron and S. Yehuda. 1996. Dead Sea mineral-based cosmetics – facts and illusions. Isr. J. Med. Sci. 32(3): 28–35.

Ma'or, Z., Y. Henis, Y. Alon, E. Orlov, K.B. Sørensen and A. Oren. 2006. Antimicrobial properties of Dead Sea black mineral mud. Int. J. Dermatol. 2006 May; 45(5): 504–11.

Ma'or, Z. 2008. The contribution of Dead Sea minerals to skin beauty. *In*: The Dead Sea – The World's Ultimate Natural Healing Resort for diseases of the skin, joints, lungs and heart, among others. The Hebrew University Magnes Press, Jerusalem, pp. 145–156.

Ma'or, Z., L. Halicz, M. Portugal-Cohen, M.Z. Russo, F. Robino, T. Vanhaecke and V. Rogiers. 2015. Safety evaluation of traces of nickel and chrome in cosmetics: The case of Dead Sea mud. Regul. Toxicol. Pharmacol. 2015 Dec; 73(3): 797–801.

Ma'or, Z. and M. Portugal-Cohen. 2021. Injecting innovation to traditional, natural cosmetic ingredients: The case of dead sea minerals. IEEE Engineering Management Review 49(2): 73–80, 1 Second quarter, June 2021.

Mennuni, G., M. Fontana, C. Perricone, S. Nocchi, R. Rosso, F. Ceccarelli and A. Fraioli. 2021. A meta-analysis of the effectiveness of mud-bath therapy on knee osteoarthritis. Clin. Ter. 2021 Jul 5; 172(4): 372–387.

Mourelle, M.L., C.P. Gómez and J.L. Legido. 2021. Microalgal peloids for cosmetic and wellness uses. Mar. Drugs 19: 666.

Portugal-Cohen, M., Y. Soroka, Z. Ma'or, M. Oron, T. Zioni, F.M. Brégégère et al. 2009. Protective effects of a cream containing Dead Sea minerals against UVB-induced stress in human skin. Exp. Dermatol. 2009 Sep. 18(9): 781–8.

Portugal-Cohen, M., D. Cohen, E. Ish-Shalom, Y. Laor-Costa and Z. Ma'or. 2019. Dead Sea minerals: new findings on skin and the biology beyond. Exp. Dermatol. 2019 May; 28(5): 585–592.

Regulation (EC) No 1223/2009 of the European Parliament and of the Council of 30 November 2009 on cosmetic products. Off. J. Eur. Union 2009, L 342/59.

Swedish Council on Health Technology Assessment. Methods of Treating Chronic Pain: A Systematic Review [Internet]. Stockholm: Swedish Council on Health Technology Assessment (SBU); 2006 Oct. SBU Yellow Report No. 177/1+2. PMID: 28876750.

Tian, X., Y. Zhang, H. Li, Y. Jiao, Q. Wang, Y. Zhang et al. 2022. Property of mud and its application in cosmetic and medical fields: a review. Environ Geochem Health. 2022 Dec; 44(12): 4235–4251.

Verhagen, A.P., S.M. Bierma-Zeinstra, M. Boers, J.R. Cardoso, J. Lambeck, R. de Bie. 2015. Balneotherapy (or spa therapy) for rheumatoid arthritis. Cochrane Database Syst. Rev. 2015 Apr. 11; 2015(4).

Zioni, T., N. Perkas, Y. Wolfus, Y. Soroka, I. Popov, M. Oron et al. 2010. Strontium hexaferrite nanomagnets suspended in a cosmetic preparation: a convenient tool to evaluate the biological effects of surface magnetism on human skin. Skin Res. Technol. 2010 Aug; 16(3): 316–24.

CHAPTER 8
Climatotherapy and the Marine Environment

Michele Antonelli,[1,*] *Marco Valussi*,[2] *Grazia Barbieri*[3] and *Davide Donelli*[4]

Introduction

The term "climatotherapy", composed of the words "climate" and "therapy", refers to temporarily or permanently relocating to areas whose climate and environmental conditions are associated with clinical improvements in some diseases (Maraver et al. 2011). Marine climatotherapy, set in seaside locations or in places with marine-like characteristics (i.e., the Dead Sea and other salt lakes), implies the body's exposure to different natural elements, such as sunlight, marine aerosol, and sometimes sand and some types of natural muds and seawater. The most important environmental components that can be harnessed for therapeutic purposes in coastal and coastal-like areas are the following:

- Thalassotherapy (from "thalassa", a Greek term standing for "sea") indicates seawater baths and therapeutic immersions, and it is sometimes recommended for patients with skin, rheumatic, or respiratory diseases associated with chronic inflammation (Munteanu and Munteanu 2019, Gomes et al. 2021).
- Psammotherapy (from Greek "psammos", "sand"), sometimes referred to as "psammatotherapy", consists of hot sand baths, possibly useful for some rheumatic diseases such as osteoarthritis (Antonelli and Donelli 2019).
- Heliotherapy (from "Helios", a Greek term standing for "the Sun") implies controlled sunlight exposure to boost the body's vitamin D supply and to treat inflammatory skin conditions such as psoriasis (Emmanuel et al. 2020, Melandri et al. 2020, Mead 2008).

[1] Department of Public Health, AUSL-IRCCS di Reggio Emilia, Reggio Emilia (Italy); Via Giovanni Amendola 2, 42122 Reggio Emilia (Italy). ORCID: 0000-0002-5941-6604
[2] Magnifica Essenza S.r.l., Verona (Italy). Email: marco@infoerbe.it; ORCID: 0000-0001-9108-848X
[3] Archilinea S.r.l., Modena (Italy). Email: grazia.barbieri.1994@gmail.com; ORCID: 0000-0003-1895-2467
[4] Cardiology Unit, University Hospital of Parma, Parma (Italy). Email: donelli.davide@gmail.com; ORCID: 0000-0002-7895-2840
* Corresponding author: michele.antonelli@ausl.re.it

- Marine climatotherapy *sensu stricto*, or exposure to coastal climatic conditions is usually within certain latitudes (for example, in the Mediterranean Basin) with pleasing mild temperatures, a variable degree of humidity, higher atmospheric pressure, relatively high wind speed and sunny weather. This also includes marine aerosol inhalation and physical interaction with biogenic compounds released in the environment by seaweed, algae and coastal vegetation. These compounds may have anti-inflammatory and antioxidant effects on the airways, and may be beneficial for patients with chronic respiratory diseases (Nurov 2010, Asselman et al. 2019). Heliotherapy can be included in the definition of climatotherapy since marine air inhalation often implies some degree of sunlight exposure. On the contrary, climatotherapy does not necessarily imply seawater, mud, or hot sand baths, even though these treatments are frequently associated in marine spa centers.

Broadly speaking, the beneficial properties of marine climatotherapy have long been known, as historical evidence of seawater uses for therapeutic purposes dates to ancient Egypt (Lucchetta et al. 2007). Hippocrates, the father of Western medicine, first underscored the importance of the environment as a powerful determinant of health, and other physicians across the centuries have highlighted the tight relationship between individual health and different environmental characteristics (Thompson 1971). Recently, the key role of natural environments in wellbeing promotion has been rediscovered, since a growing body of scientific evidence is emerging on the beneficial effects on human health of a controlled exposure to some natural elements (Antonelli et al. 2022). This new vision implies an underlying paradigm shift: the environment should not be considered as a passive background, but as a complex setting with many physical stimuli (temperature, humidity, sunlight, water, soil, vegetation, volatile compounds), which can be used to promote individual wellbeing and improve some health conditions.

This chapter will focus on the useful effects of marine climatotherapy (sunlight and air exposure) for wellbeing purposes. Implications for future research, public health, and landscape design are also briefly analyzed.

Places for marine climatotherapy

Typical places for marine climatotherapy, alone or in combination with sea balneotherapy, include the Mediterranean area, the Baltic Sea, the Canary islands, the Middle East, and California, namely coastal areas with frequently sunny and windy weather, and mild-to-hot temperatures all year long (Autio et al. 2002, Even-Paz and Efron 2003, Krzyścin et al. 2012, Alpert 2015, Melandri et al. 2020, Schuh 1993). These environmental features are well represented in the Mediterranean basin, a large geographical area with a latitude between 30° and 40°, and essentially characterized by mild wet winters and warm to hot, dry summers (Lionello et al. 2006). However, other places with specific environmental characteristics, such as the Dead Sea, are associated with similar benefits of marine climatotherapy.

The Dead Sea is a salt lake located in the Middle East, an area with Mediterranean climate (the average January temperature usually falls between 14°C and 17°C, while the average August temperature is approximately 34°C) (Asmar and Ergenzinger 1999). It is the lowest point on the Earth (415 meters below sea level) and this characteristic along with a misty atmosphere (caused by evaporating water), weakens the sun rays: in particular, UVB rays are attenuated by 12%, whereas UVA rays are reduced by 3.8% (Moses et al. 2006). This is important to mitigate the negative effects of sunlight exposure without losing the benefits of heliotherapy. The low altitude is also associated with relatively high atmospheric pressure (around 800 mmHg) and oxygen tension (7–8 mmHg higher than at sea level), thus creating a favorable environment for patients with chronic heart and respiratory diseases associated with hypoxia (Moses et al. 2006, Kramer and Godfrey 1996). Some of these benefits can be found even in coastal areas at sea level with a Mediterranean climate.

Heliotherapy

Sunlight is not only composed of visible light (VL: 380–700 nm), but also of ultraviolet radiation (subsequently divided into UVA: 320–400 nm, UVB: 290–320 nm, and UVC: 200–280 nm) and infrared radiation (IR: 770 nm^{-1} mm) (Krutmann, Morita, and Chung 2012). When sufficiently thick and undamaged by pollution, the ozone layer around the Earth can block almost all UVC radiation (the most carcinogenic) and a significant part of UVB radiation, thus partially protecting the people from UV-related skin damage (Matsumi and Kawasaki 2003).

Overall, ultraviolet radiation overexposure can accelerate skin aging and increase the risk of skin cancer: in particular, even though all UV rays can be implied in photocarcinogenesis, UVB radiation can be more dangerous than UVA radiation, the former being associated with p53 mutations, while the latter mostly causing oxidative stress by favoring reactive oxygen species (ROS) production (de Grujil 2002). Visible light can have oxidative effects too, but it is not assured whether these effects are biologically relevant for human skin (Krutmann et al. 2012). At high doses, infrared light can overheat the skin and induce ROS and hydroxyl radicals production, thus increasing oxidative stress locally (Akhalaya et al. 2014).

Nevertheless, the Sun is essential for life on Earth and also for human health, as sunlight underexposure can contribute to vitamin D deficiency and melatonin suppression, which can increase the risk of developing metabolic, psychiatric, cardiovascular, musculoskeletal, and even neoplastic diseases (Smolensky et al. 2015). In fact, data from epidemiological studies confirm that strict avoidance of sunlight exposure is a risk factor for death of a similar magnitude as smoking (Lindqvist et al. 2016).

Therapeutic and controlled sunlight exposure, or heliotherapy, is currently used for improving different health conditions, including some chronic skin diseases (psoriasis, atopic dermatitis, eczema, vitiligo) (McCullough and Lehrer 2018, Krzyścin et al. 2012, Emmanuel et al. 2020, Autio et al. 2002, Smith et al. 2019, Mrk and Wahl 2002), infections (tuberculosis) (McCullough and Lehrer 2018), chronic pain management (nonspecific low back pain, fibromyalgia) (Burgess et al. 2019, Leichtfried et al. 2014, Burgess et al. 2017) and even neuro-psychiatric illnesses (depression, mood disorders, sleep problems associated with Parkinson's disease, multiple sclerosis-related fatigue) (Terman and Terman 2005, Penders et al. 2016, Videnovic et al. 2017, Mateen et al. 2020). Additionally, sunlight therapy can be useful to prevent vitamin D deficiency and rickets (Greenhalgh and Butler 2017, Holick 2008, Rhodes et al. 2010). The most relevant clinical indications of light therapy are summarized in Table 1.

Despite being used for a wide range of health conditions, light therapy is mostly administered for skin diseases, vitamin D deficiency, and mood disorders. In a study with phototype-2 patients, a 22.50 mJ/cm^2 daily mean solar dose was proposed as a threshold for outdoor exposure to treat psoriasis, with heliotherapy being administered for 20 days and each session carried out for 2 hours around noon (Krzyścin et al. 2012). Other studies indicated that a minimum dose of 31.79 mJ/cm^2 is required to have significant anti-psoriatic effects, while lower values (at least 10.00 mJ/cm^2) are only sufficient to allow for the skin synthesis of 1000 IU of vitamin D (Czerwińska and Krzyścin 2020). In general, it has been suggested that the beneficial effect of a 20-day heliotherapy program can last a whole month and can be clinically relevant only if weather conditions permit it to reach the sunlight exposure threshold in every treatment session (Krzyścin et al. 2012). Treatment protocols can vary and, while a traditional antipsoriatic approach usually implies a sun exposure of 5–6 hours for 28 days, shorter sessions, lasting no more than 3 hours, are still associated with clinically relevant beneficial effects, especially if administered in summertime, when the lowest effective cumulative UVB dose can be as little as 170 SED (Standard Erythema Dose) (Even-Paz and Efron 2003). Other studies indicate an average optimal dose of 166 SED for a 15-day treatment protocol, while that for a regular 3-week treatment period can be around 250 SED (Osmancevic et al. 2009). The prescribed cumulative UVB irradiation dose for psoriasis in skin type 2 patients is

Table 1. Clinical indications of light therapy.

Disease type	Specific disorders	References
Skin diseases	Psoriasis	
	Atopic dermatitis	
	Eczema	
	Vitiligo	
Infections	Tuberculosis	(McCullough and Lehrer 2018)
Benign diseases associated with chronic pain	Fibromyalgia	(Burgess et al. 2019, Leichtfried et al. 2014, Burgess et al. 2017)
	Nonspecific low back pain	
Neuro-psychiatric disorders	Depression and mood disorders	(Terman and Terman 2005, Penders et al. 2016, Videnovic et al. 2017, Mateen et al. 2020)
	Sleep problems associated with Parkinson's disease	
	Multiple sclerosis-related fatigue	
Endocrine disorders	Vitamin D deficiency	(Greenhalgh and Butler 2017, Holick 2008,Rhodes et al. 2010)

25.5 MED (Minimal Erythema Dose) or 637.5 mJ/cm^2, and this dosage is achieved gradually from an initial dose of 0.4 MED (10 mJ/cm^2) on day 1 to 3.4 MED (85.0 mJ/cm^2) on day 14 (Kudish et al. 2022). Treatment protocols for vitiligo normally include 2-to-3 sessions every week, and they can be administered for months, but artificial light therapy is usually preferred in this case because the treatment dose can be better adjusted based on the patient's post-therapeutic response, namely the symptomatic erythema following phototherapy, and the selective use of narrow-band UVB (Haines et al. 2018, Myers et al. 2021). For vitiligo, the initial treatment dose is usually around 200 mJ/cm^2 and phototherapy is gradually tapered after repigmentation (Smith et al. 2019). Treatment protocols for psychiatric conditions, such as seasonal affective disorder, are often characterized by a shorter duration (for example, 30 min every morning for a week), and their efficacy seems to mostly depend on light intensity and brightness, with a recommended illuminance of 10000 Lux (this threshold is easily exceeded on sunny days) (Kogan and Guilford 1998, Terman and Terman 2005). When considering vitamin D deficiency, a 10-to-30 min sunlight exposure to one third of the skin surface area can be useful to improve the vitamin levels if repeated several times a week (Rhodes et al. 2010, Nowson et al. 2004, Cannell and Hollis 2008). So, in brief, treatment protocols largely vary depending on the disease and the patients' characteristics or skin phototype (Table 2), with longer treatments used for skin or endocrine disorders, and shorter protocols applied for psychiatric conditions.

Marine heliotherapy is also a sort of thermotherapy, since exposure to the sun in hot weather tends to increase the body's temperature, especially on its surface (Schuh 1993). The benefits of this thermal effect may be similar to that observed for baths in hot sea water and medicinal muds; however, in these hydro-therapeutic means, there are also other components (mineral elements, anti-inflammatory compounds, mud and water microbiota) beyond warmth, that are demonstrated to play a significant role in their overall pharmacological effect (Antonelli and Donelli 2018, Antonelli et al. 2021). All the same, even though hot balneotherapy is a better option some benefits may be experienced by patients with osteoarthritis, pain of rheumatic origin, and fibromyalgia even with marine climate-heliotherapy alone.

Table 2. Characteristics of different skin phototypes.

Phototype	Natural skin color	UVA MED (mJ/cm²)	UVB MED (mJ/cm²)
1	Ivory white	20–35	15–30
2	White-fair	30–45	25–40
3	White-medium	40–55	30–50
4	Beige-olive	50–80	40–60
5	Moderate brown	70–100	60–90
6	Dark brown or black	100	90-150

Adapted from "Skin phototypes 2003" by Astner and Anderson (Astner and Anderson 2004). Legends: MED=Minimal Erythema Dose.

With regard to safety, it is estimated that sunlight exposure values higher than 25.00 mJ/cm^2 (Minimal Erythema Dose) can be associated with skin redness in phototype-2 subjects (Czerwińska and Krzyścin 2020) even though this threshold can vary depending on the patients' characteristics and the body parts exposed. Patients at higher risk of developing skin neoplasms (clinical history of skin cancer, genetic predisposition, health disorders associated with tendency to cell transformation, anti-rejection therapy, and severe immunosuppression) should avoid heliotherapy, as extensive sunlight exposure in these people can further increase the chances of tumor initiation. Excessive light exposure can also trigger seizures or headache in predisposed patients, or it can worsen some eye diseases if proper sunglasses are not worn during the treatment. Sunburns can occur in subjects with skin photosensitivity (Oberman and Enticott 2018, Murphy 2001). Additionally, subjects taking photosensitizing medications must avoid heliotherapy as long as they take these drugs, including among others, amiodarone, chlorpromazine, doxycycline, hydrochlorothiazide, nalidixic acid, naproxen, piroxicam, tetracycline, thioridazine, vemurafenib, and voriconazole (Montgomery and Worswick 2022). Furthermore, in some patients undergoing light therapy for depression, symptoms of hypomania and autonomic hyperactivity may occur during the first few days of treatment (Terman and Terman 2005). While extensive use of sunscreens can markedly reduce some beneficial effects of heliotherapy, it is still important to follow the general recommendations for skin protection to sunlight exposure, especially in summertime, as sunscreen use can diminish the risk of sunlight-induced cutaneous disorders and does not compromise vitamin D synthesis (Passeron et al. 2019, Dale Wilson et al. 2012). For example, in a study with psoriatic Caucasian patients undergoing heliotherapy, the participants were advised to apply a thick layer of sunscreen all over their body only in the first day of treatment, and to restrict sunscreen use to locations that tend to easily burn in the following days (Nilsen et al. 2009): this can be a useful strategy to protect the patients' skin and, at the same time, to preserve the beneficial effects of heliotherapy. Staying well-hydrated during and after heliotherapy when the outdoor temperature is high is another important issue to consider, especially in those particularly prone to dehydration such as elderly adults or some chronic patients. In brief, since heliotherapy, as any other treatment, has its own contraindications and potential side effects, a medical checkup and supervision are necessary to ensure the patients' safety, proper eligibility, and follow-up.

In general, heliotherapy as an integrative treatment is deemed cost-effective not only in individuals with severe psoriasis, but also in those with mild-to-moderate diseases as sun therapy despite being quite expensive for travel and accommodation costs is widely accepted. It is well regarded by the patients not only for the clinical improvements associated with its administration, but also for its positive effects on psychophysical well-being, partly ascribable to nature immersion, socialization, relaxation, and limited side effects (Snellman et al. 1998). Proper sites for heliotherapy should be identified on the basis of specific climate zones, which are being modified under the influence of recent climate changes (Cui et al. 2021), and adequate models (such as the RayMan model) aimed at thoroughly predicting sunlight exposure under different environmental conditions should be used to control the treatment dose (Matzarakis et al. 2010).

Marine air composition

Marine air is rich in both non-biogenic and biogenic volatile compounds. Non-biogenic compounds are mostly sea-salt particles and ions (sodium, chlorine, iodine, bromide, sulfur) dispersed in the marine aerosol through sunlight-and temperature-induced evaporation, bubbles bursting at the sea-air interface, wind, and mechanical nebulization, especially along rocky cliffs where sea waves continuously break on coastal rocks (O'Dowd et al. 1997, Gómez Martín et al. 2022, Ooki et al. 2002, Gantt and Meskhidze 2013). It has long been known that the concentration of these inorganic substances often combined with organic compounds varies depending on different meteorological and geographic conditions, such as the wind speed and direction, the average temperature throughout the day, the site weather, altitude, latitude, and the characteristics of coastal landforms (Lewis and Schwartz 2004, Gantt and Meskhidze 2013).

Biogenic volatile organic compounds (VOCs) from vegetation are dominated by the isoprenoid chemical family, including isoprene, monoterpenes, and sesquiterpenes (Yáñez-Serrano et al. 2018). In particular, several Mediterranean plant species are known to emit a great variety of VOCs (Simon et al. 2006), and these emissions are heavily influenced by light and temperature variations (Davison et al. 2009). In an oak and pine tree coastal forest at Castel Porziano (Italy), the highest monoterpene concentrations were observed at night (Davison et al. 2009), whereas in Spanish Mediterranean forests monoterpenes tend to reach their highest concentration during daytime, especially in the morning (Peñuelas et al. 2013), with a second peak early in the afternoon (Bach et al. 2020). On the other hand, isoprene concentrations are similar regardless of the coastal site location and they reach their maximum during daytime (Davison et al. 2009, Peñuelas et al. 2013). In the Italian forest, despite the fact that monoterpenes are still mostly released under the influence of sunlight, photodegradation and specific environmental factors result in higher relative concentrations of these volatile substances at night (Davison et al. 2009). In general, isoprene emissions depend on light and temperature, while monoterpene emissions are largely a function of temperature (Peñuelas et al. 2013). The direction of the wind also plays a role in influencing biogenic VOC differences along the coastline (Guérette et al. 2019). Even though biogenic VOCs are emitted all year long, it is estimated that 81% of the yearly emissions in Mediterranean areas are produced between June and August (Simon et al. 2006). Overall, monoterpenes are more abundant than isoprene in summer, whereas both VOCs show similar levels in winter (Seco et al. 2011).

Besides isoprene, in oak and pine tree coastal forests, the most common biogenic VOCs are limonene, α-pinene, and linalool (Owen et al. 1997). In Mediterranean oak forests, the most reported volatile compounds are α-pinene, sabinene, limonene, camphene, and α-phellandrene (Bach et al. 2020). In coastal shrubland sites, volatile emissions are dominated by α-pinene, β-myrcene, β-phellandrene, limonene, and γ-terpinene (Peñuelas et al. 2013). The air of beech forests close to the Mediterranean sea is rich in α-pinene, β-pinene, sabinene, limonene, α-thujene, camphene, β-myrcene, and β-phellandrene; whereas emissions of holm oak Mediterranean forests are dominated by α-terpinene, α-phellandrene, β-pinene, β-myrcene, limonene, sabinene, and camphene (Peñuelas et al. 2013). Table 3 briefly reports the most common biogenic VOCs released in the air by different Mediterranean coastal plants, which are considered relevant for their terpene and terpenoid emissions.

Biogenic compounds in sea spray are not limited to those released in the air by coastal vegetation, but also include a range of substances produced by marine algae and microorganisms. In fact, it is demonstrated that the organic fraction accounts for 15% of the marine aerosol composition during wintertime, when the biological activity of sea microorganisms is lower, and increases up

Table 3. The most important biogenic VOCs emitted by Mediterranean plants.

Mediterranean plants	Most important VOCs emitted in the atmosphere	Citations
Helichrysum stoechas	α-pinene, Δ-3-carene	(Owen et al. 2001, Fares et al. 2009, Street et al. 1997)
Juniperus oxycedrus	α-pinene	
Lavandula stoechas	1,8-cineole, camphor	
Pinus species	α-pinene, linalool, limonene	
Quercus species	α-pinene, sabinene, β-pinene, β-myrcene, limonene, linalool	
Rosmarinus officinalis	β-pinene, 1,8-cineole	

to 63% during phytoplankton blooms (O'Dowd et al. 2004). Among these biogenic substances sampled over the ocean, primary saccharides (mannitol, glucose, and arabitol) are reported to be the dominant compounds, followed by isoprene, α-pinene, and β-caryophyllene (Fu et al. 2013).

Potential benefits of sea aerosol inhalation

Overall, marine air is characterized by some ions and biogenic organic compounds (mostly terpenes), which are believed to possess beneficial effects on human health when inhaled. For example, it is demonstrated that limonene and pinene, two relevant VOCs detectable in coastal forests, have antioxidant and anti-inflammatory properties, and they can help patients affected by chronic respiratory problems, mental fatigue, psychophysical stress, and mood disorders (Antonelli et al. 2020). A recent study also underscores that inhaling pinene during nature therapy sessions is associated with a significant anxiolytic effect (Donelli et al. 2022). Moreover, these compounds are hypothesized to improve some immune system functions, thus being potentially useful to boost immune defenses, decrease cortisol levels, and reduce psychophysical stress (Antonelli et al. 2020, Antonelli et al. 2019, Cho et al. 2017).

Apart from the volatile substances produced by coastal vegetation, biogenic compounds released in sea spray by marine microorganisms are supposed to have anti-inflammatory effects on the respiratory tract too (Asselman et al. 2019, Moore 2015). As with other nature-based interventions, these benefits may be explained by the 'old friends' hypothesis, a theory suggesting that ambient microbial species that coexist in the natural environment with humans are useful to favor a proper development and functioning of the immune system, consequently reducing the incidence of immune-related disorders (such as allergies) and boosting the immune defenses (Langgartner et al. 2019). For example, it is demonstrated that homoyessotoxin, a biochemical compound excreted by marine algae and found in sea spray, can downregulate the expression of human genes related to the mTOR pathway and proprotein convertase subtilisin/kexin type 9 (PCSK9), thus potentially having anti-inflammatory and antiproliferative effects on the respiratory airways (Asselman et al. 2019, Moore 2015, Van Acker et al. 2018). However, during algal blooms, sometimes referred to as "red tides" an excess of some of these substances can trigger bronchospasm in patients with asthma (Fleming et al. 2005). As such, it would be important for these subjects to avoid visiting coastal areas in periods of algal blooms (or to adjust their pharmacological therapy if physical avoidance is not possible for other reasons). In general, the multifaceted interplay between sea aerosol microbiota (including its byproducts) and human beings is still to be investigated in depth, and it could become a major focus for future studies on marine climatotherapy.

With regard to indirect evidence, it is suggested that inhalation therapy with nebulized high-mineral spring water, similar to sea spray, is good for rhinosinusitis and chronic respiratory diseases because it reduces local inflammation, mucus production and the incidence of respiratory infections (Khaltaev et al. 2020, Antonelli et al. 2021). Similar effects in patients with asthma and chronic obstructive pulmonary disease are hypothesized for halotherapy, namely salt therapy consisting in inhalation of tiny salt particles in natural caves or artificial chambers, even though the existing evidence is not strong enough to formulate specific recommendations yet (Rashleigh et al. 2014; Crisan-Dabija et al. 2021).

Finally, air at sea level has a relatively high atmospheric pressure (760 mmHg) and an of oxygen tension (160 mmHg) (Sharma and Hashmi 2022). As reported for the Dead Sea (Moses et al. 2006, Kramer and Godfrey 1996) these characteristics can favor blood oxygenation and improve the quality of life of patients with chronic cardiorespiratory diseases such as chronic obstructive pulmonary disease or congestive heart failure (Kramer et al. 1994, Ćurić et al. 2023, Gabizon et al. 2011).

Therefore, on the basis of direct and indirect evidence reported in this chapter, marine climatotherapy appears beneficial for ear-nose-throat diseases associated with mucus overproduction, chronic inflammation and recurrent infections; cardiorespiratory conditions characterized by chronic inflammation and hypoxia; and mental health conditions, especially mood disorders and psychophysical stress. Marine climatotherapy may also boost the immune system function and reduce the incidence of some allergic disorders, provided that red tides are avoided by those prone to bronchospasm. With regard to therapeutic programs, it could be useful to resort to those already administered in medical hydrology: an optimal set of 10–12 consecutive daily sessions, each lasting at least 30–40 min (or preferably even more), to be repeated once or twice every year (Antonelli et al. 2021).

Implications for urban health and building design in coastal areas

Being aware of how to harness the benefits of marine climatotherapy cannot only be useful to formulate therapeutic protocols but also to improve urban health and building designs in all coastal areas that are heavily subjected to anthropization because of continuous urbanization driven by the never-ending conquest of the waterfront. Property speculation, pollution, and wild urbanization of coastal areas have unfortunately increased the fragility of these natural environments (Zhan et al. 2022). However, today, thanks to new housing and material technologies all necessary tools and strategies to protect coastal environments are available. It is now possible to maximize the benefits of comfortably staying in coastal areas without negatively impacting the natural environment.

To be sustainable, buildings must necessarily respect environmental conditions, such as climate and humidity. To keep the climate in mind is an essential factor when designing new-generation green buildings; especially in conditions as particular as those occurring in coastal areas where wind, humidity, and atmospheric salts tend to corrode and quickly damage building materials (Morillas et al. 2020). Moreover, it has been demonstrated that permanence of chloride deposits on the facade drastically decreases as the altitude of the building increases. This connection, impacted by the wind, can be depicted by an exponential decay function (Meira et al. 2017). Among other environmental factors to consider, sea level rise plays a fundamental role (Samodra and Chandrabuana 2021), and should be taken into account when planning new urban areas along the coastline or restoring those that already exist.

Most studies based on urban surface geometry have revealed that, beyond weather conditions, airflow is conditioned by the city's morphology, particularly by the size, shape, and orientation of the buildings, the characteristics of the roads, and the construction density of different neighborhoods. Urban geometry, thermal and surface properties of building materials, and extension of green areas correlate with the size of the so-called 'Urban Heat Islands'. Specific research has revealed that higher temperatures tend to decrease with the increase in the building height-to-width ratio and the proximity to the sea. Providing shade has been proposed as the main strategy for lowering air and radiant temperatures (Emmanuel and Johansson 2006). Urbanization leads to a marked increase in thermal stress in hot-humid climates, due to the rise in surface and air temperatures and reduction in wind speed: in other words, heavily urbanized waterfronts reduce the strength of sea breezes (Masuda et al. 2005) while the effects of sea breezes can be more pronounced when urban areas are built farther from the waterfront. Additionally, mid-to-high-rise buildings separated by narrow streets often act as barriers along the coastline. Wind corridors perpendicular to the sea can facilitate the penetration of sea breezes deeper inside the city (Emmanuel and Johansson 2006). In fact, sea breezes are seen as an efficient source of cooling that can mitigate Urban Heat Islands, and are capable of reducing urban temperatures by about 4°C (Peng and Watanabe 2022). Additionally, once the wind speed surpasses a specific limit, it has the ability to counteract Urban Heat Islands to the extent of fully offsetting them. It is therefore necessary to increase shaded areas in urban environment without reducing the passage of air. A variation in building heights could favorably allow air movements (Shen et al. 2019). In fact, as sea breezes reduce the temperature, they contribute to heat wave dispersion in coastal areas of many cities. On the other hand, sea breeze can transport airborne particulate matter (PM2.5 and PM10) inland, and this phenomenon partially reduces the benefits described above (Kibirige et al. 2022). However, if city centers include preferential air channels that allow the breeze to reach the opposite periphery, then these atmospheric pollutants can be blown away. It follows that the city morphology, road layout and landscape design can favorably or unfavorably influence the air quality and temperature variations as they can either block or allow or redirect the passage of sea breeze in coastal areas.

In brief, planners should always keep in mind that the characteristics of the landscape play a key role in urban sites, and the basics of climatotherapy should be considered in the processes of designing new buildings in coastal areas. Conservation and promotion of coastal forests and their native vegetation can be a beneficial factor as they allow the transport of monoterpenes and other beneficial substances inland, especially in an urban context designed as described above, where the air adequately circulates across the city.

Conclusive remarks

Marine climatotherapy can have beneficial effects for health and well-being promotion, due to the diversity of volatile substances, biogenic compounds, microorganisms detectable in sea aerosol and to the properties of controlled sunlight exposure (Figure 1). Skin and cardiorespiratory diseases characterized by chronic inflammation are impacted the most in a positive way. Mental health can benefit from marine climatotherapy, especially thanks to the effects of plant-derived monoterpene inhalation and natural light exposure on mood and brain activity. Rheumatic conditions can improve as well, but the effect is greater when marine climatotherapy is coupled with balneotherapy in sea water and mud therapy. In general, marine climatotherapy is quite safe, but, as discussed above, some behavioral recommendations should be followed to avoid side effects. Current clinical evidence is quite scant, but it is enough to affirm that the benefits are more likely associated with multiple session-treatment programs, when the exposure is repeated for several days in a row, even though a single-day exposure can exert some short-lasting effects. In fact, it is estimated that these clinically relevant benefits may last for around 3 (or even more) months after completing a full thalassotherapy program implying marine climate-heliotherapy (Emmanuel et al. 2020, Schuh and Nowak 2011). Nevertheless, future research studies are recommended to better quantify the effect size of marine climatotherapy as an integrative therapy for health problems, as well as to assess any long-term benefits, risks and possible moderators of the effect.

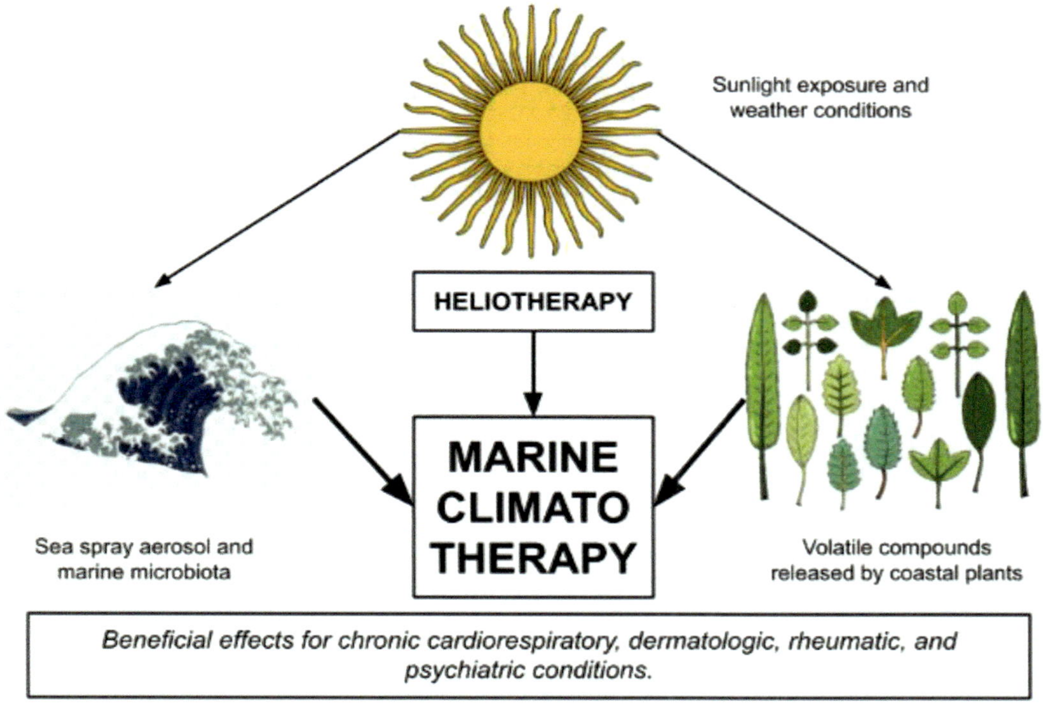

Figure 1. Infographic about the beneficial effects of marine climatotherapy for health.

References

Akhalaya, M.Y., G.V. Maksimov, A.B. Rubin, J. Lademann and M.E. Darvin. 2014. Molecular action mechanisms of solar infrared radiation and heat on human skin. Ageing Research Reviews 16(July): 1–11. http://dx.doi.org/10.1016/j.arr.2014.03.006.

Alpert, J.S. 2015. The Jeremiah Metzger Lecture: Jeremiah Metzger and the Era of Heliotherapy. Transactions of the American Clinical and Climatological Association 126: 219–226. https://www.ncbi.nlm.nih.gov/pubmed/26330679.

Antonelli, M., G. Barbieri and D. Donelli. 2019. Effects of forest bathing (shinrin-Yoku) on levels of cortisol as a stress biomarker: A systematic review and meta-analysis. International Journal of Biometeorology 63(8) (August): 1117–1134. http://dx.doi.org/10.1007/s00484-019-01717-x.

Antonelli, M., G. Barbieri and D. Donelli. 2022. Defining a new perspective in environmental health: The healing environment. International Journal of Biometeorology 66(5) (May): 1039–1044. http://dx.doi.org/10.1007/s00484-022-02251-z.

Antonelli, M. and D. Donelli. 2018. Mud therapy and skin microbiome: A review. International Journal of Biometeorology 62(11) (November): 2037–2044. http://dx.doi.org/10.1007/s00484-018-1599-y.

Antonelli, M. and D. Donelli. 2019. Hot sand baths (psammotherapy): A systematic review. Complementary Therapies in Medicine 42 (February): 1–6. http://dx.doi.org/10.1016/j.ctim.2018.10.020.

Antonelli, M., D. Donelli, G. Barbieri, M. Valussi, V. Maggini and F. Firenzuoli. 2020. Forest volatile organic compounds and their effects on human health: A state-of-the-art review. International Journal of Environmental Research and Public Health 17(18) (September 7). http://dx.doi.org/10.3390/ijerph17186506.

Antonelli, M., D. Donelli, L. Veronesi, M. Vitale and C. Pasquarella. 2021. Clinical efficacy of medical hydrology: An umbrella review. International Journal of Biometeorology (April 17). http://dx.doi.org/10.1007/s00484-021-02133-w.

Asmar, B.N. and P. Ergenzinger. 1999. Estimation of Evaporation from the Dead Sea. Hydrological Processes. https://onlinelibrary.wiley.com/doi/abs/10.1002/(SICI)1099-1085(19991215)13:17%3C2743::AID-HYP845%3E3.0.CO;2-U.

Asselman, J., E. Van Acker, M. De Rijcke, L. Tilleman, F. Van Nieuwerburgh, J. Mees et al. 2019. Marine biogenics in sea spray aerosols interact with the mTOR Signaling Pathway. Scientific Reports 9(1) (January 24): 675. http://dx.doi.org/10.1038/s41598-018-36866-3.

Astner, S. and R.R. Anderson. 2004. Skin Phototypes 2003. The Journal of Investigative Dermatology 122(2) (February): xxx–xxxi. http://dx.doi.org/10.1046/j.1523-1747.2003.22251.x.

Autio, P., P. Komulainen and H.M. Larni. 2002. Heliotherapy in atopic dermatitis: A prospective study on climatotherapy using the SCORAD Index. Acta Dermato-Venereologica 82(6): 436–440. http://dx.doi.org/10.1080/000155502762064575.

Bach, A., A.M. Yáñez-Serrano, J. Llusià, I. Filella, R. Maneja and J. Penuelas. 2020. Human breathable air in a mediterranean forest: Characterization of monoterpene concentrations under the canopy. International Journal of Environmental Research and Public Health 17(12) (June 18). http://dx.doi.org/10.3390/ijerph17124391.

Burgess, H.J., M. Park, J.C. Ong, N. Shakoor, D.A. Williams and J. Burns. 2017. Morning versus evening bright light treatment at home to improve function and pain sensitivity for women with fibromyalgia: A pilot study. Pain Medicine 18(1) (January 1): 116–123. http://dx.doi.org/10.1093/pm/pnw160.

Burgess, H.J., M. Rizvydeen, M. Kimura, M.H. Pollack, S.E. Hobfoll, K.B. Rajan et al. 2019. An open trial of morning bright light treatment among US military veterans with chronic low back pain: A pilot study. Pain Medicine 20(4) (April 1): 770–778. http://dx.doi.org/10.1093/pm/pny174.

Cannell, J.J. and B.W. Hollis. 2008. Use of vitamin D in clinical practice. Alternative Medicine Review: A Journal of Clinical Therapeutic 13(1) (March): 6–20. https://www.ncbi.nlm.nih.gov/pubmed/18377099.

Cho, K.S., Y.-R. Lim, K. Lee, J. Lee, J.H. Lee and I.-S. Lee. 2017."?/Terpenes from Forests and Human Health. Toxicology Research 33(2) (April): 97–106. http://dx.doi.org/10.5487/TR.2017.33.2.097.

Crisan-Dabija, R., I.G. Sandu, I.V. Popa, D.-V. Scripcariu, A. Covic and A. Burlacu. 2021. Halotherapy-an ancient natural ally in the management of asthma: A comprehensive review. Healthcare (Basel, Switzerland) 9(11) (November 22). http://dx.doi.org/10.3390/healthcare9111604.

Cui, D., S. Liang and D. Wang. 2021. Observed and projected changes in global climate zones based on Köppen climate classification. Wiley Interdisciplinary Reviews. Climate Change 12(3) (May). https://onlinelibrary.wiley.com/doi/10.1002/wcc.701.

Ćurić, M., O. Zafirovski and V. Spiridonov. 2023. Essentials of medical meteorology. Springer International Publishing. https://link.springer.com/book/10.1007/978-3-030-80975-1.

Czerwińska, A. and J. Krzyścin. 2020. Analysis of Measurements and Modelling of the Biologically Active UV Solar Radiation for Selected Sites in Poland – Assessment of Photo-Medical Effects. Institute of Geophysics, Polish Academy of Sciences. https://pub.igf.edu.pl/analysis-of-measurements-and-modelling-of-the-biologically-active-uv-solar-radiation-for-selected-sites-in-poland-%E2%80%93-assessment-of-photo-medical-effects,4,en,pubs,4,0,4,140.html.

Dale Wilson, B., S. Moon and F. Armstrong. 2012. Comprehensive review of ultraviolet radiation and the current status on sunscreens. The Journal of Clinical and Aesthetic Dermatology 5(9) (September): 18–23. https://www.ncbi.nlm.nih.gov/pubmed/23050030.

Davison, B., R. Taipale, B. Langford, P. Misztal, S. Fares, G. Matteucci et al. 2009. Concentrations and fluxes of biogenic volatile organic compounds above a mediterranean macchia ecosystem in Western Italy. Biogeosciences 6(8): 1655–1670. https://bg.copernicus.org/articles/6/1655/2009/.

Donelli, D., F. Meneguzzo, M. Antonelli, D. Ardissino, G. Niccoli, G. Gronchi et al. 2022. Effects of Plant-Emitted Monoterpenes on Anxiety: A Propensity-Matched Observational Cohort Study (December 31). https://www.preprints.org/manuscript/202212.0586/v1.

Emmanuel, R. and E. Johansson. 2006. Influence of Urban Morphology and Sea Breeze on Hot Humid Microclimate: The Case of Colombo, Sri Lanka. Climate Research. http://dx.doi.org/10.3354/cr030189.

Emmanuel, T., D. Lybæk, C. Johansen and L. Iversen. 2020. Effect of Dead Sea Climatotherapy on Psoriasis; A Prospective Cohort Study. Frontiers of Medicine 7 (March 18): 83. http://dx.doi.org/10.3389/fmed.2020.00083.

Even-Paz, Z. and D. Efron. 2003. Determination of solar ultraviolet dose in the dead sea treatment of psoriasis. The Israel Medical Association Journal: IMAJ 5(2) (February): 87–88. https://www.ncbi.nlm.nih.gov/pubmed/12674654.

Fares, S., S. Mereu, G. Scarascia Mugnozza, M. Vitale, F. Manes, M. Frattoni et al. 2009. The ACCENT-VOCBAS field campaign on biosphere-atmosphere interactions in a mediterranean ecosystem of Castelporziano (Rome): Site Characteristics, Climatic and Meteorological Conditions, and Eco-Physiology of Vegetation. Biogeosciences 6(6) (June 18): 1043–1058. https://bg.copernicus.org/articles/6/1043/2009/.

Fleming, L.E., B. Kirkpatrick, L.C. Backer, J.A. Bean, A. Wanner, D. Dalpra et al. 2005. Initial evaluation of the effects of aerosolized florida red tide toxins (brevetoxins) in persons with asthma. Environmental Health Perspectives 113(5) (May): 650–657. http://dx.doi.org/10.1289/ehp.7500.

Fu, P.Q., K. Kawamura, J. Chen, B. Charrière and R. Sempéré. 2013. Organic molecular composition of marine aerosols over the arctic ocean in summer: Contributions of primary emission and secondary aerosol formation. Biogeosciences 10(2) (February 1): 653–667. https://bg.copernicus.org/articles/10/653/2013/.

Gabizon, I., A. Shiyovich, V. Novack, V. Khalameizer, C. Yosefy, S.W. Moses et al. 2011. Impact of descent and stay at a dead sea resort (low Altitude) on patients with systolic congestive heart failure and an implantable cardioverter defibrillator. The Israel Medical Association Journal: IMAJ 13(7) (July): 402–407. https://www.ncbi.nlm.nih.gov/pubmed/21838181.

Gantt, B. and N. Meskhidze. 2013. The physical and chemical characteristics of marine primary organic aerosol: A review. Atmospheric Chemistry and Physics 13(8) (April 18): 3979–3996. https://acp.copernicus.org/articles/13/3979/2013/.

Gomes, C.S.F., J.V. Fernandes, F.V. Fernandes and J.B.P. Silva. 2021. Salt mineral water and thalassotherapy. Minerals Latu Sensu and Human Health: 631–656. https://link.springer.com/chapter/10.1007/978-3-030-65706-2_16.

Gómez Martín, J.C., A. Saiz-Lopez, C.A. Cuevas, A.R. Baker and R.P. Fernández. 2022. On the speciation of iodine in marine aerosol. Journal of Geophysical Research, D: Atmospheres 127(4) (February 27): e2021JD036081. http://dx.doi.org/10.1029/2021JD036081.

Greenhalgh, I. and A.R. Butler. 2017. Sanatoria Revisited: Sunlight and Health. The Journal of the Royal College of Physicians of Edinburgh 47(3) (September): 276–280. http://dx.doi.org/10.4997/JRCPE.2017.314.

de Gruijl, F.R. 2002. Photocarcinogenesis: UVA vs. UVB Radiation. Skin Pharmacology and Applied Skin Physiology 15(5) (September): 316–320. http://dx.doi.org/10.1159/000064535.

Guérette, É.-A., C. Paton-Walsh, I. Galbally, S. Molloy, S. Lawson, D. Kubistin et al. 2019. Composition of clean marine air and biogenic influences on VOCs during the MUMBA campaign. Atmosphere 10(7) (July 10): 383. https://www.mdpi.com/2073-4433/10/7/383.

Haines, R.H., K.S. Thomas, A.A. Montgomery, J.C. Ravenscroft, P. Akram, J.R. Chalmers et al. 2018. Home interventions and light therapy for the treatment of vitiligo (HI-Light Vitiligo Trial): Study Protocol for a Randomised Controlled Trial. BMJ Open 8(4) (April 3): e018649. http://dx.doi.org/10.1136/bmjopen-2017-018649.

Holick, M.F. 2008. Deficiency of Sunlight and Vitamin D. BMJ . http://dx.doi.org/10.1136/bmj.39581.411424.80.

Khaltaev, N., U. Solimene, F. Vitale and A. Zanasi. 2020. Balneotherapy and hydrotherapy in chronic respiratory disease. Journal of Thoracic Disease 12(8) (August): 4459–4468. http://dx.doi.org/10.21037/jtd-gard-2019-009.

Kibirige, G.W., C.-C. Huang, C.-L. Liu and M.C. Chen. 2022. Effects of Land-Sea Breeze Features to Long Term Pm2 . 5 Prediction in Central and Southern Taiwan Using Composite Neural Network. SSRN Electronic Journal. https://www.ssrn.com/abstract=4090304.

Kogan, A.O. and P.M. Guilford. 1998. Side Effects of Short-Term 10,000-Lux Light Therapy. The American Journal of Psychiatry 155(2) (February): 293–294. http://dx.doi.org/10.1176/ajp.155.2.293.

Kramer, M.R. and S. Godfrey. 1996. Dead sea: Natural oxygen enrichment at low altitude. Israel Journal of Medical Sciences 32 Suppl (July): S20–3. https://www.ncbi.nlm.nih.gov/pubmed/8756971.

Kramer, M.R., C. Springer, N. Berkman, E. Bar-Yishay, A. Avital, A. Mandelberg et al. 1994. Effect of natural oxygen enrichment at low altitude on oxygen-dependent patients with end-stage lung disease. Annals of Internal Medicine 121(9) (November 1): 658–662. http://dx.doi.org/10.7326/0003-4819-121-9-199411010-00005.

Krutmann, J., A. Morita and J.H. Chung. 2012. Sun exposure: What molecular photodermatology tells us about its good and bad sides. The Journal of Investigative Dermatology 132(3) Pt 2 (March): 976–984. https://linkinghub.elsevier.com/retrieve/pii/S0022202X15356062.

Krzyścin, J.W., J. Jarosławski, B. Rajewska-Więch, P.S. Sobolewski, J. Narbutt, A. Lesiak et al. 2012. Effectiveness of heliotherapy for psoriasis clearance in low and mid-latitudinal regions: A theoretical approach. Journal of Photochemistry and Photobiology. B, Biology 115(October 3): 35–41. http://dx.doi.org/10.1016/j.jphotobiol.2012.06.008.

Kudish, A.I., E.G. Evseev, G. Cohen and M. Harari. 2022. Ultraviolet measurements and photoclimatotherapy for psoriasis at the dead sea: 25 years of experience. International Journal of Environmental Research and Public Health 19(19) (September 28). http://dx.doi.org/10.3390/ijerph191912364.

Langgartner, D., C.A. Lowry and S.O. Reber. 2019. Old friends, immunoregulation, and stress resilience. Pflugers archiv: European Journal of Physiology 471(2) (February): 237–269. http://dx.doi.org/10.1007/s00424-018-2228-7.

Leichtfried, V., R. Matteucci Gothe, W. Kantner-Rumplmair, M. Mair-Raggautz, C. Bartenbach, H. Guggenbichler et al. 2014. Short-term effects of bright light therapy in adults with chronic nonspecific back pain: A randomized controlled trial. Pain Medicine 15(12) (December): 2003–2012. http://dx.doi.org/10.1111/pme.12503.

Lewis, E.R. and S.E. Schwartz. 2004. Sea salt aerosol production: mechanisms, methods, measurements, and models. geophysical monograph. Washington, D.C., DC: American Geophysical Union. http://www.agu.org/books/gm/v152/.

Lindqvist, P.G., E. Epstein, K. Nielsen, M. Landin-Olsson, C. Ingvar and H. Olsson. 2016. Avoidance of sun exposure as a risk factor for major causes of death: A competing risk analysis of the melanoma in southern sweden cohort. Journal of Internal Medicine 280(4) (October): 375–387. http://dx.doi.org/10.1111/joim.12496.

Lionello, P., P. Malanotte-Rizzoli, R. Boscolo, P. Alpert, V. Artale, L. Li et al. 2006. The Mediterranean Climate: An Overview of the Main Characteristics and Issues. In Mediterranean, 1–26. Developments in Earth and Environmental Sciences. Elsevier. https://linkinghub.elsevier.com/retrieve/pii/S1571919706800030.

Lucchetta, M.C., G. Monaco, V.I. Valenzi, M.V. Russo, J. Campanella, S. Nocchi et al. 2007. [The historical-scientific foundations of thalassotherapy: state of the art]. La Clinica terapeutica 158(6) (November): 533–541. https://www.ncbi.nlm.nih.gov/pubmed/18265720.

Maraver, F., A. Michán, C. Morer and L. Aguilera. 2011. Is thalassotherapy simply a type of climatotherapy? International Journal of Biometeorology. http://dx.doi.org/10.1007/s00484-010-0382-5.

Masuda, Y., N. Ikeda, T. Seno, N. Takahashi and T. Ojima. 2005. A basic study on utilization of the cooling effect of sea breeze in waterfront areas along tokyo bay. Journal of Asian Architecture and Building Engineering 4(2) (November): 483–487. https://www.tandfonline.com/doi/full/10.3130/jaabe.4.483.

Mateen, F.J., A.C. Vogel, T.B. Kaplan, G.C. Hotan, S.J. Grundy, K.B. Holroyd et al. 2020. Light therapy for multiple sclerosis-associated fatigue: A randomized, controlled phase II Trial. Journal of Neurology 267(8) (August): 2319–2327. http://dx.doi.org/10.1007/s00415-020-09845-w.

Matsumi, Y. and M. Kawasaki. 2003. Photolysis of atmospheric ozone in the ultraviolet region. Chemical Reviews 103(12) (December): 4767–4782. http://dx.doi.org/10.1021/cr0205255.

Matzarakis, A., F. Rutz and H. Mayer. 2010. Modelling radiation fluxes in simple and complex environments: basics of the RayMan model. International Journal of Biometeorology 54(2) (March): 131–139. http://dx.doi.org/10.1007/s00484-009-0261-0.

McCullough, P.J. and D.S. Lehrer. 2018. Vitamin D, Cod liver oil, sunshine, and phototherapy: safe, effective and forgotten tools for treating and curing tuberculosis infections - a comprehensive review. The Journal of Steroid Biochemistry and Molecular Biology 177 (March): 21–29. http://dx.doi.org/10.1016/j.jsbmb.2017.07.027.

Mead, M.N. 2008. Benefits of sunlight: A bright spot for human health. Environmental Health Perspectives 116(4) (April): A160–7. http://dx.doi.org/10.1289/ehp.116-a160.

Meira, G.R., W.T.A. Pinto, E.E.P. Lima and C. Andrade. 2017. Vertical distribution of marine aerosol salinity in a Brazilian Coastal area – the influence of wind speed and the impact on chloride accumulation into concrete. Construction and Building Materials 135 (March): 287–296. https://linkinghub.elsevier.com/retrieve/pii/S0950061816321018.

Melandri, D., V.M. Albano, M. Venturi, A. Flamigni and M. Vairetti. 2020. Efficacy of combined liman peloid baths and heliotherapy in the treatment of psoriasis at Cervia Spa, Emilia, Italy. International Journal of Biometeorology 64(7) (July): 1145–1152. http://dx.doi.org/10.1007/s00484-020-01888-y.

Montgomery, S. and S. Worswick. 2022. Photosensitizing drug reactions. Clinics in Dermatology 40(1) (January): 57–63. http://dx.doi.org/10.1016/j.clindermatol.2021.08.014.

Moore, M.N. 2015. Do Airborne biogenic chemicals interact with the PI3K/Akt/mTOR cell signalling pathway to benefit human health and wellbeing in rural and coastal environments? Environmental Research 140 (July): 65–75. http://dx.doi.org/10.1016/j.envres.2015.03.015.

Morillas, H., F.F. de Mendonça Filho, H. Derluyn, M. Maguregui, D. Grégoire and J.M. Madariaga. 2020. Decay processes in buildings close to the sea induced by marine aerosol: Salt depositions inside construction materials. The Science of the Total Environment 721 (June 15): 137687. http://dx.doi.org/10.1016/j.scitotenv.2020.137687.

Moses, S.W., M. David, E. Goldhammer, A. Tal and S. Sukenik. 2006. The dead sea, a unique natural health resort. The Israel Medical Association Journal: IMAJ 8(7) (July): 483–488. https://www.ncbi.nlm.nih.gov/pubmed/16889164.

Mrk, C. and A. Wahl. 2002. Improved quality of life among patients with psoriasis after supervised climate therapy at the canary Islands. Journal of the American Academy of Dermatology 47(2) (August): 314–316. https://linkinghub.elsevier.com/retrieve/pii/S019096220200066X.

Munteanu, C. and D. Munteanu. 2019. Thalassotherapy Today. Balneo Research Journal 10, no. Vol 104 (December 10): 440–444. http://bioclima.ro/Balneo278.pdf.

Murphy, G.M. 2001. Diseases associated with photosensitivity. Journal of Photochemistry and Photobiology. B, Biology 64(2-3) (November 15): 93–98. https://linkinghub.elsevier.com/retrieve/pii/S1011134401002287.

Myers, E., S. Kheradmand and R. Miller. 2021. An update on narrowband ultraviolet b therapy for the treatment of skin diseases. Cureus 13(11) (November): e19182. http://dx.doi.org/10.7759/cureus.19182.

Nilsen, L.T.N., E. Søyland and A.L. Krogstad. 2009. Estimated ultraviolet doses to psoriasis patients during climate therapy. Photodermatology, Photoimmunology & Photomedicine 25(4) (August): 202–208. http://dx.doi.org/10.1111/j.1600-0781.2009.00443.x.

Nowson, C.A., T.H. Diamond, J.A. Pasco, R.S. Mason, P.N. Sambrook and J.A. Eisman. 2004. Vitamin D in Australia. Issues and recommendations. Australian Family Physician 33(3) (March): 133–138. https://www.ncbi.nlm.nih.gov/pubmed/15054977.

Nurov, I. 2010. Effect of psammotherapy on cardiovascular system of patients with chronic obstructive pulmonary disease in stage of rehabilitation. Medical Principles and Practice: International Journal of the Kuwait University, Health Science Centre 2 (April 13): 8–11. http://academicpublishingplatforms.com/article.php?journal=MHSJ&number=1&article=300.

Oberman, L.M. and P.G. Enticott. 2018. Neurotechnology and brain stimulation in pediatric psychiatric and neurodevelopmental disorders. Academic Press. https://play.google.com/store/books/details?id=9Ht8DwAAQBAJ.

O'Dowd, C.D., M.C. Facchini, F. Cavalli, D. Ceburnis, M. Mircea, S. Decesari et al. 2004. Biogenically driven organic contribution to marine aerosol. Nature 431(7009) (October 7): 676–680. http://dx.doi.org/10.1038/nature02959.

O'Dowd, C.D., M.H. Smith, I.E. Consterdine and J.A. Lowe. 1997. Marine aerosol, sea-salt, and the marine sulphur cycle: A short review. Atmospheric Environment 31(1) (January): 73–80. https://linkinghub.elsevier.com/retrieve/pii/S1352231096001069.

Ooki, A., M. Uematsu, K. Miura and S. Nakae. 2002. Sources of sodium in atmospheric fine particles. Atmospheric Environment 36(27) (September): 4367–4374. https://linkinghub.elsevier.com/retrieve/pii/S1352231002003412.

Osmancevic, A., L.T. Nilsen, K. Landin-Wilhelmsen, E. Søyland, P. Abusdal Torjesen, T.A. Hagve et al. 2009. Effect of climate therapy at gran canaria on vitamin d production, blood glucose and lipids in patients with psoriasis. Journal of the European Academy of Dermatology and Venereology: JEADV 23(10) (October): 1133–1140. http://dx.doi.org/10.1111/j.1468-3083.2009.03245.x.

Owen, S., C. Boissard, R.A. Street, S.C. Duckham, O. Csiky and C.N. Hewitt. 1997. Screening of 18 mediterranean plant species for volatile organic compound emissions. Atmospheric Environment 31 (January 1): 101–117. https://www.sciencedirect.com/science/article/pii/S1352231097000782.

Owen, S.M., C. Boissard and C.N. Hewitt. 2001. Volatile organic compounds (VOCs) Emitted from 40 mediterranean plant Species: VOC speciation and extrapolation to habitat scale. Atmospheric Environment 35(32) (November 1): 5393–5409. https://www.sciencedirect.com/science/article/pii/S1352231001003028.

Passeron, T., R. Bouillon, V. Callender, T. Cestari, T.L. Diepgen, A.C. Green et al. 2019. Sunscreen photoprotection and Vitamin D status. The British Journal of Dermatology 181(5) (November): 916–931. http://dx.doi.org/10.1111/bjd.17992.

Penders, T.M., C.N. Stanciu, A.M. Schoemann, P.T. Ninan, R. Bloch and S.A. Saeed. 2016. Bright light therapy as augmentation of pharmacotherapy for treatment of depression: A systematic review and meta-analysis. The Primary Care Companion to CNS Disorders 18(5) (October 20). http://dx.doi.org/10.4088/PCC.15r01906.

Peng, S. and H. Watanabe. 2022. Analysis and mapping of sea breeze event time in coastal cities: A case study of Sendai. Atmosphere 13(9) (September 13): 1484. https://www.mdpi.com/2073-4433/13/9/1484.

Peñuelas, J., A. Guenther, F. Rapparini, J. Llusia, I. Filella, R. Seco et al. 2013. Intensive measurements of gas, water, and energy exchange between vegetation and troposphere during the MONTES campaign in a vegetation gradient from short semi-desertic shrublands to tall wet temperate forests in the NW mediterranean basin. Atmospheric Environment 75: 348–364. https://www.sciencedirect.com/science/article/pii/S1352231013002860.

Rashleigh, R., S.M.S. Smith and N.J. Roberts. 2014. A review of halotherapy for chronic obstructive pulmonary disease. International Journal of Chronic Obstructive Pulmonary Disease 9 (February 21): 239–246. http://dx.doi.org/10.2147/COPD.S57511.

Rhodes, L.E., A.R. Webb, H.I. Fraser, R. Kift, M.T. Durkin, D. Allan et al. 2010. Recommended summer sunlight exposure levels can produce sufficient (> or =20 Ng Ml(-1)) but not the proposed optimal (> or =32 Ng Ml(-1)) 25(OH)D Levels at UK latitudes. The Journal of Investigative Dermatology 130(5) (May): 1411–1418. http://dx.doi.org/10.1038/jid.2009.417.

Samodra, F.X.T.B. and C.J. Chandrabuana. 2021. Aquascape architecture 2050: A design proposal for human adaptation medium against sinking scenario. In The 5Th International Tropical Renewable Energy Conference (THE 5TH iTREC). AIP Publishing. http://aip.scitation.org/doi/abs/10.1063/5.0063486.

Schuh, A. 1993. Climatotherapy. Experientia 49(11) (November 15): 947–956. http://dx.doi.org/10.1007/BF02125641.

Schuh, A. and D. Nowak. 2011. [Evidence-based acute and long-lasting effects of climatotherapy in moderate altitudes and on the seaside]. Deutsche medizinische Wochenschrift 136(4) (January): 135–139. http://dx.doi.org/10.1055/s-0031-1272496.

Seco, R., J. Peñuelas, I. Filella, J. Llusià, R. Molowny-Horas, S. Schallhart et al. 2011. Contrasting winter and summer VOC mixing ratios at a forest site in the western mediterranean basin: The effect of local biogenic emissions. Atmospheric Chemistry and Physics 11(24) (December 21): 13161–13179. https://acp.copernicus.org/articles/11/13161/2011/.

Sharma, S. and M.F. Hashmi. 2022. Partial pressure of oxygen. In StatPearls [Internet]. StatPearls Publishing. https://www.ncbi.nlm.nih.gov/books/NBK493219/.

Shen, L., C. Zhao, Z. Ma, Z. Li, J. Li and K. Wang. 2019. Observed decrease of summer sea-land breeze in Shanghai from 1994 to 2014 and Its Association with Urbanization. Atmospheric Research 227 (October): 198–209. https://linkinghub.elsevier.com/retrieve/pii/S0169809518313826.

Simon, V., L. Dumergues, J.-L. Ponche and L. Torres. 2006. The biogenic volatile organic compounds emission inventory in France: Application to plant ecosystems in the berre-marseilles area (France). The Science of the Total Environment 372(1) (December 15): 164–182. http://dx.doi.org/10.1016/j.scitotenv.2006.08.019.

Smith, M.P., K. Ly, Q. Thibodeaux, T. Bhutani and M. Nakamura. 2019. Home phototherapy for patients with vitiligo: Challenges and solutions. Clinical, Cosmetic and Investigational Dermatology 12 (June 28): 451–459. http://dx.doi.org/10.2147/CCID.S185798.

Smolensky, M.H., L.L. Sackett-Lundeen and F. Portaluppi. 2015. Nocturnal light pollution and underexposure to daytime sunlight: Complementary mechanisms of circadian disruption and related diseases. Chronobiology International 32(8) (September 16): 1029–1048. http://dx.doi.org/10.3109/07420528.2015.1072002.

Snellman, E., T. Maljanen, A. Aromaa, A. Reunanen, T. Jyrkinen-Pakkasvirta and J. Luoma. 1998. Effect of heliotherapy on the cost of Psoriasis. The British Journal of Dermatology 138(2) (February): 288–292. http://dx.doi.org/10.1046/j.1365-2133.1998.02076.x.

Street, R.A., S. Owen, S.C. Duckham, C. Boissard and C.N. Hewitt. 1997. Effect of habitat and age on variations in volatile organic compound (VOC) Emissions from Quercus Ilex and Pinus Pinea. Atmospheric Environment 31 (January 1): 89–100. https://www.sciencedirect.com/science/article/pii/S1352231097000770.

Terman, M. and J.S. Terman. 2005. Light therapy for seasonal and nonseasonal depression: Efficacy, protocol, safety, and side effects. CNS Spectrums 10(8) (August): 647–63; quiz 672. http://dx.doi.org/10.1017/s1092852900019611.

Thompson, K. 1971. Climatotherapy in California. California Historical Quarterly 50(2) (June 1): 111–130. https://online.ucpress.edu/ch/article/50/2/111/30888/Climatotherapy-in-California.

Van Acker, E., J. Asselman, M. De Rijcke, K. De Schamphelaere and C. Janssen. 2018. Do Marine Aerosols Improve Human Health? In VLIZ Marine Science Day 2018. Vol. 81. Vlaams Instituut voor de Zee (VLIZ). https://biblio.ugent.be/publication/8578928.

Videnovic, A., E.B. Klerman, W. Wang, A. Marconi, T. Kuhta and P.C. Zee. 2017. Timed light therapy for sleep and daytime sleepiness associated with parkinson disease: A randomized clinical trial. JAMA Neurology 74(4) (April 1): 411–418. http://dx.doi.org/10.1001/jamaneurol.2016.5192.

Yáñez-Serrano, A.M., L. Fasbender, J. Kreuzwieser, D. Dubbert, S. Haberstroh, R. Lobo-do-Vale et al. 2018. Volatile diterpene emission by two mediterranean cistaceae shrubs. Scientific Reports 8(1) (May 1): 6855. http://dx.doi.org/10.1038/s41598-018-25056-w.

Zhan, Q., W. Dong, C. Fu, A. Wang, H. Yi and Z. Pan. 2022. The self-healing of marine concrete cracks based on the synergistic effect of microorganisms and inorganic minerals. Journal of Building Engineering 61(105210) (December): 105210. https://linkinghub.elsevier.com/retrieve/pii/S2352710222012165.

CHAPTER 9

Thalassotherapy and Nutrition

Manuel Portela Collazo,[1,2] *Javier Jose Prieto Troncoso*[1,2] and *Federico Mallo Ferrer*[1,2,*]

Introduction and definitions

Thalassotherapy is the combined use of marine elements (water, seaweeds, mud and climate) in a marine environment for curative and wellbeing improvement purposes (Lucchetta et al. 2007).

On the other hand, nutrition is a set of unconscious actions by which the body incorporates, transforms and uses the nutrients contained in food. The human body needs nutrients to carry out different functions covering energy needs, build and maintain body structures, regulate metabolic pathways and prevent diseases.

The combination of these two disciplines, namely thalassotherapy and nutrition, has not been previously explored. Despite the fact that there are few publications on this topic, there is evidence that thalasso-nutrition for therapeutic uses was employed for centuries. The use of seawater in nutrition has been studied in recent years, especially regarding waters upwelling from ocean depths or deep seawater.

In the following pages, we will address the links between seawater and its application in clinical nutrition, physical exercise, and gastronomy. We will also review warn against possible contraindications, and name certain precautions to be taken when using this natural resource.

Water relevance for physiology and health

Water is a fundamental component of all organisms. It is essential for cellular homeostasis and the function of our organs and systems. It accounts for 75% of body weight in children, 60%–70% in men and 50%–60% in women, and up to 55% in older adults. The water is distributed in different compartments in our body, so that approximately 65% is found in the intracellular medium and 35% in the extracellular medium. Although we obtain water from various sources such as food or even macronutrient oxidation (metabolic water), the main source is fluid intake. In fact, the main cause that motivates humans to drink is a water deficit, which activates the physiological thirst mechanism. It is essential to maintain a water balance, as well as a correct concentration of minerals that are critical for vital functions, in the human body.

[1] Laboratory of Endocrinology. Ed. Olimpia Valencia. Campus As Lagoas – Marcosende University of Vigo. 36310 – Vigo – Spain.
[2] Galicia Sur Health Research Institute. Endocrinology group. Alvaro Cunqueiro Hospital. 36213 -Vigo - Spain.
* Corresponding author: fmallo@uvigo.es, fmallof@gmail.com

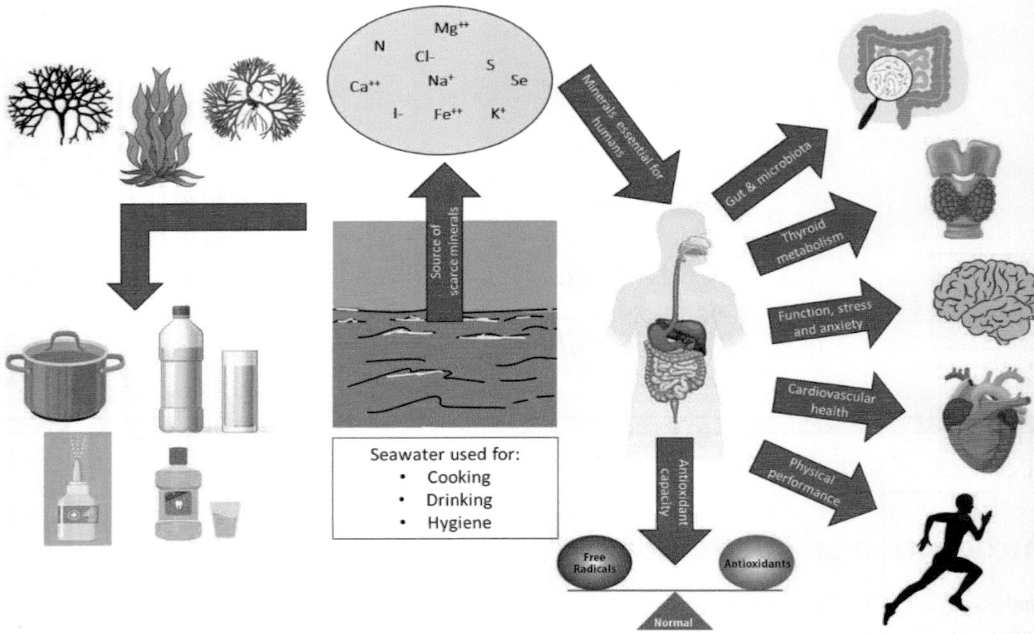

Figure 1. Uses and biological effect of seawater and seaweeds. Seawater may be used for hygiene and rinsing mouth, nose and eyes, for drinking in different dilutions, for preparing food in brine, and for boiling foods especially seafood. Algae also have a place in different culinary preparations. The minerals and other organic products from seaweeds have shown to have beneficial effects in the function of different organs including thyroid, brain, cardiovascular system, whole metabolism and gut microbiota. Seawater drinks may be used for rehydration and recovery after physical activity and other pathophysiological conditions.

The importance of these minerals is of such magnitude that their concentration in the form of solutes in biological fluids is the mechanism that the body must control to maintain a hydroelectrolytic balance. This mechanism requires the coordination of sensitive detectors (osmoceptors) in different parts of the body connected to an integrating center in the brain responsible for processing information—the hypothalamus. This system is so sensitive that it is activated by small variations (0.5%) in sodium levels and is influenced by many factors such as diuresis, natriuresis, blood pressure or body and environmental temperature. Several hormones such as vasopressin, renin, angiotensins and aldosterone which work together to maintain a delicate balance in the hydroelectrolytic control of body fluids.

Physiological activities lead to water and minerals loss. These losses have to be replaced in a proper way to maintain homeostasis and the function of the organic systems. In conditions of increased losses as sweating during physical activity and sports practice, high environmental temperature or in pathophysiological situations as diarrhea, prompt and adequate replacement becomes even more important. In such conditions, it becomes mandatory to have a hydration strategy, where the amounts of minerals in the drunk beverages is a critical factor.

Seawater nutritional properties

Composition

Seawater is a solution of water (96.5%) and salts (3.5%).

The mineral profile of seawater as containing chloride, sodium, potassium, calcium, magnesium, etc., has been known for many centuries. But it also contains more than 70 minerals, around twenty of which are present in trace amounts (*trace elements*), with iodine, zinc, bromine, cobalt and selenium standing out for their biological essentiality in the human organism. To review the complete mineral composition of seawater see Chapter 3.

To all this mineral wealth, we must add the presence of living and non-living organic matter in sea water because of the presence of plankton: phytoplankton or vegetable plankton (micro-algae) and zooplankton or animal plankton. It also contains an extraordinary number of bacteria, virus, fungi and other microorganisms which constitute the seawater microbiome (Moran 2015, Garcia 2007). Sea water also contains free amino acids, oligosaccharides, polysaccharides, fatty acids, hydrocarbons, phenols, vitamins, urea, etc. (Benner and Amon 2015).

Composition and seawater types

Seawater contains large amounts of sodium salts that makes it incompatible with our physiological needs, especially if we intend to drink it directly. Thus, the dilution of seawater is mandatory for therapeutic uses in humans except in very specific applications such as rinsing the mouth, nose or mucosa wherein it is useful because of its antiseptic properties.

Desalinated seawater

Fresh water obtained from seawater by industrial methods including filtration and reverse osmosis is desalinated seawater. It is free of salts and solvents and generally used for human purposes including drinking and as tap water in populations living in zones with very scarce sources of natural fresh water. At least 350 million people in the world currently use desalinated seawater as the exclusive drinking water source and its use is rapidly increasing in Gulf Countries, Africa, and Asia.

Isotonic seawater

The isotonic drinks are obtained by mixing seawater with fresh water to get the same concentration of the salinity as of the human plasma, which is considerably lower than that of seawater. Isotonic drinks have an approximated concentration of 9 grams of sodium salts per litre of water, and can be drunk either in its natural form or commonly with different flavorings.

Seawater taken in this form acts as a nutrients supplier. It should be drunk in small glasses and the act of drinking should be spaced out to prevent the accumulation of salt in the intestine which may have a laxative effect. It has the function of re-mineralizing and rehydrating the body in a natural way.

Hypertonic seawater

Hypertonic seawater is 100% seawater without any alteration in its composition and concentration of mineral salts. It is called *hypertonic* because it has a higher salt concentration than our body fluids do. In fact, the concentration is four times higher than ours, with about 36 grams of salts per litre of water. When orally taken it tends to act as a laxative. Ideally, it may be drunk in the morning on an empty stomach. In these cases, it acts as a purgative of the gastrointestinal system and hardly any minerals are retained. However, this practice could be dangerous as it may potentially promote deep hydroelectrolytic alterations.

Deep sea water (DSW)

DSW is water obtained from a depth of more than 200 meters from the sea surface. It has two main characteristics: it is rich in minerals and essential nutrients and it is very clean and pure as it is less exposed to pollution than surface seawater. Around the world there are few special locations where

Table 1. Minerals present in DSW. Modified using data from Mac Takahashi, M., and Huang, P. (2012). μg: micrograms, ng: nanograms, pg: picograms.

Element	[conc] μg/kg	Element	[conc] ng/kg	Element	[conc] pg/kg	Element	[conc] pg/kg
Cl	19.350.000	**Mo**	10.000	**Ti**	13.000	**Ce**	700
Na	10.780.000	**U**	3.200	**W**	10.000	**Sc**	700
Mg	1.280.000	**V**	2.000	**Re**	7.800	**Sm**	600
S	898.000	**As**	1.200	**He**	7.600	**Sn**	500
Ca	412.000	**Ni**	480	**Ti**	6.500	**Ho**	360
K	399.000	**Zn**	350	**La**	5.600	**Lu**	230
Br-	67.000	**Kr**	310	**Ge**	5.500	**Be**	210
C	27.000	**Cs**	306	**Nb**	< 5.000	**Tm**	200
N	8.720	**Cr**	212	**Hf**	3.400	**Eu**	170
Sr	7.800	**Sb**	200	**Nd**	3.300	**Tb**	170
B	4.500	**Ne**	160	**Pb**	2.700	**Hg**	140
O	2.800	**Se**	155	**Ta**	< 2.500	**Rh**	80
Si	2.800	**Cu**	150	**Ag**	2.000	**Te**	70
F	1.300	**Cd**	70	**Co**	1.200	**Pd**	60
Ar	620	**Xe**	66	**Ga**	1.200	**Pt**	50
Li	180	**Fe**	30	**Er**	1.200	**Bi**	30
Rb	120	**Al**	30	**Yb**	1.200	**Au**	20
P	62	**Mn**	20	**Dy**	1.100	**Th**	20
I	58	**Y**	17	**Gd**	900	**In**	20
Ba	15	**Zr**	15	**Pr**	700	**Ru**	5

DSW spontaneously arises to the surface—a phenomenon called 'upwelling'. At present, several such places have been identified. The best-known ones are located at North-east coast of Japan, North-west coast of Chile, North-west coast of Spain (Galicia), New Zeeland, South Africa, among others (See https://www.worldatlas.com/oceans/upwellings.html).

DSW contains more than 70 minerals and trace elements (Table 1). Its consumption could be very suitable for the replacement of useful minerals for humans and to improve homeostasis. It is important to understand that the minerals and trace elements provided by DSW cannot be synthesized *de novo* by the body, but are essential for its proper functioning, so its intake as a supplement or in recovery drinks in conditions that the optimal amounts of these components are compromised may be an important resource (Harris et al 2019).

Seawater mechanisms of biological effects

Many components of seawater may be involved in various processes within the body (See Table 1), acting by different mechanisms. Seawater may benefit health and nutrition in the following ways. By bathing with it in a similar way as other balneotherapy techniques.

- The use of saline seawater for hygiene comes from the ancient past, and currently it is recommended to soothe nasal irritations in children and adults, as a mouthwash, for hydration of the eye, to clean-up and heal small wounds, and as a general antiseptic. The mechanisms underlying these uses are in a great part related to the high salt concentration, which has remarkable antiseptic, antiviral and antifungal (García 2007) properties, which have been proven in different studies, including *in vitro* assays and randomized clinical trials (Štanfel et al. 2022).

- Seawater can provide the body with minerals and trace elements needed for proper functioning of the organism. Currently a shortage of some minerals has been observed in many human populations around the world. The use of seawater components in human nutrition may be a goodway to solve this problem, given the natural abundance of minerals in seawater.
- It can help by modifying the gut microbiome, which will increase SCFA (Short-Chain Fatty Acids) production and availability and promote a healthier metabolic profile.
- It can modify the control metabolic organs, such as the liver and the pancreas. Seawater has been shown to improve blood lipid profile and glucose metabolism, and promote body weight lost.
- It can work as an antioxidant: it slows down cellular ageing. Its antioxidant effect has been shown to extend cell life. It has an important relevance in trials with deep seawater intakes in cardiovascular diseases, protecting the system in these pathophysiological problems. This effect was also shown to improve liver pathologies (Chen et al. 2013).

Seawater bathing on metabolism and food intake

Water bathing has been used as a therapeutic tool from ancient times when the availability of therapeutic methods was limited, and that constitutes the base of later known as balneology. From the beginning of this practice, bathing at the sea showed positive results. Bathing at sea may have similar affects to bathing in fresh water which is the base of thermalism and thalassotherapy.

The application of water on the skin promotes different effects in the whole body depending on the temperature of the water. Cold water would have potent anti-inflammatory effects on the muscle-skeletal system, promoting joint and muscle pain relief, vasoconstriction of the skin, increased heart rate and blood pressure and increasing the ventilation rate of the respiratory system. After bathing in cold water, there is a reflex vasodilation of the skin and general vascular vessels, increasing organ perfusion and muscle relaxation. This kind of bathing improves cardiovascular and respiratory functions, has anti-rheumatism and anti-inflammatory effects, reduces anxiety and promotes central nervous system wellness. On the other hand, hot water bathing has an immediate vasodilation effect promoting hypotension and relaxation of muscles. The combination or "contrast" of hot and cold waters at different temperatures may generate multiple effects in different organs and its metabolism. These thermal contrasts in water have been extensively used, especially in sports to reduce muscle fatigue and improve recovery (Hou et al. 2013), and to reduce joints inflammations and pain (Higgins et al. 2017).

In addition, exposure to low temperature is known to increase the thermogenic activity of the organism, mainly by stimulating the activity of the sympathetic nervous system (SNS) and the thyroid axis. Both systems together, SNS and thyroid axis, greatly augment energy expenditure by mobilizing energy substrates, especially lipids for fueling, and increasing basal cellular metabolism. The increased basal cellular metabolism consumes substrates to oxidize at the mitochondria and produces heat to maintain the body temperature inside the homoeothermic range. Thus, repeated bathing at cold water will increase energy expenditure and contribute to weight loss by consuming lipids and reducing adipose tissue depots (Esperland et al. 2022). Subsequently, this practice may increase the appetite. On the other hand, bathing in hot waters has shown a reduction of appetite because of the reduced thermogenesis for maintaining temperature homeostasis (Chen et al. 2021).

Salts diluted in mineral waters may interact directly with the skin and mucosa including the mouth, nose and eye conjunctiva. At that level, the first effect will be to sanitize the epithelia, dragging the dirt out (Bonnomet et al. 2016, Stanfel et al. 2022). The high salt content of seawater will have a potent antiseptic activity and reduce bacteria, virus and fungi (Gelen et al. 1978, Traissac et al. 1995).

However, depending on the salt concentration, the mineral diversity and the time of immersion in the water, there may be other osmotic mechanisms by which some electrolytes may enter the body through the skin and mucosa dragging some water from the body compartments. The input electrolytes may interact with specific ionic channels in nervous cells near the skin modulating the neural functions and modify the cellular process of apoptosis/necrosis involving osmotic mechanisms in multiple cell types. These mechanisms may underlie the effectiveness of seawater bathing to reduce skin inflammation, irritation and itching (Carbajo and Maraver 2018). It has been also reported that some electrolytes and other molecules (more than 100) may abandon the body (Carbajo and Maraver 2018), then having a significant depurative effect.

Interestingly, some studies have shown that bathing in deep seawater prevents alterations in the metabolism with no changes in blood lactate and pyruvate with respect to pre-bathing values. This contrasts with the results of bathing in tap water at similar temperatures where modifications clearly occur with respect to pre-bathing values (Tsuchiya et al. 2003). Moreover, bathing in hot DSW increases circulating CD8-lymphocytes, eventually improving immunological responses in comparison with shallow seawater bathing (Tsuchiya et al. 2003b). This might be attributed to the higher amount of scarce minerals in DSW. Although promising, these findings should be validated in large well-design trials which are eagerly awaited.

Seawater oral intake benefits to organs and pathologies

The use of seawater as a therapy is an ancient method recommended by the most influential Greek medics, such as Hippocrates. The health benefits of marine cures were commonly regarded also in Roman and Arabic civilizations and had convinced defenders like Galen and Avicenna (San Martín 1995). The practice of sea baths as a proposed cure for multiple illnesses was maintained in the Middle Ages. But it was when Richard Russel (1727–1805) who lived near Brighton (England) and published the book "A dissertation *On the use of water sea in the diseases of the glands*" that sea baths became popular. The book was very successful and the first sanatoriums and health centres came up on seashores to take advantage of seawater as a sanative mean.

More recently, René Quinton (1867–1925) saved hundreds of infants from gastroenteritis and infant cholera with oral administration of sea water. This act contributed enormously to reducing infant mortality and laying the scientific foundations of seawater therapeutic potential settle in "*L'Eau de mer, milieu organique*" (1904) (San José 2004). Since then, seawater has been extensively used orally to treat several diseases. We have summarized the most relevant basic and clinical studies done in this field in the following pages (See Table 2).

The mineral and biochemical content of sea water may have some potential nutritional applications when orally given due to the salt itself, but also due to some other micronutrient supplies coming from organic materials such as plankton, microalgae zooplankton, and fish eggs and larva. In addition, seawater carries hundreds of millions of bacteria, virus and fungi, which may contribute to modifying gut microbiota, in most of the cases in a beneficial manner (Takeuchi et al. 2020).

Deep seawater deserves a special mention because of its specific characteristics, mineral richness and purity. That is the reason why many recent studies about the nutritional and health effects of seawater have been carried out using deep seawater (DSW). It must be emphasized that seawater has to be diluted in order to be well tolerated when orally administered. Eventually, in some studies seawater was offered without previous dilution, in such cases the volumes given were very small and allowed simultaneous access to fresh water.

Table 2. Major mineral components of sea water and seaweed. Summarized from cited references (Qing Zhou et al. 2022, Buturi et al. 2021, Aguirre César et al. 1997, Studer et al. 2013).

Chlorine	Control of the extracellular fluids composition Acid-base balance. Constituent of gastric acid.
Sodium	Water balance and cellular volume control Control of the extracellular fluids composition Nerve, muscle and cardiac function. Brain function
Potassium	Nerve, muscle and cardiac function. Brain function, Healing repairing Immunological system. Sense of taste. Antioxidant enzyme cofactor.
Calcium	Constituent of bone sand teeth. Coagulation. Immunologic function. Nerve, muscle and cardiac function. Nerve function
Magnesium	Key regulator of energy metabolism Controls the composition of the intracellular environment. Enzyme cofactor. ATP-ase Coagulation. Immunologic function. Nerve, muscle and cardiac function. Nerve function Constituent of bone sand teeth.
Phosphorus	Essential component of nucleic acid materials. Enzyme activation. Energy metabolism. Constituent of bone sand teeth Acid-base balance
Iron	Constituent of haemoglobin and myoglobin. Enzyme cofactor.
Iodine	Thyroid hormone synthesis
Selenium	Thyroid hormone biosynthesis and metabolism. Co-factor of deiodases Antioxidant. Immunological system.
Zinc	Gene expression. Cell division and growth. Immunological and reproductive function.
Cobalt	Vitamin B12 component. Hemoglobin composition
Copper	Antioxidant enzyme cofactor. CNS function. Haemoglobin synthesis.
Sulphur	Constituent of proteins. Enzymatic cofactor
Fluor	Maintenance of bone and teeth structure.
Silicon	Collagen and bone tissue synthesis
Manganese	Enzyme activation Development of the skeletal system Energy metabolism The function of the nervous and immune systems Reproductive function Antioxidant

Effects in metabolism and chronic diseases

Different studies, both in animal models and in humans, suggest that the minerals present in DSW once intake have a systemic impact in the body, especially at the metabolic level, with significant effects that finally contribute to improve several pathologies and their prognosis.

Deep seawater has shown to reduce body fat and body weight. *In vitro* studies have shown that deep-sea water inhibits the expression of adipocyte differentiation genes by the downregulation of adipogenetic transcription factors (Yoo et al. 2009, Ha 2014), also preventing adipocyte proliferation. Drinking DSW enhances the gene expression of specific mitochondrial proteins involved in fatty acid oxidation and phosphorylation, which priories the use of lipids as cellular fuel, promotes fat metabolic consumption and consequently weight lost (Ha 2015). Obese subjects, who usually present an imbalance in the supply of nutrients, especially specific deficiencies of minerals and trace elements, when consuming diluted seawater showed lipolysis and weight lost (Fukui et al. 2022).

Some other studies have shown the effects of DSW in hyperlipidaemia and atherosclerosis. The consumption of DSW has also been related to an upregulation of the hepatic LDL receptor and CYP7A1 and to the inhibition of Acetyl-Coa Carboxylase and HMGCoa-reductase—enzymes involved in cholesterol metabolism and fatty acid synthesis. This enzymatic downregulation promotes a reduction in total cholesterol and in LDL-c while maintaining HDL-c levels (Fu et al. 2012). Together with the activation of AMPK (Sheu et al. 2013) and Act/GSK-3B pathways (He et al. 2023) in the liver, DSW reduces lipid synthesis. It also induces a reduction in blood triglycerides, inhibits differentiation of adipocytes and a downward modification of the lipid content of hepatocytes, which also reduces lipid peroxidation and consequent liver damage. This suggests that DSW consumption could have a preventive effect on the development of hepatic steatosis induced by a high-fat diet (Ha et al. 2014). In addition, DSW consumption induces mitochondrial biogenesis by upregulating PGC-1α (Ha et al. 2015), increases adiponectin secretion and reduces the secretion of *resistin*, offering some protection against obesity and creating a more favorable metabolic environment for the health of the person Unexpectedly, because of the high salt content, deep seawater could improve cardiovascular haemodynamics and lower blood pressure. A study in hypertensive rats supplemented with deep seawater during eight weeks, revealed lower blood pressure, both diastolic and systolic blood pressure, than in the control group (Chang et al. 2011). Seawater contains high amounts of magnesium, potassium and calcium. The effect of magnesium and potassium may be the underlying cause, at least in part, of lowering blood pressure. In hypertensive subjects commonly having elevated risk of hypopotassemia, both minerals may contribute to increase potassium levels directly, but also indirectly by suppressing the adrenergic activity and promoting natriuresis—sodium elimination by urine (Itoh et al. 1997). This hypotensive effect is accompanied by a decrease in heart rate indicating a general beneficial effect on the cardiovascular system (Katsuda et al. 2008, Hsu et al. 2011). Seawater intake may also reduce the risks of cardiovascular events preventing thrombotic disorders by inhibiting the release of type-1 plasminogen activator inhibitor from human vascular endothelial cells, as shown *in vitro* (Ueshima et al. 2003). In addition, DSW greatly increases the antioxidant capacity of the organism, by promoting the up-regulation and synthesis of *Superoxide Dismutase* and *Glutathione Peroxidase*, key enzymes that reduce the damage caused by Reactive Oxygen Species (ROS). DSW reduces malondialdehyde levels (a marker of oxidative stress) and the atherogenic index (Chen et al. 2013).

DSW also has anti-diabetic effects. In diabetes, DSW induces the lowering of blood glucose, improving glucose tolerance and suppressing hyperglycemia. This effect may depend on changes occurring in different organs. Analytical techniques for studying gene expression (RT-PCR) showed an improvement in the expression of hepatic genes involved in glycogenolysis and glucose handling (Ha et al. 2014). In the pancreas, DSW improves the recovery of the size of the Langerhans´ islets and increases insulin and glucagon secretion in animal experimental models of T2DM, likely due to the additional supply of magnesium (Jo et at. 2021). In the muscles of mice consuming seawater there is a significant increase of glucose uptake and oxidation and β-oxidation of lipids also (Ha et al. 2013). These findings are consistent with other *in vivo* studies, where there is an increase in plasma levels of adiponectin, a hormone involved in the body's energy metabolism, secreted by adipose tissue and involved in glucose and fatty acid metabolism (Hwang et al. 2009). Moreover, the mechanisms underlying the effect of seawater in glucose metabolism include the activation of the AMPK pathway, as described above, and its ergogenic effect on the synthesis of GLUT4, the major inducible transporter of glucose (Ha et al. 2014) in skeletal muscle, which helps reduce insulin resistance and blood glucose levels (Mohd Nani et al. 2016).

In addition, DSW reduces fatigue and accelerates muscle recovery. Different minerals of seawater are involved in this process: they are boron, magnesium, lithium, and rubidium (Mohd Nani et al. 2016). The considerable amount of magnesium in DSW and the balance with other salts allows its use in situations of nervousness. Magnesium helps to manage this psychological situations and thus seawater helps to reduce stress and anxiety (Mohd Nani et al. 2016).

The enhancement of the anti-inflammatory and antioxidant effects provided by DSW raises the hypothesis that it may be used in the prevention and treatment of inflammatory diseases, such as atopic dermatitis (Hataguchi et al. 2005). The improvement of this pathology would also be mediated by the restoration of the proper balance of minerals in the body. DSW may reduce the symptoms of atopic dermatitis and also alleviate allergic skin reactions by the reduction of IgE and histamine released (Bak et al. 2012). This means that DSW intake also has an immunomodulatory effect, and in fact, DSW intake may increase IgA secretion in mucosa and epithelia (Shiraishi et al. 2017). Moreover, DSW consumption may be beneficial in the treatment of other diseases with an inflammatory component, such as osteoporosis, both supplying the necessary minerals for the well-functioning of bone metabolism and increasing osteoblast proliferation and thus preserving bone mineral density (Liu et al. 2013).

The benefits of DSW in regulating cellular metabolism may have application in pathologies with altered metabolism as cancer. Studies carried out in breast cancer cell lines *in vitro* indicate that the mixture of minerals present in seawater, especially magnesium, could have a potential inhibitory effect in the metastasis process (Kim et al. 2013). More research is still needed using *in vivo* models, but the reported studies seem promising.

Effects in microbiota

Seawater is a natural laxative and stimulates the peristaltic movements of the gut and emptying of the bowel. At the intestinal level, it has been observed that the consumption of DSW relieves constipation and increases the content of Short-Chain Fatty Acids (SCFA) in feces, which suggests that the consumption of seawater drinks could improve the microbiota and consequently be beneficial. Deep seawater has proven antibacterial effects. It inhibits bacterial growth and mobility and reduces the amount of *Helicobacter pylori* colonising in the stomach. In addition, it can reduce the area of stomach ulcers thanks to the increased protection provided by selenium content and its antioxidant and antiapoptotic effects (Kawada and Takeuchi 2013). Moreover, seawater can promote the antibacterial activity of yoghurt when supplemented with deep seawater. In studies with experimental animals, it was observed that a group of mice given yoghurt supplemented with seawater had a higher bacterial count in the small intestine than mice eating plain yogurt (Kang et al. 2015). The changes in the regulation of the microbiota also generate notable effects in the improvement of T2DM, increasing the efficiency of insulin (He et al. 2023). This effect is related to an increase in intestinal bacteria Lactobacillaceae and Ruminococcaceae and a reduction of Bateroides and Clostridium (He et al. 2023). Another study, also in rats, combining deep seawater with fucoidan, a substance generated from seaweed, showed improved glycaemia, insulin sensitivity and lipid metabolism in a rat model of high-fat diet-induced obesity (Chang et al. 2017).

On the other hand, it was also noted in another study that deep seawater may contain pathogenic bacteria, virus and fungi that can negatively affect the gut microbiota or have direct pathogenic activity (Chu and Zhang 2023). This has been seriously considered by the food regulatory authorities of Europe. In the last thirty years they have stablished regular strict controls and a wide web of vigilance stations along all sea coasts to promptly detect sea water contamination with biological pathogens that may threaten human nutritional health.

Sports and exercise

Those situations demanding hydration strategy can range from the practice of physical activities and sports to different pathologies and clinical conditions. That is why in recent years several studies have been carried out in order to establish the impact of seawater and especially DSW consumption and its possible benefits in improving hydration. It should be noted that the consumption of DSW is safe as long as the appropriate water treatment and consumption protocols are taken into account, just as they are for any water suitable for human consumption. On the other hand, all the investigations conclude that if the hardness of the water does not exceed 1500 parts per million, it should be safe of risk for liver or kidney damage (Mohd Nani et al. 2016).

Deep sea water and sport performance

In recent years, seawater and especially deep sea water has been seen to have great potential as a hydration drink. Research shows that it improves post-exercise acute rehydration more than a placebo does, and also more than other drinks specifically designed for athletes (Aragón-Vela 2022). On the other hand, its ergogenic capacities are interesting: *in vivo* studies have shown that sea water accelerates the recovery of aerobic capacity and the power of the thigh muscles compared to a placebo drink after an effort test. In addition, a better post-exercise inflammatory response has also been observed, which would allow the body to adapt to exercise more efficiently (Saovieng et al. 2018).

The biochemical markers of fatigue and muscle damage by exercise are also improved with the consumption of DSW. Blood lactate concentrations are lower in some studies (Pérez-Turpin et al. 2017) along with blood urea nitrogen, lactate dehydrogenase, creatine kinase and malon-dialdehyde concentrations as compared to concentrations after the consumption of a placebo. This seems to indicate that the biological stress caused by the practice of physical exercise could be attenuated by the consumption of DSW, which would favor greater performance, faster recovery and greater subsequent biological adaptations. It has also been observed that cerebral hemodynamics during physical exercise is increased by sea water consumption (Wei CY et al. 2017).

Seawater and seaweeds as a cooking ingredient

Seawater may be used for cooking and food preparation in several ways:

- It can be used for preserving foods, especially those coming from the sea. In fact, the use of saturated salt liquids or brine was one of the first uses for preserving all kind of foods in human culinary as patent in ancient Egypt, Grece, and Rome. This method is used as a final preserving method too. Bathing in brine as a first step for later sun-drying for longer preservation of fishes and meats has also been practiced. In this way many of minerals and micronutrients present in the seawater are incorporated into the final food product.

- Freezing fish and seafood in seawater is a good alternative. And in addition, the easiest way for defrosting fish and seafood is soaking it in seawater for the required period. This preparation may prevent the use of additional salt for cooking these products later.
- The ancient use of seawater for replacing common salt in cooking has been recently rediscovered. The use of deep seawater in gastronomy is still small, but it has increased in recent years. Some influencer cooks prefer seawater as a medium of cooking as it adds to the flavor of some dishes. It is a frequent component of fish and seafood broth. Other cooking uses include its use in flour masses, thus enriching them with nutrients, flavor and aroma. Interestingly, the most recommended salt for cooking is marine salt obtained from directly drying seawater under the sun. This is an old industry and is still very popular. This marine salt has plenty of minerals present in seawater in optimal proportions, and it is especially recommended as a food additive to prevent iodine-deficient diseases such as hypothyroidism. Hypothyroidism is one of the most or perhaps the most prevalent metabolic disease in the world, affecting nearly one-sixth of the whole human population of which 90% are women.

Seawater is a good resource supplement for mineral deficits or to complement the mineral intake of a normal diet. Mineral intake can be supplemented either by consuming diluted seawater directly, or by using seawater as a food ingredient.

Seawater has been used for cooking since ancient times, for example in the context of the Galician Atlantic Diet where it is still used for cooking seafood. But it can also be used to prepare broths to enrich recipes, to cook rice or pasta, giving them a very special flavor and aroma and to make delicious soups. It can also be used to cook fish especially frozen fish to make up for the loss of aroma due to defrosting. There are also known techniques of soaking meat in seawater. This increases the palatability of the meat. Several studies have been carried out on fermented foods using seawater and it has been observed that it increases the production of functional metabolites and the availability of ions—two characteristics that improve the nutritional composition of fermented foods containing seawater (Lee 2015).

The organoleptic characteristics of seawater are (Morer et al. 2016):

Colour: Seawater should be colourless, even though the sea is normally painted blue. The exact tone of the water varies greatly depending on the concentration of solutes in it, the temperature, depth and the incidence of the sun at different times of the day. Colour variations in the oceans are therefore the result of physical and biological factors.

Odour: organic "sui generis". The smell of the sea is one of those smells that anchors itself in the olfactory memory in a definitive way. It is a fresh, vital smell and unmistakable. The mixture of chemical substances from salt, fish, algae or wet rocks, within this perfume, dimethyl sulphide makes the odor of deep sea water stand out.

Taste: clearly salty and mineral. This characteristic taste is due to the high concentration of salts, not only of sodium chloride, but the other minerals too. Its powerful flavor can be softened and even refined by diluting it to make it more pleasant and enriching.

Given all the benefits associated with seawater, the food industry is already beginning to develop different product lines that include DSW at some point in the preparation or manufacture. The best-known example of this are sport drinks that include DSW in their composition or drinks composed by DSW dilutes in water, designed not necessarily only for sport practice. Another example is its use in the production of yoghurts to enhance its properties and fermentation. Its benefits and organoleptic properties depend on the concentration/dilution of the product and the way in which is used.

Seaweeds and nutrition

Within thalassotherapy, algae are used as part of topical treatments (Legido et al. 2015). However, these marine plants have been part of some traditional culinary like Japanese or European Atlantic diets, with relevant nutritional value. In Celtic and Viking times they were used to make soups as a source of nutrients. Chewing sweet seaweed prevented scurvy. Seaweed has a high vitamin C content. In Galicia (Norwest region of Spain) and some other locations of the European Atlantic coast seaweed was traditionally used for various purposes, such as fertilizing land on which vegetables were later cultivated. This practice substantially improved the mineral and oligo nutrient content of vegetables harvested in those regions. And so, given its known nutritional characteristics, seaweed may be a recommended food in today's menus and regarded as a source of nutrients. Some seaweeds are already being used for human nutritional purposes (Table 3).

Seaweeds have an outstanding mineral composition: they contain calcium, phosphorus, magnesium, sodium, potassium, sulphur, iron, iodine, cobalt, zinc, silicon, manganese, copper and selenium. In addition, same algae contain significant amounts of proteins with high biological values. However, algae are difficult to digest and absorb, for example, nori, wakame and dulse. They also contain significant quantities of vitamins, both fat-soluble (vitamin A, vitamin D and vitamin E) and water-soluble (folates, vitamin B12 and vitamin C). Algae also contain a significant amount of fiber (El-Beltagi et al. 2022), and a large variety of bioactive compounds: Ω3 & Ω6 fatty acids, lectins, flavonols, anthocyanins, diterpenes, polyphenols, pigments as chlorophyll and fucoxanthin, polysaccharides as alginate, agar and carrageen, phycobiliproteins, etc. (See Jiménez et al. 2023 for extended review).

Currently, the consumption of seaweed become popular in haute cuisine restaurants and moderate amongst the general population. The challenge is to introduce seaweed into the daily diet. As explained above seaweed has several culinary applications, as stock or soup, starters or canapés, salads, mixed with cereals, with fish and seafood etc. This may be the reason for the increasing demand of seaweed. Following the EFSA (European Food Safety Authority) guidelines the daily recommendation is limited to 50 g/person/day for a safe consumption (EFSA 2011. https://doi.org/10.2903/j.efsa.2011.2097).

The use of seaweed has potential application in therapeutic diets:

- *In menus for vegans*: Seaweed has a high supply of minerals such as calcium and iodine and good biological value in terms of protein. Together with seawater, seaweed can provide the nutrients that are basically provided by sea products such as shellfish and fish which provide scarce vitamins as vitamin B12 and vitamin D.

Table 3. Seaweeds under current exploitation. The information was collected from official regional authority's websites from Galicia (Spain). (https://mar.xunta.gal/es/el-sector/planes-de-gestion-y-biodiversidad).

Trade name	Scientific name
Sea spaghetti	*Himanthalia elongata*
Kombu	*Laminaria ochroleuca* *Laminaria sacharina*
Wakame	*Undaria pinnatifida*
Nori	*Porphyra umbilicales* *Porphyra linearis*
Agar-agar	*Gelidium sesquipedale*
Dulse seaweed	*Palmaria palmata*
Irish moss	*Chondrus crispus*
Fucus	*Fucus vesiculosus* *Fucus spiralis*

- In people treated with competitive vitamin K anticoagulants. Current land vegetables such as brassicas and other green vegetables plants contain high amounts of vitamin K and promote the growth of vitamin K producing bacteria in the bowel. This may interfere with anti-coagulant therapy and force most of the patients to avoid the intake of green vegetables. Seaweeds provide significant amounts of fiber that has no vitamin K, and could be used as substitutes for land vegetables.
- *In overweight or obese people*: Seaweed has a significant satiating capacity. It provides fiber and micronutrients, and positively modulates metabolism and should therefore be a common component of weight-loss diets.
- *As part of the diet for hypothyroidism*: Seaweed has a high amount of iodine, which is a perfect adjuvant to stimulate the thyroid gland.
- *In people with malnutrition or eating difficulties:* Being nutritious and having a special flavor helps seaweed increase the nutritional density of a menu and enhance its flavor to stimulate intake.

Seawater health risks and contraindications

Desalinated water for human drinking and hydration

Regarding the ever increasing need for water resources for human use, most countries are looking at the sea as an inexhaustible depot of water. The high content of salt in sea water requires desalination before it is fit for human consumption. Sea water purification and desalination methods are currently very efficient and becoming more affordable, thus in the near future an important proportion of drinking water in costal countries will come from desalination plants at sea sides. Already about 50% of the drinking water in Israel is from sea water and other countries like Australia and some Gulf countries are also using sea water (Schiermeier 2008). Fresh tap water contains little but significant amounts of minerals which contribute to our daily nutritional needs. Paradoxically, desalinated sea water is practically free not only of undesirable sodium salts, but also of those scarce but needed minerals, such as Mg^{2+} and Ca^{2+}, or Iodine. Some studies have shown that usual consumption of desalinated water reduces Mg^{2+} levels in blood, which might be linked to increased risks of cardiovascular events and mortality in susceptible populations (Koren 2017, Shlezinger 2016, 2018, 2019). The role of hypomagnesemia in increasing the prevalence of dementia in populations drinking desalinated water is under review (Ben Zaken 2020). Moreover, those populations currently using desalinated seawater as a major source for drinking are at higher risk of iodine-deficiency disorders and thyroid diseases (Ovadia 2016). In addition, the long-term consumption of desalinated water would give rise to hydro electrolyte disturbances including hyponatremia, hypokalaemia, hypomagnesemia and hypocalcaemia. The hydroelectrolyte imbalance would in turn trigger the mechanisms underlying cancer pathogeny as increased oncogenes expression and cellular proliferative signaling, and reduced apoptosis (Nriagu 2016). The use of desalinated water for agriculture irrigation and animal feeding may affect crop growth and cattle health. Cattle may become prone to joint issues, aggravated by the deficient supply of those oligo minerals that are essential for humans too. Certainly, all those mineral deficiencies might be corrected with specific mineral salt supplementation to the desalinated water for drinking and agricultural uses, which is the current recommendation (Avni 2013).

Seawater risks in nutrition and health

Plastics and heavy metals

Beyond desalination, sea water and sea products and foods may contain significant amounts of pollutants. In recent years, the presence of plastic particles and micro plastics have been pointed out as the major contaminants of the seas.

Plastics used by humans frequently find their way to the seas and oceans. It is certainly known that the plastic floating island in the middle of the Pacific Ocean has an extension that is three-fold higher than Spain in size and accounts for above 80.000 tons of plastics, half of which are said to be fishing nets (Lebreton 2018). In the rest of the seas the UN estimates a mean of 46.000 floating plastic pieces per square mile. A proportion of all that plastic is minced to smaller sizes and finally very fine particles of plastics are in suspension in the water of all the seas. Those particles are ingested by sea organisms and accumulate in the trophic chain including seafood and fishes consumed by humans.

Not just plastics but other pollutants may also be in equal amounts in seawater. In fact, hydrocarbons and other derivates of petroleum are also present in all seas, most of them coming from industrial and urban residuals and waste. In addition, in recent years there is an increasing concern about the amounts of heavy metals (methyl-mercury, plumb, cadmium, …) found in some sea products and fishes. Taking all this into account, alarm bells have been ringing regarding the safety of sea food products (De la Torre 2020). At present, strict food regulations established in occidental countries warranty safety of all sea foods for general human nutrition, but we have to keep in mind all these concerns when recommending seawater and algae for human intake and agricultural purposes (Rist 2018).

In this sense, deep sea water has the lowest quantity of micro plastics and pollutants, and the highest content and variety of minerals—thus it is the most recommended source for sea-water derived products.

Health contraindications to seawater

Direct consumption of seawater is physiologically prohibited except for very low daily amounts, that too in emergencies and that too for short periods. In fact, direct oral administration of undiluted seawater use can hardly be tolerated by humans and promotes immediate gastrointestinal discomfort, nausea and vomiting. On being absorbed, the salt present in seawater disturbs hydro-electrolyte control of body fluids worsening dehydration and producing cramps, muscle malfunctioning, confusion, central nervous distortion, cardiovascular and kidneys failure and finally death.

The high salt content of sea water, even when used in small amounts in the context of thalassotherapy, must be avoided in several pathophysiological conditions and diseases. For example, seawater should be always avoided if there is a risk of dehydration, kidney disease or hypertension.

On the other hand, and in the light of recent studies, desalinated sea water as a major source for drinking water might not eventually be recommended to subjects with hypothyroidism or in populations with low dietary iodine intake, as it can be a contributing factor in developing iodine-deficient diseases. At this point, it has to be highlighted that more than 1.200 million people along the world are suffering from hypothyroidism because of deficient iodine dietary intake, and most of them (90%) are women. This is a very important concern for public health politics for the World Health Organisation (de Benoist 2004). It may also be important to keep subjects who are at high risk of cardiovascular diseases and metabolic syndromes away from desalinated water. It should be completely avoided in the case of cancer patients.

Concluding remarks

The intake of seawater dilutions will have relevant benefits to different physiological conditions as exercise hydration and recovery, and pathologies as obesity and weight losing, diabetes, cardiovascular diseases, gut dysbiosis, iodine-deficient disease and some other mineral deficits, chronic inflammatory diseases and even as coadjutant in cancer nutrition. The research on its effects and benefits increases every year and we are already reaching deep points of understanding, not only to describe effects, but also to explain the action mechanisms that underlie the physiological relationships between the intake of DSW and its attributed benefits. It is likely that in the coming years we will continue to see an increase in the use of seawater drinks, both as a hydration drink and to take advantage of its multiple healthy benefits and ergogenic effects. The increasing use of desalinated seawater for tap-water in many human communities will augment the risk of mineral deficits with corresponding associated pathology. A great concern is the increasing evidence of contamination of all the seas with plastics, heavy metals, petroleum and industrial residuals that may limit the use of marine resources for human uses including drinking and agriculture. In this context the upwellings of deep sea water is an enormous valuable resource of clean, pure and reach in minerals seawater that should be reasonable exploited and consciously preserved.

References

Aragón-Vela, J., O. González-Acevedo, J. Plaza-Diaz, R.A. Casuso and J.R. Huertas. 2022. Physiological benefits and performance of sea water ingestion for athletes in endurance events: A systematic review. Nutrients 14(21): 4609.

Avni, N., M. Eben-Chaime and G. Oron. 2013. Optimizing desalinated sea water blending with other sources to meet magnesium requirements for potable and irrigation waters. Water Res. 2013 May 1; 47(7): 2164–76. doi: 10.1016/j.watres.2013.01.018.

Bak, J.P., Y.M. Kim, J. Son, C.J. Kim and E.H. Kim. 2012. Application of concentrated deep sea water inhibits the development of atopic dermatitis-like skin lesions in NC/Nga mice. BMC Complement Altern Med. 2012 Jul 26; 12: 108. doi: 10.1186/1472-6882-12-108. PMID: 22834904; PMCID: PMC3517761.

Ben Zaken, S., O. Simantov, A. Abenstein, Z. Radomysky and G. Koren. 2020. Water desalination, serum magnesium and dementia: a population-based study. J. Water Health. 2020 Oct; 18(5): 722–727. doi: 10.2166/wh.2020.132.

Benner, R. and R.M.W. Amon. 2014. The size-reactivity continuum of major bioelements in the ocean. Ann. Rev. Mar. Sci. 2015; 7: 185–205. doi: 10.1146/annurev-marine-010213-135126. Epub 2014 Jul 25. PMID: 25062478.

Bonnomet, A., E. Luczka, C. Coraux and L. de Gabory. 2016. Non-diluted seawater enhances nasal ciliary beat frequency and wound repair speed compared to diluted seawater and normal saline. Int. Forum. Allergy Rhinol. 2016 Oct; 6(10): 1062–1068. doi: 10.1002/alr.21782. Epub 2016 Apr 22. PMID: 27101776; PMCID: PMC5074274.

Carbajo, J.M. and F. Maraver. 2018. Salt water and skin interactions: new lines of evidence. Int. J. Biometeorol. 2018 Aug; 62(8): 1345–1360. doi: 10.1007/s00484-018-1545-z. Epub 2018 Apr 19. PMID: 29675710.

Chang, M.H., B.-S. Tzang, T.-Y. Yang, Y.-C. Hsiao, H.-C. Yang and Y.-C. Chen. 2011. Effects of deep-seawater on blood lipids and pressure in high-cholesterol dietary mice. Journal of Food Biochemistry 35(1): 241–259.

Chang, W.T., T.Y. Lu, M.C. Cheng, H.C. Lu, M.F. Wu and C.L. Hsu. 2017. Deep sea water improves abnormalities in lipid metabolism through lipolysis and fatty acid oxidation in high-fat diet-induced obese rats. Marine Drugs 15(12): 386. PMID: 29232925 PMCID: PMC5742846 DOI: 10.3390/md15120386.

Chen, I.S., Y.Y. Chang, C.L. Hsu, H.W. Lin, M.H. Chang, J.W. Chen et al. 2013. Alleviative effects of deep-seawater drinking water on hepatic lipid accumulation and oxidation induced by a high-fat diet. Journal of the Chinese Medical Association 76(2): 95–101.

Chen, Z., J. Yang, L. Zhu, A. Zhang, Z. Wang, P. Liu et al. 2021. Classification of typical hot springs and their relationship with health in Guizhou, China. Environ Geochem Health. 2021 Mar; 43(3): 1287–1304. doi: 10.1007/s10653-020-00724-x. Epub 2020 Sep 24. PMID: 32974884.

Chu, M. and X. Zhang. 2023. Alien species invasion of deep-sea bacteria into mouse gut microbiota. J. Adv. Res. 2023 Mar; 45: 101–115. doi: 10.1016/j.jare.2022.05.011. PMID: 35690372; PMCID: PMC10006512.

de Benoist, B., M. Andersson, I. Egli, B. Takkouche and H. Allen. 2004. Iodine status worldwide WHO Global Database on Iodine Deficiency. ISBN: 9241592001.

De-la-Torre, G.E. 2020. Microplastics: an emerging threat to food security and human health. J. Food Sci. Technol. 2020 May; 57(5): 1601–1608. doi: 10.1007/s13197-019-04138-1. Epub 2019 Oct 19. PMID: 32327770; PMCID: PMC7171031.

EFSA. 2011. Use of the EFSA comprehensive european food consumption database in exposure assessment. EFSA Journal 9(3): 2097. https://doi.org/10.2903/j.efsa.2011.2097.

El-Beltagi, H.S., A.A. Mohamed, H.I. Mohamed, K.M.A. Ramadan, A.A. Barqawi and A.T. Mansour. 2022. Phytochemical and potential properties of seaweeds and their recent applications: A review. Mar Drugs. 2022 May 24; 20(6): 342. doi: 10.3390/md20060342. PMID: 35736145; PMCID: PMC9227187.

Esperland, D., L. de Weerd and J.B. Mercer. 2022. Health effects of voluntary exposure to cold water—a continuing subject of debate. Int. J. Circumpolar Health. 2022 Dec; 81(1): 2111789. doi: 10.1080/22423982.2022.2111789. PMID: 36137565; PMCID: PMC9518606.

Fu, Z.Y., F.L. Yang, H.W. Hsu and Y.F. Lu. 2012. Drinking deep seawater decreases serum total and low-density lipoprotein–cholesterol in hypercholesterolemic subjects. Journal of Medicinal Food 15(6): 535–541.

Fukui, K., Y. Suzuki, Y. Kato, N. Takeuchi, H. Takenaka and M. Kohno. 2022. Effect of extract-added water derived from deep-sea water with different hardness on cognitive function, motor ability and serum indexes of obese mice. Nutrients. 2022 Apr 25; 14(9): 1794. doi: 10.3390/nu14091794. PMID: 35565761; PMCID: PMC9101427.

García, A. 2007. Manual del náufrago. Sobrevivencia en los océanos utilizando el agua de mar como hidratante y nutriente. Miami Florida: Resident. Alien Books.

Gelen, A.M., I.E. Mishustina, S.A. Gulevskaya, N.V. Pechnikov and L.A. Ledova. 1978. Antibacterial activity of seawater and microvibrion-predators (*Microvibrio marinus Roscoffensis*) in it. Biol. Bull. Acad. Sci. USSR. 1978 Nov.–Dec. 5(6): 690–5. PMID: 387093.

Ha, B.G., J.E. Park, H.J. Cho and Y.H. Shon. 2015. Stimulatory effects of balanced deep sea water on mitochondrial biogenesis and function. PONE. 2015 Jun 12; 10(6): e0129972. doi: 10.1371/journal.pone.0129972. PMID: 26068191; PMCID: PMC4466323.

Ha, B.G., J.E. Park, E.J. Shin and Y.H. Shon. 2014. Effects of balanced deep-sea water on adipocyte hypertrophy and liver steatosis in high-fat, diet-induced obese mice. Obesity (Silver Spring). 2014 Jul; 22(7): 1669–78. doi: 10.1002/oby.20740. Epub 2014 Mar 25. PMID: 24634394.

Ha, B.G., J.E. Park, E.J. Shin and Y.H. Shon. 2014. Modulation of glucose metabolism by balanced deep-sea water ameliorates hyperglycemia and pancreatic function in streptozotocin-induced diabetic mice. PONE. 2014 Jul 11; 9(7): e102095. doi: 10.1371/journal.pone.0102095. PMID: 25013896; PMCID: PMC4094501.

Ha, B.G., E.J. Shin, J.E. Park and Y.H. Shon. 2013. Anti-diabetic effect of balanced deep-sea water and its mode of action in high-fat diet induced diabetic mice. Mar Drugs. 2013 Oct 29; 11(11): 4193–212. doi: 10.3390/md11114193. PMID: 24172214; PMCID: PMC3853723.

Harris, P.R., D.A. Keen, E. Constantopoulos, S.N. Weninger, E. Hines, M.P. Koppinger et al. 2019. Fluid type influences acute hydration and muscle performance recovery in human subjects. Journal of the International Society of Sports Nutrition 16(1): 15.

Hataguchi, Y., H. Tai, H. Nakajima and H. Kimata. 2005. Drinking deep-sea water restores mineral imbalance in atopic eczema/dermatitis syndrome. European Journal of Clinical Nutrition 59(9): 1093–1096.

He, S., W.B. Peng, H.L. Zhou, X.J. Fu, Y.H. Sun and Z.G. Wang. 2023. A combination of deep-sea water and fucoidan alleviates T2DM through modulation of gut microbiota and metabolic pathways. Pharmaceuticals (Basel). 2023 Mar 20; 16(3): 462. doi: 10.3390/ph16030462. PMID: 36986561; PMCID: PMC10053583.

Higgins, T.R., D.A. Greene and M.K. Baker. 2017. Effects of cold water immersion and contrast water therapy for recovery from team sport: A systematic review and meta-analysis. J. Strength Cond. Res. 2017 May; 31(5): 1443–1460. doi: 10.1519/JSC.0000000000001559. PMID: 27398915.

Hou, C.W., Y.S. Tsai, W.H. Jean, C.Y. Chen, J.L. Ivy, C.Y. Huang et al. 2013. Deep ocean mineral water accelerates recovery from physical fatigue. Journal of the International Society of Sports Nutrition 10(1): 7.

Hsu, C.L., Y.Y. Chang, C.H. Chiu, K.T. Yang, Y. Wang, S.G. Fu et al. 2011. Cardiovascular protection of deep-seawater drinking water in high-fat/cholesterol fed hamsters. Food Chem. 127(3): 1146–1152.

Hwang, H.S., H.A. Kim, S.H. Lee and J.W. Yun. 2009. Anti-obesity and antidiabetic effects of deep sea water on ob/ob mice. Mar. Biotechnol. (NY). 2009 Jul–Aug; 11(4): 531–9. doi: 10.1007/s10126-008-9171-0. Epub 2008 Dec 13. PMID: 19083059.

Itoh, K., T. Kawasaki and M. Nakamura. 1997. The effects of high oral magnesium supplementation on blood pressure, serum lipids and related variables in apparently healthy Japanese subjects. British Journal of Nutrition 78(5): 737–750.

Jiménez-González, C., A.M. Torrado, F. Mallo, M.L. Rua and C. Fuciños. 2023. Red seaweed proteins: valuable marine-origin compounds with encouraging applications. Algal Res. In press. ALGAL-D-23-00425.

Jo, S.M., J. Nam, S.Y. Park, G. Park, B.G. Kim, G.H. Jeong et al. 2021. Effect of mineral-balanced deep-sea water on kidney function and renal oxidative stress markers in rats fed a high-salt diet. Int. J. Mol. Sci. 2021 Dec. 14; 22(24): 13415. doi: 10.3390/ijms222413415. PMID: 34948210; PMCID: PMC8705929.

Kang, S.M., J.W. Jhoo, J.I. Pak, I.K. Kwon, S.K. Lee and G.Y. Kim. 2015. Effect of yogurt containing deep sea water on health-related serum parameters and intestinal microbiota in mice. J. Dairy Sci. 2015 Sep; 98(9): 5967–73. doi: 10.3168/jds.2015-9492. Epub 2015 Jul 2. PMID: 26142849.

Kang, S.M., J.W. Jhoo, J.I. Pak, I.K. Kwon, S.K. Lee and G.Y. Kim. 2015. Effect of yogurt containing deep sea water on health-related serum parameters and intestinal microbiota in mice. Journal of Dairy Science 98(9): 5967–5973.

Katsuda, S., T. Yasukawa, K. Nakagawa, M. Miyake, M. Yamasaki, K. Katahira et al. 2008. Deep-sea water improves cardiovascular hemodynamics in Kurosawa and Kusanagi-Hypercholesterolemic (KHC) rabbits. Biol. Pharm. Bull. 2008 Jan; 31(1): 38–44. doi: 10.1248/bpb.31.38. PMID: 18175939.

Kim, S., S.Y. Chun, D.H. Lee, K.S. Lee and K.S. Nam. 2013. Mineral-enriched deep-sea water inhibits the metastatic potential of human breast cancer cell lines. International journal of Oncology 43(5): 1691–1700.

Koren, G., M. Shlezinger, R. Katz, V. Shalev and Y. Amitai. 2017. Seawater desalination and serum magnesium concentrations in Israel J Water Health. 2017 Apr. 15(2): 296–299. doi: 10.2166/wh.2016.164.

Lebreton, L., B. Slat, F. Ferrari, B. Sainte-Rose, J. Aitken, R. Marthouse et al. 2018. Evidence that the great pacific garbage patch is rapidly accumulating plastic. Sci. Rep. 2018 Mar 22; 8(1): 4666. doi: 10.1038/s41598-018-22939-w. PMID: 29568057; PMCID: PMC5864935.

Lee, C.L. 2015. The advantages of deep ocean water for the development of functional fermentation food. Appl. Microbiol. Biotechnol. 2015 Mar.; 99(6): 2523–31. doi: 10.1007/s00253-015-6430-7. Epub 2015 Feb 8. PMID: 25661817.

Liu, H.Y., M.C. Liu, M.F. Wang, W.H. Chen, C.Y. Tsai, K.H. Wu et al. 2013. Potential osteoporosis recovery by deep sea water through bone regeneration in SAMP8 mice. Evid Based Complement Alternat Med., vol. 2013, Article ID 161976.

Lucchetta, M.C., G. Monaco, V.I. Valenzi, M.V. Russo, J. Campanella, S. Nocchi et al. 2007. Le basi storico-scientifiche della talassoterapia: stato dell'arte [The historical-scientific foundations of thalassotherapy: state of the art]. Clin. Ter. 2007 Nov.–Dec. 158(6): 533–41. PMID: 18265720.

Kawada, M. and H. Takeuchi. 2013. Antibacterial activities of refined deep seawater on Helicobacter pylori. Journal of Genetic Syndromes & Gene Terapy s1: 2–8.

Mac Takahashi, M. and P. Huang. 2012. Novel renewable natural resource of Deep Ocean Water (DOW) and their current and future practical applications. Kuroshio. Sci. 2000, 101–13.

Mohd Nani, S.Z., F.A. Majid, A.B. Jaafar, A. Mahdzir and M.N. Musa. 2016. Potential health benefits of deep sea water: A review. Evid Based Complement Alternat. Med. 6520475. doi: 10.1155/2016/6520475. Epub 2016 Dec 26. PMID: 28105060; PMCID: PMC5221345.

Moran, M.A. 2015. The global ocean microbiome. Science. 2015 Dec. 11; 350(6266): aac8455. doi: 10.1126/science.aac8455. PMID: 26659059.

Morer, C. 2016. Talasoterapia. Bol. Soc. Esp. Hidrol. Med. 31(2): 119–146. DOI: 10.23853/bsehm.2017.0209.

Nriagu, J., F. Darroudi and B. Shomar. 2016. Health effects of desalinated water: Role of electrolyte disturbance in cancer development. Environ. Res. 2016 Oct. 150: 191–204. doi: 10.1016/j.envres.2016.05.038. Epub 2016 Jun 10.

Ovadia, Y.S., D. Gefel, D. Aharoni, S. Turkot, S. Fytlovich and A.M. Troen. 2016. Can desalinated seawater contribute to iodine-deficiency disorders? An observation and hypothesis. Public Health Nutr. 2016 Oct; 19(15): 2808–17. doi: 10.1017/S1368980016000951.

Pérez-Turpin, J.A., M. Trottini, J.J. Chinchilla-Mira and W. Cyganik. 2017. Effects of seawater ingestion on lactate response to exercise in runners. Biology of Sport 34(4): 407–412.

Rist, S., B. Carney Almroth, N.B. Hartmann and T.M. Karlsson. 2018. A critical perspective on early communications concerning human health aspects of microplastics. Sci. Total Environ. 2018 Jun 1; 626: 720–726. doi: 10.1016/j.scitotenv.2018.01.092. Epub 2018 Feb 19. PMID: 29396337.

San José, J.C. 2004. René Quinton: centenario de "L'eau de mer, milieuorganique". Bol. Soc. Esp. Hidrol. Med. 19(1): 17–22.

San Martin, J. 1995. Talasoterapia: proyección terapéutica actual. An. R. Acad. Nac. Med. (Madr) 112(2): 347–79.

Saovieng, S., J. Wu, C.Y. Huang, C.L. Kao, M.F. Higgins, R. Chuanchaiyakul et al. 2018. Deep ocean minerals minimize eccentric exercise-induced inflammatory response of rat skeletal muscle. Frontiers in Physiology 9: 1351.

Schiermeier, Q. 2008. Water: purification with a pinch of salt. Nature. 2008 Mar. 20; 452(7185): 260-1. doi: 10.1038/452260a.

Sheu, M.J., P.Y. Chou, W.H. Lin, C.H. Pan, Y.C. Chien, Y.L. Chung et al. 2013. Deep sea water modulates blood pressure and exhibits hypolipidemic effects via the AMPK-ACC pathway: an *in vivo* study. Marine Drugs 11(6): 2183–2202.

Shiraishi, H., M. Fujino, N. Shirakawa, N. Ishida, H. Funato, A. Hirata et al. 2017. Effect of minerals on intestinal IgA production using deep sea water drinks. Biological and Pharmaceutical Bulletin 40(10): 1700–1705.

Shlezinger, M., Y. Amitai, A. Akriv, H. Gabay, M. Shechter and M. Leventer-Roberts. 2018. Association between exposure to desalinated sea water and ischemic heart disease, diabetes mellitus and colorectal cancer; A population-based study in Israel. Environ Res. 2018 Oct; 166: 620–627. doi: 10.1016/j.envres.2018.06.053.

Shlezinger, M., Y. Amitai, I. Goldenberg, S. Atar and M. Shechter. 2019. Acute myocardial infarction severity, complications, and mortality associated with lack of magnesium intake through consumption of desalinated seawater. Magnes. Res. 2019 May 1; 32(2): 39–50. doi: 10.1684/mrh.2019.0449.

Shlezinger, M., Y. Amitai, I. Goldenberg and M. Shechter. 2016. Desalinated seawater supply and all-cause mortality in hospitalized acute myocardial infarction patients from the Acute Coronary Syndrome Israeli Survey 2002–2013. Int. J. Cardiol. 2016 Oct. 1; 220: 544–50. doi: 10.1016/j.ijcard.2016.06.241. Epub 2016 Jun 29. PMID: 27393841.

Štanfel, D., L. Kalogjera, S.V. Ryazantsev, K. Hlača, E.Y. Radtsig, R. Teimuraz et al. 2022. The role of seawater and saline solutions in treatment of upper respiratory conditions. Mar Drugs. 2022 May 17; 20(5): 330. doi: 10.3390/md20050330. PMID: 35621981; PMCID: PMC9147352.

Studer, J.M., W.P. Schweer, N.K. Gabler and J.W. Ross. 2022. Functions of manganese in reproduction. Anim. Reprod. Sci. 2022; 238: 106924. doi: 10.1016/j.anireprosci.2022.106924. Epub 2022 Jan 18. PMID: 35121412.

Takeuchi, H., K. Higuchi, Y. Yoshikane, R. Takagi, S. Tokuhiro, K. Takenaka et al. 2020. A randomized double-blind controlled trial. Nutrients 12(9): 2646.

Traissac, L., L. Bordenave, R. Bareille et al. 1995. *In vitro* study of the effect of sea water by-products on the respiratory mucosa. Rev. Soc. Fr ORL. 32: 43–49.

Tsuchiya, Y., T. Shimizu, T. Tazawa, K. Nakamura and M. Yamamoto. 2003. Effects of hot deep seawater bathing on the immune cell distribution in peripheral blood from healthy young men. Environ. Health Prev. Med. 2003 Nov. 8(5): 161–5. doi: 10.1007/BF02897909. PMID: 21432093; PMCID: PMC2723508.

Tsuchiya, Y., T. Shimizu, T. Tazawa, N. Shibuya, K. Nakamura and M. Yamamoto. 2003b. Changes in plasma lactate and pyruvate concentrations after taking a bath in hot deep seawater. Tohoku. J. Exp. Med. 2003b Dec. 201(4): 201–11. doi: 10.1620/tjem.201.201. PMID: 14690012.

Ueshima, S., H. Fukao, K. Okada and O. Matsuo. 2003. Suppression of the release of type-1 plasminogen activator inhibitor from human vascular endothelial cells by Hawaii deep sea water. Pathophysiology. 2003 Jan. 9(2): 103–109. doi: 10.1016/s0928-4680(02)00076-7. PMID: 14567942.

Wei, C.Y., C.Y. Chen, Y.H. Liao, Y.S. Tsai, C.Y. Huang, R. Chaunchaiyakul et al. 2017. Deep ocean mineral supplementation enhances the cerebral hemodynamic response during exercise and decreases inflammation postexercise in men at two age levels. Frontiers in Physiology 8: 1016.

Yoo, Y.G., Y.S. Chu, Y.H. Shon, K.S. Nam and J.W. Yun. 2009. Inhibitory effect of deep-sea water on differentiation of 3T3-L1 adipocytes. Mar Biotechnol (NY). 2009 Mar–Apr; 11(2): 161–8. doi: 10.1007/s10126-008-9131-8. Epub 2008 Jul 25. PMID: 18654820. Ha BG, Park JE, Cho.

Chapter 10
Introduction to Cosmeceuticals

Haresh S. Kalasariya,[1] *Carlos Eliel Maya-Ramírez,*[2,*]
Ujashkumar A. Shah[4] and *Virendra Yadav*[3]

Basics of cosmeceuticals

Cosmeceuticals represent a bridge between cosmetics and pharmaceuticals, offering enhanced skincare benefits beyond traditional cosmetic products. Scientifically, cosmeceuticals are defined as topical formulations containing active ingredients that possess therapeutic properties beneficial to the skin (Bellad et al. 2017). These products are formulated to penetrate the skin's surface, targeting specific skin concerns at a cellular level. Unlike conventional cosmetics, cosmeceuticals undergo rigorous scientific research and testing to validate their efficacy and safety (Morganti et al. 2021). The efficacy of cosmeceuticals relies on the active ingredients they contain, which may include antioxidants, peptides, retinoids, vitamins, and botanical extracts. These ingredients work synergistically to address various skin issues such as aging, hyperpigmentation, acne, and inflammation (Lin and Fisher 2007). Moreover, cosmeceuticals often incorporate advanced delivery systems to optimize the penetration of active compounds into the skin, ensuring maximum efficacy. Cosmeceuticals play a crucial role in modern skincare regimens, offering consumers the opportunity to achieve noticeable improvements in skin health and appearance. However, it's essential to note that while cosmeceuticals may provide therapeutic benefits, they are not regulated as drugs and do not undergo the same level of scrutiny by regulatory agencies. Therefore, consumers should exercise caution and choose cosmeceutical products backed by scientific evidence and manufactured by reputable companies (Chatur et al. 2021, Goyal et al. 2022, Rathod et al. 2020).

[1] Centre for Natural Products Discovery, School of Pharmacy and Biomolecular Sciences, Liverpool John Moores University, Byrom Street, Liverpool L3 3AF, UK; Faculty of Health and Life Sciences, INTI International University, Persiaran Perdana BBN, Putra Nilai, 71800 Nilai, Negeri Sembilan, Malaysia. Email: Hareshahir22@gmail.com
[2] Laboratorio de Diagnóstico molecular, Departamento de Bioquímica, Escuela Nacional de Ciencias Biológicas, Instituto Politécnico Nacional, Carpio y Plan de Ayala S/N, Colonia, Casco de Santo Tomás, C.P., 11340 Ciudad de México, México.
[3] Nootan Pharmacy College, Sankalchand Patel Vidyadham, Ambaji-Gandhinagar State Highway, Visnagar-384315. Dist: Mehsana, Gujarat, India. Email: shahujashkumar@gmail.com
[4] Department of Life Sciences, Hemchandracharya North Gujarat University, Patan, Gujarat, India.
Email: yadava94@gmail.com
* Corresponding author: emayar1800@alumno.ipn.mx

Historical background

Cosmeceuticals have a rich historical lineage deeply rooted in ancient civilizations where natural remedies and botanical extracts were utilized for skincare purposes (Kiki 2023). Historical records from ancient Egypt, Greece, and Rome demonstrate the use of various plant-derived substances and minerals in skincare preparations, indicating an early recognition of the potential therapeutic benefits of topical applications (Mahesh et al. 2019). The concept of cosmeceuticals as a distinct category began to gain prominence in the late 20th century with advancements in cosmetic science and biotechnology (Zappelli et al. 2016, Sajna et al. 2015). In the 1980s, the term "cosmeceutical" was coined to describe products that straddled the boundary between cosmetics and pharmaceuticals, offering therapeutic benefits beyond mere cosmetic enhancement (Heymann 1997, Griffiths 2010). The evolution of cosmeceuticals has been fueled by scientific discoveries elucidating the mechanisms of skin aging, damage, and disease, leading to the development of targeted formulations with specific bioactive ingredients (Zhou et al. 2021). This scientific approach has transformed skincare from a purely cosmetic endeavor to a science-driven discipline aimed at improving skin health and addressing dermatological concerns (Alves et al. 2020). Advancements in ingredient technology, such as the synthesis of retinoids, antioxidants, peptides, and growth factors have further propelled the growth of the cosmeceutical industry. These breakthroughs have expanded the repertoire of ingredients available to formulators, enabling the creation of highly efficacious products with measurable clinical benefits (Oliveira et al. 2022). Today, cosmeceuticals occupy a prominent position in the skincare market, offering consumers a wide array of products targeting various skin concerns, including aging, hyperpigmentation, acne, and sensitivity. The demand for cosmeceuticals continues to grow as consumers seek evidence-based solutions backed by scientific research and clinical validation (Tambo et al. 2016). The utilization of natural resources-derived cosmeceuticals represents a burgeoning area of interest in skincare research and product development (Mapelli-Brahm et al. 2023). These cosmeceuticals harness bioactive compounds sourced from botanicals, marine organisms, and minerals, exhibiting diverse physiological effects on the skin. Plant-derived extracts, rich in polyphenols and flavonoids, demonstrate antioxidant and anti-inflammatory properties, while botanical oils provide moisturization and replenishment of skin lipids (Dini and Laneri 2021, Korać and Khambholja 2011, Stallings and Lupo 2009). Marine extracts, containing polysaccharides and peptides contribute to skin hydration and elasticity, with potential anti-aging benefits (Kim et al. 2018). Mineral clays offer detoxification and sebum-regulating effects, promoting skin clarity and balance. Enzymes derived from fruits facilitate gentle exfoliation, promoting cellular turnover and enhancing skin radiance (Jahan and Happy 2022). The formulation of cosmeceuticals utilizing these natural resources requires rigorous scientific investigation to ascertain ingredient efficacy, safety, and stability, ensuring optimal skincare outcomes while meeting consumer demands for sustainable and eco-friendly products (Goyal and Jerold 2023, Tan et al. 2022).

The formulation of cosmeceuticals incorporating seaweed-derived ingredients necessitates meticulous consideration of various factors to ensure optimal efficacy and safety (Bilal and Iqbal 2019, Kee et al. 2023). Seaweed extracts are esteemed for their rich composition of bioactive compounds, including polysaccharides, peptides, vitamins, and minerals, which confer diverse skin benefits such as antioxidant, anti-inflammatory, and hydrating properties (Morais et al. 2021, Lomartire et al. 2021, Félix et al. 2021). A wide range of marine macroalgae species is utilized in skin care applications and formulations (Figure 1). Formulators must judiciously select seaweed-derived ingredients based on their intended functionality and concentration, taking into account their susceptibility to degradation when exposed to environmental factors like light, heat, and air (Mikušová and Mikuš 2021). To mitigate degradation and enhance stability, encapsulation or microencapsulation techniques may be employed, facilitating sustained release and protection of sensitive seaweed actives (Gunathilake et al. 2022). Additionally, the development of advanced delivery systems, such as lipid nanoparticles or nanoemulsions, enables targeted delivery of seaweed

Figure 1. (**a**–**r**) Examples of marine macroalgae. (**a**) *Palisada perforata* (red); (**b**) *Sargassum cinereum*; (**c**) *Palisada perforata* (as *Laurencia papillosa*); (**d**) *Laurencia glandulifera* (red); (**e**) *Sargassum tenerrimum* (brown); (**f**) *Sargassum tenerrimum* (brown); (**g**) *Ulva lactuca* (green); (**h**) *Laurencia* sp. (red); (**i**) *Gracilaria* sp. (red); (**j**) *Gracilaria debilis* (red); (**k**) *Udotea indica* (green); (**l**) *Champia compressa* (red); (**m**) *Tricleocarpa fragilis* (red); (**n**) *Caulerpa taxifolia* (green); (**o**) *Botryocladia leptopoda* (red); (**p**) *Centroceras clavulatum* (red); (**q**) *Scinaia moniliformis* (red); (**r**) *Padina tetrastromatica* (brown).

compounds to deeper layers of the skin, optimizing their bioavailability and therapeutic effects (Iqubal et al. 2021). Through meticulous formulation strategies, seaweed-based cosmeceuticals can unlock the full potential of these natural ingredients, offering consumers innovative and efficacious skincare solutions rooted in scientific rigor and sustainability (Achagar et al. 2024). Seaweed-derived compounds have garnered significant attention in cosmeceutical formulations due to their rich repertoire of bioactive constituents and multifaceted benefits for skin health (Pereira and Valado 2023).

These marine botanicals are abundant sources of polysaccharides, polyphenols, peptides, vitamins, and minerals, each contributing distinct therapeutic properties (Zhang et al. 2023). For instance, polysaccharides like fucoidan and ulvan exhibit potent antioxidant, anti-inflammatory, and moisturizing effects, promoting skin rejuvenation and resilience against environmental stressors (Iqbal et al. 2024). Peptides derived from seaweed proteins demonstrate remarkable anti-aging properties by stimulating collagen synthesis, improving skin elasticity, and reducing the appearance of fine lines and wrinkles (Rahman et al. 2024).

Furthermore, the presence of vitamins, such as vitamin C and E, in seaweed extracts offers photoprotective benefits, shielding the skin from harmful UV radiation and oxidative damage (López-Hortas et al. 2021). Minerals like iodine and magnesium contribute to skin nourishment and vitality, supporting cellular metabolism and tissue repair processes (Saha et al. 2021). The formulation of cosmeceuticals integrating seaweed extracts requires meticulous optimization to preserve the bioactivity of these compounds and ensure their effective delivery to target skin layers (Makgobole et al. 2024). Innovative encapsulation technologies and carrier systems enable controlled release and enhanced penetration of seaweed actives, maximizing their therapeutic potential (Desai and Mane 2024). By harnessing the diverse array of bioactive compounds present in seaweed extracts, cosmeceutical formulations can offer consumers scientifically validated skincare solutions that promote skin health, vitality, and resilience (Nagahawatta et al. 2023).

Regulatory framework

The regulatory landscape governing cosmeceuticals varies across regions and is characterized by a complex interplay of laws, regulations, and guidelines aimed at ensuring product safety, efficacy, and quality (Oliveira et al. 2022). In the United States, the Food and Drug Administration (FDA) oversees the regulation of cosmetics and cosmeceuticals under the Federal Food, Drug, and Cosmetic Act (FD&C Act) (Katz et al. 2022). Cosmeceuticals are defined by the FDA as products intended to improve the appearance of the skin through moisturizing, cleansing, toning, or altering the skin's structure or function (Rathod et al. 2020). Unlike drugs, which undergo rigorous pre-market approval, cosmeceuticals are not subject to pre-market approval by the FDA but are required to adhere to labelling and ingredient restrictions to ensure consumer safety (Meingast 1997, Ruhela et al. 2018, Gagliardi and Dorato 2007). The EU Cosmetics Regulation governs the regulation of cosmeceuticals in Europe, setting forth requirements for product safety, efficacy, and labelling. Cosmeceuticals in the EU must comply with stringent safety assessments and undergo notification or registration with the European Commission before they can be placed on the market (LODéN et al. 2007). Both regulatory frameworks emphasize the importance of conducting safety assessments, product testing, and post-market surveillance to monitor adverse events and ensure compliance with regulatory requirements (Badnjević et al. 2022). Cosmeceutical manufacturers are responsible for conducting safety and efficacy testing, including clinical trials, to substantiate product claims and demonstrate compliance with regulatory standards.

Clinical applications of cosmeceuticals

Cosmeceuticals, positioned at the intersection of cosmetics and pharmaceuticals have emerged as essential tools in dermatological practice due to their demonstrated efficacy in addressing various skin concerns (Raina et al. 2023). A robust body of clinical evidence, derived from numerous well-designed studies and trials, underscores the therapeutic potential of these formulations across diverse dermatological conditions (Dontje et al. 2024). Central to their efficacy is their ability to target key indicators of aging, including fine lines, wrinkles, and loss of skin elasticity (Naughton et al. 2023). Studies investigating the efficacy of cosmeceuticals enriched with marine algae-derived bioactive ingredients have consistently reported significant improvements in various aspects of skin aging (Alves et al. 2020). Marine algae extracts contain a plethora of bioactive compounds, including polysaccharides, peptides, and antioxidants, which contribute to their skin-rejuvenating properties (Makgobole et al. 2024). These compounds help stimulate collagen synthesis, enhance skin firmness, and protect against oxidative stress-induced damage (Liu et al. 2020). Furthermore, cosmeceuticals incorporating marine algae-derived polysaccharides, such as fucoidans and alginates, have shown promise in addressing hyperpigmentation, acne, and inflammatory skin conditions (Jesumani et al. 2019). Fucoidans, sulphated polysaccharides found in various species of brown algae, exhibit anti-inflammatory and antioxidant properties, making them effective agents for soothing irritated skin and reducing redness (Korzeniowska et al. 2018, Phull and Kim 2017, Pangestuti, Shin and Kim 2021). Alginates, another group of polysaccharides abundant in brown algae, contribute to skin hydration and barrier function, promoting overall skin health (Priyan Shanura Fernando et al. 2019). Peptides derived from marine algae have also attracted attention for their skin-rejuvenating effects. These peptides, composed of amino acid chains, stimulate collagen production, improve skin elasticity, and help reduce the appearance of fine lines and wrinkles (Cunha and Pintado 2022, Fan et al. 2014, Kim et al. 2018). Additionally, marine algae-derived antioxidants, such as vitamins and polyphenols, help protect the skin from environmental aggressors and free radical damage, contributing to a youthful and radiant complexion (Alves et al. 2020, Ferreira et al. 2021).

Safety and side effects

Ensuring the safety and tolerability of cosmeceuticals is paramount, necessitating comprehensive awareness of potential side effects and contraindications (Jaiswal et al. 2024, McAllister et al. 2013). While cosmeceuticals are generally well-tolerated, adverse reactions like allergic contact dermatitis, irritation, and photosensitivity may occur, especially among individuals with sensitive skin or underlying dermatological conditions. Reported side effects often include erythema, pruritus, and stinging sensations, typically resolving with product discontinuation or formulation adjustment. Implementing proper usage guidelines, including patch testing and gradual product introduction, minimizes the risk of adverse reactions and ensures patient safety (Fourzali and Yosipovitch 2020, Otlewska et al. 2020, Čeović et al. 2022). Dermatologists play a crucial role in educating patients about potential side effects and in guiding them in selecting suitable cosmeceutical formulations tailored to their unique skin profile. This action optimizes treatment outcomes and patient satisfaction. The cosmeceutical industry is currently undergoing a period of dynamic expansion, fueled by a convergence of shifting consumer preferences and advancements in skincare research. One prominent trend driving market growth is the escalating demand for natural and organic formulations (Ustymenko 2023, Bakri 2023). Consumers are increasingly drawn to products formulated with botanical extracts, vitamins, and minerals, perceiving them as safer and more environmentally friendly alternatives for synthetic ingredients. This preference aligns with the growing awareness of potential health and environmental impacts associated with certain synthetic compounds commonly found in traditional cosmetics (Fonseca-Santos et al. 2015, Battacchi et al. 2020). Another significant trend shaping the cosmeceutical landscape is the rise of personalized skincare solutions. With advances in technology and diagnostic tools, consumers now have access to customized skincare regimens tailored to their unique skin type, concerns, and lifestyle factors (Ustymenko 2023). Personalized formulations offer a targeted approach to addressing specific skincare issues, enhancing efficacy and consumer satisfaction (Ayu et al. 2023). Moreover, there is a growing demand for cosmeceutical products designed to target specific skincare concerns such as aging, hyperpigmentation, and environmental damage. Consumers are increasingly seeking evidence-based solutions backed by scientific research, driving innovation in cosmeceutical formulations. Ingredients with proven efficacy, such as retinoids, antioxidants, and growth factors, are in high demand due to their ability to deliver tangible results in addressing common skin concerns (Ahmed et al. 2022). Strategic marketing initiatives leveraging social media platforms, influencer endorsements, and educational content play a pivotal role in shaping consumer perceptions and driving purchasing decisions (Vrontis et al. 2021).

Future scope in cosmeceutical sector

Looking ahead, the future of cosmeceuticals holds exciting prospects for further innovation and advancement in skincare science. With ongoing research and technological developments, we can anticipate the emergence of novel ingredients and formulations designed to target specific skin concerns with greater precision and efficacy (Goyal and Jerold 2023, Rathod et al. 2021). One promising direction for future cosmeceuticals is the integration of advanced biotechnologies, such as stem cell technology and gene editing, to develop personalized skincare solutions tailored to individual genetic profiles and skin conditions. By harnessing the regenerative potential of stem cells and optimizing gene expression, researchers aim to unlock new avenues for skin rejuvenation and anti-aging therapies (Mahesh et al. 2019). Furthermore, the exploration of natural marine-derived compounds, such as those derived from algae and other marine organisms, presents a fertile ground for innovation in cosmeceutical development (Calado et al. 2018). These bioactive compounds offer unique properties, including antioxidant, anti-inflammatory, and photoprotective effects, which can be leveraged to formulate cosmeceuticals with enhanced efficacy and safety profiles (Achagar et al. 2024, Kalasariya et al. 2023). In addition to ingredient innovation, advancements in formulation technologies, such as nanotechnology and microencapsulation promise improved deliverabilty and bioavailability of active ingredients in cosmeceuticals (Oliveira et al. 2022, Kaur and Agrawal 2007). These cutting-edge approaches enable targeted delivery of key ingredients to specific layers of the skin, maximizing their therapeutic effects while minimizing potential side effects. Moreover, the integration of artificial intelligence (AI) and machine learning algorithms in skincare diagnostics and product development presents exciting opportunities for personalized skincare solutions (Koka and Burkhart 2023, Menaga and Paruvathavardhini 2022, Georgievskaya et al. 2023). By analysing vast datasets and individual skin characteristics, AI-driven platforms can recommend tailored cosmeceutical regimens optimized for each user's unique needs and preferences. Overall, the future of cosmeceuticals is characterized by a convergence of scientific innovation, technological advancement, and personalized healthcare approaches (Ostapenko 2015). As research continues to unravel the intricacies of skin biology and aging processes, we can expect a new era of cosmeceuticals that not only address existing skin concerns but also pave the way for proactive skin health management and personalized anti-aging strategies.

Conclusive remarks on cosmeceuticals

In conclusion, the concept of cosmeceuticals represents a significant advancement in skincare science, bridging the gap between cosmetics and pharmaceuticals to offer targeted solutions for various dermatological concerns. With a rich historical background rooted in the evolution of consumer preferences and scientific discoveries, cosmeceuticals have emerged as indispensable tools in modern dermatological practice. The regulatory landscape surrounding cosmeceuticals underscores the importance of ensuring product safety and efficacy, with regulatory agencies playing a crucial role in oversight and compliance. Formulation principles, guided by scientific research and technological innovation, drive the development of cosmeceuticals with optimal ingredient selection, concentration, and delivery systems. Key ingredients found in cosmeceuticals, including antioxidants, retinoids, peptides, and botanical extracts, offer diverse mechanisms of action to address specific skin concerns and promote skin health. Clinical applications of cosmeceuticals have yielded promising results, with numerous studies demonstrating their efficacy in combating aging, hyperpigmentation, acne, and other dermatological conditions. Safety considerations remain paramount in the use of cosmeceuticals, with a focus on minimizing potential side effects and allergic reactions through proper usage guidelines and patient education. Market trends reflect evolving consumer preferences for evidence-based skincare products, driving innovation and formulation of cosmeceuticals with clinically proven efficacy.

References

Achagar, R., Z. Ait-Touchente, R. El Ati, K. Boujdi, A. Thoume, A. Abdou et al. 2024. A comprehensive review of essential oil–nanotechnology synergy for advanced dermocosmetic delivery. Cosmetics 11(2): 48.

Ahmed, I.A., M.A. Mikail, N.H. Zamakshshari, M.R. Mustafa, N.M. Hashim and R. Othman. 2022. Trends and challenges in phytotherapy and phytocosmetics for skin aging. Saudi Journal of Biological Sciences 29(8): 103363.

Alves, A., E. Sousa, A. Kijjoa and M. Pinto. 2020. Marine-derived compounds with potential use as cosmeceuticals and nutricosmetics. Molecules 25(11): 2536.

Ayu, V.C., Y. Sunitiyoso and W.A. Rahayu. 2023. Validating personalized skin care needs in the indonesian market and exploring the validity of the amity skin's skin test as its marketing strategy. Riwayat: Educational Journal of History and Humanities 6(4): 2521–2526.

Badnjević, A., L.G. Pokvić, A. Deumić and L.S. Bećirović. 2022. Post-market surveillance of medical devices: A review. Technology and Health Care 30(6): 1315–1329.

Bakri, Z.F. 2023. Analyzing the influence of digital marketing strategies on business performance in the beauty industry: A comprehensive analysis of social media engagement and influencer collaborations. Journal on Economics, Management and Business Technology 2(1): 37–48.

Battacchi, D., R. Verkerk, N. Pellegrini, V. Fogliano and B. Steenbekkers. 2020. The state of the art of food ingredients' naturalness evaluation: A review of proposed approaches and their relation with consumer trends. Trends in Food Science & Technology 106: 434–444.

Bellad, K.A., B.K. Nanjwade, M.S. Kamble, T. Srichana and N.F. Idris. 2017. Development of cosmeceuticals. World. J. Pharm. Pharm. Sci. 6(4): 643–691.

Bilal, M. and H.M. Iqbal. 2019. Marine seaweed polysaccharides-based engineered cues for the modern biomedical sector. Marine Drugs 18(1): 7.

Calado, R., M.C. Leal, H. Gaspar, S. Santos, A. Marques, M.L. Nunes et al. 2018. How to succeed in marketing marine natural products for nutraceutical, pharmaceutical and cosmeceutical markets. Grand Challenges in Marine Biotechnology, pp. 317–403.

Čeović, R., L. Kovačec, Z. Bukvić Mokos and B. Marinović. 2022. Dermatologic adverse events in oncologic therapies. Acta Dermatovenerologica Croatica 30(4): 237–249.

Chatur, V.M., S.G. Walode, S.A. Awate, M.U. Gandhi and V.S. Thorat. 2021. Formulation and physical characterization of herbal face gel toner. World Journal of Advanced Research and Reviews 11(1): 138–145.

Cunha, S.A. and M.E. Pintado. 2022. Bioactive peptides derived from marine sources: Biological and functional properties. Trends in Food Science & Technology 119: 348–370.

Desai, S.S. and V.K. Mane. 2024. Biotechnological advances in cosmetic industry. Sustainable Production Innovations: Bioremediation and Other Biotechnologies, pp. 103–135.

Dini, I. and S. Laneri. 2021. The new challenge of green cosmetics: Natural food ingredients for cosmetic formulations. Molecules 26(13): 3921.

Dontje, A.E., C. Schuiling-Veninga, F.P. van Hunsel, C. Ekhart, F. Demirci and H.J. Woerdenbag. 2024. The therapeutic potential of essential oils in managing inflammatory skin conditions: A scoping review. Pharmaceuticals 17(5): 571.

Fan, X., L. Bai, L. Zhu, L. Yang and X. Zhang. 2014. Marine algae-derived bioactive peptides for human nutrition and health. Journal of Agricultural and Food Chemistry 62(38): 9211–9222.

Félix, C., R. Félix, A.M. Carmona, A.P. Januário, P.D. Dias, T.F. Vicente et al. 2021. Cosmeceutical potential of grateloupia turuturu: Using low-cost extraction methodologies to obtain added-value extracts. Applied Sciences 11(4): 1650.

Ferreira, M.S., D.I. Resende, J.M.S. Lobo, E. Sousa and I.F. Almeida. 2021. Marine ingredients for sensitive skin: market overview. Marine Drugs 19(8): 464.

Fonseca-Santos, B., M.A. Corrêa and M. Chorilli. 2015. Sustainability, natural and organic cosmetics: consumer, products, efficacy, toxicological and regulatory considerations. Brazilian Journal of Pharmaceutical Sciences, 51: 17–26.

Fourzali, K. and G. Yosipovitch. 2020. Safety considerations when using drugs to treat pruritus. Expert Opinion on Drug Safety 19(4): 467–477.

Gagliardi, L. and S. Dorato. 2007. General concepts and cosmetic legislation. Analysis of Cosmetic Products; Amparo, S., Alberto, C., Eds, pp. 3–28.

Georgievskaya, A., D. Danko, R.A. Baxter, H. Corstjens and T. Tlyachev. 2023. Artificial intelligence approaches for skin anti-aging and skin resilience research. In Artificial Intelligence for Healthy Longevity (pp. 189–214). Cham: Springer International Publishing.

Goyal, A., A. Sharma, J. Kaur, S. Kumari, M. Garg, R.K. Sindhu et al. 2022. Bioactive-based cosmeceuticals: An update on emerging trends. Molecules 27(3): 828.

Goyal, N. and F. Jerold. 2023. Biocosmetics: technological advances and future outlook. Environmental Science and Pollution Research 30(10): 25148–25169.

Griffiths, T.W. 2010. Cosmeceuticals: coming of age. British Journal of Dermatology 162(3): 469–470.

Gunathilake, T., T.O. Akanbi, H.A. Suleria, T.D. Nalder, D.S. Francis and C.J. Barrow. 2022. Seaweed phenolics as natural antioxidants, aquafeed additives, veterinary treatments and cross-linkers for microencapsulation. Marine Drugs 20(7): 445.

Heymann, L.A. 1997. The Cosmetic/Drug Dilemma: FDA Regulation of Alpha-Hydroxy Acids. Food & Drug LJ, 52: 357.

Iqbal, M.W., T. Riaz, S. Mahmood, M. Bilal, M.F. Manzoor, S.A. Qamar et al. 2024. Fucoidan-based nanomaterial and its multifunctional role for pharmaceutical and biomedical applications. Critical Reviews in Food Science and Nutrition 64(2): 354–380.

Iqubal, M.K., A. Chaudhuri, A. Iqubal, S. Saleem, M.M. Gupta, A. Ahuja et al. 2021. Targeted delivery of natural bioactives and lipid-nanocargos against signaling pathways involved in skin cancer. Current Medicinal Chemistry 28(39): 8003–8035.

Jahan, F. and A.A. Happy. 2022. Revolutionizing plant-based extracts for skin care and therapeutics. In Nanotechnology for the Preparation of Cosmetics Using Plant-Based Extracts (pp. 75–130). Elsevier.

Jaiswal, S., S. Jawade, B. Madke, S. Gupta and S.N. Gupta. 2024. Recent trends in the management of acne vulgaris: A review focusing on clinical studies in the last decade. Cureus 16(3).

Jesumani, V., H. Du, M. Aslam, P. Pei and N. Huang. 2019. Potential use of seaweed bioactive compounds in skincare—A review. Marine Drugs 17(12): 688.

Kadam Vaishali, S., G.D.K.P. Chintale Ashwini, P. Deshmukh Kshitija and N. Nalwad Digambar. 2013. Cosmeceuticals an emerging concept: A comprehensive Review. Int. J. Res. Pharm. Chem. 3(2): 308–316.

Kalasariya, H.S. and L. Pereira. 2022. Dermo-cosmetic benefits of marine macroalgae-derived phenolic compounds. Applied Sciences 12(23): 11954.

Kalasariya, H.S., L. Pereira and N.B. Patel. 2023. Comprehensive phytochemical analysis and bioactivity evaluation of *Padina boergesenii*: unveiling its prospects as a promising cosmetic component. Marine Drugs 21(7): 385.

Katz, L.M., K.M. Lewis, S.L. Spence and N. Sadrieh. 2022. Regulation of cosmetics in the United States. Dermatologic Clinics 40(3): 307–318.

Kaur, I.P. and R. Agrawal. 2007. Nanotechnology: a new paradigm in cosmeceuticals. Recent Patents on Drug Delivery & Formulation 1(2): 171–182.

Kee, P.E., S.M. Phang, J.C.W. Lan, J.S. Tan, K.S. Khoo, J.S. Chang et al. 2023. Tropical seaweeds as a sustainable resource towards circular bioeconomy: Insights and way forward. Molecular Biotechnology, pp. 1–18.

Kiki, M.J. 2023. Biopigments of microbial origin and their application in the cosmetic industry. Cosmetics 10(2): 47.

Kim, J.H., J.E. Lee, K.H. Kim and N.J. Kang. 2018. Beneficial effects of marine algae-derived carbohydrates for skin health. Marine Drugs 16(11): 459.

Koka, S.S.A. and C.G. Burkhart. 2023. Artificial intelligence in dermatology: current uses, shortfalls, and potential opportunities for further implementation in diagnostics and care. The Open Dermatology Journal, 17(1).

Korać, R.R. and K.M. Khambholja. 2011. Potential of herbs in skin protection from ultraviolet radiation. Pharmacognosy Reviews 5(10): 164.

Korzeniowska, K., B. Górka, J. Lipok and P.P. Wieczorek. 2018. Algae and their extracts in medical treatment. Algae Biomass: Characteristics and Applications: Towards Algae-based Products, pp. 73–87.
Lin, J.Y. and D.E. Fisher. 2007. Melanocyte biology and skin pigmentation. Nature 445(7130): 843–850.
Liu, T., N. Li, Y.Q. Yan, Y. Liu, K. Xiong, Y. Liu et al. 2020. Recent advances in the anti-aging effects of phytoestrogens on collagen, water content, and oxidative stress. Phytotherapy Research 34(3): 435–447.
LODéN, M.A.R.I.E., L. Ungerth and J. Serup. 2007. Changes in European legislation make it timely to introduce a transparent market surveillance system for cosmetics. Acta Dermato-venereologica 87(6): 485–492.
Lomartire, S., J.C. Marques and A.M. Gonçalves. 2021. An overview to the health benefits of seaweeds consumption. Marine Drugs 19(6): 341.
Lomartire, S., J.C. Marques and A.M. Gonçalves. 2021. An overview to the health benefits of seaweeds consumption. Marine Drugs 19(6): 341.
López-Hortas, L., N. Flórez-Fernández, M.D. Torres, T. Ferreira-Anta, M.P. Casas, E.M. Balboa et al. 2021. Applying seaweed compounds in cosmetics, cosmeceuticals and nutricosmetics. Marine Drugs 19(10): 552.
Mahesh, S.K., J. Fathima and V.G. Veena. 2019. Cosmetic potential of natural products: industrial applications. Natural Bioactive Compounds: Volume 2: Chemistry, Pharmacology and Health Care Practices, pp. 215–250.
Makgobole, M.U., S. Onwubu, A. Khathi, N. Mpofana and B. Mkhwanazi. 2024. Cosmeceuticals from marine: the prospect of marine products in skin rejuvenation and care. Egyptian Journal of Basic and Applied Sciences 11(1): 297–317.
Mapelli-Brahm, P., P. Gómez-Villegas, M.L. Gonda, A. León-Vaz, R. León, J. Mildenberger et al. 2023. Microalgae, seaweeds and aquatic bacteria, archaea, and yeasts: Sources of carotenoids with potential antioxidant and anti-inflammatory health-promoting actions in the sustainability era. Marine Drugs 21(6): 340.
McAllister, R.G., T. Franson and C.N. Kraus. 2013. The practicing clinician and regulatory safety concerns. Current Drug Safety 8(1): 17–24.
Meingast, H.S. 1997. Cosmeceuticals: Hope in a Jar. J. Med. & L., 2: 129.
Menaga, S. and J. Paruvathavardhini. 2022. AI in Healthcare. Smart Systems for Industrial Applications, pp. 115–140.
Mikušová, V. and P. Mikuš. 2021. Advances in chitosan-based nanoparticles for drug delivery. International Journal of Molecular Sciences 22(17): 9652.
Morais, T., J. Cotas, D. Pacheco and L. Pereira. 2021. Seaweeds compounds: an ecosustainable source of cosmetic ingredients? Cosmetics 8(1): 8.
Morganti, P., G. Morganti, A. Gagliardini and A. Lohani. 2021. From cosmetics to innovative cosmeceuticals—non-woven tissues as new biodegradable carriers. Cosmetics 8(3): 65.
Nagahawatta, D.P., N.M. Liyanage, T.U. Jayawardena, F. Yang, H.H.A.C.K. Jayawardena, M.J.M.S. Kurera et al. 2023. Functions and values of sulfated polysaccharides from seaweed. Algae 38(4): 217-240.
Naughton, G.K., L.I. Jiang, E.T. Makino, R. Chung, A. Nguyen, T. Cheng et al. 2023. Targeting multiple hallmarks of skin aging: preclinical and clinical efficacy of a novel growth factor-based skin care serum. Dermatology and Therapy 13(1): 169–186.
Oliveira, C., C. Coelho, J.A. Teixeira, P. Ferreira-Santos and C.M. Botelho. 2022. Nanocarriers as active ingredients enhancers in the cosmetic industry—The European and North America regulation challenges. Molecules 27(5): 1669.
Ostapenko, N. 2015. Expanding Inoherb in China: no love without herbology (a case of repositioning a domestic cosmetics brand). Journal of Marketing Development & Competitiveness, 9(1).
Otlewska, A., W. Baran and A. Batycka-Baran. 2020. Adverse events related to topical drug treatments for acne vulgaris. Expert Opinion on Drug Safety 19(4): 513–521.
Pangestuti, R., K.H. Shin and S.K. Kim. 2021. Anti-photoaging and potential skin health benefits of seaweeds. Marine Drugs 19(3): 172.
Pereira, L. and A. Valado. 2023. Harnessing the power of seaweed: Unveiling the potential of marine algae in drug discovery. Exploration of Drug Science 1(6): 475–496.
Phull, A.R. and S.J. Kim. 2017. Fucoidan as bio-functional molecule: Insights into the anti-inflammatory potential and associated molecular mechanisms. Journal of Functional Foods 38: 415–426.
Priyan Shanura Fernando, I., K.N. Kim, D. Kim and Y.J. Jeon. 2019. Algal polysaccharides: Potential bioactive substances for cosmeceutical applications. Critical Reviews in Biotechnology 39(1): 99–113.
Rahman, A., R. Rehmani, D.G. Pirvu, S.M. Huang, S. Puri and M. Arcos. 2024. Unlocking the therapeutic potential of marine collagen: A scientific exploration for delaying skin aging. Marine Drugs 22(4): 159.
Raina, N., R. Rani, V.K. Thakur and M. Gupta. 2023. New insights in topical drug delivery for skin disorders: From a nanotechnological perspective. ACS Omega 8(22): 19145–19167.
Rathod, S., S. Mali, N. Shinde and N. Aloorkar. 2020. Cosmeceuticals and beauty care products: Current trends with future prospects. Research Journal of Topical and Cosmetic Sciences 11(1): 45–51.
Rathod, S., K. Shinde, N. Shinde and N. Aloorkar. 2021. Cosmeceuticals and nanotechnology in beauty care products. Research Journal of Topical and Cosmetic Sciences 12(2): 93–101.
Ruhela, M., L. Nagar, A. Gupta and H. Popli. 2018. Cosmetics: Regulatory and market scenario for us and India. The Pharma Innovation Journal 7(6): 164–169.

Saha, S.K., N.N. Pathak, S.K. Saha and N.N. Pathak. 2021. Mineral Nutrition. Fundamentals of Animal Nutrition, pp. 113–131.

Sajna, K.V., L.D. Gottumukkala, R.K. Sukumaran and A. Pandey. 2015. White biotechnology in cosmetics. In Industrial Biorefineries & White Biotechnology (pp. 607–652). Elsevier.

Stallings, A.F. and M.P. Lupo. 2009. Practical uses of botanicals in skin care. The Journal of Clinical and Aesthetic Dermatology 2(1): 36.

Tambo, E., H. Almeer and Y. Alshamrani. 2016. Health and beauty cosmeceuticals industry and market expansion in Saudi Arabia. Pharmaceut Reg Affairs 5(172): 2.

Tan, K.X., T. Chang and X. Lin. 2022. Secretomes as an emerging class of bioactive ingredients for enhanced cosmeceutical applications. Experimental Dermatology 31(5): 674–688.

Ustymenko, R. 2023. Trends and Innovations in Cosmetic Marketing. Economics & Education 8(3): 12–17.

Vrontis, D., A. Makrides, M. Christofi and A. Thrassou. 2021. Social media influencer marketing: A systematic review, integrative framework and future research agenda. International Journal of Consumer Studies 45(4): 617–644.

Zappelli, C., A. Barbulova, F. Apone and G. Colucci. 2016. Effective active ingredients obtained through biotechnology. Cosmetics 3(4): 39.

Zhang, W., I.D. Boateng and J. Xu. 2023. Novel marine proteins as a global protein supply and human nutrition: Extraction, bioactivities, potential applications, safety assessment, and deodorization technologies. Trends in Food Science & Technology, pp. 104283.

Zhou, H., D. Luo, D. Chen, X. Tan, X. Bai, Z. Liu et al. 2021. Current advances of nanocarrier technology-based active cosmetic ingredients for beauty applications. Clinical, Cosmetic and Investigational Dermatology, pp. 867–887.

Chapter 11
Biology of Marine Macroalgae

Haresh S. Kalasariya,[1] *Carlos Eliel Maya-Ramírez,*[2] *João Cotas*[3,*] and *Leonel Pereira*[3,*]

Introduction to marine macroalgae

Marine macroalgae, colloquially known as seaweeds, stand as foundational elements within marine ecosystems. They are celebrated for their ecological significance and extensive structural diversity. They are very small, rock-like turf on coastal pebbles or tidal pools or massive kelps in deep sea. Marine macroalgae comprise the most apparent, recognizable, and possibly totemic species of the marine coastal boundary. Thriving across a broad spectrum of habitats, from the estuarine and intertidal zones to the profound abyssal depths, macroalgae orchestrate pivotal roles in driving primary productivity, facilitating nutrient cycling, and furnishing habitats within marine realms (Salehi et al. 2019, Kosek and Kukliński 2023). Taxonomically classified into three principal groups—Chlorophyta (green algae), Phaeophyceae (brown algae), and Rhodophyta (red algae)—marine macroalgae exhibit nuanced adaptations tailored to their specific ecological niches that underpin their ecological resilience and evolutionary success (Bajjouk et al. 1996). The structural complexity inherent in marine macroalgae stands as a testament to their adept adaptation to the dynamic and often challenging conditions pervasive within marine environments (Horn 1989). Anchored by the holdfast—a specialized structure affixing the alga to substrates such as rocks or coral reefs—marine macroalgae boast a plethora of morphological adaptations (Boizard 2007). The stipe, a robust axis extending from the holdfast, confers structural support and serves as a conduit for nutrient transport, ensuring the efficient delivery of essential resources to photosynthetic tissues (Croce et al. 2021). At the terminal end of the stipe, blades or fronds emerge as primary sites for photosynthesis, harnessing light energy and transforming it into chemical energy vital for algal metabolism (Dudgeon and Johnson 1992). Additionally, select species feature specialized structures such as air bladders or pneumatocysts, facilitating buoyancy regulation and optimizing light exposure in varying marine environments (Haddad et al. 2009).

[1] Centre for Natural Products Discovery, School of Pharmacy and Biomolecular Sciences, Liverpool John Moores University, Byrom Street, Liverpool L3 3AF, UK; Faculty of Health and Life Sciences, INTI International University, Persiaran Perdana BBN, Putra Nilai, 71800 Nilai, Negeri Sembilan, Malaysia. Email: hareshahir22@gmail.com

[2] Laboratorio de Diagnóstico molecular, Departamento de Bioquímica, Escuela Nacional de Ciencias Biológicas, Instituto Politécnico Nacional, Carpio y Plan de Ayala S/N, Colonia, Casco de Santo, Tomás, C.P., 11340 Ciudad de México, México. Email: emayar1800@alumno.ipn.mx

[3] Marine Resources, Conservation and Technology, Marine Algae Lab, CFE—Centre for Functional Ecology: Science for People & Planet, Department of Life Sciences, University of Coimbra, 3000-456 Coimbra, Portugal.

* Corresponding authors: jcotas@uc.pt; leonel.pereira@uc.pt

Beyond their primary role as foundational primary producers, marine macroalgae serve as indispensable habitats and food sources for a diverse array of marine organisms, spanning from microscopic invertebrates to apex predators, such as fish and marine mammals. Kelp forests, characterized by towering brown algae like *Macrocystis* and *Laminaria*, serve as thriving ecosystems teeming with biodiversity. They provide critical refuge and spawn grounds for numerous species, thereby bolstering marine ecosystem resilience, and supporting lucrative commercial fisheries (Bischof et al. 2019, Hockey and Branch 1994). Moreover, marine macroalgae contribute substantially to nutrient cycling dynamics, adeptly sequestering dissolved nutrients from the water column and liberating oxygen as a byproduct of photosynthesis. They capture carbon dioxide and thereby enhance the overall oxygenation and ecological health of marine environments (Anabtawi et al. 2024, Sudhakar et al. 2018). Human societies have historically harnessed the multifaceted benefits of marine macroalgae across diverse cultural, dietary, medicinal, agricultural, and industrial domains. Marine macroalgae, enriched with a plethora of essential vitamins, minerals, and dietary fibers, stand as dietary staples in coastal communities worldwide, furnishing valuable nutritional sustenance and culinary diversity (Öztürk et al. 2023, Figueroa et al. 2023). Furthermore, marine macroalgae harbor an array of bioactive compounds imbued with potential pharmaceutical properties, encompassing anti-inflammatory, antioxidant, and antimicrobial attributes, thus eliciting considerable interest for the development of novel therapeutic agents (Priyanka et al. 2022, Negreanu-Pirjol et al. 2022, Pereira 2022). Complementing their nutritional and medicinal value, marine macroalgae find application across an array of industrial processes, contributing to the production of versatile products such as agar, carrageenan, and alginates—polysaccharides revered for their diverse utility in food, pharmaceutical, and biotechnological sectors (Ferreira et al. 2021, Hossain et al. 2021, El-Sheekh et al. 2023) (Figure 1).

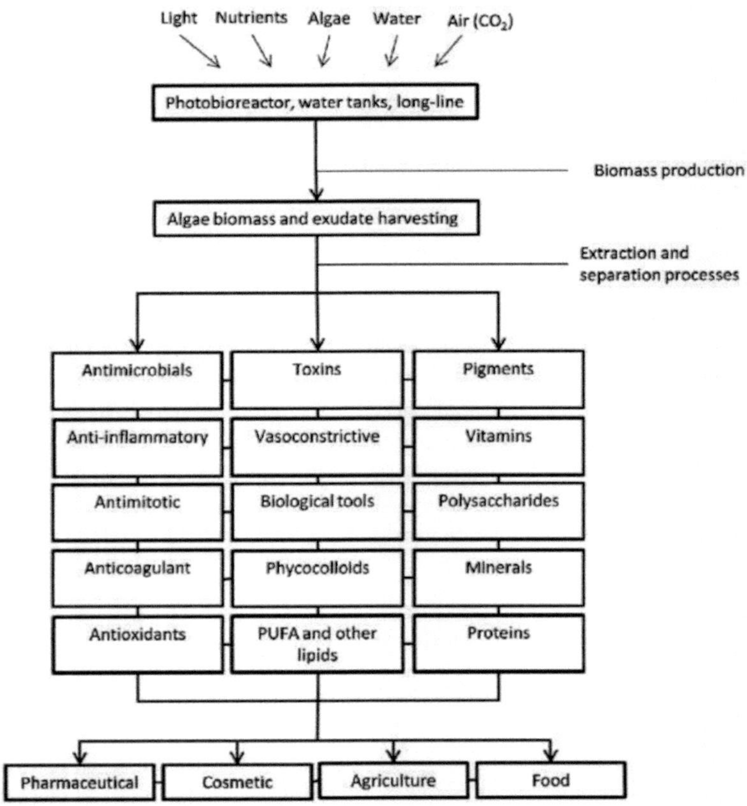

Figure 1. Schematic of the steps of an algae biotechnology production. Source: Carvalho and Pereira 2014.

Basic structure of marine macroalgae

The structural complexity of marine macroalgae is a testament to their remarkable adaptation to the dynamic and challenging marine environments they inhabit (Hurd 2000, Macreadie et al. 2017, Wahl et al. 2014). At the cellular level, these macroscopic algae display features characteristic of photosynthetic organisms, including specialized chloroplasts equipped with chlorophyll pigments, carotenoids, xanthophylls, and phycobilins. These pigments serve as the driving force behind photosynthesis apparatus, enabling the algae to capture light energy and convert it into essential chemical energy (Chapman 1974, Patel et al. 2022, Simkin et al. 2022). Macroscopically, marine macroalgae showcase a diverse array of structures finely tuned to meet their ecological needs. The holdfast, a robust anchoring structure, serves as the alga's attachment point to substrates such as rocks, shells, or other algae (Pereira 2021, Ryznar et al. 2021, Anburaj et al. 2024). From the holdfast emerges the stipe, resembling a sturdy stem-like conduit that not only provides structural support but also facilitates the transport of water and nutrients from the substrate to the photosynthetic tissues located within the blade or frond (Wernberg 2005, Wiencke and Bischof 2012). Blades, the primary photosynthetic organs of marine macroalgae, exhibit remarkable diversity in shape, size, and morphology across different species. This diversity reflects the algae's evolutionary adaptability, with blades sculpted to optimize light capture and photosynthetic efficiency (Pereira 2021, Wada et al. 2020).

Some species boast delicate fern-like fronds, while others feature robust ribbon-like structures, each tailored to thrive in specific marine habitats, from sunlit intertidal zones to dark abyssal depths. Certain marine macroalgae have evolved specialized structures to enhance buoyancy and light exposure. Air bladders or pneumatocysts, for example, serve as buoyancy regulators, allowing algae to maintain optimal positioning within the water column where light penetration is maximal (Garbary et al. 2006, Sudhakar et al. 2018, Jung et al. 2022). These adaptations are critical for sustaining the algae's photosynthetic activity and metabolic efficiency in the competitive underwater realm (Gómez et al. 2009). Reproduction in marine macroalgae is marked by a fascinating array of strategies and structures. Many species undergo alternation of generations, wherein a multicellular haploid gametophyte alternates with a multicellular diploid sporophyte. Gametophytes produce gametes (eggs and sperm) through mitosis, while sporophytes generate spores through meiosis (Thursby and Steele 2020). Reproductive structures, ranging from simple conceptacles to intricate sorus-containing receptacles, underscore the algae's reproductive prowess and evolutionary adaptability. In essence, the basic structure of marine macroalgae reflects their evolutionary resilience and ecological versatility (Clayton 1992, Littler and Littler 1980). Their morphological diversity and reproductive sophistication enable them to thrive across diverse marine habitats, playing crucial roles in marine biodiversity and ecosystem functioning. An in-depth understanding of their structural intricacies is indispensable for formulating effective conservation and management strategies aimed at preserving these invaluable components of marine ecosystems for future generations.

Physiology of marine macroalgae

Marine macroalgae, commonly known as seaweeds, are vital components of marine ecosystems, contributing significantly to primary productivity, nutrient cycling, and habitat provision. Understanding the physiological processes that govern the functioning of marine macroalgae is crucial for elucidating their ecological roles and responses to environmental change. In this section, we delve into the key physiological aspects of marine macroalgae, including photosynthesis, nutrient acquisition, and stress responses (Hurd 2000, Raven and Hurd 2012).

Photosynthesis: Photosynthesis is the primary metabolic pathway by which marine macroalgae harness solar energy to produce organic matter. Chlorophyll pigments, along with accessory pigments such as carotenoids and phycobilins, facilitate light absorption across a broad spectrum of wavelengths (Reiskind et al. 1988). Photosynthetic reactions occur within chloroplasts, where light energy is used to drive the conversion of carbon dioxide and water into carbohydrates, releasing oxygen as a byproduct (Calvin 1976). Marine macroalgae have evolved various mechanisms to optimize photosynthetic efficiency in the dynamic marine environment (Ji et al. 2016). These include the regulation of light harvesting and energy transfer within the photosystems, photoprotective mechanisms to mitigate damage from excess light energy, and the utilization of bicarbonate as a carbon source in environments with limited dissolved carbon dioxide (Mercado et al. 2020).

Nutrient acquisition: Nutrient acquisition is critical for the growth and survival of marine macroalgae, as they must contend with variable nutrient availability in seawater (Pedersen and Borum 1997). Macroalgae employ a range of strategies to acquire and assimilate nutrients, including both inorganic and organic forms. Inorganic nutrients such as nitrogen, phosphorus, and micronutrients are absorbed directly from seawater through specialized uptake mechanisms in the cell membrane (Pedersen and Borum 1997). In addition to inorganic nutrients, marine macroalgae can also utilize organic nutrients derived from dissolved organic matter or exogenous sources. Some species possess specialized structures, such as rhizoids or rhizomes, that facilitate nutrient uptake from the surrounding environment. Furthermore, macroalgae can secrete extracellular enzymes to break down complex organic molecules into simpler forms that can be readily assimilated (Maberly 1990, Morand and Merceron 2005).

Stress responses: Marine macroalgae exhibit remarkable plasticity in their physiological responses to environmental stressors such as temperature fluctuations, salinity changes, and nutrient imbalances (Duarte et al. 2018). In response to stress, macroalgae can modulate their metabolism, alter their growth rates, and adjust their biochemical composition to maintain cellular homeostasis and ensure survival (Harvey et al. 2014). One common stress response observed in marine macroalgae is the production of secondary metabolites with antioxidant properties, which help mitigate oxidative damage caused by reactive oxygen species generated during stress conditions (Tziveleka et al. 2021). Additionally, macroalgae may undergo changes in gene expression and protein synthesis to adapt to changing environmental conditions, enhancing their resilience in the face of adversity (Harley et al. 2012).

Biochemistry of marine macroalgae

The biochemistry of marine macroalgae, or seaweeds, constitutes a fascinating realm of study. It means delving into the intricate biochemical processes and compounds that underpin their physiological functions and ecological significance within marine ecosystems (Bulleri et al. 2020). Central to the biochemistry of marine macroalgae is photosynthesis, a complex metabolic pathway where chlorophyll pigments, carotenoids, and phycobilins collaborate to capture light energy. This energy is then utilized to drive the conversion of carbon dioxide and water into organic compounds, primarily carbohydrates, which serve as essential energy sources and structural components for algal growth and development. In addition to photosynthesis, marine macroalgae are prolific producers of secondary metabolites, which play diverse roles in their survival, defense, and ecological interactions (Goecke et al. 2010, Barzkar et al. 2019). These secondary metabolites encompass a wide array of compounds, including polysaccharides, polyphenolic compounds, terpenoids, and alkaloids, each with unique biochemical properties and ecological functions (Bhatla and Lal 2023, Brielmann et al. 2006). Polysaccharides such as agar, carrageenan, and alginate not only provide structural integrity to the algal cell wall but also possess bioactive properties with applications in food, pharmaceutical, and biotechnological industries (Lomartire and Gonçalves 2023, Rahmati et al. 2019, Usman et al. 2017). Polyphenolic compounds synthesized by marine macroalgae exhibit potent antioxidant properties, protecting the algal cells from oxidative damage and potentially conferring health benefits to organisms that consume them. Terpenoids and alkaloids, on the other hand, serve as chemical defenses against herbivores and pathogens, as well as allelopathic agents influencing the growth and survival of neighboring organisms in marine environments (Zolotariova et al. 2019, Fernando et al. 2016). The composition and abundance of these secondary metabolites vary significantly among different species of marine macroalgae and are influenced by various environmental factors such as light availability, temperature, nutrient availability, and biotic interactions. Furthermore, the biotechnological potential of marine macroalgae-derived bioactive compounds is gaining increasing attention, with applications ranging from pharmaceuticals and nutraceuticals to cosmetics and renewable energy production (Zubia et al. 2020, Horta et al. 2012). Understanding the biochemistry of marine macroalgae is crucial for unravelling their ecological roles, evolutionary adaptations, and responses to environmental changes. Such knowledge not only enriches our understanding of marine ecosystems but also holds significant promise for the development of sustainable biotechnological solutions and conservation strategies aimed at preserving these invaluable organisms and harnessing their bioactive compounds for the benefit of humanity and the environment (Sudhakar et al. 2018, Setthamongkol et al. 2015).

Seaweed-derived bioactive compounds

Seaweeds, or marine macroalgae are renowned for their prolific production of bioactive compounds that hold significant promise for various biomedical, pharmaceutical, and industrial applications. These bioactive compounds encompass a diverse array of chemical entities, including polysaccharides, polyphenols, terpenoids, alkaloids, and fatty acids, each exhibiting distinct biological activities and potential therapeutic benefits (Lomartire and Gonçalves 2022, Holdt and Kraan 2011, Jesumani et al. 2019, Charoensiddhi et al. 2017). Among the most studied and well-characterized bioactive compounds derived from seaweeds are polysaccharides such as agar, carrageenan, and alginate. These polysaccharides, predominantly extracted from brown and red seaweeds, possess a multitude of biological properties, including having anticoagulant, antiviral, immunomodulatory, and antitumor effects (Xie et al. 2023, Patel 2018). Furthermore, their unique physicochemical properties make them valuable ingredients in various pharmaceutical formulations, food additives, and biotechnological applications (Polat et al. 2023, Lomartire and Gonçalves 2022, Narayanan et al. 2022). Polyphenolic compounds derived from seaweeds represent another class of bioactive molecules with remarkable antioxidant properties. These compounds, including phlorotannins, flavonoids, bromophenols and phenolic acids, scavenge free radicals and inhibit oxidative stress, thereby conferring protective effects against various oxidative damage-related diseases, such as cardiovascular disorders, neurodegenerative diseases, and cancer (Agregan et al. 2017, Cotas et al. 2020, Subbiah et al. 2023). Terpenoids and alkaloids synthesized by seaweeds serve as chemical defenses against herbivores and pathogens, as well as allelopathic agents influencing the growth and survival of neighboring organisms in marine ecosystems. Some of these compounds exhibit potent antimicrobial, antifungal, and anti-inflammatory activities, holding promise for the development of novel therapeutic agents against infectious diseases and inflammatory conditions (Rengasamy et al. 2020, Vera et al. 2011, Rocha et al. 2018). Additionally, seaweed-derived fatty acids, particularly omega-3 fatty acids such as eicosapentaenoic acid (EPA) and docosahexaenoic acid (DHA), are recognized for their profound health benefits, including cardiovascular protection, anti-inflammatory effects, and cognitive enhancement. These fatty acids are abundant in certain species of marine macroalgae, making seaweeds valuable dietary sources of essential nutrients for human health and wellbeing (Jaworowska and Murtaza 2022, Peñalver et al. 2020, Padam and Chye 2020). Overall, seaweed-derived bioactive compounds represent a vast and largely untapped reservoir of natural products with significant biomedical, pharmaceutical, and industrial potential. Further exploration of the chemical diversity and biological activities of these compounds holds promise for the discovery of novel therapeutics, functional ingredients, and biotechnological applications, that will contribute to human health, environmental sustainability, and economic prosperity.

Biological actions of seaweed-derived compounds

Seaweed-derived compounds exhibit a plethora of biological actions, owing to their diverse chemical structures and bioactive properties. These compounds exert multifaceted effects on biological systems, including antioxidant, anti-inflammatory, antimicrobial, antiviral, antitumor, immunomodulatory, and neuroprotective activities, among others.

Antioxidant activity: Seaweed-derived compounds, including polyphenols, phlorotannins, and carotenoids, exhibit profound antioxidant properties by neutralizing free radicals and reactive oxygen species (ROS). These compounds scavenge ROS, preventing oxidative damage to biomolecules such as proteins, lipids, and DNA (Hermund 2018, Pradhan et al. 2022, Echave et al. 2022). By inhibiting lipid peroxidation and protein oxidation, seaweed antioxidants help maintain cellular integrity and function, thereby mitigating oxidative stress-related pathologies such as aging, inflammation, cardiovascular diseases, and neurodegenerative disorders (Babakhani et al. 2016, Raja et al. 2016, Senevirathne et al. 2006).

Anti-inflammatory Effects: Certain seaweed-derived compounds possess potent anti-inflammatory activities, and thus modulate immune responses and inflammatory signaling pathways. Polysaccharides, polyphenols, and fatty acids found in seaweeds inhibit the production of pro-inflammatory cytokines (e.g., TNF-α, IL-6) and enzymes (e.g., COX-2, iNOS), thereby attenuating inflammatory processes (Juárez-Portilla et al. 2019, Saraswati et al. 2019, Jaworowska and Murtaza 2022, Lomartire and Gonçalves 2022). These compounds also suppress the activation of NF-κB and MAPK pathways, key regulators of inflammation, leading to reduced inflammation and tissue damage in conditions such as arthritis, inflammatory bowel diseases, and asthma (Meinita et al. 2022, Olsthoorn et al. 2021, Khalid et al. 2018).

Antimicrobial and antifungal actions: Seaweed-derived compounds exhibit broad-spectrum antimicrobial and antifungal properties, targeting a wide range of pathogenic microorganisms. These compounds disrupt microbial cell membranes, inhibit essential metabolic enzymes, and interfere with microbial adhesion and biofilm formation (Singh et al. 2015, Lomartire and Gonçalves 2022, Shukla et al. 2021). Moreover, certain seaweed compounds exhibit synergistic effects with conventional antibiotics, enhancing their efficacy against multidrug-resistant pathogens. By combating microbial infections, seaweed-derived compounds offer promising alternatives for the treatment of infectious diseases and the prevention of antibiotic resistance (Lee et al. 2014, Kulshreshtha et al. 2020).

Antiviral activity: Seaweed-derived compounds demonstrate potent antiviral activities against various viruses, including influenza, herpes simplex virus (HSV), human immunodeficiency virus (HIV), and severe acute respiratory syndrome coronavirus 2 (SARS-CoV-2). These compounds inhibit viral attachment, entry, replication, or release by targeting viral envelope proteins, viral enzymes, or host cell receptors (Lomartire and Gonçalves 2022, Ray et al. 2021). Some seaweed compounds also stimulate innate antiviral immune responses, enhancing host defense mechanisms against viral infections. Their multifaceted antiviral actions make them attractive candidates for the development of novel antiviral therapeutics and preventive strategies (Singh and Kumar 2022, Rana and Chahal 2023).

Antitumor properties: Seaweed-derived compounds exhibit promising antitumor activities, targeting multiple hallmarks of cancer progression, including cell proliferation, apoptosis evasion, angiogenesis, and metastasis. These compounds induce apoptosis (programmed cell death) in cancer cells, inhibit cell cycle progression, and suppress tumor angiogenesis by targeting vascular endothelial growth factor (VEGF) signaling pathways (Agena et al. 2023, Katiyar et al. 2023, El-Sheekh et al. 2022). Moreover, certain seaweed compounds modulate immune responses against cancer cells, enhancing cytotoxic T cell activity and natural killer (NK) cell-mediated cytotoxicity. Their multifaceted antitumor effects hold potential for the development of novel anticancer therapies and adjuvant treatments to complement conventional chemotherapy and radiotherapy (Nuzzo et al. 2022, Hwang et al. 2022).

Immunomodulatory effects: Seaweed-derived compounds exert immunomodulatory effects by modulating innate and adaptive immune responses, enhancing defense mechanisms of the host against pathogens and tumors. These compounds stimulate immune cell proliferation, activation, and cytokine production, promoting both cellular and humoral immune responses (Hwang et al. 2022, Pradhan et al. 2023, Jaworowska and Murtaza 2022). Moreover, seaweed compounds regulate the balance between pro-inflammatory and anti-inflammatory cytokines, maintaining immune homeostasis and dampening excessive inflammation. Their immunomodulatory properties offer therapeutic potential for immune-related disorders, autoimmune diseases, and cancer immunotherapy (Flórez-Fernández et al. 2023, Gothai et al. 2016).

Neuroprotective actions: Certain seaweed-derived compounds possess neuroprotective effects and safeguard neuronal function and promote brain health. These compounds mitigate oxidative stress-induced neuronal damage by scavenging free radicals and enhancing antioxidant defense mechanisms (Schepers et al. 2020, Alghazwi et al. 2016). They also modulate neuroinflammatory responses by inhibiting microglial activation and pro-inflammatory cytokine production, thereby attenuating neuroinflammation and neurodegeneration. Additionally, seaweed compounds promote neurogenesis and synaptic plasticity, enhancing cognitive function and neuronal repair in neurodegenerative diseases and brain injuries (Pereira and Valado 2021, Jannat et al. 2023). Their neuroprotective actions hold promise for the development of novel therapeutic strategies for neurological disorders and age-related cognitive decline.

Hydration and moisturization: Polysaccharides extracted from seaweeds, notably carrageenan and alginate, exhibit exceptional water retention properties, primarily through the formation of occlusive layers on the skin's surface. This moisture-locking mechanism operates via several scientifically elucidated processes, contributing to enhanced skin hydration and the fortification of a supple, resilient skin barrier (Choi et al. 2013, López-Hortas et al. 2021, Jesumani et al. 2019). At a molecular level, carrageenan and alginate molecules possess hydrophilic properties, characterized by their affinity for water molecules. Upon application to the skin, these polysaccharides form hydrogen bonds with water molecules present in the skin's uppermost layers, effectively trapping moisture within the epidermis (Yu et al. 2019, Roh and Shin 2006). Furthermore, the polysaccharide chains within carrageenan and alginate polymers possess large surface areas and considerable flexibility, facilitating their arrangement into dense, cohesive networks when deposited onto the skin. This network formation creates an occlusive barrier that retards the evaporation of water from the skin's surface, a process known as trans-epidermal water loss (TEWL) (Kumar et al. 2004, Kanlayavattanakul and Lourith 2015). Additionally, the occlusive film formed by seaweed-derived polysaccharides serves as a physical barrier against environmental stressors such as wind, cold air, and pollutants, further minimizing water loss and maintaining optimal hydration levels within the skin (Martins and Marto 2023). Moreover, carrageenan and alginate may interact with natural moisturizing factors (NMFs) present in the skin, enhancing their water-binding capacity and promoting the retention of moisture within the epidermal layers. This synergistic action contributes to the restoration of the skin's natural hydration balance and the alleviation of dryness and discomfort (Nieto 2016).

Antioxidant protection: Seaweed-derived polyphenolic compounds, including phlorotannins and flavonoids, manifest potent antioxidant properties through a sophisticated mechanism of action, rooted in their molecular structure and biochemical interactions (Mateos et al. 2020, El Khalloufi and Oudra 2022). These compounds function as radical scavengers, effectively neutralizing free radicals and reactive oxygen species (ROS) within the skin's microenvironment, thereby mitigating oxidative stress and its deleterious effects on skin health (Xing et al. 2020, Pereira 2018). At a molecular level, phlorotannins and flavonoids possess aromatic ring structures with conjugated double bonds, conferring them with electron-rich properties. When exposed to oxidative stressors such as UV radiation, pollution, and metabolic processes, organic molecules within the skin may undergo oxidative damage, resulting in the generation of free radicals and ROS (Cruces et al. 2017). In response, seaweed-derived polyphenolic compounds intercept and quench these reactive species through a process known as hydrogen atom transfer (HAT) or single electron transfer (SET). By donating hydrogen atoms or electrons to free radicals, phlorotannins and flavonoids effectively neutralize their reactivity, halting the chain reaction of lipid peroxidation, protein oxidation, and DNA damage associated with oxidative stress (Del Mondo et al. 2021, Wang et al. 2019). Moreover, polyphenolic compounds exhibit metal-chelating properties, sequestering transition metal ions such as iron and copper that catalyze the generation of harmful hydroxyl radicals via Fenton chemistry. By inhibiting metal-catalyzed oxidation reactions, phlorotannins and flavonoids mitigate oxidative stress and reduce the risk of oxidative damage to cellular components within the skin (Lakey-Beitia et al. 2021, Cho et al. 2007, Begum et al. 2021). Furthermore, seaweed-derived polyphenols stimulate endogenous antioxidant defense mechanisms within skin cells, upregulating the expression of antioxidant enzymes such as superoxide dismutase (SOD), catalase, and glutathione peroxidase. By bolstering the cellular antioxidant defense system, these compounds enhance the skin's capacity to counteract oxidative stress and maintain redox homeostasis (Begum et al. 2021, Pangestuti et al. 2021).

Anti-inflammatory effects: Bioactive compounds such as fucoidans and laminarins harness profound anti-inflammatory properties through intricate molecular interactions and modulation of key signaling pathways within cutaneous tissues (Nagahawatta et al. 2023, Duan et al. 2023). These compounds, derived from seaweeds, exert a multifaceted approach to inflammation attenuation, culminating in the mitigation of skin redness, swelling, and discomfort associated with inflammatory skin conditions (Costa et al. 2023). At a mechanistic level, fucoidans and laminarins interact with various cellular receptors and signaling molecules implicated in the initiation and propagation of inflammatory responses. Notably, these compounds target toll-like receptors (TLRs) and pattern recognition receptors (PRRs) expressed on immune cells and keratinocytes, thereby modulating the activation of innate immune responses to exogenous and endogenous stimuli (Li et al. 2023). Furthermore, fucoidans and laminarins interfere with pro-inflammatory signaling cascades, including nuclear factor-kappa B (NF-κB) and mitogen-activated protein kinase (MAPK) pathways, which orchestrate the expression of inflammatory mediators such as cytokines, chemokines, and adhesion molecules (Vaishnudevi and Viswanathan 2016). By inhibiting the activation and nuclear translocation of NF-κB and the phosphorylation of MAPKs, these compounds attenuate the transcriptional upregulation of pro-inflammatory genes, dampening the inflammatory cascade. Moreover, fucoidans and laminarins modulate cytokine production within cutaneous tissues by influencing the activity of immune cells such as macrophages, dendritic cells, and T lymphocytes (Tunon et al. 2009, Surh et al. 2001). These compounds suppress the secretion of pro-inflammatory cytokines, including tumor necrosis factor-alpha (TNF-α), interleukin-1 beta (IL-1β), interleukin-6 (IL-6), and interleukin-8 (IL-8), while promoting the release of anti-inflammatory cytokines such as interleukin-10 (IL-10). This cytokine balance shift contributes to the resolution of inflammation and the restoration of tissue homeostasis (Coutinho-Wolino et al. 2022, Juárez-Portilla et al. 2019). Furthermore, fucoidans and laminarins attenuate the activation and recruitment of inflammatory cells to the site of injury or inflammation by inhibiting the expression of adhesion molecules on endothelial cells and leukocytes. This impediment in leukocyte-endothelial cell interactions reduces the infiltration of immune cells into inflamed tissues, mitigating the inflammatory response and alleviating associated symptoms such as erythema and edema.

Collagen synthesis and elasticity enhancement: Seaweed-derived peptides and amino acids orchestrate a sophisticated cascade of molecular events to stimulate collagen synthesis and enhance skin elasticity and firmness. Through their intricate interactions with cellular receptors and signaling pathways, these bioactive compounds fortify the skin's extracellular matrix, resulting in a reduction of fine lines and sagging, and promoting a rejuvenated, more resilient skin texture (Hayes et al. 2015, Olsthoorn et al. 2021). At a mechanistic level, seaweed-derived peptides and amino acids interact with specific cell surface receptors, including integrins and growth factor receptors, present on fibroblasts—the primary cells responsible for collagen synthesis. Binding these bioactive compounds to their respective receptors triggers intracellular signaling cascades, activating transcription factors such as transforming the growth factor-beta (TGF-β), which plays a pivotal role in regulating collagen gene expression (Tous et al. 2011, Zhou et al. 2022). Furthermore, seaweed-derived peptides and amino acids stimulate the production and secretion of collagen precursors, such as procollagen type I and procollagen type III, within fibroblasts. These precursors are subsequently processed and cross-linked in the extracellular matrix by enzymes such as collagen prolyl hydroxylase and lysyl oxidase, leading to the formation of mature collagen fibrils (Hayes et al. 2015, Gelse et al. 2003). Moreover, seaweed-derived bioactive compounds modulate the activity of matrix metalloproteinases (MMPs) and tissue inhibitors of metalloproteinases (TIMPs), which regulate collagen turnover and degradation in the skin. By inhibiting MMP activity and upregulating TIMP expression, these compounds help maintain the structural integrity of the extracellular matrix, preventing excessive collagen degradation and preserving skin elasticity (Costa et al. 2023, Fonseca et al. 2023). Additionally, seaweed-derived peptides and amino acids enhance skin hydration and moisture retention, creating an optimal microenvironment for collagen synthesis and deposition (Echave et al. 2022). By promoting the synthesis of hyaluronic acid and glycosaminoglycans within the extracellular matrix, these compounds improve skin plumpness and suppleness, further enhancing the appearance of fine lines and wrinkles (Mohiuddin 2019).

Skin regeneration and repair: Essential vitamins (e.g., vitamin C, vitamin E) and minerals (e.g., zinc, magnesium) derived from seaweeds orchestrate a complex array of molecular mechanisms to promote skin regeneration and wound healing. Through their multifaceted actions on cellular signaling pathways and biochemical processes, these micronutrients bolster epidermal turnover, augment barrier function, and expedite tissue repair processes, culminating in a revitalized, healthier skin complexion (Ashokkumar et al. 2022, Calder et al. 2022). At a mechanistic level, vitamin C (ascorbic acid) serves as a crucial cofactor for enzymes involved in collagen biosynthesis, such as prolyl and lysyl hydroxylases. By enhancing hydroxylation reactions essential for collagen triple helix formation, vitamin C facilitates the synthesis and deposition of collagen fibrils within the extracellular matrix, promoting skin strength and integrity (Boo 2022, Ronchetti et al. 1996, Reilly and Lozano 2021). Moreover, vitamin C exerts antioxidant properties, scavenging free radicals and reactive oxygen species (ROS) that can impair cellular function and impede wound healing processes (Comino-Sanz et al. 2021). Similarly, vitamin E (tocopherol) functions as a potent lipid-soluble antioxidant, protecting cellular membranes and lipoproteins from oxidative damage induced by ROS. By neutralizing lipid peroxyl radicals and inhibiting lipid peroxidation chain reactions, vitamin E safeguards the structural integrity of cell membranes and preserves the viability of keratinocytes and fibroblasts essential for skin regeneration and repair (Niki 2014, Azzi 2018, Butt et al. 2019). Furthermore, minerals such as zinc and magnesium play essential roles in various biochemical processes implicated in wound healing, including cell proliferation, migration, and extracellular matrix synthesis. Zinc, for instance, acts as a cofactor for enzymes involved in DNA synthesis and cell division, promoting keratinocyte proliferation and epithelialization of wounds (Dalisson and Barralet 2019, Schwartz et al. 2005, Lansdown et al. 2007, Skrajnowska and Bobrowska-Korczak 2019).

Additionally, zinc exhibits anti-inflammatory properties, modulating cytokine expression and immune cell function within the wound microenvironment (Wessels et al. 2017). Magnesium, on the other hand, participates in ATP-dependent processes crucial for cellular energy metabolism and protein synthesis (Pasternak et al. 2010 Gröber et al. 2015). By facilitating ATP production and ribosomal activity, magnesium supports the synthesis of structural proteins and growth factors necessary for tissue regeneration and remodeling (Zhou et al. 2021). Moreover, magnesium modulates calcium signaling pathways involved in keratinocyte differentiation and epidermal barrier function, further enhancing skin regeneration and wound closure (Chandrasekaran et al. 2016, Proksch et al. 2005).

UV protection and photoprotection: Seaweed-derived compounds, notably mycosporine-like amino acids (MAAs) and carotenoids, employ sophisticated mechanisms to confer photoprotective properties, safeguarding the skin from the harmful effects of ultraviolet (UV) radiation (Vega et al. 2021). Through their interactions with UV photons and cellular components, these bioactive compounds form a photoprotective shield that effectively absorbs and scatters UV radiation, mitigating UV-induced DNA damage, sunburn, and oxidative stress, and preserving skin health (Dunaway et al. 2018, Archoo et al. 2022). At a mechanistic level, mycosporine-like amino acids (MAAs) serve as potent UV-absorbing molecules, capable of absorbing UV radiation in the UVA and UVB spectra. Structurally resembling amino acids, MAAs contain conjugated double bonds and functional groups that enable them to absorb UV photons and dissipate the absorbed energy harmlessly as heat, preventing UV-induced photodamage to cellular biomolecules such as DNA, proteins, and lipids (Shick and Dunlap 2002, Raj et al. 2021). Furthermore, MAAs exhibit antioxidant properties, scavenging reactive oxygen species (ROS) generated by UV radiation exposure and quenching oxidative stress within the skin. By neutralizing free radicals and ROS, MAAs mitigate oxidative damage to cellular components and prevent the propagation of oxidative chain reactions, thereby minimizing UV-induced skin aging and inflammation (Singh et al. 2021). Similarly, carotenoids, such as β-carotene and astaxanthin, function as photoprotective agents by virtue of their ability to absorb UV radiation and dissipate excess energy as heat. These pigments, which impart the characteristic yellow, orange, or red hues to seaweeds, possess conjugated double bonds that enable them to absorb UV photons in the UVA and UVB spectra, thereby shielding underlying skin tissues from direct UV exposure (Flieger et al. 2024, Salian et al. 2021). Moreover, carotenoids exhibit antioxidant properties, scavenging free radicals generated by UV radiation and lipid peroxidation processes. By donating electrons to reactive oxygen species (ROS), carotenoids neutralize their reactivity and prevent oxidative damage to cellular membranes and organelles, thereby preserving skin integrity and mitigating UV-induced photodamage (Dewanjee et al. 2021, El-Beltagi and Mohamed 2013, Edge and Truscott 2018).

Anti-aging and anti-wrinkle effects: Seaweed-derived bioactive compounds orchestrate a multifaceted approach to combat skin aging by targeting key mechanisms underlying the aging process. Through their synergistic actions on collagen synthesis, inflammation inhibition, and oxidative stress neutralization, these compounds diminish the appearance of wrinkles, fine lines, and age spots, fostering a more youthful, luminous skin complexion. At a mechanistic level, seaweed-derived bioactive compounds stimulate collagen synthesis within the skin by activating signaling pathways involved in collagen production and deposition (López-Hortas et al. 2021). These compounds, which include peptides, polysaccharides, and growth factors derived from seaweeds, interact with cell surface receptors on fibroblasts—the primary cells responsible for collagen synthesis (Pereira 2018). The binding of these compounds to their receptors triggers intracellular signaling cascades, leading to the upregulation of collagen gene expression and the synthesis of collagen precursors such as procollagen type I and procollagen type III (Meinita et al. 2021). Furthermore, seaweed-derived bioactive compounds exhibit potent anti-inflammatory properties, mediated by their ability to modulate pro-inflammatory signaling pathways and cytokine production within the skin. By inhibiting nuclear factor-kappa B (NF-κB) activation and downregulating pro-inflammatory cytokines such as tumor necrosis factor-alpha (TNF-α) and interleukin-6 (IL-6), these compounds mitigate chronic inflammation—a key driver of skin aging and wrinkle formation (Sakthivel and Devi 2019, Jaworowska and Murtaza 2022). Additionally, some bioactive compounds, such as fucoidans and laminarins, suppress the activity of matrix metalloproteinases (MMPs) involved in collagen degradation, further preserving skin elasticity and firmness (Agatonovic-Kustrin and Morton 2018). Moreover, seaweed-derived bioactive compounds possess potent antioxidant properties, counteracting oxidative stress induced by environmental factors such as UV radiation, pollution, and metabolic processes (Menaa et al. 2021). These compounds scavenge free radicals and reactive oxygen species (ROS) within the skin's microenvironment, preventing oxidative damage to cellular components such as lipids, proteins, and DNA. By preserving cellular integrity and functionality, antioxidant compounds mitigate the signs of aging, including wrinkles, fine lines, and age spots, and promote a more youthful skin appearance (Polat et al. 2023, Menaa et al. 2021).

Skin whitening effects: Seaweed or marine macroalgae have garnered attention in skincare for their potential to promote skin whitening and brightening effects. These effects are attributed to various bioactive compounds present in seaweeds, including pigments, vitamins, minerals, and phenolic compounds, which exert their influence through different mechanisms targeting melanin production, distribution, and skin pigmentation (Pereira 2018, Kalasariya et al. 2021, Hayes et al. 2015). One mechanism by which seaweed-derived compounds contribute to skin whitening is by inhibiting melanin synthesis. Melanin is the pigment responsible for skin coloration, and its overproduction can lead to hyperpigmentation and uneven skin tone (Braun-Falco et al. 2000). Certain compounds found in seaweeds, such as phlorotannins and fucoidans, have been shown to inhibit key enzymes involved in melanin synthesis, such as tyrosinase. By blocking these enzymes, seaweed-derived compounds can reduce melanin production, leading to a lighter and more even skin tone (Zhao et al. 2022). Additionally, seaweed-derived compounds exhibit antioxidant properties that help protect the skin from oxidative stress and damage caused by environmental factors such as UV radiation and pollution (Polat et al. 2023). Oxidative stress can trigger melanin production and exacerbate hyperpigmentation. By scavenging free radicals and reducing oxidative damage, seaweed-derived antioxidants can help prevent or mitigate skin darkening and promote a brighter complexion. Furthermore, some seaweed-derived compounds possess exfoliating properties that help remove dead skin cells and promote skin renewal (Chen et al. 2021, Diehl 2014). By facilitating the shedding of pigmented skin cells and promoting the turnover of new, lighter-coloured skin cells, these compounds can contribute to a more radiant and even skin tone. Moreover, certain vitamins and minerals present in seaweeds, such as vitamin C and zinc, play essential roles in skin health and pigmentation regulation (Jablonski 2004). Vitamin C, for example, is involved in the regulation of melanin production and can help fade existing hyperpigmentation. Zinc contributes to the inhibition of tyrosinase activity and the regulation of melanin distribution within the skin (Mohiuddin 2019).

Inhibition of melanin synthesis: Seaweed-derived compounds, such as phlorotannins and fucoidans, exert their skin-whitening effects by inhibiting melanin synthesis through various mechanisms. One key target is the enzyme tyrosinase, which plays a central role in melanin production by catalyzing the conversion of tyrosine to melanin precursors (Pangestuti et al. 2021, Ando et al. 2007, Singh et al. 2013). Compounds like phlorotannins and fucoidans inhibit tyrosinase activity, thereby reducing the production of melanin and leading to a lighter skin tone. Additionally, these compounds may interfere with other steps in the melanin synthesis pathway, further suppressing melanin production and promoting skin whitening (Wijesinghe and Jeon 2011, Qian et al. 2020).

Antioxidant protection: Seaweed-derived antioxidants, including phenolic compounds and vitamins, help protect the skin from oxidative stress induced by environmental factors such as UV radiation and pollution. Oxidative stress can stimulate melanin synthesis and contribute to hyperpigmentation (Chen et al. 2021). By scavenging free radicals and reducing oxidative damage, seaweed-derived antioxidants mitigate melanin production and promote a brighter complexion. Furthermore, antioxidants may regulate signalling pathways involved in melanin synthesis, providing an additional mechanism for their skin-whitening effects (Begum et al. 2021).

Exfoliation and skin renewal: Some seaweed-derived compounds possess exfoliating properties that promote the removal of dead skin cells and stimulate skin renewal. By facilitating the shedding of pigmented skin cells and promoting the turnover of new, lighter-colored skin cells, these compounds contribute to a more even skin tone and enhance the effectiveness of other skin-whitening agents (Nagahawatta et al. 2023). Exfoliation also improves the penetration of skin-whitening ingredients, enhancing their efficacy in reducing hyperpigmentation and promoting skin brightening (Mohiuddin 2019).

Regulation of pigmentation processes: Certain vitamins and minerals present in seaweeds, such as vitamin C and zinc, play essential roles in regulating pigmentation processes within the skin. Vitamin C, for example, inhibits the activity of tyrosinase and other enzymes involved in melanin synthesis, leading to reduced melanin production and a lighter skin tone (Yi et al. 2009, Thawabteh et al. 2023). Zinc contributes to the regulation of melanin distribution within the skin, helping to achieve a more even complexion. By modulating these pigmentation processes, seaweed-derived vitamins and minerals promote skin whitening and brightening effects (Kaczmarek, Wełniak, and Czajkowski 2023; Vigneshwaran and Shaju 2024).

Conclusive note on biological characteristics of macroalgae

In summation, the investigation into marine macroalgae entails a meticulous examination of their ecological, physiological, and biochemical intricacies, thereby unravelling their profound significance within marine ecosystems and beyond. From their structural diversity and evolutionary resilience to their multifaceted physiological processes and synthesis of bioactive compounds, marine macroalgae emerge as pivotal constituents of coastal and oceanic habitats, exerting a profound influence on ecosystem dynamics and fostering biodiversity. By delving into the nuanced intricacies of marine macroalgae, we not only deepen our comprehension of their ecological functions but also glean invaluable insights into their biotechnological potentials and contributions to human welfare. Moving forward, sustained research endeavors and conservation initiatives are imperative for safeguarding these indispensable organisms and unlocking their diverse array of benefits, ensuring their perpetuation for future generations and the sustainable coexistence of humanity and marine ecosystems. Moreover, the comprehensive exploration into seaweed-derived compounds unveils a wide spectrum of biological actions, each carrying significant implications for skincare and human health. Ranging from antioxidant defense mechanisms and anti-inflammatory pathways to the augmentation of collagen synthesis and photoprotective capabilities, these bioactive constituents epitomize the multifaceted potential inherent in marine macroalgae for promoting skin vitality, resilience, and overall well-being. Furthermore, transcending mere cosmetic applications, these compounds offer promising therapeutic modalities, encompassing wound healing facilitation, immune system modulation, and antiviral interventions. In essence, the scientific inquiry into seaweed-derived compounds underscores their pivotal role in propelling the frontier of skincare formulations and therapeutic interventions. This heralds novel approaches to addressing a myriad of dermatological challenges, empowering individuals to attain and sustain healthier, more radiant skin. As our understanding of seaweed biochemistry and its intricate interplay with cutaneous physiology deepens, the prospects for transformative breakthroughs and applications in skincare and allied disciplines remain immensely promising. Thus, the ongoing endeavor to harness the full potential of seaweed-derived compounds represents an enticing frontier in the pursuit of beauty, wellness, and sustainable healthcare solutions, poised to yield enduring benefits for humanity.

References

Agatonovic-Kustrin, S. and D.W. Morton. 2018. The cosmeceutical properties of compounds derived from marine Algae. Mar. Macro-Microalgae Overv, 198.

Agena, R., A. de Jesús Cortés-Sánchez, H. Hernández-Sánchez and M.E. Jaramillo-Flores. 2023. Pro-apoptotic activity of bioactive compounds from seaweeds: Promising sources for developing novel anticancer drugs. Marine Drugs 21(3): 182.

Agregan, R., P.E. Munekata, D. Franco, R. Dominguez, J. Carballo and J.M. Lorenzo. 2017. Phenolic compounds from three brown seaweed species using LC-DAD–ESI-MS/MS. Food Research International 99: 979–985.

Alghazwi, M., Y.Q. Kan, W. Zhang, W.P. Gai, M.J. Garson and S. Smid. 2016. Neuroprotective activities of natural products from marine macroalgae during 1999–2015. Journal of Applied Phycology 28: 3599–3616.

Anabtawi, H.M., W.H. Lee, A. Al-Anazi, M.M. Mohamed and A. Aly Hassan. 2024. Advancements in biological strategies for controlling Harmful Algal Blooms (HABs). Water 16(2): 224.

Anburaj, R., K. Kathiresan and C. Prasannakumar. 2024. Exploring the depths: A comprehensive review of marine algae. Recent Trends In, p. 176.

Ando, H., H. Kondoh, M. Ichihashi and V.J. Hearing. 2007. Approaches to identify inhibitors of melanin biosynthesis via the quality control of tyrosinase. Journal of Investigative Dermatology 127(4): 751–761.

Archoo, S., S.H. Naikoo and S.A. Tasduq. 2022. Role of herbal products as therapeutic agents against ultraviolet radiation-induced skin disorders. In Herbal Medicines (pp. 345–360). Academic Press.

Ashokkumar, V., S. Jayashree, G. Kumar, S.A. Sharmili, M. Gopal, S. Dharmaraj et al. 2022. Recent developments in biorefining of macroalgae metabolites and their industrial applications-A circular economy approach. Bioresource Technology 359: 127235.

Azzi, A. 2018. Many tocopherols, one vitamin E. Molecular Aspects of Medicine 61: 92–103.

Babakhani, A., K.S. Farvin and C. Jacobsen. 2016. Antioxidative effect of seaweed extracts in chilled storage of minced Atlantic mackerel (Scomber scombrus): Effect on lipid and protein oxidation. Food and Bioprocess Technology 9: 352–364.

Bajjouk, T., B. Guillaumont and J. Populus. 1996. Application of airborne imaging spectrometry system data to intertidal seaweed classification and mapping. In Fifteenth International Seaweed Symposium: Proceedings of the Fifteenth International Seaweed Symposium held in Valdivia, Chile, in January 1995 (pp. 463–471). Springer Netherlands.

Barzkar, N., S.T. Jahromi, H.B. Poorsaheli and F. Vianello. 2019. Metabolites from marine microorganisms, micro, and macroalgae: Immense scope for pharmacology. Marine Drugs 17(8): 464.

Begum, R., S. Howlader, A.N.M. Mamun-Or-Rashid, S.M. Rafiquzzaman, G.M. Ashraf, G.M. Albadrani et al. 2021. Antioxidant and signal-modulating effects of brown seaweed-derived compounds against oxidative stress-associated pathology. Oxidative Medicine and Cellular Longevity, 2021.

Bhatla, S.C. and M.A. Lal. 2023. Secondary metabolites. In Plant physiology, development and metabolism (pp. 765–808). Singapore: Springer Nature Singapore.

Bischof, K., C. Buschbaum, S. Fredriksen, F.J. Gordillo, S. Heinrich, C. Jiménez et al. 2019. Kelps and environmental changes in Kongsfjorden: Stress perception and responses. The Ecosystem of Kongsfjorden, Svalbard, pp. 373–422.

Boizard, S.M.D.S.V. 2007. The ecology and anchorage mechanics of kelp holdfasts (Doctoral dissertation, University of British Columbia). Retrieved from https://open.library.ubc.ca/collections/ubctheses/831/items/1.0100425.

Boo, Y.C. 2022. Ascorbic acid (vitamin C) as a cosmeceutical to increase dermal collagen for skin antiaging purposes: Emerging combination therapies. Antioxidants 11(9): 1663.

Braun-Falco, O., G. Plewig, H.H. Wolff, W.H. Burgdorf, O. Braun-Falco, G. Plewig et al. 2000. Disorders of melanin pigmentation. Dermatology, pp. 1013–1042.

Brielmann, H.L., W.N. Setzer, P.B. Kaufman, A. Kirakosyan and L.J. Cseke. 2006. Phytochemicals: The chemical components of plants. Natural Products from Plants 2: 1–49.

Bulleri, F., S. Batten, S.D. Connell, L. Benedetti-Cecchi, M. Gibbons, M.M. Nugues et al. 2020. Human pressures and the emergence of novel marine ecosystems. In Oceanography and Marine Biology (pp. 456–535). CRC Press.

Butt, H., A. Mehmood, M. Ali, S. Tasneem, M.N. Tarar and S. Riazuddin. 2019. Vitamin E preconditioning alleviates *in vitro* thermal stress in cultured human epidermal keratinocytes. Life Sciences 239: 116972.

Calder, P.C., E.F. Ortega, S.N. Meydani, Y. Adkins, C.B. Stephensen, B. Thompson et al. 2022. Nutrition, immunosenescence, and infectious disease: An overview of the scientific evidence on micronutrients and on modulation of the gut microbiota. Advances in Nutrition 13(5): S1–S26.

Calvin, M. 1976. Photosynthesis as a resource for energy and materials. Photochemistry and Photobiology 23(6): 425–444.

Carvalho, L.G. and L. Pereira. 2014. Review of marine algae as source of bioactive metabolites: A marine biotechnology approach. Marine Algae, pp. 195–227.

Chandrasekaran, N.C., W.Y. Sanchez, Y.H. Mohammed, J.E. Grice, M.S. Roberts and R.T. Barnard. 2016. Permeation of topically applied Magnesium ions through human skin is facilitated by hair follicles. Magnesium Research 29(2): 35–42.

Chapman, A.R.O. 1974. The ecology of macroscopic marine algae. Annual Review of Ecology and Systematics 5(1): 65–80.

Charoensiddhi, S., M.A. Conlon, C.M. Franco and W. Zhang. 2017. The development of seaweed-derived bioactive compounds for use as prebiotics and nutraceuticals using enzyme technologies. Trends in Food Science & Technology 70: 20–33.

Chen, J., Y. Liu, Z. Zhao and J. Qiu. 2021. Oxidative stress in the skin: Impact and related protection. International Journal of Cosmetic Science 43(5): 495–509.

Cho, S.H., S.E. Kang, J.Y. Cho, A.R. Kim, S.M. Park, Y.K. Hong et al. 2007. The antioxidant properties of brown seaweed (*Sargassum siliquastrum*) extracts. Journal of Medicinal Food 10(3): 479–485.

Choi, J.S., W.S. Moon, J.N. Choi, K.H. Do, S.H. Moon, K.K. Cho et al. 2013. Effects of seaweed Laminaria japonica extracts on skin moisturizing activity *in vivo*. J. Cosmet. Sci. 64(3): 193–205.

Clayton, M.N. 1992. Propagules of marine macroalgae: Structure and development. British Phycological Journal 27(3): 219–232.

Comino-Sanz, I.M., M.D. López-Franco, B. Castro and P.L. Pancorbo-Hidalgo. 2021. The role of antioxidants on wound healing: A review of the current evidence. Journal of Clinical Medicine 10(16): 3558.

Costa, J.P., L. Custódio and C.P. Reis. 2023. Exploring the potential of using marine-derived ingredients: From the extraction to cutting-edge cosmetics. Marine Drugs 21(12): 620.

Cotas, J., A. Leandro, P. Monteiro, D. Pacheco, A. Figueirinha, A.M. Gonçalves et al. 2020. Seaweed phenolics: From extraction to applications. Marine Drugs 18(8): 384.

Coutinho-Wolino, K.S., P.P. Almeida, D. Mafra and M.B. Stockler-Pinto. 2022. Bioactive compounds modulating Toll-like 4 receptor (TLR4)-mediated inflammation: Pathways involved and future perspectives. Nutrition Research 107: 96–116.

Croce, M.E., M.C. Gauna, C. Fernández, A.M. Poza and E.R. Parodi. 2021. Biology and ecology of the benthic algae. The Bahía Blanca Estuary: Ecology and Biodiversity, pp. 113–151.

Cruces, E., R. Rautenberger, Y. Rojas-Lillo, V.M. Cubillos, N. Arancibia-Miranda, E. Ramírez-Kushel et al. 2017. Physiological acclimation of *Lessonia spicata* to diurnal changing PAR and UV radiation: differential regulation among down-regulation of photochemistry, ROS scavenging activity and phlorotannins as major photoprotective mechanisms. Photosynthesis Research 131: 145–157.

Dalisson, B. and J. Barralet. 2019. Bioinorganics and wound healing. Advanced Healthcare Materials 8(18): 1900764.

Del Mondo, A., A. Smerilli, L. Ambrosino, A. Albini, D.M. Noonan, C. Sansone et al. 2021. Insights into phenolic compounds from microalgae: Structural variety and complex beneficial activities from health to nutraceutics. Critical Reviews in Biotechnology 41(2): 155–171.

Dewanjee, S., N. Bhattacharjee, P. Chakraborty and S. Bhattacharjee. 2021. Carotenoids as antioxidants. Carotenoids: Structure and Function in the Human Body, pp. 447–473.

Diehl, C. 2014. Melanocytes and oxidative stress. Pigmentary Disorders 1(127): 2.

Duan, X., O.T. Agar, C.J. Barrow, F.R. Dunshea and H.A. Suleria. 2023. Improving potential strategies for biological activities of phlorotannins derived from seaweeds. Critical Reviews in Food Science and Nutrition, pp. 1–23.

Duarte, B., I. Martins, R. Rosa, A.R. Matos, M.Y. Roleda, T.B. Reusch et al. 2018. Climate change impacts on seagrass meadows and macroalgal forests: an integrative perspective on acclimation and adaptation potential. Frontiers in Marine Science 5: 190.

Dudgeon, S.R. and A.S. Johnson. 1992. Thick vs. thin: Thallus morphology and tissue mechanics influence differential drag and dislodgement of two co-dominant seaweeds. Journal of Experimental Marine Biology and Ecology 165(1): 23–43.

Dunaway, S., R. Odin, L. Zhou, Y. Zhang and A.L. Kadekaro. 2018. Natural antioxidants: Multiple mechanisms to protect skin from solar radiation. Frontiers in Pharmacology 9: 358049.

Echave, J., P. Otero, P. Garcia-Oliveira, P.E. Munekata, M. Pateiro, J.M. Lorenzo et al. 2022. Seaweed-derived proteins and peptides: promising marine bioactives. Antioxidants 11(1): 176.

Edge, R. and T.G. Truscott. 2018. Singlet oxygen and free radical reactions of retinoids and carotenoids—a review. Antioxidants 7(1): 5.

El Khalloufi, F. and B. Oudra. 2022. Seaweeds as a New Source of Bioactive Compounds and Potent Biotechnological Applications. In Seaweed Biotechnology (pp. 229–287). Apple Academic Press.

El-Beltagi, H.S. and H.I. Mohamed. 2013. Reactive oxygen species, lipid peroxidation and antioxidative defense mechanism. Notulae Botanicae Horti Agrobotanici Cluj-Napoca 41(1): 44–57.

El-Sheekh, M.M., M. Nassef, E. Bases, S.E. Shafay and R. El-Shenody. 2022. Antitumor immunity and therapeutic properties of marine seaweeds-derived extracts in the treatment of cancer. Cancer Cell International 22(1): 267.

Fernando, I.S., M. Kim, K.T. Son, Y. Jeong and Y.J. Jeon. 2016. Antioxidant activity of marine algal polyphenolic compounds: A mechanistic approach. Journal of Medicinal Food 19(7): 615–628.

Ferreira, M., C. Teixeira, H. Abreu, J. Silva, B. Costas, V. Kiron et al. 2021. Nutritional value, antimicrobial and antioxidant activities of micro-and macroalgae, single or blended, unravel their potential use for aquafeeds. Journal of Applied Phycology 33(6): 3507–3518.

Figueroa, V., M. Farfán and J.M. Aguilera. 2023. Seaweeds as novel foods and source of culinary flavors. Food Reviews International 39(1): 1–26.

Flieger, J., M. Raszewska-Famielec, E. Radzikowska-Büchner and W. Flieger. 2024. Skin protection by carotenoid pigments. International Journal of Molecular Sciences 25(3): 1431.

Flórez-Fernández, N., A. Rodríguez-Coello, T. Latire, N. Bourgougnon, M.D. Torres, M. Buján et al. 2023. Anti-inflammatory potential of ulvan. International Journal of Biological Macromolecules, p. 126936.
Fonseca, S., M.N. Amaral, C.P. Reis and L. Custódio. 2023. Marine natural products as innovative cosmetic ingredients. Marine Drugs 21(3): 170.
Garbary, D.J., A. Brackenbury, A.M. McLean and D. Morrison. 2006. Structure and development of air bladders in Fucus and *Ascophyllum* (Fucales, Phaeophyceae). Phycologia 45(5): 557–566.
Gelse, K., E. Pöschl and T. Aigner. 2003. Collagens—structure, function, and biosynthesis. Advanced Drug Delivery Reviews 55(12): 1531–1546.
Goecke, F., A. Labes, J. Wiese and J.F. Imhoff. 2010. Chemical interactions between marine macroalgae and bacteria. Marine Ecology Progress Series 409: 267–299.
Gómez, I., A. Wulff, M.Y. Roleda, P. Huovinen, U. Karsten, M.L. Quartino et al. 2009. Light and temperature demands of marine benthic microalgae and seaweeds in polar regions. Botanica Marina 52(6): 593–608. https://doi.org/10.1515/BOT.2009.073.
Gothai, S., P. Ganesan, S.Y. Park, S. Fakurazi, D.K. Choi and P. Arulselvan. 2016. Natural phyto-bioactive compounds for the treatment of type 2 diabetes: Inflammation as a target. Nutrients 8(8): 461.
Gröber, U., J. Schmidt and K. Kisters. 2015. Magnesium in prevention and therapy. Nutrients 7(9): 8199–8226.
Haddad, Jr., V., O. Lupi, J.P. Lonza and S.K. Tyring. 2009. Tropical dermatology: marine and aquatic dermatology. Journal of the American Academy of Dermatology 61(5): 733–750.
Harley, C.D., K.M. Anderson, K.W. Demes, J.P. Jorve, R.L. Kordas, T.A. Coyle et al. 2012. Effects of climate change on global seaweed communities. Journal of Phycology 48(5): 1064–1078.
Harvey, B.P., B. Al-Janabi, S. Broszeit, R. Cioffi, A. Kumar, M. Aranguren-Gassis et al. 2014. Evolution of marine organisms under climate change at different levels of biological organisation. Water 6(11): 3545–3574.
Hayes, M., M. García-García, C. Fitzgerald and T. Lafarga. 2015. Seaweed and milk derived bioactive peptides and small molecules in functional foods and cosmeceuticals. Biotechnology of Bioactive Compounds: Sources and applications, pp. 669–691.
Hermund, D.B. 2018. Antioxidant properties of seaweed-derived substances. In Bioactive seaweeds for food applications (pp. 201–221). Academic Press.
Hockey, P.A. and G.M. Branch. 1994. Conserving marine biodiversity on the African coast: Implications of a terrestrial perspective. Aquatic conservation: Marine and Freshwater Ecosystems 4(4): 345–362.
Holdt, S.L. and S. Kraan. 2011. Bioactive compounds in seaweed: Functional food applications and legislation. Journal of Applied Phycology 23: 543–597.
Horn, M.H. 1989. Biology of marine herbivorous fishes. Oceanogr. Mar. Biol. 27: 167–272.
Horta, P.A., T. Vieira-Pinto, C.D. Martins, M.N. Sissini, F. Ramlov, C. Lhullier et al. 2012. Evaluation of impacts of climate change and local stressors on the biotechnological potential of marine macroalgae: A brief theoretical discussion of likely scenarios. Revista Brasileira de Farmacognosia, 22: 768–774.
Hossain, M.T., A.A.M. Sohag, M.N. Haque, M. Tahjib-Ul-Arif, R. Dash, M.T.H. Chowdhury et al. 2021. Nutritional value, phytochemical profile, antioxidant property and agar yielding potential of macroalgae from Coasts of Cox's Bazar and St. Martin's Island of Bangladesh. Journal of Aquatic Food Product Technology 30(2): 217–227.
Hurd, C.L. 2000. Water motion, marine macroalgal physiology, and production. Journal of Phycology 36(3): 453–472.
Hwang, J., D. Yadav, P.C. Lee and J.O. Jin. 2022. Immunomodulatory effects of polysaccharides from marine algae for treating cancer, infectious disease, and inflammation. Phytotherapy Research 36(2): 761–777.
Jablonski, N.G. 2004. The evolution of human skin and skin color. Annu. Rev. Anthropol., 33: 585–623.
Jannat, K., R. Balakrishnan, J.H. Han, Y.J. Yu, G.W. Kim and D.K. Choi. 2023. The neuropharmacological evaluation of seaweed: A potential therapeutic source. Cells 12(22): 2652.
Jaworowska, A. and A. Murtaza. 2022. Seaweed derived lipids are a potential anti-inflammatory agent: A review. International Journal of Environmental Research and Public Health 20(1): 730.
Jesumani, V., H. Du, M. Aslam, P. Pei and N. Huang. 2019. Potential use of seaweed bioactive compounds in skincare—A review. Marine Drugs 17(12): 688.
Ji, Y., Z. Xu, D. Zou and K. Gao. 2016. Ecophysiological responses of marine macroalgae to climate change factors. Journal of Applied Phycology 28: 2953–2967.
Juárez-Portilla, C., T. Olivares-Bañuelos, T. Molina-Jiménez, J.A. Sánchez-Salcedo, D.I. Del Moral, T. Meza-Menchaca et al. 2019. Seaweeds-derived compounds modulating effects on signal transduction pathways: A systematic review. Phytomedicine 63: 153016.
Jung, S., T.V. Chau, M. Kim and W.B. Na. 2022. Artificial seaweed reefs that support the establishment of submerged aquatic vegetation beds and facilitate ocean macroalgal afforestation: A review. Journal of Marine Science and Engineering 10(9): 1184.
Kaczmarek, M.E., A. Wełniak and R. Czajkowski. 2023. The role of zinc in skin physiology–relevance for the pathogenesis and treatment of vitiligo. Dermatology Review/Przegląd Dermatologiczny 110(2): 142–150.
Kalasariya, H.S., V.K. Yadav, K.K. Yadav, V. Tirth, A. Algahtani, S. Islam et al. 2021. Seaweed-based molecules and their potential biological activities: An eco-sustainable cosmetics. Molecules 26(17): 5313.

Kanlayavattanakul, M. and N. Lourith. 2015. Biopolysaccharides for skin hydrating cosmetics. Polysaccharides: Bioactivity and Biotechnology; Springer International Publishing: New York, NY, USA, pp. 1867–1892.

Katiyar, S., X.H. Jin and D. Yadav. 2023. Red seaweed-derived compounds: A Desired Approach for Treating Cancer. Current Pharmaceutical Design 29(22): 1729–1740.

Khalid, S., M. Abbas, F. Saeed, H. Bader-Ul-Ain and H.A.R. Suleria. 2018. Therapeutic potential of seaweed bioactive compounds. London, UK: IntechOpen.

Kosek, K. and P. Kukliński. 2023. Impact of kelp forest on seawater chemistry–A review. Marine Pollution Bulletin 196: 115655.

Kulshreshtha, G., A. Critchley, B. Rathgeber, G. Stratton, A.H. Banskota, J. Hafting et al. 2020. Antimicrobial effects of selected, cultivated red seaweeds and their components in combination with tetracycline, against poultry pathogen Salmonella enteritidis. Journal of Marine Science and Engineering 8(7): 511.

Kumar, M.R., R. Muzzarelli, C. Muzzarelli, H. Sashiwa and A.J. Domb. 2004. Chitosan chemistry and pharmaceutical perspectives. Chemical Reviews 104(12): 6017–6084.

Lakey-Beitia, J., A.M. Burillo, G. La Penna, M.L. Hegde and K.S. Rao. 2021. Polyphenols as potential metal chelation compounds against Alzheimer's disease. Journal of Alzheimer's Disease 82(s1): S335–S357.

Lansdown, A.B., U. Mirastschijski, N. Stubbs, E. Scanlon and M.S. Ågren. 2007. Zinc in wound healing: Theoretical, experimental, and clinical aspects. Wound Repair and Regeneration 15(1): 2–16.

Lee, J.H., S.H. Eom, E.H. Lee, Y.J. Jung, H.J. Kim, M.R. Jo et al. 2014. *In vitro* antibacterial and synergistic effect of phlorotannins isolated from edible brown seaweed *Eisenia bicyclis* against acne-related bacteria. Algae 29(1): 47–55.

Li, H., Y. Liu, Y. Teng, Y. Zheng, M. Zhang, X. Wang et al. 2023. Enhancement of seaweed polysaccharides (fucoidan and laminarin) on the phagocytosis of macrophages via activation of intelectin in blunt snout bream (*Megalobrama amblycephala*). Frontiers in Marine Science 10: 1124880.

Littler, M.M. and D.S. Littler. 1980. The evolution of thallus form and survival strategies in benthic marine macroalgae: Field and laboratory tests of a functional form model. The American Naturalist 116(1): 25–44.

Lomartire, S. and A.M. Gonçalves. 2022. An overview of potential seaweed-derived bioactive compounds for pharmaceutical applications. Marine Drugs 20(2): 141.

Lomartire, S. and A.M. Gonçalves. 2022. Antiviral activity and mechanisms of seaweeds bioactive compounds on enveloped viruses—A review. Marine Drugs 20(6): 385.

Lomartire, S. and A.M. Gonçalves. 2023. Algal phycocolloids: Bioactivities and pharmaceutical applications. Marine Drugs 21(7): 384.

López-Hortas, L., N. Flórez-Fernández, M.D. Torres, T. Ferreira-Anta, M.P. Casas, E.M. Balboa et al. 2021. Applying seaweed compounds in cosmetics, cosmeceuticals and nutricosmetics. Marine Drugs 19(10): 552.

Maberly, S.C. 1990. Exogenous sources of inorganic carbon for photosynthesis by marine macroalgae 1. Journal of Phycology 26(3): 439–449.

Macreadie, P.I., J. Jarvis, S.M. Trevathan-Tackett and A. Bellgrove. 2017. Seagrasses and macroalgae: Importance, vulnerability and impacts. Climate change impacts on fisheries and aquaculture: A Global Analysis 2: 729–770.

Martins, A.M. and J.M. Marto. 2023. A sustainable life cycle for cosmetics: From design and development to post-use phase. Sustainable Chemistry and Pharmacy 35: 101178.

Mateos, R., J.R. Pérez-Correa and H. Domínguez. 2020. Bioactive properties of marine phenolics. Marine Drugs 18(10): 501.

Meinita, M.D.N., D. Harwanto and J.S. Choi. 2022. Seaweed exhibits therapeutic properties against chronic diseases: An overview. Applied Sciences 12(5): 2638.

Meinita, M.D.N., D. Harwanto, J.H. Sohn, J.S. Kim and J.S. Choi. 2021. Hizikia fusiformis: Pharmacological and nutritional properties. Foods 10(7): 1660.

Menaa, F., U. Wijesinghe, G. Thiripuranathar, N.A. Althobaiti, A.E. Albalawi, B.A. Khan et al. 2021. Marine algae-derived bioactive compounds: A new wave of nanodrugs? Marine Drugs 19(9): 484.

Mercado, J.M., S. Salles and D. Cortés. 2020. Ecophysiology of marine algae: Factors regulating primary production and growth in natural environments. Encyclopedia of Marine Biotechnology 1: 341–357.

Mohiuddin, A. 2019. Skin aging and modern age anti—Aging strategies. Int. J. Clin. Dermatol. Res. 7: 209–240.

Mohiuddin, A.K. 2019. Skin lightening & management of hyperpigmentation. Pharma Sci. Anal. Res. J. 2: 180020–180068.

Morand, P. and M. Merceron. 2005. Macroalgal population and sustainability. Journal of Coastal Research 21(5): 1009–1020.

Nagahawatta, D.P., N.M. Liyanage, T.U. Jayawardena, F. Yang, H.H.A.C.K. Jayawardena, M.J.M.S. Kurera et al. 2023. Functions and values of sulfated polysaccharides from seaweed. Algae 38(4).

Narayanan, M., S. Kandasamy, Z. He, S. Hemaiswarya, R. Raja and I.S. Carvalho. 2022. Algae biotechnology for nutritional and pharmaceutical applications. In Biotechnology in Healthcare (pp. 177–194). Academic Press.

Negreanu-Pirjol, B.S., T. Negreanu-Pirjol, D.R. Popoviciu, R.E. Anton and A.M. Prelipcean. 2022. Marine bioactive compounds derived from macroalgae as new potential players in drug delivery systems: A review. Pharmaceutics 14(9): 1781.

Nieto, M.B. 2016. Edible film and packaging using gum polysaccharides. Edible Food Packaging Materials and Processing Technologies. CRC Press, Boca Raton, pp. 9–79.

Niki, E. 2014. Role of vitamin E as a lipid-soluble peroxyl radical scavenger: *in vitro* and *in vivo* evidence. Free Radical Biology and Medicine 66: 3–12.

Nuzzo, G., G. Senese, C. Gallo, F. Albiani, L. Romano, G. d'Ippolito et al. 2022. Antitumor potential of immunomodulatory natural products. Marine Drugs 20(6): 386.

Olsthoorn, S.E., X. Wang, B. Tillema, T. Vanmierlo, S. Kraan, P.J. Leenen et al. 2021. Brown seaweed food supplementation: Effects on allergy and inflammation and its consequences. Nutrients 13(8): 2613.

Öztürk, M., K.R. Sridhar, M. Sarwat, V. Altay and F.M. Huerta-Martínez [eds.]. 2023. Ethnic knowledge and perspectives of medicinal plants: Volume 1: Curative Properties and Treatment Strategies. CRC Press.

Padam, B.S. and F.Y. Chye. 2020. Seaweed components, properties, and applications. In Sustainable seaweed technologies (pp. 33–87). Elsevier.

Pangestuti, R., K.H. Shin and S.K. Kim. 2021. Anti-photoaging and potential skin health benefits of seaweeds. Marine Drugs 19(3): 172.

Pasternak, K., J. Kocot and A. Horecka. 2010. Biochemistry of magnesium. Journal of Elementology 15(3): 601–616.

Patel, A.K., F.P.J.B. Albarico, P.K. Perumal, A.P. Vadrale, C.T. Nian, H.T.B. Chau et al. 2022. Algae as an emerging source of bioactive pigments. Bioresource Technology 351: 126910.

Patel, S. 2018. Seaweed-derived sulfated polysaccharides: scopes and challenges in implication in health care. In Bioactive seaweeds for food applications (pp. 71–93). Academic Press.

Pedersen, M.F. and J. Borum. 1997. Nutrient control of estuarine macroalgae: Growth strategy and the balance between nitrogen requirements and uptake. Marine Ecology Progress Series 161: 155–163.

Peñalver, R., J.M. Lorenzo, G. Ros, R. Amarowicz, M. Pateiro and G. Nieto. 2020. Seaweeds as a functional ingredient for a healthy diet. Marine Drugs 18(6): 301.

Pereira, L. and A. Valado. 2021. The seaweed diet in prevention and treatment of the neurodegenerative diseases. Marine Drugs 19(3):128.

Pereira, L. 2018. Biological and therapeutic properties of the seaweed polysaccharides. International Biology Review, 2(2).

Pereira, L. 2018. Seaweeds as source of bioactive substances and skin care therapy—cosmeceuticals, algotheraphy, and thalassotherapy. Cosmetics 5(4): 68.

Pereira, L. 2021. Macroalgae. Encyclopedia 1(1): 177–188.

Pereira, L. 2022. Therapeutic and nutritional uses of marine algae: A pharmacy in the ocean. Traditional Medicine Research 7: 30.

Polat, S., M. Trif, A. Rusu, V. Šimat, M. Čagalj, G. Alak et al. 2023. Recent advances in industrial applications of seaweeds. Critical reviews in Food Science and Nutrition 63(21): 4979–5008.

Pradhan, B., P.P. Bhuyan and J.S. Ki. 2023. Immunomodulatory, antioxidant, anticancer, and pharmacokinetic activity of Ulvan, a seaweed-derived sulfated polysaccharide: An updated comprehensive review. Marine Drugs 21(5): 300.

Pradhan, B., P.P. Bhuyan, S. Patra, R. Nayak, P.K. Behera, C. Behera et al. 2022. Beneficial effects of seaweeds and seaweed-derived bioactive compounds: Current evidence and future prospective. Biocatalysis and Agricultural Biotechnology 39: 102242.

Priyanka, K.R., R. Rajaram and S.R. Sivakumar. 2022. A critical review on pharmacological properties of marine macroalgae. Biomass Conversion and Biorefinery, pp. 1–25.

Proksch, E., H.P. Nissen, M. Bremgartner and C. Urquhart. 2005. Bathing in a magnesium-rich Dead Sea salt solution improves skin barrier function, enhances skin hydration, and reduces inflammation in atopic dry skin. International Journal of Dermatology 44(2): 151–157.

Qian, W., W. Liu, D. Zhu, Y. Cao, A. Tang, G. Gong et al. 2020. Natural skinwhitening compounds for the treatment of melanogenesis. Experimental and Therapeutic Medicine 20(1): 173–185.

Rahmati, M., Z. Alipanahi and M. Mozafari. 2019. Emerging biomedical applications of algal polysaccharides. Current Pharmaceutical Design 25(11): 1335–1344.

Raj, S., A.M. Kuniyil, A. Sreenikethanam, P. Gugulothu, R.B. Jeyakumar and A.K. Bajhaiya. 2021. Microalgae as a source of mycosporine-like amino acids (MAAs); advances and future prospects. International Journal of Environmental Research and Public Health 18(23): 12402.

Raja, R., S. Hemaiswarya, K. Arunkumar and I.S. Carvalho. 2016. Antioxidant activity and lipid profile of three seaweeds of Faro, Portugal. Brazilian Journal of Botany 39: 9–17.

Rana, B. and K. Chahal. 2023. Promising seaweeds against human viruses: A review. Antiviral Therapy 14:16.

Raven, J.A. and C.L. Hurd. 2012. Ecophysiology of photosynthesis in macroalgae. Photosynthesis Research 113: 105–125.

Ray, B., I. Ali, S. Jana, S. Mukherjee, S. Pal, S. Ray et al. 2021. Antiviral strategies using natural source-derived sulfated polysaccharides in the light of the COVID-19 pandemic and major human pathogenic viruses. Viruses 14(1): 35.

Reilly, D.M. and J. Lozano. 2021. Skin collagen through the lifestages: Importance for skin health and beauty. Plast Aesthet Res. 8(2): 10–20517.

Reiskind, J.B., P.T. Seamon and G. Bowes. 1988. Alternative methods of photosynthetic carbon assimilation in marine macroalgae. Plant Physiology 87(3): 686–692.

Rengasamy, K.R., M.F. Mahomoodally, M.Z. Aumeeruddy, G. Zengin, J. Xiao and D.H. Kim. 2020. Bioactive compounds in seaweeds: An overview of their biological properties and safety. Food and Chemical Toxicology 135: 111013.

Rocha, D.H., A.M. Seca and D.C. Pinto. 2018. Seaweed secondary metabolites *in vitro* and *in vivo* anticancer activity. Marine Drugs 16(11): 410.

Roh, Y.H. and C.S. Shin. 2006. Preparation and characterization of alginate–carrageenan complex films. Journal of applied Polymer Science 99(6): 3483–3490.

Ronchetti, I.P., D. Quaglino Jr., and G. Bergamini. 1996. Ascorbic acid and connective tissue. Subcellular biochemistry: Ascorbic acid: biochemistry and Biomedical Cell Biology, pp. 249–264.

Ryznar, E.R., P. Fong and C.R. Fong. 2021. When form does not predict function: Empirical evidence violates functional form hypotheses for marine macroalgae. Journal of Ecology 109(2): 833–846.

Sakthivel, R. and K.P. Devi. 2019. Antioxidant, anti-inflammatory and anticancer potential of natural bioactive compounds from seaweeds. Studies in Natural Products Chemistry 63: 113–160.

Salehi, B., J. Sharifi-Rad, A.M. Seca, D.C. Pinto, I. Michalak, A. Trincone et al. 2019. Current trends on seaweeds: Looking at chemical composition, phytopharmacology, and cosmetic applications. Molecules 24(22): 4182.

Salian, A., S. Dutta and S. Mandal. 2021. A roadmap to UV-protective natural resources: Classification, characteristics, and applications. Materials Chemistry Frontiers 5(21): 7696–7723.

Saraswati, Giriwono, P.E., D. Iskandriati, C.P. Tan and N. Andarwulan. 2019. Sargassum seaweed as a source of anti-inflammatory substances and the potential insight of the tropical species: A review. Marine Drugs 17(10): 590.

Schepers, M., N. Martens, A. Tiane, K. Vanbrabant, H.B. Liu, D. Lütjohann et al. 2020. Edible seaweed-derived constituents: An undisclosed source of neuroprotective compounds. Neural Regeneration Research 15(5): 790–795.

Schwartz, J.R., R.G. Marsh and Z.D. Draelos. 2005. Zinc and skin health: Overview of physiology and pharmacology. Dermatologic Surgery 31: 837–847.

Senevirathne, M., S.H. Kim, N. Siriwardhana, J.H. Ha, K.W. Lee and Y.J. Jeon. 2006. Antioxidant potential of ecklonia cavaon reactive oxygen species scavenging, metal chelating, reducing power and lipid peroxidation inhibition. Food Science and Technology International 12(1): 27–38.

Setthamongkol, P., S. Tunkijjanukij, K. Satapornvanit and J. Salaenoi. 2015. Growth and nutrients analysis in marine macroalgae. Agriculture and Natural Resources 49(2): 211–218.

Shick, J.M. and W.C. Dunlap. 2002. Mycosporine-like amino acids and related gadusols: Biosynthesis, accumulation, and UV-protective functions in aquatic organisms. Annual review of Physiology 64(1): 223–262.

Shukla, P.S., T. Borza, A.T. Critchley and B. Prithiviraj. 2021. Seaweed-based compounds and products for sustainable protection against plant pathogens. Marine Drugs 19(2): 59.

Simkin, A.J., L. Kapoor, C.G.P. Doss, T.A. Hofmann, T. Lawson and S. Ramamoorthy. 2022. The role of photosynthesis related pigments in light harvesting, photoprotection and enhancement of photosynthetic yield in planta. Photosynthesis Research 152(1): 23–42.

Singh, A., M. Čížková, K. Bišová and M. Vitova. 2021. Exploring mycosporine-like amino acids (MAAs) as safe and natural protective agents against UV-induced skin damage. Antioxidants 10(5): 683.

Singh, R.P., P. Kumari and C.R.K. Reddy. 2015. Antimicrobial compounds from seaweeds-associated bacteria and fungi. Applied Microbiology and Biotechnology 99: 1571–1586.

Singh, S. and M.S. Kumar. 2022. Antiviral applications of macroalgae. In Sustainable Global Resources of Seaweeds Volume 2: Food, Pharmaceutical and Health Applications (pp. 455–471). Cham: Springer International Publishing.

Singh, S., A.G. Malhotra, A. Pandey and K.M. Pandey. 2013. Computational model for pathway reconstruction to unravel the evolutionary significance of melanin synthesis. Bioinformation 9(2): 94.

Skrajnowska, D. and B. Bobrowska-Korczak. 2019. Role of zinc in immune system and anti-cancer defense mechanisms. Nutrients 11(10): 2273.

Subbiah, V., C. Xie, F.R. Dunshea, C.J. Barrow and H.A. Suleria. 2023. The quest for phenolic compounds from seaweed: Nutrition, biological activities and applications. Food Reviews International 39(8): 5786–5813.

Sudhakar, K., R. Mamat, M. Samykano, W.H. Azmi, W.F.W. Ishak and T. Yusaf. 2018. An overview of marine macroalgae as bioresource. Renewable and Sustainable Energy Reviews 91: 165–179.

Surh, Y.J., K.S. Chun, H.H. Cha, S.S. Han, Y.S. Keum, K.K. Park et al. 2001. Molecular mechanisms underlying chemopreventive activities of anti-inflammatory phytochemicals: Down-regulation of COX-2 and iNOS through suppression of NF-κB activation. Mutation Research/Fundamental and Molecular Mechanisms of Mutagenesis 480: 243–268.

Thawabteh, A.M., A. Jibreen, D. Karaman, A. Thawabteh and R. Karaman. 2023. Skin pigmentation types, causes and treatment—a review. Molecules 28(12): 4839.

Thursby, G.B. and R.L. Steele. 2020. Sexual reproduction tests with marine seaweeds (macroalgae). In Fundamentals of Aquatic Toxicology (pp. 171–188). CRC Press.

Tous, E., B. Purcell, J.L. Ifkovits and J.A. Burdick. 2011. Injectable acellular hydrogels for cardiac repair. Journal of Cardiovascular Translational Research, 4: 528–542.

Tunon, M.J., M.V. Garcia-Mediavilla, S. Sanchez-Campos and J. González-Gallego. 2009. Potential of flavonoids as anti-inflammatory agents: Modulation of pro-inflammatory gene expression and signal transduction pathways. Current Drug Metabolism 10(3): 256–271.

Tziveleka, L.A., M.A. Tammam, O. Tzakou, V. Roussis and E. Ioannou. 2021. Metabolites with antioxidant activity from marine macroalgae. Antioxidants 10(9): 1431.
Usman, A., S. Khalid, A. Usman, Z. Hussain and Y. Wang. 2017. Algal polysaccharides, novel application, and outlook. In Algae Based Polymers, Blends, and Composites (pp. 115–153). Elsevier.
Vaishnudevi, D. and P. Viswanathan. 2016. Seaweed polysaccharides-new therapeutic insights against the inflammatory response in diabetic nephropathy. Anti-Inflammatory & Anti-Allergy Agents in Medicinal Chemistry (Formerly Current Medicinal Chemistry-Anti-Inflammatory and Anti-Allergy Agents) 15(3): 178–190.
Vega, J., G. Schneider, B.R. Moreira, C. Herrera, J. Bonomi-Barufi and F.L. Figueroa. 2021. Mycosporine-like amino acids from red macroalgae: UV-photoprotectors with potential cosmeceutical applications. Applied Sciences 11(11): 5112.
Vera, J., J. Castro, A. Gonzalez and A. Moenne. 2011. Seaweed polysaccharides and derived oligosaccharides stimulate defense responses and protection against pathogens in plants. Marine Drugs 9(12): 2514–2525.
Vigneshwaran, L.V. and A.M. Shaju. 2024. Potential cosmeceuticals from marine constituents. European Journal of Medical Genetics and Clinical Biology 1(1): 14–20.
Wada, S., K.I. Ishida, M. Noda and H. Abe. 2020. Marine Algae and Plants. Japanese Marine Life: A Practical Training Guide in Marine Biology, pp. 49–64.
Wahl, M., M. Molis, A.J. Hobday, S. Dudgeon, R. Neumann, P.D. Steinberg et al. 2014. The responses of brown macroalgae to environmental change from local to global scales: Direct versus ecologically mediated effects. Perspectives in Phycology 2(1): 11–29.
Wang, A., W. Wang, H. Cheng and C. Zhou [eds.]. 2019. Novel approaches to design eco-friendly Materials based on natural nanomaterials. Frontiers Media SA.
Wernberg, T. 2005. Holdfast aggregation in relation to morphology, age, attachment and drag for the kelp Ecklonia radiata. Aquatic Botany 82(3): 168–180.
Wessels, I., M. Maywald and L. Rink. 2017. Zinc as a gatekeeper of immune function. Nutrients 9(12): 1286.
Wiencke, C. and K. Bischof. 2012. Seaweed biology. Ecological Studies, 219.
Wijesinghe, W.A.J.P. and Y.J. Jeon. 2011. Biological activities and potential cosmeceutical applications of bioactive components from brown seaweeds: A review. Phytochemistry Reviews 10: 431–443.
Xie, C., Z.J. Lee, S. Ye, C.J. Barrow, F.R. Dunshea and H.A. Suleria. 2023. A review on seaweeds and seaweed-derived polysaccharides: Nutrition, chemistry, bioactivities, and applications. Food Reviews International, pp. 1–36.
Xing, M., Q. Cao, Y. Wang, H. Xiao, J. Zhao, Q. Zhang et al. 2020. Advances in research on the bioactivity of alginate oligosaccharides. Marine Drugs 18(3): 144.
Yi, W., X. Wu, R. Cao, H. Song and L. Ma. 2009. Biological evaluations of novel vitamin C esters as mushroom tyrosinase inhibitors and antioxidants. Food Chemistry 117(3): 381–386.
Yu, F., T. Cui, C. Yang, X. Dai and J. Ma. 2019. κ-Carrageenan/Sodium alginate double-network hydrogel with enhanced mechanical properties, anti-swelling, and adsorption capacity. Chemosphere 237: 124417.
Zhao, W., A. Yang, J. Wang, D. Huang, Y. Deng, X. Zhang et al. 2022. Potential application of natural bioactive compounds as skin-whitening agents: A review. Journal of Cosmetic Dermatology 21(12): 6669–6687.
Zhou, H., B. Liang, H. Jiang, Z. Deng and K. Yu. 2021. Magnesium-based biomaterials as emerging agents for bone repair and regeneration: From mechanism to application. Journal of Magnesium and Alloys 9(3): 779–804.
Zhou, X., Y. Guo, K. Yang, P. Liu and J. Wang. 2022. The signaling pathways of traditional Chinese medicine in promoting diabetic wound healing. Journal of Ethnopharmacology 282: 114662.
Zolotariova, Y.K., V.M. Mokrosnop and S.S. Stepanov. 2019. Polyphenol compounds of macroscopic and microscopic algae. International Journal on Algae 21(1).
Zubia, M., O.P. Thomas, S. Soulet, M. Demoy-Schneider, D. Saulnier, S. Connan et al. 2020. Potential of tropical macroalgae from French Polynesia for biotechnological applications. Journal of Applied Phycology 32: 2343–2362.

Chapter 12
Valuable Compounds from Marine Resources in Cosmeceuticals

Julie Queffelec, Noelia Flórez-Fernández, Herminia Domínguez and *María Dolores Torres**

Introduction

In the last decades, a growing concern about health and physical looks have promoted the daily use of cosmeceuticals leading to a changing and competitive market. Safe natural active compounds and environmentally sustainable processes have gained attention in their development with special emphasis on marine resources, especially macro and micro algae, sponges and shells (Brunt et al. 2018, Pimentel et al. 2018).

Cosmetics have been conventionally formulated with different additives and excipients of animal origin, which could be replaced by those that have a marine origin (Brunt et al. 2018). Marine sources are rich in active compounds (phenolics, carotenoids, vitamins, peptides, polysaccharides, oils…) which can exert a number of potential actions on cosmeceuticals. One of the best-known actions is that of an antioxidant, which is important both as a bioactive agent and as a stabilizing agent—phenolic compounds being one of the chemical structures most broadly analyzed for this use (Sathasivam et al. 2018). Marine active compounds also include other beneficial biological and technological properties such as anti-ageing, anti-browning, photo-protective, odor control, anti-inflammatory, antimicrobial, anti-acne, emollient, moisturizer, anti-wrinkle, texturizer, perfuming or hydrating properties (Daudt et al. 2015). Nevertheless, it is dramatically relevant to optimize the maximum content of natural extract used in order to avoid potential allergies as well as the generation of reactive oxygen species (Gunia-Krzyżak et al. 2018). Among the most studied compounds are polyphenolic and carotenoid extracts due to their beneficial biological features (Sathasivam et al. 2018). *In vitro* analysis with dermal fibroblasts or epidermal keratinocytes proved their antioxidant and photoprotective activities and inhibition of dermal proteases (Mirata et al. 2023). Additionally, some few clinical works established an anti-aging action of polyphenols on the skin after topical use (Zillich et al. 2018).

CINBIO, Universidade de Vigo, Department of Chemical Engineering, 32004 Ourense, Spain.
* Corresponding author: matorres@uvigo.gal

This chapter summarizes some illustrative active compounds found in a number of representative marine sources, which can have a relevant action on cosmeceutical applications. It also covers emerging extraction methodologies for the recovery of the corresponding compounds.

Marine biodiversity as a source of compounds for cosmetics applications

Even if seaweeds, fish and crustacean seem familiar to us, are we aware of the wide set of bioactive compounds they can synthetize. Moreover, these well-known species only represent a minor part of the marine biodiversity. The principal sources of bioactive compounds with potential use in cosmetics are summarized in Figure 1, and are succinctly described in this section.

Macroalgae

Macroalgae are visible multicellular photosynthetic marine organisms, classified into 3 types: Rhodophyta (red), Phaeophyta (brown) and Chlorophyta (green); as a function of their pigment composition. They have been used in traditional medicine and consumed as food since ages, but their interesting bioactive compounds with uses in the cosmetic, food and health industry are currently drawing the attention of researchers and the industry. Among these compounds are polysaccharides, phenolic compounds, pigments, amino acids, tannins (Michalak et al. 2020)

Microalgae are unicellular photosynthetic organisms able to live in different kinds of aquatic environment such as oceans, lakes, rivers or even wastewater. Microalgae can be prokaryotic and thus similar to cyanobacteria or eukaryotic and similar to green algae (Khan et al. 2018). Interest in microalgae has grown for their wide number of bioactive compounds such as polysaccharides, pigments, fatty acids, phenols, proteins, peptides, amino acids, vitamin E and the resulting potential application in food, health, cosmetics, and biofuels. Nowadays, microalgae are cultivated in industrial farms.

Fish/shellfish

The dictionary defines fishes as aquatic cold-bloody vertebrate animals, including bony fishes or cartilaginous and jawless fishes. They are mainly caught to be eaten as they represent, according to the FAO, around 17% of the world consumption of animal proteins. But they are now also exploited for their collagen, polysaccharides, and fatty acids (Alves et al. 2017, Claverie et al. 2020, Mitura et al. 2020, Sionkowska et al. 2020).

Shellfish are aquatic invertebrates, classified into crustaceans which have a segmented body and a hard shell to protect them, like shrimp, crabs and lobsters and mollusks like squids or clams that have a soft unsegmented body usually protected by a calcareous shell (Fehrenbach et al. 2022).

Shellfish represent an important part of human diet. To support this consumption, breeding shellfish came into being and shellfish farms are now their main source. But the interest in shellfish goes beyond their gustatory qualities. The main component of their exoskeleton is a nitrogen-containing polysaccharide called chitin. Chitin and a derivative of its deacetylation, chitosan are produced at industrial level because, as well as forming hydrogel and being thickening agents, they present anti-aging, photo-protective and antioxidant properties. In addition, shellfish also synthetizes astaxanthin, a carotenoid pigment with several valuable properties (Casadidio et al. 2019, Claverie et al. 2020).

Marine bacteria/cyanobacteria

Marine bacteria are the most abundant of marine microorganisms. They are microscopic unicellular ubiquitous microorganisms (procaryotes) that are essential to the preservation of the ecosystem (Srinivasan et al. 2021). Some of them live in such harsh habitat that they are called extremophilic bacteria.

Bacteria and cyanobacteria (also called blue algae, a sub-class of bacteria) are considered as an amazing source of a wide variety of bioactive compounds. Among others, exopolysaccharides, enzymes, peptides and amino acids, pigments and phytohormones are increasingly in demand (Corinaldesi et al. 2017, Geraldes and Pinto 2021, Mourelle et al. 2017, Pagels et al. 2022, Resende et al. 2021, Srinivasan et al. 2021).

Phytoplankton

Phytoplankton are diverse aquatic microscopic organisms. Most of them are single-cell organisms (microalgae), but some are bacteria or protists. Some of the specific enzymes, amino acids, pigments, and vitamins they synthetize have drawn special interest for cosmetic applications (Kim 2014).

Halophyte

Halophytes are highly resistant plants that tolerate saline environments, drought, extreme temperature, and UV radiation. To survive in such hostile conditions, they have developed physiological and morphological adaptations. Halophytes are thus an interesting source of phenolic compounds, organic acids, essential oils, and carbohydrates, which can provide them antioxidant, anti-inflammatory, antitumoral and antimicrobial activities (Pereira 2022).

Sea mud/water

Due to their ionic composition and concentration, sea water and mud can exhibit therapeutic properties for damaged or ill skins. For example, baths in the Dead Sea, which has a salinity of about 300% and a specific ionic composition, significantly improve the symptoms of skin diseases such as psoriasis or atopic dermatitis (Dai et al. 2023).

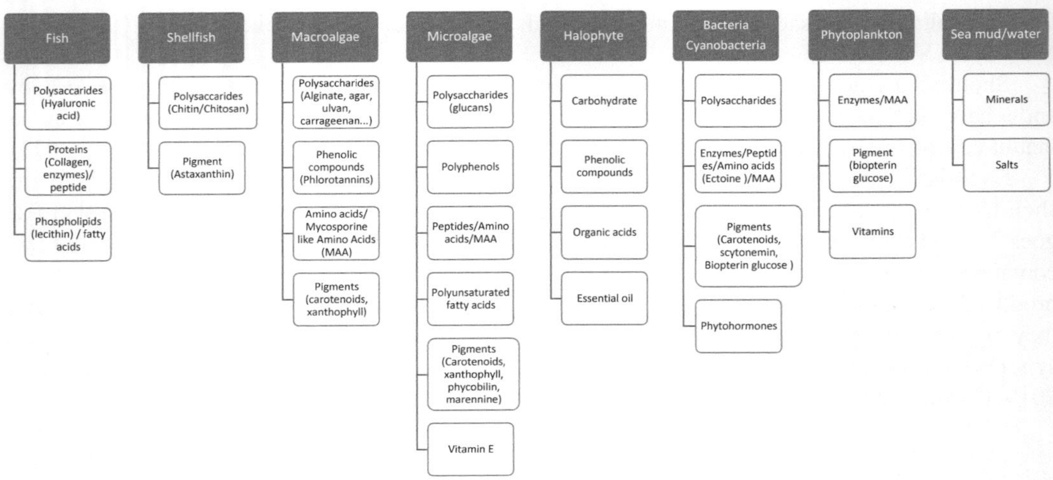

Figure 1. Example of marine resources and some of their compounds of interest for cosmetics.

Indeed, sea water is rich in vital minerals such as sodium, potassium, magnesium, calcium, sulfates, and chlorides. These minerals, which have a strong affinity with the skin can refresh, replenish and hydrate it. Sea salts also have cosmetically benefits as deodorants, cleansers, or air conditioners (Kim 2014).

Sea mud, which contains nutrients and minerals, can also be added in skin care products for its beneficial effects on skin diseases and for its anti-aging, anti-bacterial and hydration properties (Kim 2014).

Different uses of marine natural ingredients in cosmetics

According to the Chapter 2 of the European Regulation 123/2009, marine ingredients are qualified as cosmetics: "Any substance or mixture intended to be placed in contact with the external parts of the human body (epidermis, hair system, nails, lips and external genital organs) or with the teeth and the mucous membranes of the oral cavity with a view exclusively or mainly to cleaning them, perfuming them, changing their appearance, protecting them, keeping them in good condition or correcting body odors".

Scandals of recent years bring to light the toxicity (allergens, endocrine disruptors, cancerogenic) of some chemicals in cosmetic formulations and the current environmental crisis where plastics and chemical products are polluting the Earth. This has made people demand products that are more natural and less toxic.

The diversity of compounds present in marine organisms and suffering from hard living conditions (such as high UV exposure or on the contrary lack of light, high salt concentration, predators…) opens up a wide range of potentially new applications as less toxic, more biocompatible and more effective adjuvants, additives and active substances for cosmetics.

Although the term cosmetics is quite well defined and regulated, there is no legal definition for "natural cosmetics" in Europe. The certification "natural or organic" is delivered by three agencies (Soil association, BDIH and ECOCERT) according to the Cosmetics Organic and Natural Standard. The ISO 16128-2:2017 also provides some guidelines about these natural compounds.

Even if the definition and regulation of "natural products" remains weak, they represent a growing market. The cosmetic market in 2022 was USD 262.21 billion with an expected annual growth of 4.2% from 2023 to 2030. The market valued of global marine-based skincare ingredients was at USD 575.64 billion and expected to grow up to USD 959.86 billion by 2030.

Active substances

Since several years, we find the term "cosmeceutical" in papers and used by the cosmetic industry. It refers to a mix between cosmetics and pharmaceuticals and names cosmetics that have active substances that should provide medical benefits. However, care is needed when using the term 'active substance', as it is normally used for drugs and authorized by the FDA (Food and Drugs Administration) once the safety and efficiency of the molecule has been proven.

In the case of cosmetic products, the term 'active substance' names the cosmetic ingredient that is supposed to help solve a specific problem, for example skin dryness or redness, but that is not a regulated drug. The most common benefits of cosmetic active substances are described below and listed in Table 1.

Moisturizing

The skin, divided into the epidermis, the dermis, and the hypodermis, is a barrier that reduces transepidermal water loss and pathogenic invasion through the upper layer of the epidermis, called the stratum corneum. It is essential in this protective process. It is made up of natural moisturizing factors which are a set of amino acids and other hygroscopic and water-soluble molecules. Ceramides

Table 1. Illustrative examples of active substances from marine resources and the corresponding cosmetic potential.

Application	Compounds	References
Active substance		
Hydration	Polysaccharides (chitin/chitosan, hyaluronic acid, carrageenan) Oligosaccharides Fatty acids (linoleic acid) Proteins (collagen/hydrolyzed collagen) Amino acids (ex: serine, ectoine), Carotenoids	Brunt and Burgess 2018, Casadidio et al. 2019, Guillerme et al. 2017, Michalak et al. 2020, Mitura et al. 2020, Mourelle et al. 2017, Pagels et al. 2022, Resende et al. 2021, Sionkowska et al. 2020
Anti-aging	Bacterial/algal polysaccharides (hyaluronic acid, chitin/chitosan, β-glucan) Sulfated polysaccharides Polyunsaturated fatty acids Phenolic compounds (phlorotannins) and Phloroglucinol derived polyphenols Carotenoids (β-carotene, saproxanthin, myxol, astaxanthin, fucoxanthin,) Inhibitors of matrix metalloproteinase (fish peptide, phenolic compound derivates (eckol, dieckol)) Mycosporine-like amino acids Phytohormones Sea water minerals	Brunt and Burgess 2018, Casadidio et al. 2019, Couteau and Coiffard 2020, Guillerme et al. 2017, Kim 2014, Michalak et al. 2020, Mourelle et al. 2017, Resende et al. 2021
Skin photo protection	Pigments (Scytonemin, Carotenoids (fucoxanthin, astaxanthin, zeaxanthin, β-carotene, lutein), Biopterin glucose…) Mycosporines and mycosporine-like amino acids (MAA) chitin/chitosan Gadusol	Brunt and Burgess 2018, Casadidio et al. 2019, Couteau and Coiffard 2020, Guillerme et al. 2017, Kim 2014, Michalak et al. 2020, Mourelle et al. 2017, Resende et al. 2021
Whitening	Fucoidan Carotenoids (Zeaxanthin, astaxanthin, fucoxanthin) Phenolic compounds (phloroglucinol, phlorotannin (7-phloroeckol), gallic acid, Flavonoid (epicatechin))	Brunt and Burgess 2018, Guillerme et al. 2017, Kim 2014, Mourelle et al. 2017
Antioxidant	Polysaccharides (Chitosan/chitin) Oligosaccharides Phenolic compounds (e.g., phlorotannin) Pigments (ex: astaxanthin) Terpenoids, Fucosterol	Brunt and Burgess 2018, Casadidio et al. 2019, Kim 2014, Michalak et al. 2020, Mourelle et al. 2017, Pagels et al. 2022, Quitério et al. 2021, Resende et al. 2021
Anti-inflammatory	Sulfated polysaccharides (fucoidan, carrageenan, ulvan) Phenolic compounds (phlorotannin) Protein (lectin) Phytosterols Phytohormone (abscisic acid)	Cheung et al. 2015, Kim 2014, Michalak et al. 2020, Mourelle et al. 2017
Emollient	Polysaccharides Proteins Vitamins (A, B_1) Minerals	Michalak et al. 2020

Table 1 contd. ...

...Table 1 contd.

Application	Compounds	References
	Preservative, adjuvant, & additive	
Preservatives (antioxidant/anti-microbial)	Polysaccharides (chitosan/chitin) Sulfated polysaccharide (agar, carrageenan, fucoidan) Polyphenols Proteins	Casadidio et al. 2019, Couteau and Coiffard 2020, Guillerme et al. 2017, Mitura et al. 2020, Resende et al. 2021
Thickening agents Emulsifier	Polysaccharides (alginates, carrageenan, agar, fucoidan, chitosan) Proteins (Collagen) Lipids/Lecithin	Alhajj et al. 2020, Kim 2014, Sionkowska et al. 2020
Encapsulation systems	Biopolymers	Barbosa et al. 2019, Casadidio et al. 2019, Mitura et al. 2020
Dyes	Pigments (phycobiliproteins, marennine, carotenoids, chlorophyll)	Couteau and Coiffard 2020, Guillerme et al. 2017, Hosikian et al. 2010, Mourelle et al. 2017, Pagels et al. 2022
Fragrance	Essential oil Proteins Fatty acid derivatives Terpenoids Pigments (Carotenoids, Chlorophyll) Sulfur compounds	Couteau and Coiffard 2020, Guillerme et al. 2017, Hosikian et al. 2010

(big lipid molecules whose precursors are linoleic acid) and glycosaminoglycan (such as hyaluronic acid) also compose the dermal skin matrix, making it possible to retain water and have elasticity. (Brunt and Burgess 2018). A combination of environmental, nutritional, physiological, and aging factors can lead to skin dehydration (Camilion et al. 2022).

Compounds capable of attracting and retaining water in the epidermis are therefore sought after in cosmetic products designed to moisturize the skin. Polysaccharides such as chitin, chitosan (shellfish), small molecular weight hyaluronic acid (fish cartilage, deep-vent bacteria) and carrageenan (macroalgae), with water retention, anti-inflammatory and antimicrobial properties have attracted increased interest (Claverie et al. 2020, Corinaldesi et al. 2017). In addition to polysaccharides, fatty acids, astaxhanthin, amino acids (serine), and proteins such as fish collagen/ hydrolyzed collagen,, may decrease water loss and increase skin moisture.

Research is also oriented to marine bacterial polysaccharides as they exhibit great water absorbing properties and seem to stimulate hyaluronic acid and keratinocytes production (Brunt and Burgess 2018).

Photo-protection

It is well known that UV radiations cause a large range of problems from sunburn and skin aging to skin cell DNA mutation and cancer. Even though many photo-protective products are available on the market, the controversies about the side effects of some of their compounds (avobenzone, octocrylene, oxybenzone, para-aminobenzoic acid, Titanium dioxide nanoparticles) for both environment (marine life destruction) and human health (allergen, cancerogenic, endocrine disruptors) urge us to find some new photo-protective molecules.

Marine biodiversity, due to its high contact with UV radiations, represents an interesting source of photo-protective molecules. Sun damages are mainly due to the production of reaction by-products: reactive oxygen species (ROS) such as superoxide radicals, hydroxyl radicals, hydrogen peroxide, lipids peroxides and singlet oxygen. At low concentration, ROS maintain the proper functioning of various biological processes but in excess these highly unstable and reactive compounds cause oxidative stress with potential DNA, protein and lipid damage (Anbualakan et al. 2022, de Jager et al. 2017).

Molecules that play a UV filter role or those that exhibit strong antioxidant activity reacting with these ROS could protect human skin from sun damage. Particularly active are pigments such as scytonemin, carotenoids which include fucoxanthin, astaxanthin, zeaxanthin, β-carotene, and lutein, molecules like mycosporines and mycosporines like amino acids (MAA) and phenolic compounds. (Brunt and Burgess 2018, Casadidio et al. 2019, Couteau and Coiffard 2020, Guillerme et al. 2017, Kalasariya et al. 2021, Khan et al. 2018, Kim et al. 2008, Michalak et al. 2020, Mourelle et al. 2017, Pagels et al. 2022, Stahl and Sies 2012).

This kind of pigments and MAA are found in macro and microalgae as well as in marine bacteria, cyanobacteria, and phytoplankton. They also synthesize phenolic compounds that can also be found in halophyte and fish.

Anti-ageing: antioxidant, whitening, anti-wrinkle

Hyperpigmentation, wrinkles, skin elasticity loss, dehydration, skin laxity are all signs of skin aging. They are the consequences of oxidative stress and dermal protein damage due to environmental, nutritional and physiological factors. Ultraviolet radiation can induce cancer, skin aging and is responsible for the stimulation of the matrix metalloproteinase, which deteriorates collagen, hyaluronic acid and elastin, thus decomposing the extracellular matrix and causing skin aging (Fonseca et al. 2023).

Sulfated polysaccharides from macro and microalgae or marine bacteria are attracting interest due to their elastase and collagenase inhibition properties. Enzymes and peptides from salmon eggs and extremophile bacteria show anti-aging properties. Some compounds such as marine collagen (which demonstrates a lower cytotoxicity than the bovine one) may act on the aging signs attracting water and filling the wrinkles.

Tyrosinase is the enzyme that catalyzes important steps of the melanin synthesis. Its inhibition would thus limit melanin production and fight hyperpigmentation. Marine phenolic extracts from halophytes (tannins and flavonoid) or from brown macroalgae (phlorotannin) and carotenoids (zeaxanthin, astaxanthin) show tyrosinase inhibition properties. Fucoidan from brown seaweeds also decreases the melanin synthesis through the activation of the extracellular signal-related kinase (ERK) pathway (Brunt and Burgess 2018). These properties are even more interesting than current commercial tyrosinase inhibitors (kojic acid, hydroquinone), controversial for their toxicity.

Moreover, antioxidant compounds, with their direct action on ROS, make it possible, to a certain extent to prevent skin aging. Compounds exhibiting antioxidant activity are phenolic compounds (phlorotannin), pigments (astaxanthin), terpenoids, sulfated polysaccharides/polysaccharides, fucosterol… (Brunt and Burgess 2018, Casadidio et al. 2019, Kim, 2014, Michalak et al. 2020, Mourelle et al. 2017, Pagels et al. 2022, Resende et al. 2021).

Emollient (adoucissant)

Compounds that soothe irritated or reactive skin are also often added in cosmetics products. Indeed, inflammatory diseases, from redness and itching to more severe diseases such as atopic dermatitis or psoriasis are common dermatological problems. Cosmetics with anti-inflammatory ingredients tempt to reduce these dermatological issues, thus improving skin comfort and aspect.

Many marine compounds contain anti-inflammatory and soothing properties. For example, phenolic compounds such as phlorotannin from brown seaweeds, sulfated polysaccharides (fucoidan, ulvan), algal and cyanobacterial lectin and phytohormones demonstrated anti-inflammatory activity

(Cheung et al. 2015, Kim 2014, Michalak et al. 2020, Mourelle et al. 2017). Polysaccharides, vitamins A and B1 and minerals are also introduced in cosmetic formulations as emollient agents (Michalak et al. 2020).

Adjuvant and additive

In cosmetic products, some adjuvants are added to enhance the role of the active substance. Additives are also needed to improve both preservation and organoleptic properties. Nowadays, many of them are controversial and suspected to cause dermatological problems or to be endocrine disruptors.

Thickening agent, emulsifier, encapsulation systems

In cosmetics, texture is fundamental for a product's appeal to consumers and for production issues. Thickening agents are compounds that can regulate moisture content, improve structure, rheological properties, and the stability of a cosmetic product (Himashree et al. 2022). In addition to their action as active ingredients, marine polysaccharides from macro/micro algae like fucoidan, agar, carrageenan, alginate, chitosan can improve the texture of a product by thickening or stabilizing it. Alginate also stabilizes and thickens emulsions at low pH (Kim 2014). Some proteins like collagen also have these properties but due to their high cost, they are more often used as active ingredient instead of adjuvants (Sionkowska et al. 2020).

Compounds made up of lipids in all their forms (phospholipids, glycolipids, fatty acids, triglycerides, sterols) such as lecithin from marine animals, stabilize emulsions by forming liposomes (Alhajj et al. 2020).

Some active molecules cannot be directly added in cosmetic formulations due to their heat, pH, light or oxidation sensitivity. Encapsulation of these compounds in an adequate medium is currently used to solve this problem. More than just protecting the compound of interest, the encapsulation systems make it possible to control its release when applied. Hydrogels made of marine polysaccharides such as chitosan, sodium alginate or fucoidan represent great encapsulation systems for bioactive compounds thanks to their biocompatibility and biodegradability (Barbosa et al. 2019, Casadidio et al. 2019, Mitura et al. 2020).

Preservatives (antimicrobial/antioxidant)

Parabens were until today additives used to prevent microbial and fungal product contamination but their safety is now controversial. It is thus important to find new preservative molecules that exhibit antioxidant and antimicrobial activity. Polysaccharides (chitosan and chitin) as well as sulfated polysaccharide from brown (fucoidan), red (agar, carrageenan) and green (ulvan) seaweeds demonstrate antibacterial and antioxidant activity and can help preserve cosmetic products. Polyphenols from macro/microalgae also present these properties (Casadidio et al. 2019, Couteau and Coiffard 2020, Resende et al. 2021). Some preservative properties of algal extract were also correlated to their high protein content (Couteau and Coiffard 2020).

Sensory agents (dyes and fragrance)

Natural dyes from marine sources drew the attention of the industry because some synthetic dyes were found to be allergenic. Cyanobacteria and micro/macroalgae synthetize phycobiliproteins, chlorophyll and carotenoids (ex: canthaxanthin) pigments that exhibit a wide range of colors from yellow, orange, and red to blue, green (Couteau and Coiffard 2020; Hosikian et al. 2010; Mourelle et al. 2017). Green colors can be obtained by the marennine pigment, synthetized by microalgae (Guillerme et al. 2017).

Marine compounds are also used to give interesting smells to cosmetic products. Algae and cyanobacteria produce terpenoids, carotenoids, fatty acid derivates and sulfur compounds. For example, geosmin from cyanobacteria has earthy or musty smells whereas b-ionone produced by green algae has a woody and fruity odor and polyunsaturated fatty acids derivatives give fishy, rancid and cucumber odors (Couteau and Coiffard 2020).

Some halophyte essential oils are also known for their lemony and musty smell (Guillerme et al. 2017). Lastly, the deodorizing function of chlorophyll from microalgae is already commercially exploited in toothpastes and deodorants (Hosikian et al. 2010).

Extraction of marine compounds

Different methodologies are suitable for obtaining bioactive compounds—such as polysaccharides, oils, peptides, vitamins or oils from different raw materials (Table 2). Nowadays, extraction technologies and techniques have advanced to include sustainability. In this context, non-conventional, green or ecofriendly extraction technologies have been developed to obtain our necessities without harming the environment. Another important factor in this section is the solvent that is used. Whereas in conventional extraction organic solvents have a special role, in green extraction technologies cleaner and safer alternatives are preferred. In terms of advantages or disadvantages, simplicity is associated with conventional extraction, among the disadvantages are the low selectivity, long extraction time, high energy consumption, high temperature and the need of organic toxic solvents. In the case of ecofriendly extraction technologies, reduction of energy consumption and shorter times, improve solubility, reproducibility, scalability, and the possibility of using compact equipment are some of the advantages. In this case, the main disadvantage is the investment as usually these equipment are sophisticated and require special necessities.

The most popular conventional methodologies are maceration and hydrodistillation. These technologies are commonly used as a homemade preparation to develop liquors, essential oils or unguent for the traditional medicine. Before extraction, the raw material needs proper drying and particle size reduction to enhance the surface area and favor the contact with solvent.

The main steps of these methodologies are summarized below:

Maceration requires placing the prepared raw material and the solvent in a closed vessel with adequate mixing to maintain a homogeneous suspension and to increase mass transfer. The next step is filtration to separate the liquid and the solid phase. The solvent in the liquid phase must be separated from the solute, which would be further purified (Azmir et al 2013, Giacosa et al. 2023, Paissoni et al. 2020).

Hydrodistillation can be performed by different methodologies. Currently, there are three different possibilities, water distillation, water and steam distillation and direct steam distillation. In these methods the quantity of solvent (water) is an important parameter to enhance the distillation process. These methodologies allow the use of raw vegetable materials such as leaves and flowers. They can be employed for the extraction of phenolic compounds, antioxidants, among other compounds with interest in industries. According to the target compounds, the type of extraction must be carefully selected (Azmir et al. 2013, Tăbărașu et al. 2023).

Nowadays concerns regarding the environment, sustainability and green chemistry are fundamental to design extraction technologies or processes. In this context, sustainability is based on keeping social, economic and ecologic points in view. The concept of green chemistry has been defined as a model targeted to design processes and products from a point of view of being ecofriendly (Chemat et al. 2012). Some of the most used ecofriendly extraction technologies are cited below:

Ultrasound assisted extraction is based on the application of ultrasonic waves causing a cavitation phenomenum, where the energy supplied depends on frequency and intensity influencing the bubble sizes. Besides, the solvent media, system and solid geometry, also influence the effects of the ultrasounds (Santha Kumar et al. 2023). Several authors have used this extraction technology

Table 2. Several marine raw materials are used to obtain bioactive compounds with ecofriendly extraction technologies.

Extraction technologies	Raw material	References
Ultrasound assisted extraction	*Caulerpa racemosa*	Sherley et al. 2023
	Pandalus borealis	Dave et al. 2020
	Nizamuddinia zanardinii	Alboofetileh et al. 2019
	Enteromorpha tubulosa	Hu et al. 2018
	Sea cucumber	Liu et al. 2015
Microwave assisted extraction	*Ulva pertusa*	Le et al. 2019
	Solieria filiformis	Peñuela et al. 2018
	Acaudina Molpadioides	Jin et al. 2019
	Holothuria atra	Yuniati and Sulardiono 2019
	Apostichopus japonicus	Liu et al. 2022
Pressurized liquid extraction	*Durvillaea antarctica*	Ruiz-Domínguez et al. 2023
	Kappaphycus alvarezii	Rudke et al. 2022
	Fucus vesiculosus	Getachew et al. 2022, Sumampouw et al. 2021
	Spirulina	Zhou et al. 2022
	Xestospongia	Bayona et al. 2018
Supercritical fluid extraction	*Undaria pinnatifida*	Yin et al. 2022
	Tuna	Taati et al. 2017
	Cladophora glomerata	Fabrowska et al. 2017
	Laminaria digitata	Ivahnov et al. 2016
	Dunaliella salina	Mouahid et al. 2016

to obtain bioactive compounds from marine sources, such as seaweeds (Chen et al. 2023, Tarman et al. 2023). It has also been used to recover collagen (Lu et al. 2023, Pezeshk et al. 2022).

Microwave assisted extraction is an emerging technology based on the rapid transfer of energy to the solution, by two processes: dipole rotation and ionic conduction generating a temperature increase in the medium. In addition, in the solid matrix, the cell wall rupture induced by the rapid water vaporization facilitates the release and of the target molecules to the media (Kadam et al. 2013, Quitério et al. 2022, Zhang et al. 2022). Besides, it is a suitable technology to recover marine phenolic compounds (Getachew et al. 2022) and also polysaccharides from seaweeds (Dobrinčić et al. 2020).

Pressurized liquid extraction is operated in the range of the extraction temperature from 50°C to 200°C and pressures from 3.5 to 20 MPa. The combination of pressure and temperature improves the extraction by increasing the solubility of the solutes, and by enhancing the mass transfer phenomena due to changes in viscosity and surface tension of the solvent (Kadam et al. 2013). Phenolic compounds from brown seaweeds (Mekinić et al. 2019, Shanmugam et al. 2021), lipids (Zhou et al. 2022) and pigments (Derwenskus et al. 2020) have been recovered using this technology.

Supercritical fluid extraction is based on an operation above the critical point of the solvent, thus exhibiting a density similar to a liquid, a viscosity similar to a gas, and diffusivity between the values of liquid and gases. These features improve the transport properties compared to those of a conventional liquid solvent (Kadam et al. 2013). Some works have used this extraction technology to obtain lipids from microalgae (Zhou et al. 2022) carotenoids and other compounds form marine sources (Ballesteros-Vivas et al. 2022).

Future trend in cosmetic marine products

Even if marine biodiversity is an almost infinite source of bioactive compounds with potential cosmeceutical applications, many of these molecules that seem to exhibit outstanding anti-aging, photo-protective and antioxidant properties, are not yet included in cosmeceutical products because of their instability, reactivity and because of a lack of information about them. Some examples could be cited. They are the phytohormones that are promising anti-aging, the mycosporine-like amino acids or even planktonic and cyanobacteria pigments such as biopterin glucose, which is an UV-A absorbing chromophore (Brunt and Burgess 2018). Finding adequate formulations for these biomolecules would open up the way to even more efficient cosmetic products.

Beyond innovation in formulations and extraction techniques, the cosmeceutical industry also works on the reduction of its environmental impact, looking for new material replacing plastic and non-biodegradable ones in traditional packaging. Biopolymers from marine source such as chitosan, alginate, carrageenan, thanks to their filming properties, can be considered as interesting candidate as bioplastics for packaging. In addition to their mechanical properties, they also present antimicrobial and antioxidant activity which could enhance product conservation. Adding antioxidant compounds in biomatrices in order to increase product shelf-life is also currently being studied for the food industry but could be interesting for cosmetics too (Bhullar et al. 2015, Lomartire et al. 2022, Kanatt et al. 2015).

According to the National Oceanic and Atmospheric Administration, 91% of ocean species are still unclassified and more than 80% of the ocean is unmapped, unobserved, and unexplored, which represents just as many opportunities to discover new bioactive compounds to improve the cosmeceutical field.

References

Alboofetileh, M., M. Rezaei, M. Tabarsa and S.G. You. 2019. Ultrasound-assisted extraction of sulfated polysaccharide from *Nizamuddinia zanardinii*: Process optimization, structural characterization, and biological properties. Journal of Food Process Engineering 42(2): e12979. https://doi.org/10.1111/jfpe.12979.

Alhajj, M.J., N. Montero, C.J. Yarce and C.H. Salamanca. 2020. Lecithins from vegetable, land, and marine animal sources and their potential applications for cosmetic, food, and pharmaceutical sectors. Cosmetics 7(4), Article 4. https://doi.org/10.3390/cosmetics7040087.

Alves, A.L., A.L.P. Marques, E. Martins, T.H. Silva and R.L. Reis. 2017. Cosmetic potential of marine fish skin collagen. Cosmetics 4(4), Article 4. https://doi.org/10.3390/cosmetics4040039.

Anbualakan, K., N.Q. Tajul Urus, S. Makpol, A. Jamil, E.S. Mohd Ramli, S.H. Md Pauzi et al. 2022. A scoping review on the effects of carotenoids and flavonoids on skin damage due to ultraviolet radiation. Nutrients 15(1): 92. https://doi.org/10.3390/nu15010092.

Azmir, J., I.S.M. Zaidul, M.M. Rahman, K.M. Sharif, A. Mohamed, F. Sahena et al. 2013. Techniques for extraction of bioactive compounds from plant materials: A review. Journal of Food Engineering 117(4): 426–436. https://doi.org/10.1016/j.jfoodeng.2013.01.014.

Ballesteros-Vivas, D., J.P. Ortega-Barbosa, F. Parada-Alfonso, S.R.S. Ferreira and A. del Pilar Sánchez-Camargo. 2022. Chapter 13 - Supercritical fluid extraction of lipids, carotenoids, and other compounds from marine sources. pp. 277–317. *In*: Garcia-Vaquero, M. and G. Rajauria [eds.]. Innovative and Emerging Technologies in the Bio-marine Food Sector. Academic Press. https://doi.org/10.1016/B978-0-12-820096-4.00016-X.

Barbosa, A.I., A.J. Coutinho, S.A. Costa Lima and S. Reis. 2019. Marine polysaccharides in pharmaceutical applications: Fucoidan and chitosan as key players in the drug delivery match field. Marine Drugs 17(12), Article 12. https://doi.org/10.3390/md17120654.

Bayona, L.M., M. Videnova and Y.H. Choi. 2018. Increasing metabolic diversity in marine sponges extracts by controlling extraction parameters. Marine Drugs 16(10): 393. https://doi.org/10.3390/md16100393.

Bhullar, S.K., B. Kaya and M.B.-G. Jun. 2015. Development of bioactive packaging structure using melt electrospinning. Journal of Polymers and the Environment 23(3): 416–423. https://doi.org/10.1007/s10924-015-0713-z.

Brunt, E.G. and J.G. Burgess. 2018. The promise of marine molecules as cosmetic active ingredients. Int. J. Cosmet. Sci. 2018 Feb. 40(1): 1–15. doi: 10.1111/ics.12435. Epub 2017 Dec 4. PMID: 29057483.

Brunt, E.G. and J.G. Burgess. 2018. The promise of marine molecules as cosmetic active ingredients. International Journal of Cosmetic Science 40(1): 1–15. https://doi.org/10.1111/ics.12435.

Camilion, J.V., S. Khanna, S. Anasseri, C. Laney and H.N. Mayrovitz. 2022. Physiological, Pathological, and Circadian Factors Impacting Skin Hydration. Cureus. https://doi.org/10.7759/cureus.27666.

Casadidio, C., D.V. Peregrina, M.R. Gigliobianco, S. Deng, R. Censi and P. Di Martino. 2019. Chitin and chitosans : characteristics, eco-friendly processes, and applications in cosmetic science. Marine Drugs 17(6), Article 6. https://doi.org/10.3390/md17060369.

Chemat, F., M.A. Vian and G. Cravotto. 2012. Green extraction of natural products: Concept and principles. International Journal of Molecular Sciences 13(7): 8615–8627. https://doi.org/10.3390/ijms13078615.

Chen, Z., N. Shen, X. Wu, J. Jia, Y. Wu, H. Chiba et al. 2023. Extraction and quantitation of phytosterols from edible brown seaweeds: Optimization, validation, and application. Foods 12(2): 244. https://doi.org/10.3390/foods12020244.

Cheung, R.C.F., J.H. Wong, W. Pan, Y.S. Chan, C. Yin, X. Dan et al. 2015. Marine lectins and their medicinal applications. Applied Microbiology and Biotechnology 99(9): 3755–3773. https://doi.org/10.1007/s00253-015-6518-0.

Claverie, M., C. McReynolds, A. Petitpas, M. Thomas and S.C.M. Fernandes. 2020. Marine-derived polymeric materials and biomimetics: An overview. Polymers 12(5): 1002. https://doi.org/10.3390/polym12051002.

Corinaldesi, C., G. Barone, F. Marcellini, A. Dell'Anno and R. Danovaro. 2017. Marine microbial-derived molecules and their potential use in cosmeceutical and cosmetic products. Marine Drugs 15(4): 118. https://doi.org/10.3390/md15040118.

Couteau, C. and L. Coiffard. 2020. Phycocosmetics and other marine cosmetics, specific cosmetics formulated using marine resources. Marine Drugs 18(6), Article 6. https://doi.org/10.3390/md18060322.

Dai, D., X. Ma, X. Yan and X. Bao. 2023. The biological role of dead sea water in skin health : A review. Cosmetics 10(1), Article 1. https://doi.org/10.3390/cosmetics10010021.

Daudt, R.M., P.I. Back, N.S.M. Cardozo, L.D.F. Marczak and I.C. Külkamp-Guerreiro. 2015. Pinhão Starch and Coat Extract as New Natural Cosmetic Ingredients: Topical Formulation Stability and Sensory Analysis. Carbohydr. Polym. 134: 573–580.

Dave, D., Y. Liu, J. Pohling, S. Trenholm and W. Murphy. 2020. Astaxanthin recovery from Atlantic shrimp (*Pandalus borealis*) processing materials. Bioresource Technology Reports 11: 100535. https://doi.org/10.1016/j.biteb.2020.100535.

de Jager, T.L., A.E. Cockrell and S.S. Du Plessis. 2017. Ultraviolet light induced generation of reactive oxygen species. Advances in Experimental Medicine and Biology 996: 15–23. https://doi.org/10.1007/978-3-319-56017-5_2.

Derwenskus, F., B. Schäfer, J. Müller, K. Frick, A. Gille, K. Briviba et al. 2020. Coproduction of EPA and fucoxanthin with *P. tricornutum*—a promising approach for up- and downstream processing. Chemie-Ingenieur-Technik 92(11): 1780–1789. https://doi.org/10.1002/cite.202000046.

Dobrinčić, A., S. Balbino, Z. Zorić, S. Pedisić, D.B. Kovačević, I.E. Garofulić et al. 2020. Advanced technologies for the extraction of marine brown algal polysaccharides. Marine Drugs 18(3): 168. https://doi.org/10.3390/md18030168.

Fabrowska, J., A. Kapuscinska, B. Leska, K. Feliksik-Skrobich and I. Nowak. 2017. *In vivo* studies and stability of *Cladophora glomerata* extract as a cosmetic active ingredient. Acta Poloniae Pharmaceutica 74(2): 633–641.

Fehrenbach, G.W., R. Pogue, F. Carter, E. Clifford and N. Rowan. 2022. Implications for the seafood industry, consumers and the environment arising from contamination of shellfish with pharmaceuticals, plastics and potentially toxic elements: A case study from Irish waters with a global orientation. Science of The Total Environment, 844: 157067. https://doi.org/10.1016/j.scitotenv.2022.157067.

Fonseca, S., M.N. Amaral, C.P. Reis and L. Custódio. 2023. Marine natural products as innovative cosmetic ingredients. Marine Drugs 21(3), Article 3. https://doi.org/10.3390/md21030170.

Geraldes, V. and E. Pinto. 2021. Mycosporine-Like Amino Acids (MAAs) : Biology, chemistry and identification features. Pharmaceuticals 14(1), Article 1. https://doi.org/10.3390/ph14010063.

Getachew, A.T., S.L. Holdt, A.S. Meyer and C. Jacobsen. 2022. Effect of extraction temperature on pressurized liquid extraction of bioactive compounds from *Fucus vesiculosus*. Marine Drugs 20(4): 263. https://doi.org/10.3390/md20040263.

Giacosa, S., L. Ferrero, M.A. Paissoni, S. Río Segade, V. Gerbi and L. Rolle. 2023. Grape skin anthocyanin extraction from red varieties during simulated maceration: Influence of grape seeds and pigments adsorption on their surface. Food Chemistry 424: 136463. https://doi.org/10.1016/j.foodchem.2023.136463.

Guillerme, J.-B., C. Couteau and L. Coiffard. 2017. Applications for marine resources in cosmetics. Cosmetics 4(3), Article 3. https://doi.org/10.3390/cosmetics4030035.

Gunia-Krzyżak A., K. Słoczyńska, J. Popiół, P. Koczurkiewicz, H. Marona and E. Pękala. 2018. Cinnamic acid derivatives in cosmetics: current use and future prospects Int. J. Cosmetic Sci. 40(4): 356–366.

Himashree, P., A.S. Sengar and C.K. Sunil. 2022. Food thickening agents : Sources, chemistry, properties and applications - A review. International Journal of Gastronomy and Food Science 27: 100468. https://doi.org/10.1016/j.ijgfs.2022.100468.

Hosikian, A., S. Lim, R. Halim and M.K. Danquah. 2010. Chlorophyll extraction from microalgae : A review on the process engineering aspects. International Journal of Chemical Engineering, 1–11. https://doi.org/10.1155/2010/391632.

Hu, Z., P. Hong, Y. Cheng, M. Liao and S. Li. 2018. Polysaccharides from *Enteromorpha tubulosa*: Optimization of extraction and cytotoxicity. Journal of Food Processing and Preservation 42(1): e13373. https://doi.org/10.1111/jfpp.13373.

Ivahnov, A.D., T.E. Skrebets and K.G. Bogolitsyn. 2016. Supercritical fluid extraction of chlorophylls and carotenoids from White Sea algae. Russian Journal of Physical Chemistry B 10(8): 1244–1247. https://doi.org/10.1134/S1990793116080133.

Jin, H.X., H.P. Xu, Y. Li, Q.W. Zhang and H. Xie. 2019. Preparation and evaluation of peptides with potential antioxidant activity by microwave assisted enzymatic hydrolysis of collagen from sea cucumber *Acaudina molpadioides* obtained from Zhejiang province in China. Marine Drugs 17(3): 169. https://doi.org/10.3390/md17030169.

Kadam, S.U., B.K. Tiwari and C.P. O'Donnell. 2013. Application of novel extraction technologies for bioactives from marine algae. Journal of Agricultural and Food Chemistry 61(20): 4667–4675. https://doi.org/10.1021/jf400819p.

Kalasariya, H.S., V.K. Yadav, K.K. Yadav, V. Tirth, A. Algahtani, S. Islam et al. 2021. Seaweed-based molecules and their potential biological activities : An eco-sustainable cosmetics. Molecules 26(17), Article 17. https://doi.org/10.3390/molecules26175313.

Khan, M.I., J.H. Shin and J.D. Kim. 2018. The promising future of microalgae : Current status, challenges, and optimization of a sustainable and renewable industry for biofuels, feed, and other products. Microbial Cell Factories 17, 36. https://doi.org/10.1186/s12934-018-0879-x.

Kim, S.-K. 2014. Marine cosmeceuticals. Journal of Cosmetic Dermatology 13(1): 56–67. https://doi.org/10.1111/jocd.12057.

Kim, S.-K., D. Ravichandran, S. Khan and Y. Kim. 2008. Prospective of the cosmeceuticals derived from marine organisms. Biotechnology and Bioprocess Engineering 13: 511–523. https://doi.org/10.1007/s12257-008-0113-5.

Le, B., K.S. Golokhvast, S.H. Yang and S. Sun. 2019. Optimization of microwave-assisted extraction of polysaccharides from *Ulva pertusa* and evaluation of their antioxidant activity. Antioxidants 8: 129. https://doi.org/10.3390/antiox8050129.

Liu, G., X. Wang, K. Ge, J. Wu, Z. Wang, J. Du, L. Song and Z. Zhou. 2022. Isolation, identification, and quantitative determination of Saponin in *Apostichopus japonicus* by HPLC-DAD. Journal of Ocean University of China 21(2): 473–478. https://doi.org/10.1007/s11802-022-4854-0.

Liu, Y., J. Xu, Y. Xue, J. Wang, Y. Chang and C. Xue. 2015. Determination of trace vanadium in sea cucumbers by ultrasound-assisted cloud point extraction and graphite furnace atomic absorption spectrometry. International Journal of Environmental Analytical Chemistry 95(3): 258–270. https://doi.org/10.1080/03067319.2014.1002494.

Lomartire, S., J.C. Marques and A.M.M. Gonçalves. 2022. An overview of the alternative use of seaweeds to produce safe and sustainable bio-packaging. Applied Sciences 12(6), Article 6. https://doi.org/10.3390/app12063123.

Lu, W.-C., C.-S. Chiu, Y.-J. Chan, A.T. Mulio and P.-H. Li. 2023. Characterization and biological properties of marine by-product collagen through ultrasound-assisted extraction. Aquaculture Reports, 29: 101514. https://doi.org/10.1016/j.aqrep.2023.101514.

Mekinić, I.G., D. Skroza, V. Šimat, I. Hamed, M. Čagalj and Z.P. Perković. 2019. Phenolic content of brown algae (Pheophyceae) species: Extraction, identification, and quantification. Biomolecules 9(6): 244. https://doi.org/10.3390/biom9060244.

Michalak, I., A. Dmytryk and K. Chojnacka. 2020. Algae cosmetics. In Encyclopedia of Marine Biotechnology (p. 65–85). John Wiley & Sons, Ltd. https://doi.org/10.1002/9781119143802.ch3.

Mirata, S., V. Asnaghi, M. Chiantore, A. Salis, M. Benvenuti, G. Damonte et al. 2023. Photoprotective and anti-aging properties of the apical frond extracts from the mediterranean seaweed *Ericaria amentacea*. Mar. Drugs 21: 306. https://doi.org/10.3390/md21050306.

Mitura, S., A. Sionkowska and A. Jaiswal. 2020. Biopolymers for hydrogels in cosmetics: Review. Journal of Materials Science: Materials in Medicine 31(6): 50. https://doi.org/10.1007/s10856-020-06390-w.

Mouahid, A., C. Crampon, S.-A.A. Toudji and E. Badens. 2016. Effects of high water content and drying pre-treatment on supercritical CO_2 extraction from *Dunaliella salina* microalgae: Experiments and modelling. The Journal of Supercritical Fluids 116: 271–280. https://doi.org/10.1016/j.supflu.2016.06.007.

Mourelle, M.L., C.P. Gómez and J.L. Legido. 2017. The potential use of marine microalgae and cyanobacteria in cosmetics and thalassotherapy. Cosmetics 4(4), Article 4. https://doi.org/10.3390/cosmetics4040046.

Pagels, F., C. Almeida, V. Vasconcelos and A.C. Guedes. 2022. Cosmetic potential of pigments extracts from the marine *Cyanobacterium cyanobium* sp. Marine Drugs 20(8), Article 8. https://doi.org/10.3390/md20080481.

Paissoni, M.A., S. Río Segade, C. Carrero-Carralero, C. Montanini, S. Giacosa and L. Rolle. 2020. Role of anthocyanin traits on the impact of oenological tannins addition in the first stage of red winegrape skin simulated maceration. Food Chemistry 320: 126633. https://doi.org/10.1016/j.foodchem.2020.126633.

Peñuela, A., D. Robledo, N. Bourgougnon, G. Bedoux, E. Hernández-Nú ez and Y. Freile-Pelegrín. 2018. Environmentally friendly valorization of *Solieria filiformis* (Gigartinales, rhodophyta) from IMTA using a biorefinery concept. Marine Drugs 16(12): 487. https://doi.org/10.3390/md16120487.

Pereira, C.G. 2022. Special issue on advances in marine biotechnology: exploitation of halophyte plants. Applied Sciences 12(20), Article 20. https://doi.org/10.3390/app122010573.

Pezeshk, S., M. Rezaei and M. Abdollahi. 2022. Impact of ultrasound on extractability of native collagen from tuna by-product and its ultrastructure and physicochemical attributes. Ultrasonics Sonochemistry 89: 106129. https://doi.org/10.1016/j.ultsonch.2022.106129.

Pimentel, F.B., R.C. Alves, F. Rodrigues and M.B.P.P. Oliveira. 2018. Macroalgae-derived ingredients for cosmetic industry—An update. Cosmetics 5: 2. https://doi.org/10.3390/cosmetics5010002.

Quitério, E., C. Grosso, R. Ferraz, C. Delerue-Matos and C. Soares. 2022. A critical comparison of the advanced extraction techniques applied to obtain health-promoting compounds from seaweeds. Marine Drugs 20(11): 677. https://doi.org/10.3390/md20110677.

Quitério, E., C. Soares, R. Ferraz, C. Delerue-Matos and C. Grosso. 2021. Marine health-promoting compounds : Recent trends for their characterization and human applications. Foods 10(12): 3100. https://doi.org/10.3390/foods10123100.

Kanatt, S.R., P. Lahare, S. Chawla and A. Sharma. 2015. Kappaphycus alvarezii : Its antioxidant potential and use in bioactive packaging films. Journal of Microbiology, Biotechnology and Food Sciences 5(1): 1–6. https://doi.org/10.15414/jmbfs.2015.5.1.1-6.

Resende, D.I.S.P., M. Ferreira, C. Magalhães, J.M. Sousa Lobo, E. Sousa and I.F. Almeida. 2021. Trends in the use of marine ingredients in anti-aging cosmetics. Algal Research 55: 102273. https://doi.org/10.1016/j.algal.2021.102273.

Rudke, A.R., M. da Silva, C.J. de Andrade, L. Vitali and S.R.S. Ferreira. 2022. Green extraction of phenolic compounds and carrageenan from the red alga *Kappaphycus alvarezii*. Algal Research 67: 102866. https://doi.org/10.1016/j.algal.2022.102866.

Ruiz-Domínguez, M.C., J.A. Mendiola, J.D. Sánchez-Martínez, M. Bueno, P. Cerezal-Mezquita and E. Ibáñez. 2023. Evaluation of the antioxidant and neuroprotective activity of the seaweed *Durvillaea antarctica* (cochayuyo) extracts using pressurized liquids. Journal of Applied Phycology 35(2): 835–847. https://doi.org/10.1007/s10811-022-02901-0.

Santha Kumar, A.R.S., A. Padmakumar, U. Kalita, S. Samanta, A. Baral, N.K. Singha et al. 2023. Ultrasonics in polymer science: applications and challenges. Progress in Materials Science 136: 101113. https://doi.org/10.1016/j.pmatsci.2023.101113.

Shanmugam, M., A.R. Ganesan and G. Rajauria. 2021. Extraction technologies to recover dietary polyphenols from macro- and microalgae. In Recent Advances in Micro and Macroalgal Processing (pp. 163–187). Wiley. https://doi.org/10.1002/9781119542650.ch7.

Sherley, Hikmawanti, N.P.E. and M. Agustin. 2023. The effect of extraction methods on total phenolics and antioxidant activities in *Caulerpa racemosa* (Forsskal) J. Agardh extracts. Vitae 30(2): 349405. https://doi.org/10.17533/udea.vitae.v30n2a349405.

Sionkowska, A., K. Adamiak, K. Musiał and M. Gadomska. 2020. Collagen based materials in cosmetic applications: A review. Materials 13(19), Article 19. https://doi.org/10.3390/ma13194217.

Srinivasan, R., A. Kannappan, C. Shi and X. Lin. 2021. Marine bacterial secondary metabolites: A treasure house for structurally unique and effective antimicrobial compounds. Marine Drugs 19(10): 530. https://doi.org/10.3390/md19100530.

Stahl, W. and H. Sies. 2012. β-Carotene and other carotenoids in protection from sunlight. The American Journal of Clinical Nutrition 96(5): 1179S–1184S. https://doi.org/10.3945/ajcn.112.034819.

Sumampouw, G.A., C. Jacobsen and A.T. Getachew. 2021. Optimization of phenolic antioxidants extraction from *Fucus vesiculosus* by pressurized liquid extraction. Journal of Applied Phycology 33(2): 1195–1207. https://doi.org/10.1007/s10811-020-02362-3.

Taati, M.M., B. Shabanpour and M. Ojagh. 2017. Extraction of oil from tuna by-product by supercritical fluid extraction (SFE) and comparison with wet reduction method. AACL Bioflux 10(6): 1546–1553.

Tăbăraşu, A.M., D.N. Anghelache, I. Găgeanu, Biriş, S. Ştefan and N.V. Vlăduţ. 2023. Considerations on the use of active compounds obtained from lavender. Sustainability (Switzerland) 15(11): 8879. https://doi.org/10.3390/su15118879.

Tarman, K., M.W.U. Zuhai and R.F. Pari. 2023. Ultrasound-assisted depolymerization of carrageenan from *Kappaphycus alvarezii* hydrolized by a marine fungus. IOP Conference Series: Earth and Environmental Science 1137(1): 012048. https://doi.org/10.1088/1755-1315/1137/1/012048.

Yin, S., L. Niu, M. Shibata, Y. Liu and T. Hagiwara. 2022. Optimization of fucoxanthin extraction obtained from natural by-products from *Undaria pinnatifida* stem using supercritical CO_2 extraction method. Frontiers in Nutrition 9: 981176. https://doi.org/10.3389/fnut.2022.981176.

Yuniati, R. and B. Sulardiono. 2019. Exploration of the collagen of non commercial sea cucumber *Holothuria atra* and commercial sea cucumber *Stichopus vastus* in the Karimunjawa Islands, Indonesia. Indo Pacific Journal of Ocean Life 3(1): 18–23. https://doi.org/10.13057/oceanlife/o030103.

Zhang, Y., X. Duan, T. Wassie, H.H. Wang, T. Li, C. Xie and X. Wu. 2022. *Enteromorpha prolifera* polysaccharide-zinc complex modulates the immune response and alleviates LPS-induced intestinal inflammation via inhibiting the TLR4/NF-κB signaling pathway. Food and Function 13(1): 52–63. https://doi.org/10.1039/d1fo02171k.

Zhou, J., M. Wang, J.A. Saraiva, A.P. Martins, C.A. Pinto, M.A. Prieto et al. 2022. Extraction of lipids from microalgae using classical and innovative approaches. Food Chemistry 384: 132236. https://doi.org/10.1016/j.foodchem.2022.132236.

Zillich, O.V., U. Schweiggert-Weisz, P. Eisner and M. Kerscher. 2018. Int. J. Cosmetic Sci. 37(5): 455-464 (2015). Sathasivam R., Ki J.-S. Mar. Drugs 16(1), Article number 26.

CHAPTER 13

Significance of Algal Lipids in Cosmeceutical Sector

Khevna Pandya,[1] *Mittal Thakkar*[2,*] and *Anupama Shrivastav*[3,*]

Introduction

Cosmeceuticals, which includes physiologically active substances with therapeutic effect is a phrase that combines the terms cosmetics and pharmaceuticals. Cosmeceuticals are made up of a variety of chemical compounds, which can be sourced naturally from plants, animals, algae, and minerals. However, some compounds such as sodium lauryl sulphate, PVP, and ethyl paraben are artificially produced (Couteau et al. 2020). Due to the active substances present in cosmetics, their goal is to improve the skin's look without changing its structure or function (Mourelle et al. 2017, Vermeer et al. 1996). Contrarily, the term "cosmeceuticals" is used in the cosmetics business to describe cosmetics that contain substances that have therapeutic or drug-like properties (Mourelle et al. 2017, Kligman 2005). The bulk of cosmeceutical products are topical lotions or creams. Dermatologists recommend 30–40% of these products even though the wording may be unclear and the cosmeceuticals are not officially recognized by the European Union or the US Food and Drug Administration (US-FDA) (Pandey et al. 2021). Algae have been intensively researched as a potential replacement source for botanical cosmeceuticals and are popular and offer a number of advantages (Spolaore et al. 2006, Chakdar et al. 2012, Stolz et al. 2005).

Algal products have been utilized in the cosmetics industry as antioxidants, moisturizing agents, skin sensitizers, thickening agents and sunscreens, to enhance the skin's resilience against abrasions, tanning, and other factors (Fig. 1) (Wang et al. 2015). Algae are primitive unicellular or multicellular eukaryotes that utilize energy from sunlight to generate chemical energy for the creation of organic molecules like sugars. They are primary producers of chlorophyll, a green pigment found in some types of algae which is essential for the photosynthesis process. The pigments aid in the transfer of energy from the light source to the reaction centers of photosystems I and II. These pigments can be classified into two categories: Chlorophyll a and Chlorophyll b. This process requires the

[1] Department of Microbiology, Faculty of Applied Science, Parul University, Ahmedabad, Gujarat, India.
[2] Department of Chemical Engineering, Faculty of Technology, Dharmsinh Desai University, Nadiad-387001, Gujarat, India.
[3] Department of Microbiology, Faculty of Life Health and Allied Sciences, ITM Vocational University, Vadodara, Gujarat, India.
* Corresponding authors: mittald.thakkar777@gmail.com, mittalthakkar.ch@ddu.ac.in; anupamashrivastav@gmail.com

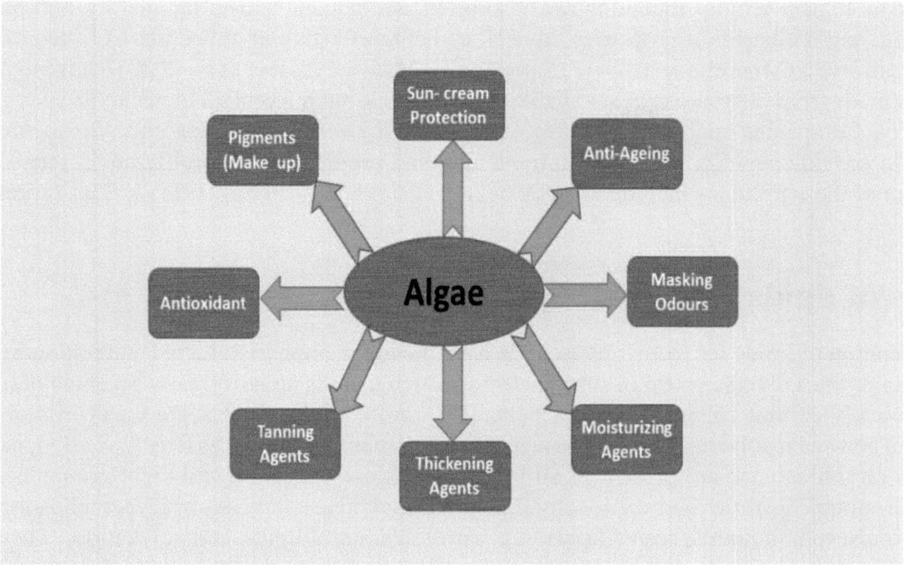

Figure 1. Potential benefits of algae in cosmeceuticals.

conversion of carbon dioxide, water, and sunlight into sugars like glucose/starch and biomass (Croce et al. 2014). Algal species can endure and thrive in harsh environmental extremes like as pH, temperature, osmotic pressure, salinity, UV light exposure, and anaerobiosis. When faced with intruders, algae can protect their biological components by producing primary metabolites such as oleic acids, vitamin E, vitamin B12, lutein, and zeaxanthin (Morais et al. 2015). In harsh conditions, secondary metabolites may also be produced. Algae's metabolites have antibiotic and antimicrobial properties, making them effective against pathogenic fungi and viruses (Wang et al. 2014). Algae are divided into two categories based on their size into microalgae and macroalgae.

Microalgae, also known as blue-green algae or cyanobacteria, are tiny, unicellular prokaryotic algae with a diameter ranging from 1 to 50 m. Although some can grow heterotrophically, most are phototrophic. They use photosynthesis releasing oxygen, which is like that of terrestrial plants, to power their metabolism (Wolkers et al. 2011). They can exist as solitary cells, in pairs, or in chains. Microalgae are rich in nutrients such as phosphorus, calcium, iron, vitamins A, B, C, and E, pantothenic acid, biotin, beta-carotene, folic acid and vitamin B12 (Fabregas and Herrero 1990). Certain microalgal species have the ability to substitute phospholipids for non-phosphorus membrane lipids to adapt to changes when the environmental phosphorus supply is depleted (Bonachela et al. 2011). Seaweeds are macro-algae that are eukaryotic, macroscopic, and multicellular. Marine or seawater with the ideal amount of light is the habitat for macroalgal organisms (Milledge 2014). Macroalgae are benthic plants, and their survival depends on their attachment to the seafloor or a hard layer of rock below them. Unlike ordinary terrestrial plants, which typically have complex tissue and organ organization, macroalgae have distinct visual characteristics such as the thallus, holdfast, lamina, and frond sorus (Adhithya et al. 2023).

Algae produce and release significant metabolites as a result of ongoing exposure to a variety of stressors, including high or low temperatures, photo-oxidation, osmotic pressure, high salinity, and UV light (Gauthier et al. 2020). Algae have garnered significant biotechnological attention due to their abundant metabolites, which can be utilized for various purposes in industries such as medical, business, energy, agriculture, and food. As stress can enhance or diminish the production of many valuable substances, scientific research is focused on identifying the optimal conditions for algae cultivation to maximize industrial output (Patel et al. 2019). Algal metabolites exhibit an immense biological activity, including UV absorption, antioxidant, anti-inflammatory, anti-blemish, anti-aging, and antibacterial properties, as well as being effective at collecting CO_2, reducing the greenhouse effect (Mourelle et al. 2017, Sartori et al. 2021, Levasseur et al. 2020). Currently, a few products made from microalgae are available in the market, but microalgal biotechnology is gaining popularity. Companies such as DSM use extracts from *Dunaliella salina* and *Nannochloropsis oculata* to provide benefits such as skin tightening and promoting cell proliferation, respectively. Examples of these products include PEPHA®-TIGHT CB and PEPHA®-CTIVE CB (Yarkent et al 2020).

Basics of lipids with their classification

The generation of lipids for industrial uses has been the main emphasis of algal cultivation in recent years. Lipids are a diverse group of substances for which there is no universally accepted definition. They typically exhibit chemical heterogeneity. They are compounds that are water-insoluble and soluble in non-polar solvents. Hydrophobic building components known as fatty acyl/alkyl, isoprene fractions or sphingosine are present in all lipids. Lipids are classified into eight groups based on their hydrophobic qualities and chemically functional backbones: fatty acyls, glycerophospholipids, glycerolipids, sphingolipids, prenol lipids, saccharolipids, sterol lipids, and polyketides. These lipid types are further classified into lipid classes and subclasses (Baumann et al. 2002, Fahy et al. 2011). A variety of crucial biological activities include lipids. Because of their hydrophobic properties, lipids are fundamental components of the biological membranes, and as such, lipids serve as the structural basis for all living organisms. Another function that lipids carry out is the preservation of extra energy for subsequent use. Finally, because they transmit signals and increase regulatory cascades lipids also contribute to extracellular and intracellular signaling processes. Lipids play a role in the photosynthetic processes in plants (Zullig et al. 2020, Li-beisson et al. 2016). In addition to their significance in biology, lipids are used for a variety of commercial endeavors. Lipids are a crucial component in the food industry, as they are used in the production of food and dietary supplements for both humans and animals, among other important applications. Detergents, paints, biofuels, inks, hydraulic fluids, lubricants and phytochemical compounds require oils (Tao et al. 2007). Increasing customer demand is gravitating towards vegetable oils. That have lipids. Lipid susceptibility is a crucial feature to take into account since it is heavily influenced by a variety of factors. Lipids are particularly susceptible to oxidation which impairs their biological functions. There are numerous internal and external factors that affect oxidative stability. First, different lipid classes exhibit varying degrees of oxidative stability. The susceptibility of lipids to oxidation is determined by the amount of fatty acid unsaturation and the spatial orientation within the triglyceride molecules (Fahy et al. 2011).

Algal lipids

Phospholipids, glycolipids (glycosylglycerides), non-polar glycerolipids (neutral lipids), betaine and a few other uncommon lipids that are present in a particular type of algae species or genus are among the several types of lipids found in algae. They differ significantly from higher plants in terms of chain length and unsaturation. The hydrophobic acyl groups of glycerolipids are esterified to the glycerol framework at the sn-1 and sn-2 locations, which is a physiologically-derived compound from glycerol 3-phosphate. Based on the presence of a phosphate group at the sn-3 position, which is further connected to a hydrophilic head group, certain phospholipid molecules can be classified. The main phospholipids present in algae are phosphatidic acid (PA), phosphatidylinositol (PI), phosphatidylserine (PS), phosphatidylethanolamine (PE), and phosphatidylglycerol (PG). Different distinctive head groups, including serine, glycerol, ethanolamine, choline, and myo-inositol, are present in these phospholipids (Figure 2) (Macabuhay et al. 2022).

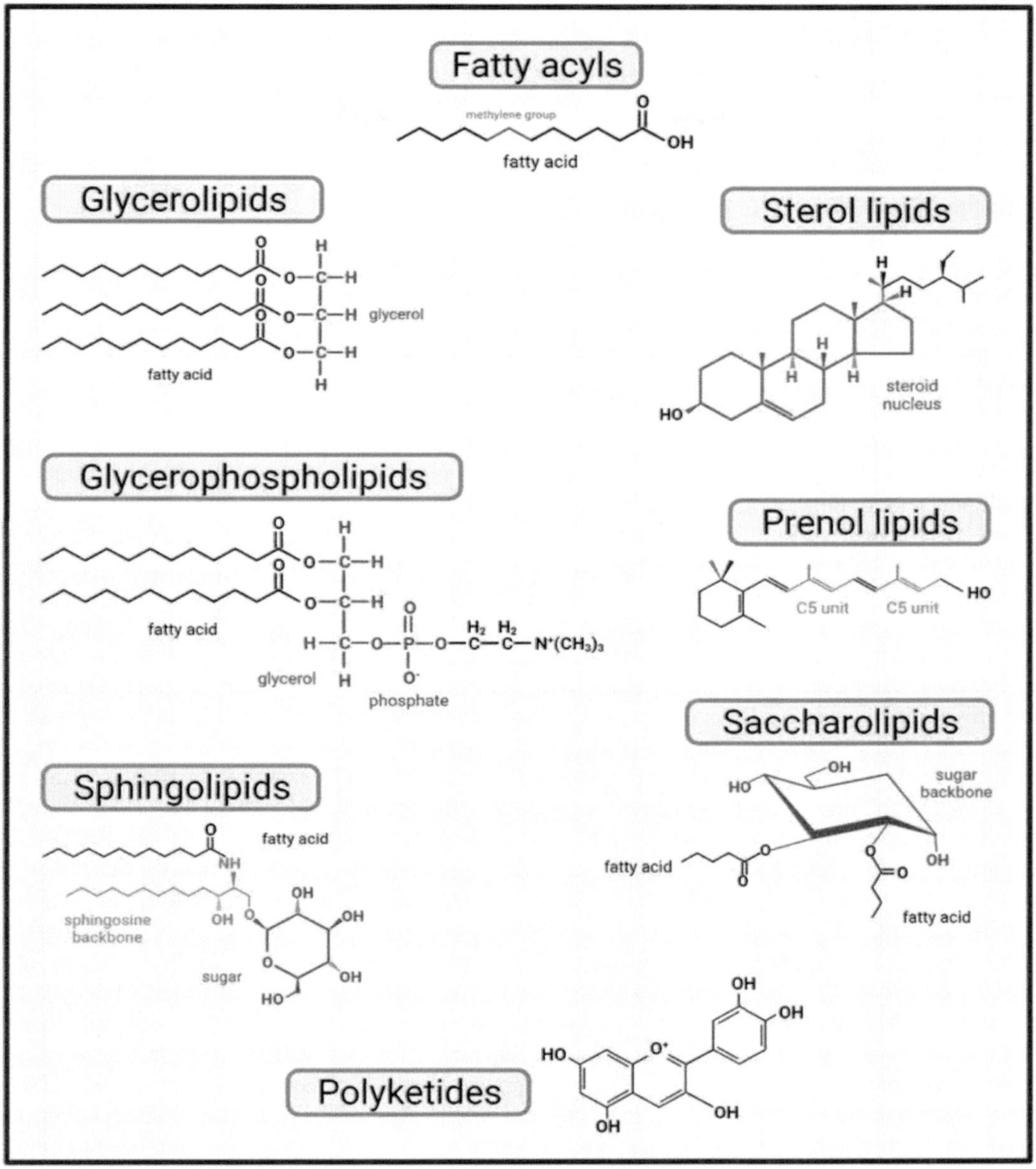

Figure 2. Structured of common lipid molecular.

Lipids are involved in a wide range of important biological functions. They are the major components of biological membranes due to their hydrophobic qualities, and as such, they provide the physical foundation for all living organisms. Another role that lipids perform is the storage of excess energy for later use. Finally, lipids contribute to extracellular and intracellular signaling processes by transmitting signals and increasing regulatory cascades. Plant photosynthetic processes are aided by lipids (Zullig et al. 2020, Li-beisson et al. 2016). Plant-derived lipids, particularly those generated from oilseeds such as palm, soy, rapeseed, and sunflower oils, often meet the large industrial demand. There were various concerns expressed by the rising lipid demand, including decreased availability and the amount of available land for farming and resource use.

A sustainable alternative to lipid sources is microorganisms. It is known that different prokaryotic and eukaryotic microbes create lipids in varying amounts and compositions (Stevens et al. 2004). They are categorized as "oleaginous" when their lipid content surpasses 20% of their dry biomass. Oleaginous microorganisms fall into three major family groups: algae, fungus (molds and yeast), and bacteria (Lopes et al. 2019).

Lipids for cosmetic purposes

Cosmetics have long been important for society. They are exact blends of active principles, excipients, and additives, each of which is responsible for the stated aesthetic activity. They have, other compounds as well that assure proper pharmaceutical dosages. They are material mixes that are stable and homogeneous. In addition to the additives used to preserve the product and increase its organoleptic properties (Couteau et al. 2020). Cosmetics are applied to clean, smell, protect, or change the appearance of the skin, hair, nails, lips, teeth, and mucous membranes of the mouth, among other parts of the human body (Guillerme et al. 2017). The uses of cosmetics include anything from beauty products like perfumes and makeup to common hygiene items like soap, shampoo, deodorants and toothpaste. In order to protect consumers, cosmetics production is strictly regulated (European commission 2009). The cosmetics industry has gradually increased the usage of several synthetic chemicals over time replacing the natural substances that were formerly used (Panico et al. 2019). Examples of these include mineral oils and waxes. They are stable, dermatologically safe substances and are utilized to control formulation viscosity as well as for their lubricating and protecting qualities. Natural crude petroleum oil is refined by several processes, such as distillation, extraction, crystallization, and purification to eliminate potentially hazardous elements, to create mineral oils and waxes. Additionally, mineral oils are extremely stable, non-allergenic, and resistant to rancidity and oxidation (Chuberre et al. 2019). Although there are strict restrictions in place, there are significant worries about the likelihood of finding traces of dangerous solvents in finished chemical formulations and the potential implications this poses for human health. The proposal to replace them with vegetable oils has not yet been completely adopted since chemicals provide stronger economic competitiveness. Numerous worries are emerging regarding the use of chemicals. Synthetic compounds can lack safety information, which puts consumers at risk of adverse reactions like hypersensitivity, anaphylaxis, deadly poisonings, or long-lasting effects. Testing for genetic toxicity, phototoxicity, photogenotoxicity, toxicokinetics, and carcinogenicity of these compounds can be done in a systematic manner. Additionally, fresh research may disclose different toxicity data (Petry et al. 2017, Morais et al. 2021, Yarkent et al. 2020). In addition, using many cosmetic items on a regular basis exposes the customer to various chemicals continuously. As a result, the presence of chemicals may cause additive action and synergistic interaction of several compounds, which are identical components utilized in multiple products. These factors make the need to replace chemicals more and more urgent, driving the cosmetics industry to constantly search for new technologies. Finding natural ingredients to satisfy customers' increasingly frequent requests, which are driven by growing awareness of the significance of using high-quality, ecologically friendly products, is one of the current issues (Birjiandi et al. 2020, Wang et al. 2015). Macroalgae-based cosmetics are currently available in the market and are gradually replacing their synthetic equivalents. These products might comprise macroalgal extracts, using several bioactive compounds in their purest form. Macroalgae are mostly used as thickening and gelling agents; the active substances that lead to their use in cosmetics are diverse (proteins, phospholipids, polysaccharides, peptides, amino acids, phenolic compounds, fatty acids, sterols, glycolipids, colors, vitamins). Even if some substances made from microalgae are currently available on the market, the world of microalgae is still mostly unexplored (Lourenço-Lopes et al. 2020, Morocho-Jácome 2020). Aside from lipids with chemical origins, many kinds of vegetable and animal oils and fats can be employed as neutral bases and bioactive substances; nevertheless, lipids and their derivatives are now acquired in cosmetic formulations from plants or biotechnology. Lipids that are often utilized in cosmetics include glycerophospholipids, triacyl glycerides, waxes, ceramides, sterols, hydrogenated, esterified, and oxidized lipids (Hernandez et al. 2020, Bonnet 2018, Rabasco et al. 2000).

Lipids serve distinct purposes in cosmetic compositions. Through several processes, they act as moisturizing agents to prevent water loss. The first method is occlusion, which is achieved by covering the skin with a water-resistant layer to postpone the surface evaporation of moisture. Hydrocarbons, fatty acids, fatty alcohols, vegetal waxes, phospholipids, and sterols are typical substances employed for these purposes. Putting humectant chemicals to the skin's surface to draw water is another method . A way of moisturizing the skin is by utilizing hydrophilic matrices, which physically coat the skin with a protective layer to stop evaporation. Therefore, photoprotection is seen as a sort of indirect moisturization that uses sunscreen chemicals to stop cellular damage and the resulting dehydration. The emollient and softening properties of lipids give skin its smoothness and suppleness. Thin, oily substances have the capacity to briefly deposit between corneocytes, smoothing out their edges, and this is what gives skin its smoothness and softness. Since they depend on how much light the skin surface reflects, skin's brightness and vibrancy are highly correlated with the smoothness of the skin surface, boosting the appearance of the skin (Draelos 2018).

Lipids are utilized in goods as emulsifiers and surfactants to keep water and oil mixed and to minimize the contact between the skin and the material. Surfactants reduce surface tension and make it simpler for different densities of liquids to mix to form emulsions. Due to the hydrophobic and hydrophilic moieties in their chemical structure, they have polarity. Detergency, wetting, emulsifying, solubilizing, dispersing and foaming effects are carried out via their amphiphilicity. Polyethylene glycol ethers are the most frequently used conventional commercial surfactants in cosmetics. On the basis of chemical structures, biosurfactants are typically classified as fatty acids, glycolipids, lipopeptides, polymeric substances, and phospholipids. Glycolipids are biosurfactants widely used in cosmetics and self-care products because of their physicochemical properties, biocompatibility, biological activity, and biodegradability. Additionally, they serve as multipurpose components in the creation of cosmetics. Sophorolipids, rhamnolipids, and mannosylerythritol lipids are the most well-known glycolipids used in cosmetic and pharmaceutical science (Lourith et al. 2009, Varvaresou at el. 2015). In order to improve spreadability and consistency, lipids are utilized as texturizers to give gel-like products consistency, and as carriers for colour and aroma (Hernandez et al. 2020). Two key factors that influence customer choice of cosmetics are colour and smell. Examples of lipid-based, colourful compounds used in industry include pigments (Morocho-Jácome et al. 2020). Essential oils add flavor to a variety of products, behave as active substances and preservatives along with offering the skin a range of health advantages. Lipids function as transporters for preservatives and can inhibit the development of microorganisms (Hernandez et al. 2020). Preservatives are used in cosmetic compositions to reduce the possibility of bacterial contamination during use. Lipids have been used as a preservation system in cosmetics in a number of instances. Lipids have historically been utilized in cosmetics as penetration enhancers, They have traditionally been utilized in cosmetics as absorption enhancers, notably as nanoparticles for the transfer of bioactive substances. Nanostructured lipid carriers, liposomes, nanosomes, and solid lipid nanospheres are examples of these nanoparticles (Hammer et al. 2011).

Although the majority of the study on microalgal lipids is on using them to make biofuels, many other businesses, including the food and feed industries, can profit from the unique characteristics of microalgal lipids. In these regions, lipids with omega-3 and omega-6 fatty acids are most often employed. As a sustainable energy source, transesterification is needed to produce fatty acid alkyl or methyl esters, sometimes referred to as biodiesel, from lipids (Khan et al. 2018). The interest in lipids with beneficial properties, such as phytosterols, carotenoids, and polyunsaturated fatty acids (PUFAs), is also growing in the beauty and pharmaceutical sectors. Additionally, oils are frequently employed in dermatology as delivery agents, manufactured into creams or emulsions, as already mentioned to give a more uniform effective administration and conveyance of active agents

(Zerres et al. 1865, Cezare-Gomes et al. 2019). Fatty acyls are the most often used biological lipids due to their structure that acts as the primary lipid building block of complex lipids. Fatty acyls are further classified into conjugates and fatty acids, fatty alcohols and esters, eicosanoids, etc. Fatty acyls from oil seeds are the key source of bio-based surfactant feedstock in the cosmetics industry (Hayes and Smith 2019). Microalgae have a high PUFA content. They are fatty acids with multiple double bonds throughout the carbon chain and have a variety of beneficial properties. Microalgae may generate both the PUFAs -3 and -6 families, which include linoleic acid (LA), -linolenic acid (GLA), and arachidonic acid (ARA). Eicosapentaenoic acid (EPA), docosahexaenoic acid (DHA), and -linolenic acid (ALA) are also present. Because of their obvious benefits to tissue integrity and health, they are vital ingredients in dietary supplements, cosmetics, and feeds (Maltsev et al. 2021).

Glycerophospholipids, often known as phospholipids, are essential components of all organisms' bilayers of fatty acids including microalgae. They act as a barrier that can be selectively penetrated, insulate the cell from its surroundings and aid in the division of the various internal organelles. Glycerophospholipids, which are found in membrane components, can affect how a cell transmits information by acting as second messengers or their derivatives, as well as protein binding sites (Fahy et al. 2005). Phospholipids are effective emulsifiers because of their amphiphilic character, as they have hydrophilic head groups and hydrophobic tails. Consequently, they have the capability to stabilize oil-water emulsions, making them suitable for use as carriers for drugs, food, cosmetics, and components of medications (Li et al. 2019).

The sterol family of compounds, which perform a variety of biological functions is distinguished by a distinctive fused ring structure. Particularly for the optimum maintenance of membrane fluidity they are an essential component of the lipids that make up membranes. Both free and conjugated versions of the sterols are found in microalgae and several of them are rare in higher terrestrial plants (Stonik et al. 2018). Plants contain a large number of sterols, also known as phytosterols. Several studies have found that phytosterols and the total lipid fraction isolated from microalgae are responsible for anti-inflammatory activity because they increase anti-inflammatory cytokines while decreasing pro-inflammatory cytokines. As a result, their inclusion in animal feed may be able to modulate immune responses during inflammation, reducing the requirement for antibiotics. Their biological properties make them helpful in the cosmetics industry (Ciliberti et al. 2019).

Algae's challenges for cosmetics

According to research by Klaschka (Klaschka 2016). The International Nomenclature of Cosmetic substances (INCI) listed 1358 natural chemicals, 369 of which were designated as dangerous substances (That is 257 for the skin, 182 for the aquatic environment, and 53 as carcinogenic). Clearly, all the ingredients used in cosmetics are not entirely safe for use. Therefore, there has been a shift in customer preferences towards natural products due to their perceived safety as compared to chemical-based cosmetics. The various effects of bioactive metabolites generated from marine algae have sparked substantial interest in the cosmetic sector (Wang et al. 2015, Thomas et al. 2013, Guillerme et al. 2017). Algae are regarded as safe because they exhibit minimal cytotoxicity when evaluated on human cell lines (Alvarez et al. 2019) and because they may be bioprocessed utilizing cutting-edge, environmentally friendly extraction methods (Fabrowska et al. 2015). This collection includes a variety of fatty acids and bioactive compounds that have use not only in the production of biofuels, sewage treatment, CO_2 sequestration, and the release of oxygen into the atmosphere, but also in the beauty sector. These applications work together to reduce the greenhouse effect. Algal extracts are also used in the food and pharmaceutical industries to provide nutritional supplements to patients suffering from neurological illnesses (Olasehinde et al. 2017). The byproducts of these businesses may potentially represent a viable source of compounds with biological properties suited for topical use (Ribeiro et al. 2015). Algae could emerge as a highly effective and economically feasible alternative to manmade products as a result of these applications. However, there are some hurdles to using algae as a cosmeceutical ingredient, such as (i) creating effective biomass culturing procedures, (ii) improving metabolite extraction methods, and (iii) assuring quality assurance and regulatory compliance.

The ideal circumstances for large-scale algae cultivation include light intensity, pH regulation, contamination-free environments, nutrients that are readily available, the presence of CO2, salinity, inorganic carbon, temperature, and nutrients (Fabrowska et al. 2015). Additionally, the sort of bioreactor utilized for growing algae is created depending on the kind of algae and the goal of the culture. In Asia, for example, *Nostoc* sp., algae are typically gathered from their native habitats or grown under specific conditions (Khan et al. 2018). Usually open cultivation, like that in tanks or ponds, takes place on land next to the sea. Although it is cheap to produce algae, it can be easily contaminated when grown in an open system. It is possible to cultivate algae using seawater rather than freshwater, which boosts the production of lipids and other nutrients while lowering costs (Fabrowska et al. 2015).

There are two types of biometabolites extraction methods: traditional and novel techniques. Traditional extraction techniques include saponification, maceration, Soxhlet extraction, and traditional solvent extraction, while more recent techniques include enzyme-assisted extraction (EAE). The three techniques for assisting extraction are ultrasound, supercritical fluid, and microwave. The use of new approaches is recommended because they have a number of benefits, such as quick extraction times, little need for solvents, and higher yields and higher-quality extractions (Fabrowska et al. 2015).

For the security, efficacy, and high level of excellence in cosmetic goods, it is critical to prioritize quality control and standardization of the fundamental ingredients utilized in their composition. Some algae are important sources of raw materials for cosmetic stuff. Therefore, it becomes imperative to thoroughly assess the samples for potential allergens, toxins, heavy metals (such as arsenic, mercury, lead, and cadmium), pesticides (including organochlorines), as well as other chemical contaminants. This meticulous evaluation is essential to ensure product safety and quality. The World Health Organization and the US Food and Drug Administration have established minimal levels of contamination, such as lead and arsenic at 10 parts per million, mercury at 1 part per million, and cadmium at 0.3 parts per million (Fabrowska et al. 2015). When developing cosmetic goods, it is important to keep in mind that some drugs have phototoxic adverse effects. Phytophotodermatitis, also known as phytophoto contact dermatitis, is caused by the activation of a phototoxin or photoallergen upon skin contact, followed by light exposure. Several of these compounds produce singlet oxygen and other reactive oxygen species (ROS) (Collett et al. 2019). Consumers are frequently unaware that natural-based cosmetic products contain a complex blend of chemical ingredients that may have negative effects on human health. Cyanotoxins, a substance produced by algae, have been shown to have lethal effects on immunological and brain cells, even though they are used as pesticides in agriculture (Takser et al.2016, Singh et al. 2017).

Conclusions and prospects for the future

Due to public demand and the fact that synthetic cosmetic products clearly have a negative impact on people, the cosmetic industry is researching novel compounds derived from natural materials. Algae present an intriguing source of bioactive substances with potential industrial uses. There are numerous different types of lipids that are beneficial in many fields. They are employed in the production of biodiesel, the pharmaceutical business, the cosmetic industry, and the improvement of food and animal feed's nutritional content. A significant amount of algal lipid is becoming more and more popular in the cosmetic sector (Jacob-Lopes et al. 2019). The usage of algae in cosmetics is an intriguing method to expand the hunt for novel natural components from biomass that is environmentally friendly. It is possible to create a mixture of valuable chemicals from whole cells. On the other hand, as total microalgae biomass is challenging to include in cosmetic compositions, isolated compounds could also be suggested to focus on specific benefits (Morocho-Jácome et al. 2020). Lipid products offer formulas for cosmetics and are beneficial. Different lipid extraction techniques are being used today. However the extraction of oil from algae has often only been done on a small scale in the lab. Testing novel extraction and purification techniques that could become green technologies is therefore important. The choice of a suitable extraction method is a key consideration. A biocompatible buffer that doesn't affect the isolated compounds' bioactivity must be used. Today's traditional extraction methods use dry biomass as a starting material and organic solvents that must be used over an extended period of time. New methods are being created that shorten extraction times, increase extraction yields, and do not use hazardous solvents (Imbimbo et al. 2020).

Algae could be used to produce different bioactive compounds in addition to lipids. The cosmetic industry is constantly looking for fresh components for cutting-edge product compositions. Seaweeds currently account for a large portion of the manufacturing of algal extracts used in cosmetics. Prior to the 2010s, the utilization of microalgae in beauty treatments was largely secretive, with limited awareness and demand. However, with the significant surge in demand for natural beauty treatments the usage of microalgae has become more prominent. It is anticipated that the market for cosmetics made with microalgae will expand further. A lot of focus has been placed on cosmeceutical goods recently, which are defined as cosmetics with biologically active chemicals that make claims of therapeutic or drug-like advantages. This trend has resulted in a blurring of the line between pharmaceuticals and cosmetics. Lipids are already employed in the cosmetics industry for their role as moisturizing agents or surfactants (Rumin et al. 2020). While algae show great potential as a long-term resource for the development of lipids and various other bioactive compounds, several challenges must be addressed before scaling up lipid production from algae on a large scale (Sun et al. 2020). In addition to strain genetic engineering, algae lipid production can be paired with other high-value metabolic products to make algal industrial usage economically viable. Hydrophilic substances, such as proteins and carbohydrates, can be recovered from defatted biomass, whereas hydrophobic substances, such as lipids, can be extracted concurrently and purified afterwards. Lipids can also be produced as a consequence of employing algae for environmental objectives, such as wastewater treatment (Bellou et al. 2014).

References

Adhithya, S., R. Ramkumar, S.M.S.V. Arjunan and K. Abirami. 2022. A review on application and future prospects of algae in pharmaceutical and food industry. Int. J. Res. Appl. Sci. Eng. Technol. 10(October): 1469–1475. doi: 10.22214/ijraset.2022.47221.

Álvarez-Gómez, F., N. Korbee, V. Casas-Arrojo, R.T. Abdala-Díaz and F.L. Figueroa. 2019. UV photoprotection, cytotoxicity and immunology capacity of red algae extracts. Molecules (Basel, Switzerland) 24(2): 341. doi:10.3390/molecules24020341.

Baumann, N.A. and A.K. Menon. 2002. Chapter 2: Lipid modifications of proteins. New Comprehensive Biochemistry, 37–54. (ISBN 9780444511386). doi:10.1016/S0167-7306(02)36004-6.

Bellou, S., M.N. Baeshen, A.M. Elazzazy, D. Aggeli, F. Sayegh and G. Aggelis. 2014. Microalgal lipids biochemistry and biotechnological perspectives. Biotechnol. Adv. 32(8): 1476–1493. doi:10.1016/j.biotechadv.2014.10.

Birjandi Nejad, H., L. Blasco, B. Moran, J. Cebrian, J. Woodger, E. Gonzalez et al. 2020. Bio-based algae oil: An oxidation and structural analysis. Int. J. Cosmet. Sci. 42: 237–247. doi:10.1111/ics.12606.

Bonachela, J.A., M. Raghib and S.A. Levin. 2011. Dynamic model of flexible phytoplankton nutrient uptake. Proc. Natl. Acad. Sci. U.S.A. 108(51): 20633–20638. doi:10.1073/pnas.1118012108.

Bonnet, C. 2018. Lipids, a natural raw material at the heart of cosmetics innovation. OCL 25(5), D501. doi:10.1051/ocl/2018055.

Cezare-Gomes, E.A., Mejia-da-Silva, L. del C., L.S. Pérez-Mora, M.C. Matsudo, L.S. Ferreira-Camargo et al. 2019. Potential of microalgae carotenoids for industrial application. Appl. Biochem. Biotechnol. 188(3): 602–634. doi:10.1007/s12010-018-02945-4.

Chakdar, H., S.D. Jadhav, D.W. Dhar and S. Pabbi. 2012. Potential applications of blue green algae. J. Sci. Ind. Res. 71: 13–20.

Chuberre, B., E. Araviiskaia, T. Bieber and A. Barbaud. 2019. Mineral oils and waxes in cosmetics: An overview mainly based on the current European regulations and the safety profile of these compounds. J. Eur. Acad. Dermatol. Venereol. 33(S7): 5–14. doi:10.1111/jdv.15946.

Ciliberti, M.G., M. Albenzio, M. Francavilla, G. Neglia, L. Esposito and M. Caroprese. 2019. Extracts from microalga chlorella sorokiniana exert an anti-proliferative effect and modulate cytokines in sheep peripheral blood mononuclear cells. Animals 9(2): 45. doi:10.3390/ani9020045.

Ciliberti, M., M. Francavilla, S. Intini, M. Albenzio, R. Marino, A. Santillo et al. 2017. Phytosterols from Dunaliella tertiolecta reduce cell proliferation in sheep fed flaxseed during post partum. Mar. Drugs 15(7): 216. doi:10.3390/md15070216.

Collett, M.G. 2019. Photosensitisation diseases of animals: Classification and a weight of evidence approach to primary causes. Toxicon: X, 3: 100012. doi:10.1016/j.toxcx.2019.100012.

Couteau, C. and L. Coiffard. 2020. Phycocosmetics and other marine cosmetics, specific cosmetics formulated using marine resources. Mar. Drugs 18: 322. doi:10.3390/md18060322.

Croce, R. and H.V. Amerongen. 2014. Natural strategies for photosynthetic light harvesting. Nat. Chem. Biol. 10(7): 492–501. doi:10.1038/nchembio.1555.

de Morais, M.G., B. da Silva Vaz, E.G. de Morais and J.A. Vieira Costa. 2015. Biologically active metabolites synthesized by microalgae. Biomed. Res. Int. 15: 1-15. doi:10.1155/2015/835761.

Draelos, Z.D. 2018. The science behind skin care: Moisturizers. J. Cosmet. Dermatol. 17(2): 138–144. doi:10.1111/jocd.12490.

Fabregas, J. and C. Herrero. 1990. Vitamin content of four marine microalgae. Potential use as source of vitamins in nutrition. J. Ind. Microbiol. 5(4): 259–263. doi:10.1007/bf01569683.

Fabrowska, J., B. Łęska, G. Schroeder, B. Messyasz and M. Pikosz. 2015. Biomass and extracts of algae as material for cosmetics. *In*: Kim, S. and K. Chojnacka [eds.]. Marine algae extracts, processes, products, and applications. Wiley-VCH-Verl; Weinheim, Germany: 2015, 681–706. doi:10.1002/9783527679577.ch38.

Fahy, E., D. Cotter, M. Sud and S. Subramaniam. 2011. Lipid classification, structures and tools. Biochim. Biophys. Acta. 1811(11): 637–647. doi:10.1016/j.bbalip.2011.06.009.

Gauthier, M.R., G.N.A. Senhorinho and J.A. Scott. 2020. Microalgae under environmental stress as a source of antioxidants. Algal Res. 52: 102104, 1–15. doi:10.1016/j.algal.2020.102104.

Guillerme, J.-B., C. Couteau and L. Coiffard. 2017. Applications for marine resources in cosmetics. Cosmetics 4(3): 35. doi:10.3390/cosmetics4030035.

Hammer, K.A. and C.F. Carson. 2011. Antibacterial and antifungal activities of essential oils. Lipids and essential oils as antimicrobial agents. pp. 255–306. *In*: Thormar, H. [ed.]. lipids and essential oils as antimicrobial agents. John Wiley & Sons, London, UK.

Hayes, D.G. and G.A. Smith. 2019. Biobased Surfactants: Overview and industrial state of the art. pp. 3–38. *In*: Hayes, D.G., D.K.Y. Solaiman and R.D. Ashby [eds.]. biobased surfactants synthesis, properties, and applications. Elsevier: Amsterdam, The Netherlands, ISBN 9780128127056.

Hernandez, E.M. 2020. Pharmaceutical and cosmetic use of lipids. In Bailey's industrial oil and fat products, 1–28. doi:10.1002/047167849x.bio068.pub2.

Imbimbo, P., L. D'Elia, D. Liberti, G. Olivieri and D.M. Monti. 2020. Towards green extraction methods from microalgae learning from the classics. Appl. Microbiol. Biotechnol. 104(21): 9067–9077. doi:10.1007/s00253-020-10839-x.

Jacob-Lopes, E., M.M. Maroneze, M.C. Deprá, R.B. Sartori, R.R. Dias and L.Q. Zepka. 2018. Bioactive food compounds from microalgae: An innovative framework on industrial biorefineries. Curr. Opin. Food Sci. 25: 1–7. doi:10.1016/j.cofs.2018.12.003.

Khan, M.I., J.H. Shin and J.D. Kim. 2018. The promising future of microalgae: current status, challenges, and optimization of a sustainable and renewable industry for biofuels, feed, and other products. Microb. Cell. Fact. 17: 1–21. doi:10.1186/s12934-018-0879-x.

Klaschka, U. 2016. Natural personal care products-analysis of ingredient lists and legal situation. Environ. Sci. Eur. 28(1): 8. doi: 10.1186/s12302-016-0076-7.

Kligman, A. 2005. The future of cosmeceuticals: an interview with Albert Kligman, MD, PhD. Interview by Zoe Diana Draelos. Dermatologic surgery: official publication for American Society for Dermatologic Surgery [et al.], 31(7 Pt 2), 890–891.

Levasseur, W., P. Perré and V. Pozzobon. 2020. A review of high value-added molecules production by microalgae in light of the classification. Biotechnol. Adv. 41: 107545. doi:10.1016/j.biotechadv.2020.107.

Li, J., Y. He, S. Anankanbil and Z. Guo. 2019. Phospholipid-based surfactants. In biobased surfactants (Second Edi, Issue January). Elsevier. doi:10.1016/B978-0-12-812705-6.00007-1.

Li-Beisson, Y. 2016. Subcellular biochemistry: Lipids in plant and algae development. Springer. Berlin, Germany, 86, ISBN 978-3-319-25977-2.

Lourenço-Lopes, C., M. Fraga-Corral, C. Jimenez-Lopez, A.G. Pereira, P. Garcia-Oliveira, M. Carpena et al. 2020. Metabolites from macroalgae and its applications in the cosmetic industry: A circular economy approach. Resources 9(9): 101. doi:10.3390/resources9090101.

Lourith, N. and M. Kanlayavattanakul. 2009. Natural surfactants used in cosmetics: glycolipids. Int. J. Cosmet. Sci. 31(4): 255–261. doi:10.1111/j.1468-2494.2009.00493.x.

Lopes da Silva, T., P. Moniz, C. Silva and A. Reis. 2019. The dark side of microalgae biotechnology: a heterotrophic biorefinery platform directed to ω-3 rich lipid production. Microorganisms 7(12): 670. doi:10.3390/microorganisms7120670.

Macabuhay, A., B. Arsova, R. Walker, A. Johnson and M. Watt. 2022. Modulators or facilitators? Roles of lipids in plant root – microbe interactions. Trends Plant Sci. 27(2): 180–190. doi:10.1016/j.tplants.2021.08.004.

Maltsev, Y. and K. Maltseva. 2021. Fatty acids of microalgae: Diversity and applications. Rev. Environ. Sci. Biotechnol. 20. 515–547. doi:10.1007/s11157-021-09571-3.

Milledge, J.J., B. Smith, P.W. Dyer and P. Harvey. 2014. Macroalgae-derived biofuel: A review of methods of energy extraction from seaweed biomass. Energies 7: 7194–7222. doi:10.3390/en7117194.

Morais, T., J. Cotas, D. Pacheco and L. Pereira. 2021. Seaweeds compounds: An ecosustainable source of cosmetic ingredients? Cosmetics 8, 8: 1–28. doi:10.3390/cosmetics8010008.

Morocho-Jácome, A.L., N. Ruscinc, R.M. Martinez, J.C.M. de Carvalho, T. Santos de Almeida, C. Rosado et al. 2020. (Bio)Technological aspects of microalgae pigments for cosmetics. Appl. Microbiol. Biotechnol. 104(22): 9513–9522. doi:10.1007/s00253-020-10936-x.

Mourelle, M., C. Gómez and J. Legido. 2017. The potential use of marine microalgae and cyanobacteria in cosmetics and thalassotherapy. Cosmetics 4(4): 46. doi:10.3390/cosmetics4040046.

Olasehinde, T.A., A.O. Olaniran and A.I. Okoh. 2017. Therapeutic potentials of microalgae in the treatment of Alzheimer's disease. Molecules (Basel, Switzerland) 22(3): 480. doi:10.3390/molecules22030480.

Pandey, A., G.K. Jatana and S. Sonthalia. 2021. Cosmeceuticals. In Stat Pearls; StatPearls Publishing: Treasure Island, FL, USA, 2021. Available online: https://www.ncbi.nlm.nih.gov/books/NBK544223/ (accessed on 29 December 2023).

Panico, A., F. Serio, F. Bagordo, T. Grassi, A. Idolo, M. DE Giorgi et al. 2019. Skin safety and health prevention: an overview of chemicals in cosmetic products. J. Prev. Med. Hyg. 60(1): E50–E57. doi:10.15167/2421-4248/jpmh2019.60.1.1080.

Patel, A., L. Matsakas, U. Rova and P. Christakopoulos. 2019. A perspective on biotechnological applications of thermophilic microalgae and cyanobacteria. Bioresour. Technol., 278: 424–434. doi:10.1016/j.biortech.2019.01.063.

Petry, T., D. Bury, R. Fautz, M. Hauser, B. Huber, A. Markowetz et al. 2017. Review of data on the dermal penetration of mineral oils and waxes used in cosmetic applications. Toxicol. Lett. 280: 70–78. doi:10.1016/j.toxlet.2017.07.899.

Rabasco Alvarez, A.M. and M.L. González Rodríguez. 2000. Lipids in pharmaceutical and cosmetic preparations. Grasas y Aceites. 51: 74–96. doi:10.3989/gya.2000.v51.i1-2.409.

Ribeiro, A.S., M. Estanqueiro, M.B. Oliveira and J.M. Sousa Lobo. 2015. Main benefits and applicability of plant extracts in skin care products. Cosmetics 2: 48–65. doi:10.3390/cosmetics2020048.

Rumin, J., E. Nicolau, R. Gonçalves de Oliveira Junior, C. Fuentes-Grünewald and L. Picot. 2020. Analysis of scientific research driving microalgae market opportunities in Europe. Mar. Drugs 18: 264. doi:10.3390/md18050264.

Sartori, R.B., S.F. Siqueira, M.M. Maroneze, M.B. Fagundes, R. Wagner, L.Q. Zepka et al. 2021. Microalgal secondary metabolites: effect of climatic variables, seasons, and photocycles on the biogeneration of volatile organic compounds (VOCs). J. Appl. Phycol. 33(3): 1457–1472. doi:10.1007/s10811-021-02391-6 Microalgal.

Singh, R., P. Parihar, M. Singh, A. Bajguz, J. Kumar, S. Singh et al. 2017. Uncovering potential applications of cyanobacteria and algal metabolites in biology, agriculture and medicine: Current status and future prospects. Front. Microbiol. 8: 515. doi:10.3389/fmicb.2017.00515.

Spolaore, P., C. Joannis-Cassan, E. Duran and A. Isambert. 2006. Commercial applications of microalgae. J. Biosci. Bioeng. 101(2): 87–96. doi: 10.1263/jbb.101.87.

Stevens, C.V. and R. Verhé. [eds.]. 2004. Renewable bioresources: scope and modification for non-food applications. John Wiley & Sons. ISBN 0470021047.

Stolz, P. and B. Obermayer. 2005. Manufacturing microalgae for skin care. Cosmet. Toilet. 120(3): 99–106.

Stonik, V.A. and I.V. Stonik. 2018. Sterol and sphingoid glycoconjugates from microalgae. Mar. Drugs. 16(12): 514. doi: 10.3390/md16120514.

Sun, X.M., L.J. Ren, Q.Y. Zhao, X.J. Ji and H. Huang. 2019. Enhancement of lipid accumulation in microalgae by metabolic engineering. Biochim. Biophys. Acta Mol. Cell Biol. Lipids 1864(4): 552–566. doi:10.1016/j.bbalip.2018.10.004.

Takser, L., N. Benachour, B. Husk, H. Cabana and D. Gris. 2016. Cyanotoxins at low doses induce apoptosis and inflammatory effects in murine brain cells: Potential implications for neurodegenerative diseases. Toxicol. Rep. 3: 180–189. doi:10.1016/j.toxrep.2015.12.008.

Tao, B.Y. 2007. Industrial applications for plant oils and lipids. In book: Bioprocessing for value-added products from renewable resources, 611–627. doi:10.1016/B978-044452114-9/50025-6. ISBN 9780444521149.

Thomas, N.V. and S.K. Kim. 2013. Beneficial effects of marine algal compounds in cosmeceuticals. Mar. Drugs 11(1): 146–164. doi:10.3390/md11010146.

Tibocha-Bonilla, J.D., C. Zuñiga, R.D. Godoy-Silva and K. Zengler. 2018. Advances in metabolic modeling of oleaginous microalgae. Biotechnol. Biofuels 11: 241. doi:10.1186/s13068-018-1244-3.

Union E. Regulation (EU) No 1223/2009 of the European Parliament and of the Council of 30 November 2009 on cosmetic products. Off. J. Eur. Union. L. 2009, 342: 59–209.

Varvaresou, A. and K. Iakovou. 2015. Biosurfactants in cosmetics and biopharmaceuticals. Lett. Appl. Microbiol. 61(3): 214–223. doi:10.1111/lam.12440.

Vermeer, B.J. and B.A. Gilchrest. 1996. Cosmeceuticals. A proposal for rational definition, evaluation, and regulation. Arch. Dermatol. 132(3): 337–340. doi:10.1001/archderm.132.3.337.

Wang, H.D., C.C. Chen, P. Huynh and J.S. Chang. 2015. Exploring the potential of using algae in cosmetics. Bioresour. Technol. 184: 355–362. doi:10.1016/j.biortech.2014.12.001.

Wolkers, H., M.J. Barbosa, D.M.M. Kleinegris, R. Bosma, R.H. Wijffels and P.F.H. Harmsen. 2011. Microalgae: the green gold of the future? Large-scale sustainable cultivation of microalgae for the production of bulk commodities. Wageningen UR Food & Biobased Research. Available online: https://edepot.wur.nl/170781 (accessed on 29 December 2023).

Yarkent, C., C. Gürlek and S.S. Oncel. 2020. Potential of microalgal compounds in trending natural cosmetics: A review. Sustain. Chem. Pharm. 17: 100304. doi:10.1016/j.scp.2020.100304.

Zerres, S. and W. Stahl. 2020. Carotenoids in human skin. Biochim. Biophys. Acta Mol. Cell Biol. Lipids 1865(11): 158588. doi:10.1016/j.bbalip.2019.158588.

Züllig, T., M. Trötzmüller and H.C. Köfeler. 2020. Lipidomics from sample preparation to data analysis: A primer. Anal. Bioanal. Chem. 412(10): 2191–2209. doi:10.1007/s00216-019-02241-y.

CHAPTER 14

Carbohydrates and Proteins from Seaweed

A Promising Frontier in Skin Cosmeceuticals

Haresh S. Kalasariya,[1] *Leonel Pereira*[2,*] and *João Cotas*[2,*]

Introduction to cosmeceuticals

Consumer preferences for green and eco-friendly products have witnessed a notable upsurge in recent years (Soulioti et al. 2013, Pangestuti et al. 2021). This inclination extends to the cosmetics industry, which operates within a fiercely competitive and swiftly evolving global market. The demand for natural, safe, and effective ingredients has become paramount for the development of innovative skincare products (Pimentel et al. 2017, Resende et al. 2021, Aslam et al. 2021). The term "cosmeceutical" denotes products that are actively developed and rigorously tested by the cosmetics industry to enhance skin appearance and effectively address various dermatological conditions (Preetha and Karthika 2009). These products offer a diverse range of functionalities (Preetha and Karthik 2009, Draelos 2014). A plethora of active ingredients, including vitamins, phytochemicals, enzymes, antioxidants, and essential oils are considered and can be incorporated into formulations such as creams, lotions, ointments, or masks (Querellou et al. 2010). The use of cosmeceuticals has experienced a significant surge in recent years, catering not only to traditional consumer segments but also to less-conventional populations like men and children (Brandt et al. 2011, Pham and Dinulos 2014, de Lacerda et al. 2013, Draelos 2018). Ensuring safety, efficacy, and favourable sensorial qualities is imperative for both cosmetics and cosmeceuticals (Preetha and Karthika 2009, Lin 2010). Additionally, nutricosmetics demand optimal characteristics for their development. Collaboration across various disciplines, including biotechnology, is crucial for the optimal advancement of these products.

Marine resources emerge as a readily available and promising reservoir of unique and active compounds with the potential to fuel the production of cosmetics, cosmeceuticals, and nutricosmetics. Collaboration with biotechnology can unlocks the full potential of marine resources for skincare

[1] Centre for Natural Products Discovery, School of Pharmacy and Biomolecular Sciences, Liverpool John Moores University, Byrom Street, Liverpool L3 3AF, UK. Email: hareshahir22@gmail.com

[2] Marine Resources, Conservation and Technology, Marine Algae Lab, CFE—Centre for Functional Ecology; Science for People & Planet, Department of Life Sciences, University of Coimbra, 3000-456 Coimbra, Portugal.

* Corresponding authors: leonel.pereira@uc.pt; jcotas@uc.pt

product development. Among natural resources, seaweeds stand out as a sustainable and renewable source, garnering increasing attention for various applications (Pereira 2018, Sotelo et al. 2021). Moreover, the valorisation of waste from seaweed processing has become a focus area. Regardless of their origin or type, there is a growing demand for the development of environmentally friendly extraction methods that prioritize low extraction time, minimal solvent usage, higher extraction yields, and quality outcomes.

The utilization of seaweed-derived ingredients in cosmetic products has seen a surge in recent years, driven by scientific studies showcasing the skincare potential of seaweed bioactive compounds (Jesumani et al. 2019). These bioactive compounds encompass a diverse array of biologically active molecules, including carotenoids, fatty acids, polysaccharides, phlorotannins, vitamins, sterols, tocopherol, phycobilins, and phycocyanins (Hamed et al. 2015, Michalak and Chojnacka 2015, Shahidi and Rahman 2018). The rich compositions of seaweeds have positioned them as potential ingredients in classical cosmetics, with applications ranging from solid soaps to replacements for sodium lauryl sulphate/sodium laureth sulphate (Wahyuni 2021). Furthermore, algal extracts are being used in nutritional supplements, cosmetics, and alternative medicines recommended for various skin-related diseases (Kharkwal et al. 2012). In such formulations, seaweed-derived ingredients serve as active agents, offering a multitude of benefits, including photoprotection, moisturization, antioxidant properties, anti-melanogenic effects, anti-allergic, anti-inflammatory, anti-acne, anti-wrinkling, antimicrobial, anti-aging, and whitening properties (Senevirathne and Kim 2013, Vo et al. 2015, Salehi et al. 2019). Additionally, these ingredients exhibit low cytotoxicity and allergen contents, further enhancing their appeal for skincare applications (Morais et al. 2021). Thus, this chapter focuses on the most common compounds found in seaweeds, polysaccharides, and proteins; exploiting their bioactivities for the cosmetic area and focusing on future roads.

Seaweed derived polysaccharides in skin benefits

Seaweeds are rich in polysaccharides, which constitute an essential carbohydrate fraction forming part of their cell walls. These polysaccharides vary depending on the type of algae: brown algae contain alginate, laminaran, and fucoidan; green algae contain ulvan; and red algae contain agar and carrageenan, with carrageenan being the most prominent (Shanura Fernando et al. 2018). Polysaccharides have garnered increasing attention due to their biofunctional and physicochemical characteristics (Shanura Fernando et al. 2018). Sulphated polysaccharides, in particular, have been noted for their health benefits and biological activities (Morya, Kim and Kim 2012, Casas et al. 2016, Ruocco et al. 2016, Lee et al. 2017, Sanjeewa et al. 2018, Fernando et al. 2019, Cikoš et al. 2021).

The activity of polysaccharides is closely related to their composition and structure, particularly their molecular weight. Therefore, depolymerization is often proposed to enhance their activity (Fournière et al. 2021), although other structural modifications can also be employed. Simple hydrophobization reactions, such as esterification, acylation, alkylation, amidation, or cross-linking reactions on native hydroxyl-, amine-, or carboxylic acid functions, have been shown to enhance bioactivity (Wong et al. 2021).

Alginates, composed of chains of D-mannuronic acid and L-guluronic acid, are prevalent in brown seaweeds. These compounds exhibit various properties relevant to cosmetics and wellness products, particularly anti-allergic properties (Yu et al. 2020). This anti-allergic action has also been observed in formulations of hydrogels containing alginate (Szekalska et al. 2020) and has been linked to potential benefits in preventing obesity (Holdt and Kraan 2011, Xing et al. 2020).

Laminarin is characterized by its inability to form viscous solutions and is known for its prebiotic properties (Deville et al. 2007, Shannon et al. 2021). It also exhibits antioxidant activity (Choi et al. 2012, Gómez-Mascaraque et al. 2021, Huang et al. 2021, Ozanne et al, 2020) and possesses anti-photoaging and regenerative properties (Kadam et al. 2015). Novel hydrogel systems have been developed based on the wound healing properties of laminarin sulphate (Costa et al. 2020), showing

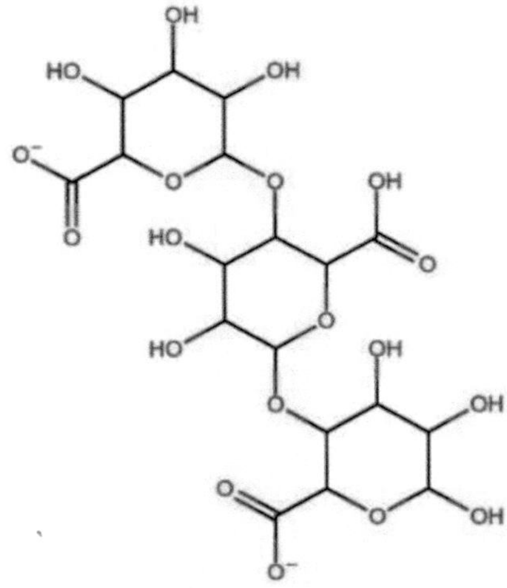

Alginate

promise in biomedical applications such as tissue engineering and cancer therapies (Li et al. 2019, Tümen Erden et al. 2021). Additionally, laminarin has demonstrated anti-inflammatory properties (Zargarzadeh et al. 2020).

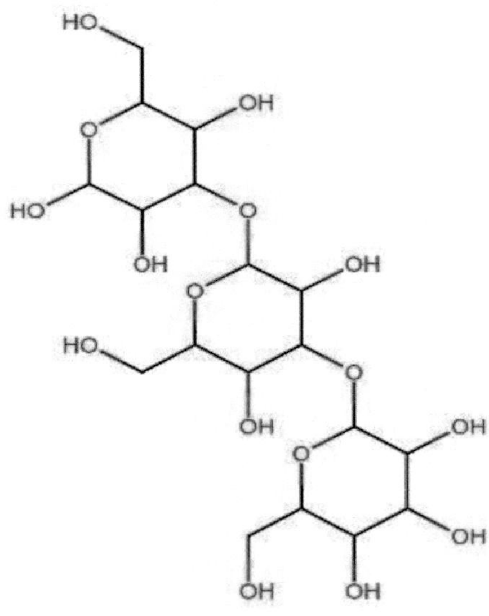

Laminarin

Fucoidans, heteropolysaccharides primarily composed of fucose and other monosaccharides, have gained attention for their potential as cosmetic ingredients (Baweja et al. 2016, Li et al. 2005). They are recognized for their non-toxic, biodegradable, and biocompatible nature (Citkowska et al. 2019, Jiao et al. 2011) and exhibit a wide range of biological properties (Wijesinghe and Jeon 2012, Kartik et al. 2021). Fucoidans also possess antioxidant and antiradical properties (Xue et al. 2001, Kim et al. 2011), with their effectiveness depending on their molecular weight and sulphate content (Li et al. 2019). Fucoidans have demonstrated benefits in preventing and treating skin photoaging. They inhibit UVB-induced collagenase and gelatinase activities, as well as elastase activity in human skin (Moon et al. 2009, Su et al. 2020, Senni et al. 2011). Furthermore, fucoidans inhibit wrinkle-related enzymes and enhance collagen synthesis in human dermal fibroblasts (Wang et al. 2020). They also exhibit anti-inflammatory effects, particularly in relation to extracellular matrix degradation by matrix metalloproteinases (Wang et al. 2021, Ali Karami et al. 2021).

Sulphated polysaccharides derived from green algae, such as rhamnans, arabinogalactans, galactans, and mannans, exhibit diverse compositions and structures, with certain properties being strongly influenced by molecular weight, particularly in terms of antiradical and chelating properties (Costa et al. 2011, Qiao et al. 2016, Lin et al. 2020, Adrien et al. 2017). Ulvans, complex and variable sulphated polysaccharides found in ulvales, are primarily composed of rhamnose, xylose, glucose, glucuronic acid, iduronic acid, and sulphate (Lahaye and Robic 2007, Cindana Mo'o et al. 2020). Ulvans exhibit various activities, including gelling, anti-aging, anti-hyperlipidemic, and antiherpetic properties (Lahaye and Robic, 2007, Kidgell et al. 2021, Yu et al. 2020, Li et al. 2020).

Agar, composed mainly of β-D-galactopyranose and 3,6-anhydro-α-L-galactopyranose units with variable sulphate, pyruvate, and uronate substituents, finds applications in pharmaceutical and industrial cosmetics, serving as a thickener and ingredient for tablets or capsules for drug delivery (Chen et al. 2021, Aziz et al. 2020). Agar is also suggested for use in creams, as an emulsifier, stabilizer, moisturizer, and in various cosmetic products such as lotions, deodorants, anti-aging treatments, and exfoliants (Balboa et al. 2015).

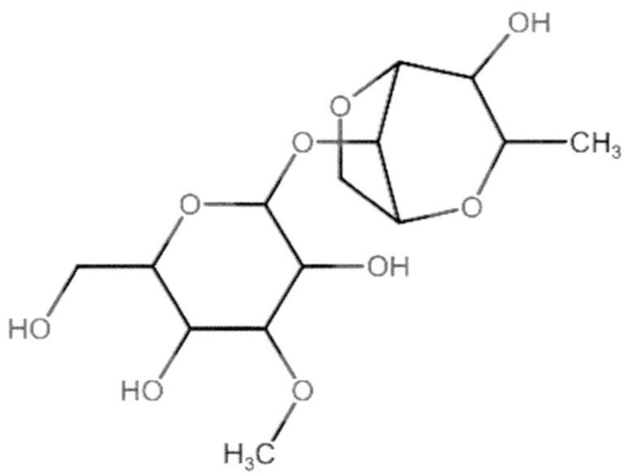

Agar

Carrageenans, generally recognized as safe (GRAS) and approved for food applications, are high-molecular-weight sulphated linear polysaccharides with a backbone of alternating 3-β-D-galactopyranose and 4-α-D-galactopyranose with anhydrogalactose residues (Laurienzo 2010, Ghanbarzadeh et al. 2018).

Porphyran, a complex sulphated galactan found in *Porphyra* sp., possesses therapeutic properties such as antioxidant activity, antiallergic effects, tyrosinase inhibitory activity, protection against ultraviolet B radiation, anti-inflammatory and antitumoral activity, and promotion of beneficial bacteria growth in intestinal microbiota (Isaka et al. 2015, Beaumont et al. 2021, Kwon and Nam 2006, Pacheco-Quito et al. 2020, Qiu et al. 2021, Xie et al. 2020, Thevanayagam et al. 2014, de Araújo et al. 2011, Zhang et al. 2020). Agarooligosaccharides (AOS) and carrageenan-oligosaccharides (COS) exhibit enhanced biological properties compared to their native counterparts, including prebiotic, antitumoral, and antioxidant actions, attributed to their chemical structure, molecular weight, degree of polymerization, and glycosidic linkage flexibility (Cheong et al. 2018).

Polysaccharides are pivotal compounds found in macroalgae, renowned for their diverse biological activities beneficial for skin health. Seaweeds encompass a spectrum of polysaccharides, including chitin, fucoidans, agar, carrageenan, alginates, ulvans, terpenoids, and tocopherol (Usov and Zelinsky 2013, Ahmed and Taha 2011, Godlewska et al. 2016, Paduch et al. 2007, Cha et al. 2011,

Murugan and Iyer 2013). In the realm of skin cosmeceuticals, marine algae have demonstrated myriad activities such as anti-melanogenesis, antioxidant, anti-skin-aging, anti-inflammation, anti-atopic dermatitis, anti-skin-cancer, and repair of UV-induced damage (Leandro et al. 2019, Buono et al. 2012, Kang et al. 2015, Pimentel et al. 2017, Kim et al. 2014, Nurjanah et al. 2019). Many macroalgal species, including *Kappaphycus alvarezii*, *Porphyra umbilicalis* (Rhodophyta), *Sargassum polycystum*, *Padina boryana*, *Fucus vesiculosus* (Phaeophyceae), etc., boast a substantial carbohydrate content. Polysaccharides exhibit versatility in applications such as photoprotection, moisturizing, wound-healing, thickening, emulsifying, and preservation (Quah et al. 2014, Campos et al. 2014, Kim 2014, Jutur et al. 2016, Percival 1979, Song et al. 2015). Contemporary skincare product developers prioritize compounds that regulate tyrosinase inhibition, collagenase and elastase inhibition, reduction of matrix metalloproteinase (MMP) activity, attenuation of reactive oxygen species (ROS) and antioxidant activity (Shao et al. 2015, Fujimura et al. 2000, Holtkamp et al. 2009, Fujimura et al. 2002, Teixeira and Hellewell 1997, Chen et al. 2004). Holtkamp et al. (2009) also underscore the efficacy of fucoidan, particularly for its skin-protecting, antioxidant, anti-aging, antiviral, anti-inflammatory, antitumor, and anticoagulant properties, supported by epidemiological and experimental studies.

Table 1 delineates the various potential skin beneficial effects of seaweed polysaccharides. Notably, Fujimura et al. (2002) expounded on the application of purified fucoidan extracts from *Fucus vesiculosus* in creams and lotions imparting antiaging and anti-wrinkle benefits. They also highlighted its efficacy in reducing collagenase expression, exerting anti-inflammatory activity, and inhibiting matrix enzymes such as hyaluronidase, heparinase, tyrosine kinase, and phospholipase A2. Furthermore, Fujimura et al. (2002) demonstrated a notable reduction in skin thickness and enhanced elasticity with a gel formulation containing 1% *Fucus* extract. Additionally, Teixeira and Hellewell (1997) elucidated the utility of fucoidan as a topical anti-inflammatory agent for cosmetic applications targeting after-sun damage, allergic conditions, or postsurgical formulations.

Polysaccharides have also been extensively studied for their antioxidant, antiviral, anticoagulant, and antitumor properties in commercial products. For instance, the lipid-soluble fraction of the edible red alga *Gelidium amansii* induced apoptosis in cancer cells *in vitro* (Chen et al. 2004). Similar to fucoidan, carrageenan facilitates the creation of various textures for creams, lotions, sticks, sprays, and foams formulations (Campo et al. 2009). It exhibits a range of biological activities and finds applications in pharmaceutical formulations, including antitumor, immunomodulation, anti-hyperlipidemic, and anticoagulant activities (Munaf et al. 2009). Another noteworthy phenolic compound is alginate, found in the cell walls of brown algae (Phaeophyceae), predominantly in species like *Laminaria*, *Macrocystis pyrifera*, *Ascophyllum nododsum*, *Ecklonia maxima*, *Lessonia nigrescens*, *Durvillea antarctica*, and *Sargassum* sp. (Mišurcová et al. 2012, Ibrahim et al. 2017). Podkorytova et al. (2007) also utilized alginates in cosmetics for face masks and body washes due to their beneficial effects on skin structure and function. Additionally, Skjåk-Bræk et al. (1989) suggested the inhomogeneity of low-molecular-weight alginate in gel formulations with a low concentration of Ca^{+2} and the absence of non-gelling ions Na^{+}.

A diverse array of polysaccharides including agar, alginic acid, carrageenan, porphyrin, laminarin, fucoidan, and ulvan found in seaweeds offer valuable applications in skin cosmetics. Agarophytes algae genera like *Gelidium* sp., *Gracilaria* sp., *Gelidiela* sp., *Pterocladiella* sp. (Rhodophyta), among others, are renowned for producing agar-agar (Kanlayavattanakul and Lourith 2015, Priyan Shanura et al. 2019, Chen et al. 2021). Alginic acid, akin to agar, is derived from various brown algal species such as Fucales, Laminariales, *Ascophyllum* sp., *Durvillaea* sp., *Ecklonia* sp. (Phaeophyceae), and others (Pereira 2018, Sreekumar and Bindhu 202, Fabrowska et al. 2015).

Fabrowska et al. (2015) highlighted its utilization as an ingredient in beauty masks or facial packs in the formulation of skin-protective creams for dermatitis treatment. Additionally, kappa-, iota-, lambda-, and beta-carrageenan are extracted from various carrageenophytes like *Betaphycus*

Table 1. Seaweed derived polysaccharides in skin benefits.

No.	Name of Macroalgae	Polysaccharides	Cosmetic Benefits	References
1	*Ulva lactuca* (C)	SP (Ulvan)	Antioxidant, Moisturizer, Photoprotective	(Robic et al. 2009)
	Pyropia yezoensis (R)	Porphyran	Antiinflammation	(Jiang et al. 2012, Isaka et al. 2015)
2	*Porphyridium* sp. (R), *Costaria costata* (P), *Ulva lactuca* (C)	Sulphated polysaccharides	Antioxidant, Anti-inflammatory, Antiaging	(Raposo et al. 2013)
3	*Fucus vesiculosus* (P)	Fucoidans	Antiaging, Antiwrinkle	(Kim et al. 2008)
4	*Ascophyllum nododum*, *Chnoospora minima*, *Sargassum fusiforme*, *Saccharina japonica*, *Sargassum polycystum*, *S. vachellianum*, *S. hemiphyllum* (P)	Fucoidans	Photoprotection, Anti photoaging, Anti-inflammatory, Anti-elastase, Anti-collagenase, Skin whitening	(Pangestuti et al. 2021; Jesumani et al. 2019, Morya et al. 2012, Sanjeewa et al. 2018)
5	*Fucus vesiculosus* (P)	Fucoidan	Anticoagulant, Antioxidant, Enhancer of Skin fibroblast formation	(Rupérez et al. 2002)
6	*Pyropia haitanensis* (R)	Porphyran	Antioxidant	(Zhang et al. 2003; Zhang et al. 2004)
7	*Saccharina latissima* (formerly *Saccharina longicruris*) (P)	Laminaran	Anti-inflammation, Antioxidant, Reconstruction of dermis	(Ayoub et al. 2015, Ozanne et al. 2020)
8	*Saccharina latissima* (formerly *Saccharina longicruris*) (P)	Galactofucans	Enhance fibroblast formation, Increase synthesis of MMP complex and collagen-1	(Rioux et al. 2013)
9	*Eucheuma denticulatum* (R)	Carrageenan	Antioxidant, photoprotection	(Thevanayagam et al. 2014)
10	*Gelidium* sp. (R)	Agar	Thickener	(Wells et al. 2017)
11	*Ascophyllum* sp., *Fucus* sp., *Sargassum sp.*, *Undaria* sp. (P)	Laminaran	Anticellulite	(Pereira et al. 2013)
12	*Saccharina cichorioides* (P)	Fucoidan	Anti-atopic dermatitis	(Yang 2012)
13	*Corallina officinalis* (R)	Sulphated polysaccharides	Antioxidant	(Yang et al. 2011)
14	*Ulva australis* (C)	Ulvan	Antiaging	(Fournière et al. 2021, Jiang et al. 2020)
15	*Acanthophora muscoides* (R)	Sulphated polysaccharides-Carrageenan	Anticoagulant, Antinociceptive, Anti-inflammation, Gel agents	(Quinderé et al. 2013, Rodrigues et al. 2016, Rodrigues et al. 2016)
16	*Chondrus crispus* (R)	Carrageenan	Gel and Thickening agent, Skin moisturizer	(Stengel and Connan 2015)
17	*Ulva rigida* (Figure 3m), *U. pseudorotundata* (C)	Sulphated polysaccharides	Antioxidant, Chelators, Gel agents, Moisturizer	(Lahaye and Robic 2007)
18	*Ascophyllum nodosum* (P)	Fucoidan	Anti-inflammation, Antiviral, Antiaging, Anti elastase, Photoprotective, Tyrosinase inhibition, Anticellulite	(Wijesinghe and Jeon 2012)
19	*Gracilaria* sp. (R)	Agar	Thickener	(Dita and Triastuti 2020)
20	*Padina boergesenii* (P)	Sulphated polysaccharides	Formation of collagen	(Kordjazi et al. 2013)
21	*Macrocystis* sp., *Lessonia* sp., *Laminaria* sp. (P)	Alginate	Gelling and Stabilizing agent, Moisturizer, Chelator	(Malinowska 2011, Tønnesen and Karlsen 2002)
22	*Kjellmaniella crassifolia* (P)	Fucoidan	Antiaging, Antiwrinkle	(Mizutani et al. 2010)

Table 1 contd. ...

...Table 1 contd.

No.	Name of Macroalgae	Polysaccharides	Cosmetic Benefits	References
23	Brown algae (P)	Alginate	Thickening agent, Gelling agent	(Morais et al. 2021)
24	*Sargassum vachellianum* (P)	Polysaccharides	Skin moisturizer and protectors	(Jesumani et al. 2020)
25	*Fucus vesiculosus*, *Laminaria digitata*, *Undaria pinnatifida* (P)	Fucoidan	Antioxidant, Antiaging, Anticoagulant, Increase skin fibroblast formulation	(Rupérez 2002, Je et al. 2009)
26	*Ascophyllum nodosum* (P)	Fucoidan	Anti-elastase, gelatinase A inhibition, Inhibition of interleukin-1 beta in fibroblast cells	(Pereira 2018)
27	*Ecklonia cava* (P)	Phlorotannins	Photoprotectors against UV-B	(Heo et al. 2009, Ko et al. 2011)
28	*Pyropia haitanensis*, *Gracilaria parvispora* (formerly *Gracilaria chouae*), *G. blodgetti* (R)	Agar	Antioxidant, Thickeners, Antitumor, Radiation protector, Antiaging	(Fernando et al. 2019, Xue et al. 1998)
29	*Turbinaria conoides* (P)	Laminarin, Alginate, Fucoidan	Antioxidant	(Berthon et al. 2017)

(Note: Abbreviations used: (R) - Rhodophyta, (P) - Phaeophyceae, (C) - Chlorophyta)

gelatinus, *Chondrus crispus*, *Eucheuma denticulatum*, *Gigartina* sp., and others from the Rhodophyta, finding applications in cosmetology ranging from lotions to sunray protectors (Fernando et al. 2018, Pereira et al. 2009, Charlier and Chaineux 2009).

Furthermore, porphyran, derived from aqueous extracts of red algae *Porphyra* sp. and *Bangia* sp., exhibits promising cosmeceutical applications including skin whitening and anti-inflammatory properties (Villarroel and Zanlungo 1981, Aziz et al. 2020). Laminaran, prevalent in brown seaweed species like *Laminaria* sp., *Saccharina* sp., *Ascophyllum* sp., *Fucus* sp., and *Undaria* sp. (Phaeophyceae), offers various benefits such as antitumor and antioxidant properties (Lourenço et al. 2020, Kim et al. 2013, Fauziee et al. 2021).

Moon et al. (2011) highlighted fucoidan's role in inhibiting matrix metalloproteinase induced by UVB radiation, suggesting its potential in preventing skin photoaging, while also acting as a tyrosinase inhibitor for skin-whitening formulations (Kim et al. 2020, Wijesinghe and Jeon 2012). Ulvan, suggested by Carvalho and Pereira (2015) and Pereira (2018), stands out as a desirable raw material for cosmeceuticals due to its moisturizing and antioxidative properties (Cindana Mo'o et al. 2020, Ruslan et al. 2021). Additionally, Yaich et al. (2017) elaborated on the skin protective and bioactive effects of rhamnose and fucose against skin aging. The distinctive chemical and physiochemical characteristics of polysaccharides render them appealing candidates for novel functional and biologically active polymers in cosmeceuticals.

Marine algae serve as rich reservoirs of polysaccharides, with distinct types prevalent across various algae types: fucoidans in brown algae, ulvans in green algae, and carrageenans in red algae (Ngo and Kim, 2013). The advantageous impacts of polysaccharides from marine algae on skin health are outlined in Table 2, delineating the marine algal species, their biological functions, and the associated benefits for the skin.

Proteins, peptides and amino acids

Proteins, fundamental biological macromolecules composed of amino acids, are ubiquitous in living organisms, serving as essential components in numerous cellular processes. They manifest in various forms, including enzymes, hormones, vitamins, and pigments (Lahaye and Robic 2007, Stengel and Connan 2015). Macroalgae exhibit a rich diversity of amino acids, encompassing aliphatic, hydroxyl-group-containing, aromatic, and mycosporine amino acids, as detailed in Table 2. Amino acid compositions have been documented in several macroalgae species such as *Palmaria palmata*, *Chondrus crispus*, *Porphyra* sp. (Rhodophyta), *Undaria pinnatifida*, *Ulva* sp., and *Euchema* sp. (Phaeophyceae), highlighting their potential utility in cosmeceutical formulations (Černá 2011, Harnedy and Fitzgerald 2013, Cotas et al. 2020). These amino acids contribute to various cosmeceutical activities, including anti-inflammatory, antioxidant, antiaging, and photoprotective effects (Holdt and Kraan 2011, Yuan et al. 2009).

Research by Galland-Irmouli et al. (1999) and Samarakoon and Jeon (2012) underscores the multifaceted benefits of proteins and amino acids in cosmetics, including anti-inflammatory, antioxidant, antitumor, antiaging, and moisturizing properties. Essential amino acids like histidine and taurine have been identified in *Ulva australis* by Houston (2005), while *Palmaria palmata* (Rhodophyta) and *Himanthalia elongata* (Phaeophyceae) are rich sources of serine, alanine, and glutamic acid, as demonstrated by Galland-Irmouli et al. (1999) and Martínez–Hernández et al. (2018).

Detection of mycosporines, specifically mycosporine-like amino acids (MAAs), in various red macroalgae species (Rhodophyta) such as *Chondrus crispus*, *Palmaria palmata*, *Gelidium* sp., *Porphyra* sp. (Rhodophyta), and others, has been reported by Reef, Kaniewska and Hoegh-Guldberg (2009) and Pereira (2018). The role of MAAs as UV protectors and promoters of cell proliferation in cosmetics has been elucidated by Nowruzi, Sarvari and Blanco (2020) and Chrapusta et al. (2017). Notably, mycosporine-like amino acids Porphyra-334 and Shinorine, derived from the red alga *Pyropia elongata* (formerly *Porphyra rosengurttii*), exhibit remarkable photostability and photoprotective properties against radiation exposure (Conde, Churio and Previtali 2000). These MAAs have been found to mitigate sunburn cell formation and prevent skinfold thickening in the epidermis/dermis of mice following UV radiation exposure (De la Coba et al. 2009). Seaweed-derived peptides and amino acids, which are reported to have various skin cosmeceutical benefits, are summarized in Table 3.

Certain seaweeds serve as abundant protein sources, with cultivation offering a higher protein yield per unit area (2.5–7.5 tons/Ha/year) compared to terrestrial crops, although successful extraction is often influenced by the presence of polysaccharides like alginates in brown seaweed or carrageenans in red seaweed (Bleakley and Hayes 2017). Seasonal variations and habitat conditions impact the protein, peptides, and amino acids contents in seaweed, with red algae (Rhodophyta) generally exhibiting higher contents (up to 47%) compared to green (Chlorophyta) (9–26%) and brown (Phaeophyceae) seaweeds (3–15%) (Fleurence 2004, Gressler et al. 2011, Cheung et al. 2015). Proteins in these macroalgae groups contain essential and non-essential amino acids (Qasim 1991, Wong and Cheung 2001, Matanjun et al. 2009).

Proteins and bioactive peptides derived from seaweed offer numerous health benefits and exhibit high antioxidant properties, particularly in low molecular weight molecules, which are often considered safer than their synthetic counterparts and have reduced side effects (Admassu et al. 2018, Meisel 2004, Murray and FitzGerald 2007, Harnedy and FitzGerald 2011). Bioactive peptides typically comprise 3–20 amino acid residues, with their composition and sequence influencing activities such as antioxidant and antimicrobial effects, among other pharmacological interests (Indumathi and Mehta 2016, Jo et al. 2017, Lafarga et al. 2020, Sridhar et al. 2021). Peptides like carnosine, glutathione, and taurine exhibit antioxidant and chelating properties (Tiwari and Troy 2015).

Taurine, lacking a carboxyl group, is not considered a "true" amino acid but possesses various health-promoting properties, often accumulating in the thalli of certain red algae species such as *Ahnfeltia plicata*, *Euthora cristata*, and *Ceramium virgatum* (Rhodophyta) (Yanshin et al. 2021). The peptide PPY1, consisting of five amino acids and obtained via enzymatic hydrolysis from *Pyropia yezoensis*, demonstrates anti-inflammatory effects by suppressing inflammatory cytokines (Lee et al. 2015).

Table 2. Skin benefits and biological actions of marine algae derived polysaccharides.

Marine Algae	Bioactive Compounds	Skin Benefits	Biological Actions	References
Fucus vesiculosus	Fucoidan	Anti-melanogenesis	Activation of ERK pathway, reduction in melanin content	(Song et al. 2015)
Sargassum tenerrimum, Turbinaria conoides	Fucoidan	Antioxidant	- DPPH radical scavenging - Suppression of superoxide radicals - High total antioxidant capacity and FRAP ability	(Rupérez et al.2002, Wang et al. 2008, Wang et al. 2009, Marudhupandi et al. 2014)
Costaria costata	Fucoidan	Skin anti-aging	- Induction of UVB-induced mRNA and protein expression of MMP-1 and type 1 pro-collagen, along with activation of ERK and JNK pathways, demonstrated *in vitro* using HS68 cells	(Moon et al. 2008, Moon et al. 2009)
Costaria costata	Fucoidan	Skin anti-aging	- Stimulation of MMP-1 expression and type 1 pro-collagen production observed *in vitro* with HaCaT cells	(Moon et al. 2009)
Undaria pinnatifida (P) (Mekabu)	Fucoidan	Improved skin hydration and elasticity; Enhanced skin barrier function	- Reduction of UVB-induced edema and thickness of the prickle cell layer, as well as modulation of MMP-1 activity and expression, and IFN- levels demonstrated *in vivo* studies	(Maruyama et al. 2015)
Ascophyllum nodosum (P)	Fucoidan	Anti-aging effects; Reduction of skin redness and irritation	- Inhibition of IL-1-induced MMP-9 and MMP-3 expression/secretion, and upregulation of TIMP-1 shown *in vitro* using HDFs	(Senni et al. 2006)
Saccharina japonica (P)	Fucoidan	Moisturizing Higher moisture-absorption and moisture-retention ability than HA	attract and retain moisture in the skin	(Wang et al. 2013)
Saccharina cichorioides (P)	Fucoidan (water soluble)	Anti-skin cancer	- Inhibition of EGF or TPA-induced neoplastic cell transformation, and interference with EGF and EGFR binding demonstrated in vitro using JB6 Cl41 cells	(Lee et al. 2008)
Porphyra sp. (R)	Porphyran	Antioxidant	- *in vitro* through ferrous ion chelation, reduction power, and scavenging of DPPH and superoxide radicals	(Kuda et al. 2005)
Pyropia haitanensis (R)	Porphyran fraction F1 fraction F2	Antioxidant	- Enhancement of antioxidant enzyme activity, reduction of lipid peroxidation, and total antioxidant capacity observed *in vivo* in Kumming mice	(Zhang et al. 2003, Zhang et al. 2004)

Table 2 contd. ...

...Table 2 contd.

Marine Algae	Bioactive Compounds	Skin Benefits	Biological Actions	References
P. haitanensis	Porphyran with different molecular weights (including LMW Porphyran SD, AD, PD, BD)	Antioxidant	- *in vitro* through scavenging of DPPH and hydroxyl radicals, and inhibition of superoxide generation	(Zhao et al. 2006, Zhang et al. 2009)
P. haitanensis (R)	Porphyran	Anti-inflammation	- Inhibition of LPS-induced NO production, iNOS levels, NF-_B activation, TNF-production, and nuclear translocation of p65, along with degradation of I_B-demonstrated in vitro using RAW264.7 cells	(Jiang et al. 2012, Isaka et al. 2015)
Pyropia yezoensis (R)	Carrageenan	Anti-melanogenesis	- Reduction in melanosome levels observed *in vivo* in guinea pigs	(Takematsu and Seiji 1984)
P. yezoensis	Commercial iota (II)-Carrageenan	Antioxidant Photoprotective	- Antioxidant and photoprotective effects demonstrated *in vitro*, including prevention of UVB-induced cell death, reduction of intracellular ROS, and scavenging of DPPH radicals	(Thevanayagam et al. 2014)
Eucheuma denticulatum (R) *Kappaphycus alvarezii* (R)	iota (V)-Carrageenan; kappa (III)-Carrageenan; carrageenan-derived oligosaccharides (37.7 kDa) Commercial l-Carrageenan; Commercial i(II)-Carrageenan;	Antioxidant	- Antioxidant activity shown *in vitro* through scavenging of superoxide and hydroxyl radicals, as well as DPPH radical scavenging and reducing power	(Sun et al. 2015)

(Note: Abbreviations used: (R) - Rhodophyta, (P) - Phaeophyta, (C) - Chlorophyta)

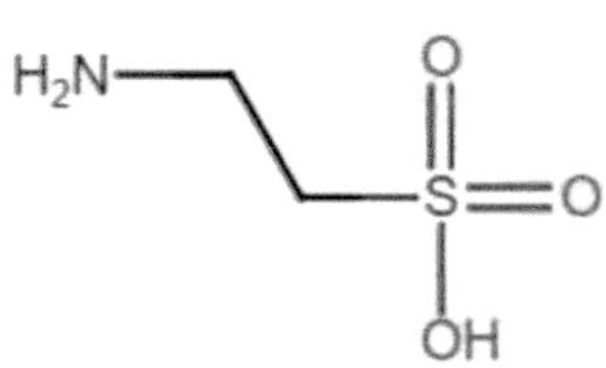

Taurine

The peptides PYP1-5 and Porphyra 334, derived from *Porphyra yezoensis* f. *coreana* Ueda, have demonstrated the ability to enhance elastin and collagen production while reducing the expression of matrix metalloproteinases (MMP) (Ryu et al. 2014). Ultrasound-assisted enzymatic hydrolysis has been suggested for the effective extraction of iodinated amino acids from *Palmaria palmata* and *Porphyra umbilicalis* (red seaweeds) (Romarís-Hortas et al. 2013). Mycosporine-like amino acids (MAAs) are secondary metabolites synthesized by seaweeds for protection against solar radiation (Shick et al. 2000. Bedoux et al. 2014). These compounds, consisting of cyclohexenone or cyclohexenimine chromophores with various amino acids, predominantly glycine

Table 3. Seaweed derived peptides and amino acids for skin benefits.

No.	Name of Macroalgae	Compounds	Cosmetic Benefits	References
1	*Gracilaria vermiculophylla* (R)	Porphyra-334, Palythine, Asterina-330, Shinorine	Antioxidant, UV protector	(Barceló-Villalobos et al. 2017)
2	*Ulva lactuca* (C), *Asparagopsis armata* (R)	MAAs, Amino acids	Antiaging, Anti wrinkles, Improves collagen formation	(Leandro et al. 2019)
3	*Gracilaria chilensis*, *Pyropia plicata*, *Champia novae-zelandiae* (R)	MAAs	Anti UV, Antioxidant	(Orfanoudaki et al. 2019)
4	*Ulva lactuca* (C)	Arginine, Aspartic acid, Glycine	Enhance collagen and elastin synthesis	(Orfanoudaki et al. 2019)
5	*Porphyra umbilicalis* (R)	MAAs, (2:1 ratio of Porphyra-334 and Shinorine)	Antiaging	(Guglielmo and Montanari 2008)
6	*Palmaria palmata* (R), *Catenella caespitosa* (R)	MAAs	UV and UV-A protection	(Daniel et al. 2004)
7	*Porphyra* sp., *Catenella caespitosa* (R), *Padina crassa*, *Desmarestia aculeata* (P)	MAAs such as Aminocyclohexenone-type, Aminocyclohexene imine-type	Photoprotection, Antiaging, Anti-inflammatory, Antioxidant	(Hartmann et al. 2015)
8	*Curdiea racovitzae*, *Iridaea cordata* (R)	Palythine, asterina-330	Antioxidant, Anti-UV, Antiaging	(Gao and Garcia-Pichel 2011)
9	*Palmaria* sp., *Porphyra* sp. (R)	High amounts of Glycine and Arginine	Natural moisturizing factor	(Fleurence 2004)
10	*Chondrus crispus* (R), *Mastocarpus stellatus* (R), *Palmaria palmata* (R)	Palythine, Usujirene, Porphyra-334, Shinorine, Asterina, palythinol	Antioxidant, Anti-proliferation	(Athukorala et al. 2016)
11	*Pyropia yezoensis* (R)	Peptide PPy1	Anti-inflammatory	(Lee, Kim and Nam 2015)
12	*Palmaria palmata* (R)	MAAs	UV protector	(Yuan et al. 2009)
13	*Porphyra umbilicalis* (R)	Porphyra-334, Shinorine	Moisturization, Skin protector, Antiwrinkle, Protect against roughness	(Ryu et al. 2014)
14	*Porphyra yezoensis* f. *coreana* (R)	Peptides, PYP1-5, porphyra-334	Enhance Elastin and collagen formation, reduce MMP expression	(Vega et al. 2021)
15	*Palmaria palmata* (R), *Porphyra umbilicalis* (R)	MAAs	Antiaging, Collagenase inhibition	(Vega et al. 2021, Nishida et al. 2020)

(Note: Abbreviations used: (R) - Rhodophyta, (P) - Phaeophyta, (C) - Chlorophyta)

or iminoalcohol groups, exhibit antioxidant and photoprotective properties (Dunlap and Yamamoto 1995, Bandaranayake 1998, Conde et al. 2000, Gröniger et al. 2000, Adams and Shick 2001, Singh et al. 2008).

Common MAAs found in seaweeds include shinorine, porphyra-334, palythine, asterina-330, mycosporine-glycine, palythinol, and palythene, with their concentration's dependent on geographic, seasonal, and bathymetric factors, peaking during summer and decreasing with water depth (Karsten et al. 1998, Rastogi et al. 2017, Orfanoudaki et al. 2019). A multifunctional cosmetic liposome formulation containing UV filters, vitamins (A, C, and E), Ginkgo biloba extract (rich in quercetin), and *Porphyra umbilicalis* extract (rich in proteins, vitamins, minerals, and MAAs porphyra-334 and shinorine) has shown efficacy against aging signs by improving hydration and reducing wrinkles and skin roughness (Gianeti and Maia Campos 2014). Leandro, Pereira and Gonçalves (2019) incorporated an extract of *Asparagopsis armata* (ASPAR'AGE™) containing MAA molecules in lotions with anti-aging properties, along with a hydrolysed extract Aosaine® (derived from *Ulva lactuca*, rich in amino acids similar to those promoting skin elasticity) presenting anti-aging, anti-wrinkle, and collagen stimulation properties. Additionally, an extract from *Gelidium corneum* (rich in minerals, trace elements, and amino acids) improves skin softness and elasticity. MAAs exhibit various properties such as natural sunscreens, antioxidants, anti-inflammatory, and anti-aging

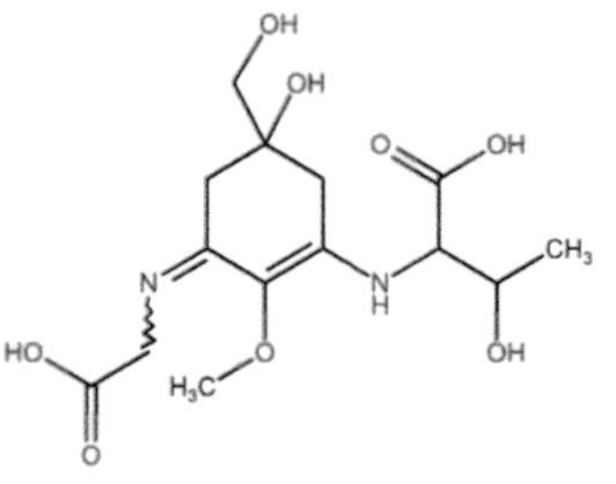

porphyra-334

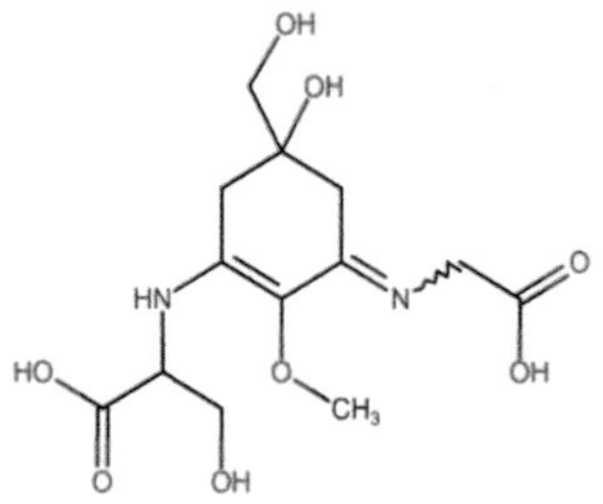

Shinorine

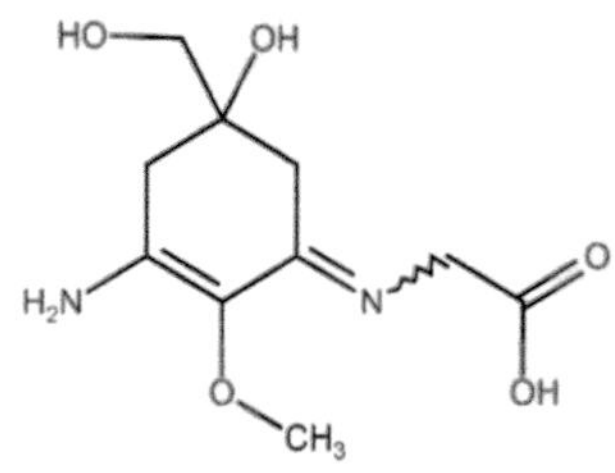

Palythine

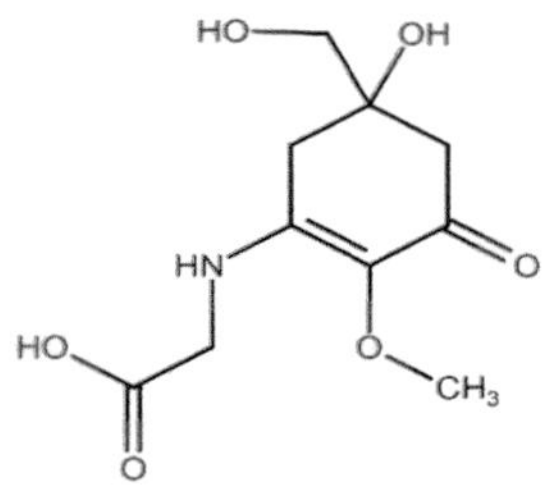

mycosporine-glycine

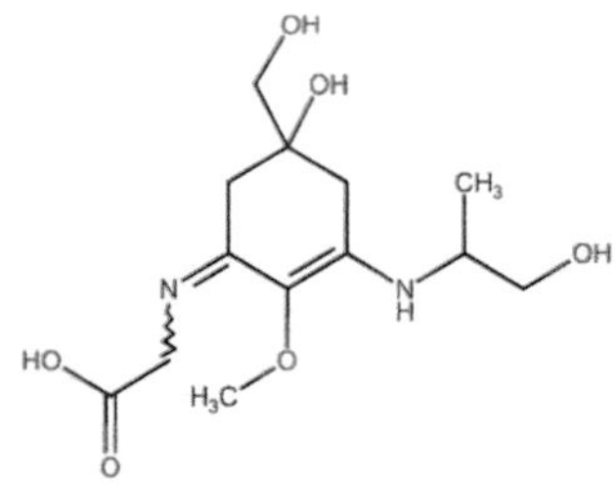

Palythinol

effects, and can stimulate skin renewal and cell proliferation, making them promising and safe options for pharmaceutical and cosmetic applications (Chrapusta et al. 2017).

Given the toxic effects of several synthetic dyes and the growing demand for natural colors in food, pharmaceuticals, cosmetics, and textiles, there is increasing interest in using phycobiliproteins such as C-phycocyanin and R-phycoerythrin. Phycobiliproteins, consisting of water-soluble compounds composed of proteins covalently bound to linear tetrapyrroles (phycobilins), possess fluorescent properties and high molecular weights, making them suitable for reddish colourings (Arad and Yaron 1992, Bermejo et al. 2000, Chronakis 2000, Rossano et al. 2003). B-phycoerythrin exhibits resistance to changes in pH and possesses antioxidant properties (Sekar and Chandramohan 2008), making it suitable for use as a pink or purple dye in cosmetics (Viskari and Colyer 2003). Phycobilins, including phycoerythrins (red) and phycocyanins and allophycocyanins (blue), contribute to the characteristic colours of microalgae and red algae (Rhodophyta). Phycocyanin is the predominant pigment in microalgae such as *Spirulina* sp., while the red colour of Rhodophyta is attributed to both phycoerythrin and phycocyanin pigments.

Extracts of phycobiliproteins from *Gracilaria gracilis*, particularly R-phycoerythrin, have been found to exhibit high antioxidant and radical scavenging activities, especially when harvested during winter (Francavilla et al. 2013). Extraction methods utilizing aqueous solutions of ionic liquids, such as cholinium chloride, have been successful in yielding up to 46.5% of R-phycoerythrin from fresh algal biomass (Martins et al. 2016). Saluri et al. (2020) conducted a study on *Furcellaria lumbricalis* and *Coccotylus truncatus*, revealing an exponential correlation between R-phycoerythrin and allophycocyanin concentrations and collection depth. Additionally, the contents of phycoerythrin and phycocyanin were found to be slightly higher and lower, respectively, in dried commercial *Porphyra* sp. extracts compared to *Spirulina* sp. (Osório et al. 2020).

Towards the seaweed polysaccharides and proteins into cosmetic products

Natural ingredients in cosmetic substances are becoming more crucial in terms of sustainability and environmental stewardship. Polysaccharides and proteins produced by seaweeds constitute a treasure trove of bioactive substances with unrivalled promise in skincare formulation. Their multipurpose capabilities, which include hydration, antioxidant protection, photoprotection, and cosmetic improvement, render them ideal tools in the search for fresh glowing skin. The next steps include producing specified types of seaweed (under controlled cultivation in specific regions) which decrease the ecological footprint, improve component stabilization and quality level, and assure a consistent supply network.

Seaweed extracts may be employed in formulations tailored to certain skin types or concerns, resulting in a more targeted and effective approach. Biotechnological processes, such as genetic modification and bioengineering, might be utilized to boost the synthesis of valuable chemicals in seaweeds. This could result in the development of seaweed varieties featuring better properties for cosmetic usage.

Ongoing studies on unique green extraction technologies will result in more efficient isolation of specific substances using grown seaweed. This improves the production, purity, and safety of chosen elements, ensuring their efficacy in cosmetic compositions (by decreasing contamination or impurities). Formulators must explore chemical and biological interactions between seaweed extracts/compounds and other natural ingredients known for their skincare benefits. Combining seaweed with plant extracts, vitamins, and antioxidants can produce strong, multipurpose cosmetic products.

Beyond skin applications, the skincare sector and research and development teams might begin to explore the use of seaweed extracts in functional meals and nutricosmetic products. Seaweed-based foods can assist to enhance skin wellness and attractiveness by altering the human metabolic process. Nevertheless, it requires the creation and development of rigorous clinical trials to verify their safety and effectiveness. To conclude, clinical research will be critical for assessing the benefits of seaweed for skin health in a green future. In addition, this environmental future development should involve greater efforts to educate the public about the benefits and drawbacks of seaweed-based cosmetics.

The future development of seaweed-based skincare products is likely to be characterized by an intricate combination of technological advances, industrial research, pharmaceutical safety approaches, concerns over the environment, consumer demands, and a commitment towards secure, trustworthy, and ethical sourcing.

Concluding remark on potential of carbohydrates and proteins in skin benefits

In conclusion, the utilization of seaweed-derived polysaccharides in skincare products presents a compelling avenue for the cosmetics and cosmeceutical industries to leverage natural bioactive compounds for enhanced skin health and wellness. Through rigorous scientific inquiry and technological advancement, researchers and industry experts have elucidated the diverse array of beneficial properties inherent in these marine-derived polysaccharides.

The rich composition of seaweeds, encompassing polysaccharides such as alginate, laminarin, fucoidans, ulvans, agar, carrageenans, and porphyran, provides a wealth of bioactive molecules with demonstrated skincare benefits. These polysaccharides exhibit various biological activities, including antioxidant, anti-inflammatory, moisturizing, and wound-healing properties, making them valuable ingredients for skincare formulations. Scientific studies have underscored the efficacy of seaweed polysaccharides in addressing various skin concerns. For example, fucoidans have been shown to inhibit UV-induced collagenase and elastase activities, thereby protecting against skin photoaging. Similarly, laminarin exhibits antioxidant activity and promotes wound healing, while ulvans demonstrate anti-inflammatory and anti-hyperlipidemic properties. Moreover, the structural characteristics of seaweed polysaccharides, such as molecular weight and sulphate content, play a crucial role in determining their bioactivity. Depolymerization and other structural modifications have been explored to enhance the efficacy of these polysaccharides, further highlighting the importance of scientific research in optimizing their skincare applications.

Collaboration between biotechnology and the cosmetics industry holds promise for unlocking the full potential of seaweed-derived polysaccharides. By leveraging advanced extraction techniques and biotechnological approaches, researchers can develop innovative skincare products that harness the biofunctional properties of these natural compounds. Furthermore, the sustainability of seaweed-derived ingredients aligns with consumer preferences for eco-friendly and ethically sourced skincare products. Responsible harvesting practices and environmentally friendly extraction methods ensure the preservation of marine ecosystems while meeting the growing demand for sustainable beauty solutions.

Moreover, the utilization of seaweed-derived proteins and amino acids in cosmetic formulations represents a promising avenue for the skincare industry, driven by scientific research elucidating their diverse bioactive properties. Seaweeds—abundant sources of proteins and peptides—offer a rich repository of essential and non-essential amino acids, including histidine, taurine, serine, and glutamic acid, among others. These compounds contribute to various cosmeceutical activities such as anti-inflammatory, antioxidant, anti-aging, and photoprotective effects. Research findings underscore the multifaceted benefits of seaweed-derived proteins and amino acids in skincare, including collagen stimulation, wrinkle reduction, and UV protection. Mycosporine-like amino acids (MAAs) found in red macroalgae demonstrate remarkable photostability and photoprotective properties against UV radiation, making them valuable ingredients for sunscreen formulations. Furthermore, bioactive peptides derived from seaweed exhibit antioxidant and antimicrobial effects, contributing to skin health and vitality. The sustainable cultivation and extraction of proteins and peptides from seaweed offer significant advantages over traditional terrestrial sources, with seaweed cultivation yielding higher protein yields per unit area. Advanced extraction techniques, such as ultrasound-assisted enzymatic hydrolysis, enable the efficient extraction of bioactive compounds from seaweed, ensuring the preservation of their beneficial properties. Moreover, the integration of seaweed-derived proteins and peptides into cosmetic formulations aligns with consumer preferences for natural and environmentally friendly skincare products. Future research directions may involve exploring biotechnological approaches to enhance the synthesis of valuable compounds in seaweed and investigating the synergistic effects of seaweed extracts with other natural ingredients for improved skincare formulations.

Overall, the integration of seaweed-derived polysaccharides as well as peptides and amino acids into skincare formulations represents a scientifically sound approach to meeting consumer demands for safe, effective, and environmentally friendly skincare products. Through continued research and collaboration, the cosmetics industry can harness the full potential of these natural compounds to innovate and advance the field of skincare. The integration of seaweed-derived bioactives into skincare products holds immense potential for addressing various skin concerns while meeting consumer demands for safe, effective, and sustainable cosmetics. Continued scientific research and technological innovation are essential for unlocking the full potential of seaweed as a valuable resource in the skincare industry.

References

Adams, N.L. and J.M. Shick. 2001. Mycosporine-like amino acids prevent UVB-induced abnormalities during early development of the green sea urchin *Strongylocentrotus droebachiensis*. Marine Biology 138: 267–280.

Admassu, H., M.A.A. Gasmalla, R. Yang and W. Zhao. 2018. Bioactive peptides derived from seaweed protein and their health benefits: antihypertensive, antioxidant, and antidiabetic properties. Journal of Food Science 83(1): 6–16.

Adrien, A., A. Bonnet, D. Dufour, S. Baudouin, T. Maugard and N. Bridiau. 2017. Pilot production of ulvans from *Ulva* sp. and their effects on hyaluronan and collagen production in cultured dermal fibroblasts. Carbohydrate Polymers 157: 1306–1314.

Ahmed, A. and R. Taha. 2011. Marine phytochemical compounds and their cosmeceutical applications. pp. 51–61. *In*: Kim, S. [ed.]. Marine Cosmeceuticals: Trends and Prospects. CRC Press: Boca Raton, FL, USA.

Ali Karami, M., B. Sharif Makhmalzadeh, M. Pooranian and A. Rezai. 2021. Preparation and optimization of silibinin-loaded chitosan–fucoidan hydrogel: An in vivo evaluation of skin protection against UVB. Pharmaceutical Development and Technology 26(2): 209–219.

Arad, S.M. and A. Yaron. 1992. Natural pigments from red microalgae for use in foods and cosmetics. Trends in Food Science & Technology 3: 92–97.

Aslam, A., A. Bahadar, R. Liaquat, M. Saleem, A. Waqas and M. Zwawi. 2021. Algae as an attractive source for cosmetics to counter environmental stress. Science of The Total Environment 772: 144905.

Athukorala, Y., S. Trang, C. Kwok and Y.V. Yuan. 2016. Antiproliferative and antioxidant activities and mycosporine-like amino acid profiles of wild-harvested and cultivated edible Canadian marine red macroalgae. Molecules 21(1): 119.

Ayoub, A., J.M. Pereira, L.E. Rioux, S.L. Turgeon, M. Beaulieu and V.J. Moulin. 2015. Role of seaweed laminaran from *Saccharina longicruris* on matrix deposition during dermal tissue-engineered production. International Journal of Biological Macromolecules 75: 13–20.

Aziz, E., R. Batool, M.U. Khan, A. Rauf, W. Akhtar, M. Heydari et al. 2020. An overview on red algae bioactive compounds and their pharmaceutical applications. Journal of Complementary and Integrative Medicine 17(4): 20190203.

Balboa, E.M., M.L. Soto, D.R. Nogueira, N. González-López, E. Conde, A. Moure et al. 2014. Potential of antioxidant extracts produced by aqueous processing of renewable resources for the formulation of cosmetics. Industrial Crops and Products 58: 104–110.

Balboa, E.M., E. Conde, M.L. Soto, L. Pérez-Armada and H. Domínguez. 2015. Cosmetics from marine sources. pp. 1015–1042. *In*: Kim, S.-K. [ed.]. Springer Handbook of Marine Biotechnology. Springer: Berlin/Heidelberg, Germany. ISBN 978-3-642-53970-1.

Bandaranayake, W. 1998. Mycosporines: are they nature's sunscreens? Natural Product Reports 15(2): 159–172.

Barceló-Villalobos, M., F.L. Figueroa, N. Korbee, F. Álvarez-Gómez and M.H. Abreu. 2017. Production of mycosporine-like amino acids from *Gracilaria vermiculophylla* (Rhodophyta) cultured through one year in an integrated multi-trophic aquaculture (IMTA) system. Marine Biotechnology 19: 246–254.

Baweja, P., S. Kumar, D. Sahoo and I. Levine. 2016. Biology of Seaweeds in Seaweed. pp. 41–106. *In*: Fleurence, J. and I. Levine [eds.]. Health and Disease Prevention. Nikki Levy Elsevier Inc.: Amsterdam, The Netherlands.

Beaumont, M., R. Tran, G. Vera, D. Niedrist, A. Rousset, R. Pierre et al. 2021. Hydrogel-forming algae polysaccharides: From seaweed to biomedical applications. Biomacromolecules 22(3): 1027–1052.

Bedoux, G., K. Hardouin, C. Marty, L. Taupin, L. Vandanjon and N. Bourgougnon. 2014. Chemical characterization and photoprotective activity measurement of extracts from the red macroalga *Solieria chordalis*. Botanica Marina 57(4): 291–301.

Bermejo, R., E.M. Talavera and J.M. Alvarez-Pez. 2000. C-phycocyanin incorporated into reverse micelles: A fluorescence study. Colloids and Surfaces B: Biointerfaces 18(1): 51–59.

Berthon, J.Y., R. Nachat-Kappes, M. Bey, J.P. Cadoret, I. Renimel and E. Filaire. 2017. Marine algae as attractive source to skin care. Free Radical Research 51(6): 555–567.

Bleakley, S. and M. Hayes. 2017. Algal proteins: extraction, application, and challenges concerning production. Foods 6(5): 33.

Brandt, F.S., A. Cazzaniga and M. Hann. 2011, September. Cosmeceuticals: current trends and market analysis. In Seminars in cutaneous medicine and surgery. WB Saunders 30(3): 141–143.

Buono, S., A.L. Langellotti, A. Martello, M. Bimonte, A. Tito, A. Carola et al. 2012. Biological activities of dermatological interest by the water extract of the microalga *Botryococcus braunii*. Archives of Dermatological Research 304: 755–764.

Campo, V.L., D.F. Kawano, D.B. da Silva Jr. and I. Carvalho. 2009. Carrageenans: Biological properties, chemical modifications and structural analysis–A review. Carbohydrate Polymers 77(2): 167–180.

Campos, P.M.B.G.M., M.O. Melo and F.B. Camargo Junior. 2014. Effects of polysaccharide-based formulations on human skin. Polysaccharides; Springer: Cham, Switzerland, pp. 1–8.

Carvalho, L.G. and L. Pereira. 2015. Review of marine algae as source of bioactive metabolites. *In*: Pereira, L. and J.M. Neto [Eds.]. Marine Algae—Biodiversity, Taxonomy, Environmental Assessment and Biotechnology, 1st ed; CRC Press: Boca Raton, FL, USA, ISBN 9781466581678.

Casas, M.P., V. Rodríguez-Hermida, P. Pérez-Larrán, E. Conde, M.T. Liveri, D. Ribeiro et al. 2016. *In vitro* bioactive properties of phlorotannins recovered from hydrothermal treatment of *Sargassum muticum*. Separation and Purification Technology 167: 117–126.

Černá, M. 2011. Seaweed proteins and amino acids as nutraceuticals. Advances in Food and Nutrition Research 64: 297–312.

Cha, S.H., S.C. Ko, D. Kim and Y.J. Jeon. 2011. Screening of marine algae for potential tyrosinase inhibitor: Those inhibitors reduced tyrosinase activity and melanin synthesis in zebrafish. The Journal of Dermatology 38(4): 354–363.

Charlier, R.H. and M.C.P. Chaineux. 2009. The healing sea: A sustainable coastal ocean resource: Thalassotherapy. Journal of Coastal Research 25(4): 838–856.

Chen, X., X. Fu, L. Huang, J. Xu and X. Gao. 2021. Agar oligosaccharides: A review of preparation, structures, bioactivities and application. Carbohydrate Polymers 265: 118076.

Chen, Y.H., C.J. Tu and H.T. Wu. 2004. Growth-inhibitory effects of the red alga *Gelidium amansii* on cultured cells. Biological and Pharmaceutical Bulletin 27(2): 180–184.

Cheong, K.L., H.M. Qiu, H. Du, Y. Liu and B.M. Khan. 2018. Oligosaccharides derived from red seaweed: Production, properties, and potential health and cosmetic applications. Molecules 23(10): 2451.

Cheung, R.C.F., T.B. Ng and J.H. Wong. 2015. Marine peptides: Bioactivities and applications. Marine Drugs 13(7): 4006–4043.

Choi, J.S., Y.M. Ha, C.U. Joo, K.K. Cho, S.J. Kim and I.S. Choi. 2012. Inhibition of oral pathogens and collagenase activity by seaweed extracts. Journal of Environmental Biology 33(1): 115.

Chrapusta, E., A. Kaminski, K. Duchnik, B. Bober, M. Adamski and J. Bialczyk. 2017. Mycosporine-like amino acids: Potential health and beauty ingredients. Marine Drugs 15(10): 326.

Chronakis, I.S. 2000. Biosolar proteins from aquatic algae. In Developments in food science (Vol. 41, pp. 39–75). Elsevier.

Cikoš, A.M., I. Jerković, M. Molnar, D. Šubarić and S. Jokić. 2021. New trends for macroalgal natural products applications. Natural Product Research 35(7): 1180–1191.

Cindana Mo'o, F.R., G. Wilar, H.P. Devkota and N. Wathoni. 2020. Ulvan, a polysaccharide from macroalga *Ulva* sp.: A review of chemistry, biological activities and potential for food and biomedical applications. Applied Sciences 10(16): 5488.

Citkowska, A., M. Szekalska and K. Winnicka. 2019. Possibilities of fucoidan utilization in the development of pharmaceutical dosage forms. Marine Drugs 17(8): 458.

Conde, F.R., M.S. Churio and C.M. Previtali. 2000. The photoprotector mechanism of mycosporine-like amino acids. Excited-state properties and photostability of porphyra-334 in aqueous solution. Journal of Photochemistry and Photobiology B: Biology 56(2-3): 139–144.

Conde, F.R., M.S. Churio and C.M. Previtali. 2000. The photoprotector mechanism of mycosporine-like amino acids: excited-state properties and photostability of porphyra-334 in aqueous solution. Journal of Photochemistry and Photobiology B: Biology 56(2-3): 139–144. https://doi.org/10.1016/S1011-1344(00)00163-0.

Costa, A.M., J.M. Rodrigues, M.M. Perez-Madrigal, A.P. Dove and J.F. Mano. 2020. Modular functionalization of laminarin to create value-added naturally derived macromolecules. Journal of the American Chemical Society 142(46): 19689–19697.

Costa, L.S., G.P. Fidelis, C.B.S. Telles, N. Dantas-Santos, R.B.G. Camara, S.L. Cordeiro et al. 2011. Antioxidant and antiproliferative activities of heterofucans from the seaweed *Sargassum filipendula*. Marine Drugs 9(6): 952–966.

Cotas, J., A. Leandro, D. Pacheco, A.M. Gonçalves and L. Pereira. 2020. A comprehensive review of the nutraceutical and therapeutic applications of red seaweeds (Rhodophyta). Life 10(3): 19.

Daniel, S., S. Cornelia and Z. Fred. 2004. UV-A sunscreen from red algae for protection against premature skin aging. Cosmet. Toilet. Manuf. Worldw, pp. 139–143.

de Araújo, I.W.F., E.D.S.O. Vanderlei, J.A.G. Rodrigues, C.O. Coura, A.L.G. Quinderé, B.P. Fontes et al. 2011. Effects of a sulfated polysaccharide isolated from the red seaweed *Solieria filiformis* on models of nociception and inflammation. Carbohydrate Polymers 86(3): 1207–1215.

de Jesus Raposo, M.F., R.M.S.C. De Morais and A.M.M.B. de Morais. 2013. Bioactivity and applications of sulphated polysaccharides from marine microalgae. Marine Drugs 11(1): 233–252.

De la Coba, F., J. Aguilera, M.V. De Galvez, M. Alvarez, E. Gallego, F.L. Figueroa et al. 2009. Prevention of the ultraviolet effects on clinical and histopathological changes, as well as the heat shock protein-70 expression in mouse skin by topical application of algal UV-absorbing compounds. Journal of Dermatological Science 55(3): 161–169.

de Lacerda, D., D. Thioly-Bensoussan and K. Burke. 2013. Cosmeceuticals for men. Cutis, pp. 6–12.

Deville, C., M. Gharbi, G. Dandrifosse and O. Peulen. 2007. Study on the effects of laminarin, a polysaccharide from seaweed, on gut characteristics. Journal of the Science of Food and Agriculture 87(9): 1717–1725.

Dita, L.R. and J. Triastuti. 2020, February. Utilization of agar *Gracilaria* sp. as a natural thickener on liquid bath soap formulation. In IOP Conference Series: Earth and Environmental Science (Vol. 441, No. 1, p. 012021). IOP Publishing.

Dolorosa, M.T., S. Purwaningsih and E. Anwar. 2020. Utilization of *Kappaphycus alvarezii* and *Sargassum plagyophyllum* from Banten as cosmetic creams. In IOP Conference Series: Earth and Environmental Science (Vol. 404, No. 1, p. 012008). IOP Publishing.

Draelos, Z.D. 2014. Cosmeceuticals: efficacy and influence on skin tone. Dermatologic Clinics 32(2): 137–143.

Draelos, Z.D. 2018. Cosmeceuticals for male skin. Dermatologic Clinics 36(1): 17–20.

Dunlap, W.C. and Y. Yamamoto. 1995. Small-molecule antioxidants in marine organisms: antioxidant activity of mycosporine-glycine. Comparative Biochemistry and Physiology Part B: Biochemistry and Molecular Biology 112(1): 105–114.

Fabrowska, J., B. Łęska, G. Schroeder, B. Messyasz and M. Pikosz. 2015. Biomass and extracts of algae as material for cosmetics. Marine algae extracts: Processes, Products, and Applications, pp. 681–706.

Fauziee, N.A.M., L.S. Chang, W.A.W. Mustapha, A.R.M. Nor and S.J. Lim. 2021. Functional polysaccharides of fucoidan, laminaran and alginate from Malaysian brown seaweeds (*Sargassum polycystum*, *Turbinaria ornata* and *Padina boryana*). International Journal of Biological Macromolecules 167: 1135–1145.

Félix, R., A.M. Carmona, C. Félix, S.C. Novais and M.F. Lemos. 2020. Industry-friendly hydroethanolic extraction protocols for *Grateloupia turuturu* UV-shielding and antioxidant compounds. Applied Sciences 10(15): 5304.

Fernando, I.P.S., K.N. Kim, D. Kim and Y.J. Jeon. 2019. Algal polysaccharides: Potential bioactive substances for cosmeceutical applications. Critical Reviews in Biotechnology 39(1): 99–113.

Fernando, I.S., K.A. Sanjeewa, S.Y. Kim, J.S. Lee and Y.J. Jeon. 2018. Reduction of heavy metal (Pb2+) biosorption in zebrafish model using alginic acid purified from *Ecklonia cava* and two of its synthetic derivatives. International Journal of Biological Macromolecules 106: 330–337.

Fleurence, J. 2004. Seaweed Proteins. pp. 197–213. *In*: Yada, R.Y. [Ed.]. Proteins in Food Processing. Woodhead Publishing Limited: Cambridge, UK.

Fournière, M., G. Bedoux, N. Lebonvallet, R. Leschiera, C. Le Goff-Pain, N. Bourgougnon et al. 2021. Poly-and oligosaccharide *ulva* sp. Fractions from enzyme-assisted extraction modulate the metabolism of extracellular matrix in human skin fibroblasts: Potential in anti-aging dermo-cosmetic applications. Marine Drugs 19(3): 156.

Francavilla, M., M. Franchi, M. Monteleone and C. Caroppo. 2013. The red seaweed *Gracilaria gracilis* as a multi products source. Marine Drugs 11(10): 3754–3776.

Fujimura, T., K. Tsukahara, S. Moriwaki, T. Kitahara, T. Sano and Y. Takema. 2002. Treatment of human skin with an extract of *Fucus vesiculosus* changes its thickness and mechanical properties. Journal of Cosmetic Science 53(1): 1–9.

Fujimura, T., K. Tsukahara, S. Moriwaki, T. Kitahara and Y. Takema. 2000. Effects of natural product extracts on contraction and mechanical properties of fibroblast populated collagen gel. Biological and Pharmaceutical Bulletin 23(3): 291–297.

Galland-Irmouli, A.V., J. Fleurence, R. Lamghari, M. Luçon, C. Rouxel, O. Barbaroux et al. 1999. Nutritional value of proteins from edible seaweed *Palmaria palmata* (Dulse). The Journal of Nutritional Biochemistry 10(6): 353–359.

Gao, Q. and F. Garcia-Pichel. 2011. Microbial ultraviolet sunscreens. Nature Reviews Microbiology 9(11): 791–802.

Ghanbarzadeh, M., A. Golmoradizadeh and A. Homaei. 2018. Carrageenans and carrageenases: Versatile polysaccharides and promising marine enzymes. Phytochemistry Reviews 17(3): 535–571.

Gianeti, M.D. and P.M. Maia Campos. 2014. Efficacy evaluation of a multifunctional cosmetic formulation: The benefits of a combination of active antioxidant substances. Molecules 19(11): 18268–18282.

Godlewska, K., I. Michalak, Ł. Tuhy and K. Chojnacka. 2016. Plant growth biostimulants based on different methods of seaweed extraction with water. BioMed research international, 5973760.

Gómez-Mascaraque, L.G., M. Martínez-Sanz, R. Martínez-López, A. Martínez-Abad, B. Panikuttira, A. López-Rubio et al. 2021. Characterization and gelling properties of a bioactive extract from *Ascophyllum nodosum* obtained using a chemical-free approach. Current Research in Food Science 4: 354–364.

Gressler, V., M.T. Fujii, A.P. Martins, P. Colepicolo, J. Mancini-Filho and E. Pinto. 2011. Biochemical composition of two red seaweed species grown on the Brazilian coast. Journal of the Science of Food and Agriculture 91(9): 1687–1692.

Gröniger, A., R.P. Sinha, M. Klisch and D.P. Häder. 2000. Photoprotective compounds in cyanobacteria, phytoplankton and macroalgae—a database. Journal of Photochemistry and Photobiology B: Biology 58(2-3): 115–122.

Guglielmo, M. and D. Montanari. 2008. Cosmetic composition with a lifting effect for sustaining relaxed tissues. Patent WO2008146116 A, 2: 2008.

Hamed, I., F. Özogul, Y. Özogul and J.M. Regenstein. 2015. Marine bioactive compounds and their health benefits: A review. Comprehensive Reviews in Food Science ad Food Safety 14(4): 446–465.

Harnedy, P.A. and FitzGerald, R.J., 2011. Bioactive proteins, peptides, and amino acids from macroalgae 1. Journal of Phycology, 47(2): 218-232.

Harnedy, P.A. and R.J. Fitzgerald. 2013. Bioactive proteins and peptides from macroalgae, fish, shellfish and marine processing waste. Marine Proteins and Peptides: Biological Activities and Applications: 5–39.

Hartmann, A., K. Becker, U. Karsten, D. Remias and M. Ganzera. 2015. Analysis of mycosporine-like amino acids in selected algae and cyanobacteria by hydrophilic interaction liquid chromatography and a novel MAA from the red alga *Catenella repens*. Marine Drugs 13(10): 6291–6305.

Heo, S.J., S.C. Ko, S.H. Cha, D.H. Kang, H.S. Park, Y.U. Choi et al. 2009. Effect of phlorotannins isolated from *Ecklonia cava* on melanogenesis and their protective effect against photo-oxidative stress induced by UV-B radiation. Toxicology *in vitro* 23(6): 1123–1130.

Holdt, S.L. and S. Kraan. 2011. Bioactive compounds in seaweed: functional food applications and legislation. Journal of Applied Phycology 23: 543–597.

Holtkamp, A.D., S. Kelly, R. Ulber and S. Lang. 2009. Fucoidans and fucoidanases—focus on techniques for molecular structure elucidation and modification of marine polysaccharides. Applied Microbiology and Biotechnology 82: 1–11.

Houston, M.C. 2005. Nutraceuticals, vitamins, antioxidants, and minerals in the prevention and treatment of hypertension. Progress in Cardiovascular Diseases 47(6): 396–449.

Huang, Y., H. Jiang, X. Mao and F. Ci. 2021. Laminarin and laminarin oligosaccharides originating from brown algae: Preparation, biological activities, and potential applications. Journal of Ocean University of China 20: 641–653.

Ibrahim, M., M. Salman, S. Kamal, S. Rehman, A. Razzaq and S.H. Akash. 2017. Algae-based biologically active compounds. In Algae Based Polymers, Blends, and Composites (pp. 155–271). Elsevier.

Indumathi, P. and A. Mehta. 2016. A novel anticoagulant peptide from the Nori hydrolysate. Journal of Functional Foods 20: 606–617.

Isaka, S., K. Cho, S. Nakazono, R. Abu, M. Ueno, D. Kim et al. 2015. Antioxidant and anti-inflammatory activities of porphyran isolated from discolored nori (*Porphyra yezoensis*). International Journal of Biological Macromolecules 74: 68–75.

Je, J.Y., P.J. Park, E.K. Kim, J.S. Park, H.D. Yoon, K.R. Kim et al. 2009. Antioxidant activity of enzymatic extracts from the brown seaweed *Undaria pinnatifida* by electron spin resonance spectroscopy. LWT-Food Science and Technology 42(4): 874–878.

Jesumani, V., H. Du, M. Aslam, P. Pei and N. Huang. 2019. Potential use of seaweed bioactive compounds in skincare—A review. Marine Drugs 17(12): 688.

Jesumani, V., H. Du, P. Pei, M. Aslam and N. Huang. 2020. Comparative study on skin protection activity of polyphenol-rich extract and polysaccharide-rich extract from *Sargassum vachellianum*. PLoS One 15(1): e0227308.

Jesumani, V., H. Du, P. Pei, C. Zheng, K.L. Cheong and N. Huang. 2019. Unravelling property of polysaccharides from *Sargassum* sp. as an anti-wrinkle and skin whitening property. International Journal of Biological Macromolecules 140: 216–224.

Jiang, N., B. Li, X. Wang, X. Xu, X. Liu, W. Li et al. 2020. The antioxidant and antihyperlipidemic activities of phosphorylated polysaccharide from *Ulva pertusa*. International Journal of Biological Macromolecules 145: 1059–1065.

Jiang, Z., Y. Hama, K. Yamaguchi and T. Oda. 2012. Inhibitory effect of sulphated polysaccharide porphyran on nitric oxide production in lipopolysaccharide-stimulated RAW264. 7 macrophages. The Journal of Biochemistry 151(1): 65–74.

Jiao, G., G. Yu, J. Zhang and H.S. Ewart. 2011. Chemical structures and bioactivities of sulfated polysaccharides from marine algae. Marine Drugs 9(2): 196–223.

Jo, C., F.F. Khan, M.I. Khan and J. Iqbal. 2017. Marine bioactive peptides: Types, structures, and physiological functions. Food Reviews International 33(1): 44–61.

Jutur, P.P., A.A. Nesamma and K.M. Shaikh. 2016. Algae-derived marine oligosaccharides and their biological applications. Frontiers in Marine Science 3: 83.

Kadam, S.U., B.K. Tiwari and C.P. O'Donnell. 2015. Extraction, structure and biofunctional activities of laminarin from brown algae. International Journal of Food Science & Technology 50(1): 24–31.

Kang, H., C.H. Lee, J.R. Kim, J.Y. Kwon, S.G. Seo, J.G. Han et al. 2015. *Chlorella vulgaris* attenuates dermatophagoides farinae-induced atopic dermatitis-like symptoms in NC/Nga mice. International Journal of Molecular Sciences 16(9): 21021–21034.

Kanlayavattanakul, M. and N. Lourith. 2015. Biopolysaccharides for skin hydrating cosmetics. Polysaccharides: Bioactivity and Biotechnology; Springer International Publishing: New York, NY, USA: 1867–1892.

Karsten, U., T. Sawall and C. Wiencke. 1998. A survey of the distribution of UV-absorbing substances in tropical macroalgae. Phycological Research 46(4): 271–279.

Kartik, A., D. Akhil, D. Lakshmi, K.P. Gopinath, J. Arun, R. Sivaramakrishnan et al. 2021. A critical review on production of biopolymers from algae biomass and their applications. Bioresource Technology 329: 124868.

Kharkwal, H., D.D. Joshi, P.R.E.E.T.I. Panthari, M.K. Pant and A.C. Kharkwal. 2012. Algae as future drugs. Asian J. Pharm. Clin. Res, 5: 1-4.

Kidgell, J.T., Carnachan, S.M., Magnusson, M., Lawton, R.J., Sims, I.M., Hinkley, S.F., de Nys, R. and Glasson, C.R., 2021. Are all ulvans equal? A comparative assessment of the chemical and gelling properties of ulvan from blade and filamentous Ulva. Carbohydrate Polymers, 264, p.118010.

Kim, K.B.W.R., Jeong, S.M., Kim, M.J. and Ahn, D.H., 2020. Tyrosinase Inhibitory Effects of Sargachromanol G, Sargachromanol I and Mojabanchromanol b isolated from Myagropsis myagroides. Indian Journal of Pharmaceutical Sciences 82(1).

Kim, M.S., G.H. Oh, M.J. Kim and J.K. Hwang. 2013. Fucosterol inhibits matrix metalloproteinase expression and promotes type-1 procollagen production in UVB-induced HaCaT cells. Photochemistry and Photobiology 89(4): 911–918.

Kim, S., D.H. You, T. Han and E.M. Choi 2014. Modulation of viability and apoptosis of UVB-exposed human keratinocyte HaCaT cells by aqueous methanol extract of laver (*Porphyra yezoensis*). Journal of Photochemistry and Photobiology B: Biology 141: 301–307.

Kim, S.K. 2014. Marine cosmeceuticals. Journal of Cosmetic Dermatology 13(1): 56–67.

Kim, S.K., Y.D. Ravichandran, S.B. Khan and Y.T. Kim. 2008. Prospective of the cosmeceuticals derived from marine organisms. Biotechnology and Bioprocess Engineering 13: 511–523.

Kim, S.M., K. Kang, J.S. Jeon, E.H. Jho, C.Y. Kim, C.W. Nho et al. 2011. Isolation of phlorotannins from *Eisenia bicyclis* and their hepatoprotective effect against oxidative stress induced by tert-butyl hyperoxide. Applied Biochemistry and Biotechnology 165: 1296–1307.

Ko, S.C., S.H. Cha, S.J. Heo, S.H. Lee, S.M. Kang and Y.J. Jeon. 2011. Protective effect of *Ecklonia cava* on UVB-induced oxidative stress: *In vitro* and *in vivo* zebrafish model. Journal of Applied Phycology 23: 697–708.

Kordjazi, M., B. Shabanpour, E. Zabihi, M.A. Faramarzi, F. Feizi, H.A. Gavlighi et al. 2013. Sulfated polysaccharides purified from two species of Padina improve collagen and epidermis formation in the rat. International Journal of Molecular and Cellular Medicine 2(4): 156.

Kuda, T., M. Tsunekawa, H. Goto and Y. Araki. 2005. Antioxidant properties of four edible algae harvested in the Noto Peninsula, Japan. Journal of Food Composition and Analysis 18(7): 625–633.

Kwon, M.J. and T.J. Nam. 2006. Porphyran induces apoptosis related signal pathway in AGS gastric cancer cell lines. Life Sciences 79(20): 1956–1962.

Lafarga, T., F.G. Acién-Fernández and M. Garcia-Vaquero. 2020. Bioactive peptides and carbohydrates from seaweed for food applications: Natural occurrence, isolation, purification, and identification. Algal Research 48: 101909.

Lahaye, M. and A. Robic. 2007. Structure and functional properties of ulvan, a polysaccharide from green seaweeds. Biomacromolecules 8(6): 1765–1774.

Laurienzo, P. 2010. Marine polysaccharides in pharmaceutical applications: An overview. Marine Drugs 8(9): 2435–2465.

Leandro, A., L. Pereira and A.M. Gonçalves. 2019. Diverse applications of marine macroalgae. Marine Drugs 18(1): 17.

Lee, H.A., I.H. Kim and T.J. Nam. 2015. Bioactive peptide from *Pyropia yezoensis* and its anti-inflammatory activities. International Journal of Molecular Medicine 36(6): 1701–1706.

Lee, N.Y., S.P. Ermakova, H.K. Choi, M.I. Kusaykin, N.M. Shevchenko, T.N. Zvyagintseva et al. 2008. Fucoidan from *Laminaria cichorioides* inhibits AP-1 transactivation and cell transformation in the mouse epidermal JB6 cells. Molecular Carcinogenesis: Published in cooperation with the University of Texas MD Anderson Cancer Center 47(8): 629–637.

Lee, Y.E., H. Kim, C. Seo, T. Park, K.B. Lee, S.Y. Yoo et al. 2017. Marine polysaccharides: Therapeutic efficacy and biomedical applications. Archives of Pharmacal Research 40: 1006–1020.

Li, B., H. Xu, X. Wang, Y. Wan, N. Jiang, H. Qi and X. Liu. 2020. Antioxidant and antihyperlipidemic activities of high sulfate content purified polysaccharide from *Ulva pertusa*. International Journal of Biological Macromolecules 146: 756–762.

Li, J., C. Cai, C. Yang, J. Li, T. Sun and G. Yu. 2019. Recent advances in pharmaceutical potential of brown algal polysaccharides and their derivatives. Current Pharmaceutical Design 25(11): 1290–1311.

Lin, G.P., D.S. Wu, X.W. Xiao, Q.Y. Huang, H.B. Chen, D. Liu et al. 2020. Structural characterization and antioxidant effect of green alga *Enteromorpha prolifera* polysaccharide in *Caenorhabditis elegans* via modulation of microRNAs. International Journal of Biological Macromolecules 150: 1084–1092.

Lin, T.J. 2010. Evolution of cosmetics: Increased need for experimental clinical medicine. Journal of Experimental & Clinical Medicine 2(2): 49–52.

Lourenço-Lopes, C., M. Fraga-Corral, C. Jimenez-Lopez, A.G. Pereira, P. Garcia-Oliveira, M. Carpena et al. 2020. Metabolites from macroalgae and its applications in the cosmetic industry: A circular economy approach. Resources 9(9): 101.

Malinowska, P. 2011. Algae extracts as active cosmetic ingredients. Zeszyty Naukowe/Uniwersytet Ekonomiczny w Poznaniu (212): 123–129.

Martínez–Hernández, G.B., N. Castillejo, M.D.M. Carrión–Monteagudo, F. Artés and F. Artés-Hernández. 2018. Nutritional and bioactive compounds of commercialized algae powders used as food supplements. Food Science and Technology International 24(2): 172–182.

Martins, M., F.A. Vieira, I. Correia, R.A. Ferreira, H. Abreu, J.A. Coutinho et al. 2016. Recovery of phycobiliproteins from the red macroalga *Gracilaria* sp. using ionic liquid aqueous solutions. Green Chemistry 18(15): 4287–4296.

Marudhupandi, T., T.T. Kumar, S.L. Senthil and K.N. Devi. 2014. *In vitro* antioxidant properties of fucoidan fractions from *Sargassum tenerrimum*. Pak. J. Biol. Sci. 17(3): 402–7.

Maruyama, H., H. Tamauchi, F. Kawakami, K. Yoshinaga and T. Nakano. 2015. Suppressive effect of dietary fucoidan on proinflammatory immune response and MMP-1 expression in UVB-irradiated mouse skin. Planta Medica 81(15): 1370–1374.

Matanjun, P., S. Mohamed, N.M. Mustapha and K. Muhammad. 2009. Nutrient content of tropical edible seaweeds, *Eucheuma cottonii*, *Caulerpa lentillifera* and *Sargassum polycystum*. Journal of Applied Phycology 21: 75–80.

Meisel, H. 2004. Multifunctional peptides encrypted in milk proteins. Biofactors 21(1-4): 55–61.

Michalak, I. and K. Chojnacka. 2015. Algae as production systems of bioactive compounds. Engineering in Life Sciences 15(2): 160–176.

Mišurcová, L., S. Škrovánková, D. Samek, J. Ambrožová and L. Machů. 2012. Health benefits of algal polysaccharides in human nutrition. Advances in Food and Nutrition Research 66: 75–145.

Mizutani, S., S. Deguchi, E. Kobayashi, E. Nishiyama, H. Sagawa and I. Kato. 2010. Takara Bio Inc, 2010. Fucoidan-containing cosmetics. U.S. Patent 7,678,368.

Moon, H.E., M.N. Islam, B.R. Ahn, S.S. Chowdhury, H.S. Sohn, H.A. Jung et al. 2011. Protein tyrosine phosphatase 1B and α-glucosidase inhibitory phlorotannins from edible brown algae, *Ecklonia stolonifera* and *Eisenia bicyclis*. Bioscience, Biotechnology, and Biochemistry 75(8): 1472–1480.

Moon, H.J., S.H. Lee, M.J. Ku, B.C. Yu, M.J. Jeon, S.H. Jeong et al. 2009. Fucoidan inhibits UVB-induced MMP-1 promoter expression and down regulation of type I procollagen synthesis in human skin fibroblasts. European Journal of Dermatology 19(2): 129–134.

Moon, H.J., S.R. Lee, S.N. Shim, S.H. Jeong, V.A. Stonik, V.A. Rasskazov et al. 2008. Fucoidan inhibits UVB-induced MMP-1 expression in human skin fibroblasts. Biological and Pharmaceutical Bulletin 31(2): 284–289.

Moon, H.J., K.S. Park, M.J. Ku, M.S. Lee, S.H. Jeong, T.I. Imbs et al. 2009. Effect of *Costaria costata fucoidan* on expression of matrix metalloproteinase-1 promoter, mRNA, and protein. Journal of Natural Products 72(10): 1731–1734.

Morais, T., J. Cotas, D. Pacheco and L. Pereira. 2021. Seaweeds compounds: An ecosustainable source of cosmetic ingredients? Cosmetics 8(1): 8.

Morya, V.K., J. Kim and E.K. Kim. 2012. Algal fucoidan: Structural and size-dependent bioactivities and their perspectives. Applied Microbiology and Biotechnology 93: 71–82.

Munaf, E., R. Zein, A. Dharma, L.W. Lim and T. Takeuchi. 2009. Optimation study of carrageenan extraction from red algae (*Eucheuma cottonii*). Jurnal Riset Kimia 2(2): 120–120.

Murray, B.A. and R.J. FitzGerald. 2007. Angiotensin converting enzyme inhibitory peptides derived from food proteins: Biochemistry, bioactivity and production. Current Pharmaceutical Design 13(8): 773–791.

Murugan, K. and V.V. Iyer. 2013. Differential growth inhibition of cancer cell lines and antioxidant activity of extracts of red, brown, and green marine algae. *In Vitro* Cellular & Developmental Biology-Animal 49: 324–334.

Narayanaswamy, R., B.W. Jo, S.K. Choi and I.S. Ismail. 2013. Fucoidan: Versatile cosmetic ingredient. An overview. Journal of Applied Cosmetology 31(3/4): 131–138.

Ngo, D.H. and S.K. Kim. 2013. Sulfated polysaccharides as bioactive agents from marine algae. International Journal of Biological Macromolecules 62: 70–75.

Nishida, Y., Y. Kumagai, S. Michiba, H. Yasui and H. Kishimura. 2020. Efficient extraction and antioxidant capacity of mycosporine-like amino acids from red alga dulse *Palmaria palmata* in Japan. Marine Drugs 18(10): 502.

Nowruzi, B., G. Sarvari and S. Blanco. 2020. The cosmetic application of cyanobacterial secondary metabolites. Algal Research 49: 101959.

Nurjanah, Luthfiyana, N., T. Hidayat, M. Nurilmala and E. Anwar. 2019, May. Utilization of seaweed porridge *Sargassum* sp. and *Eucheuma cottonii* as cosmetic in protecting skin. In IOP Conference Series: Earth and Environmental Science (Vol. 278, No. 1, p. 012055). IOP Publishing.

Orfanoudaki, M., A. Hartmann, U. Karsten and M. Ganzera. 2019. Chemical profiling of mycosporine-like amino acids in twenty-three red algal species. Journal of Phycology 55(2): 393–403.

Osório, C., S. Machado, J. Peixoto, S. Bessada, F.B. Pimentel, C. Alves, R. et al. 2020. Pigments content (chlorophylls, fucoxanthin and phycobiliproteins) of different commercial dried algae. Separations 7(2): 33.

Ozanne, H., H. Toumi, B. Roubinet, L. Landemarre, E. Lespessailles, R. Daniellou et al. 2020. Laminarin effects, a β-(1, 3)-glucan, on skin cell inflammation and oxidation. Cosmetics 7(3): 66.

Pacheco-Quito, E.M., R. Ruiz-Caro and M.D. Veiga. 2020. Carrageenan: Drug delivery systems and other biomedical applications. Marine Drugs 18(11): 583.

Paduch, R., M. Kandefer-Szerszeń, M. Trytek and J. Fiedurek. 2007. Terpenes: substances useful in human healthcare. Archivum immunologiae et therapiae experimentalis 55: 315–327.

Pangestuti, R., K.H. Shin and S.K. Kim. 2021. Anti-photoaging and potential skin health benefits of seaweeds. Marine Drugs 19(3): 172.

Percival, E. 1979. The polysaccharides of green, red and brown seaweeds: Their basic structure, biosynthesis and function. British Phycological Journal 14(2): 103–117.

Pereira, L. 2016. Edible Seaweeds of the World, 1st ed.; Taylor & Francis Group: Boca Raton, FL, USA, p. 453. ISBN 9781498730471.

Pereira, L. 2018. Biological and therapeutic properties of the seaweed polysaccharides. International Biology Review 2(2): 1–50.

Pereira, L. 2018. Seaweeds as source of bioactive substances and skin care therapy—cosmeceuticals, algotheraphy, and thalassotherapy. Cosmetics 5(4): 68.

Pereira, L., A.M. Amado, A.T. Critchley, F. Van de Velde and P.J. Ribeiro-Claro. 2009. Identification of selected seaweed polysaccharides (phycocolloids) by vibrational spectroscopy (FTIR-ATR and FT-Raman). Food Hydrocolloids 23(7): 1903–1909.

Pereira, L., S.F. Gheda and P.J. Ribeiro-Claro. 2013. Analysis by vibrational spectroscopy of seaweed polysaccharides with potential use in food, pharmaceutical, and cosmetic industries. International Journal of Carbohydrate Chemistry.

Pham, A.K. and J.G. Dinulos. 2014. Cosmeceuticals for children: should you care? Current Opinion in Pediatrics 26(4): 446–451.

Pimentel, F.B., R.C. Alves, F. Rodrigues and M.B. PP Oliveira. 2017. Macroalgae-derived ingredients for cosmetic industry—An update. Cosmetics 5(1): 2.

Podkorytova, A.V., L.H. Vafina, E.A. Kovaleva and V.I. Mikhailov. 2007. Production of algal gels from the brown alga, *Laminaria japonica* Aresch., and their biotechnological applications. Journal of Applied Phycology 19: 827–830.

Preetha, J.P. and K. Karthika. 2009. Cosmeceuticals–an evolution. International Journal of ChemTech Research 1(4): 1217–1223.

Priyan Shanura Fernando, I., K.N. Kim, D. Kim and Y.J. Jeon. 2019. Algal polysaccharides: Potential bioactive substances for cosmeceutical applications. Critical Reviews in Biotechnology 39(1): 99–113.

Qasim, R. 1991. Amino acid composition of some common seaweeds. Pakistan Journal of Pharmaceutical Sciences 4(1): 49–54.

Qiao, L., Y. Li, Y. Chi, Y. Ji, Y. Gao, H. Hwang et al. 2016. Rheological properties, gelling behavior and texture characteristics of polysaccharide from Enteromorpha prolifera. Carbohydrate Polymers 136: 1307–1314.

Qiu, Y., H. Jiang, L. Fu, F. Ci and X. Mao. 2021. Porphyran and oligo-porphyran originating from red algae Porphyra: Preparation, biological activities, and potential applications. Food Chemistry 349: 129209.

Quah, C.C., K.H. Kim, M.S. Lau, W.R. Kim, S.H. Cheah and R. Gundamaraju. 2014. Pigmentation and dermal conservative effects of the astonishing algae *Sargassum polycystum* and *Padina tenuis* on guinea pigs, human epidermal melanocytes (HEM) and Chang cells. African Journal of Traditional, Complementary and Alternative Medicines 11(4): 77–83.

Querellou, J., T. Børresen, C. Boyen, A. Dobson, M. Höfle, A. Ianora et al. 2010. Marine biotechnology: Realising the full potential of Europe. VLIZ Special Publication 47: 21.

Quinderé, A.L., B.P. Fontes, E. de SO Vanderlei, I.N. de Queiroz, J.A. Rodrigues, I.W. de Araújo et al. 2013. Peripheral antinociception and anti-edematogenic effect of a sulfated polysaccharide from *Acanthophora muscoides*. Pharmacological Reports 65(3): 600–613.

Rastogi, R.P., R.R. Sonani and D. Madamwar. 2017. UV photoprotectants from algae—synthesis and bio-functionalities. In Algal green chemistry (pp. 17–38). Elsevier.

Reef, R., P. Kaniewska and O. Hoegh-Guldberg. 2009. Coral skeletons defend against ultraviolet radiation. PLoS One 4(11): e7995.

Resende, D.I., M. Ferreira, C. Magalhães, J.S. Lobo, E. Sousa and I.F. Almeida. 2021. Trends in the use of marine ingredients in anti-aging cosmetics. Algal Research 55: 102273.

Rioux, L.E., V. Moulin, M. Beaulieu and S.L. Turgeon. 2013. Human skin fibroblast response is differentially regulated by galactofucan and low molecular weight galactofucan. Bioactive Carbohydrates and Dietary Fibre 1(2): 105–110.

Robic, A., C. Rondeau-Mouro, J.F. Sassi, Y. Lerat and M. Lahaye. 2009. Structure and interactions of ulvan in the cell wall of the marine green algae *Ulva rotundata* (Ulvales, Chlorophyceae). Carbohydrate Polymers 77(2): 206–216.

Rodrigues, J.A.G., I.N.L. de Queiroz, A.L.G. Quinderé, N.M.B. Benevides, A.M.F. Tovar and P.A. de Souza Mourão. 2016. Extraction and structural properties of *Acanthophora muscoides* (Rhodophyceae) extracellular matrix sulfated polysaccharides and their effects on coagulation. Acta Scientiarum. Technology 38(3): 273–282.

Rodrigues, J.A.G., I.N.L. de Queiroz, A.L.G. Quinderé, N.M.B. Benevides, A.M.F. Tovar and P.A. de Souza Mourão. 2016. Mild-acid hydrolysis of a native polysulfated fraction from *Acanthophora muscoides* generates sulfated oligosaccharides displaying in vitro thrombin generation inhibition. Acta Scientiarum. Biological Sciences 38(1): 7–15.

Romarís-Hortas, V., P. Bermejo-Barrera and A. Moreda-Pineiro. 2013. Ultrasound-assisted enzymatic hydrolysis for iodinated amino acid extraction from edible seaweed before reversed-phase high performance liquid chromatography–inductively coupled plasma-mass spectrometry. Journal of Chromatography a, 1309: 33–40.

Rossano, R., N. Ungaro, A. D'Ambrosio, G.M. Liuzzi and P. Riccio. 2003. Extracting and purifying R-phycoerythrin from Mediterranean red algae *Corallina elongata* Ellis & Solander. Journal of Biotechnology 101(3): 289–293.

Ruocco, N., S. Costantini, S. Guariniello and M. Costantini. 2016. Polysaccharides from the marine environment with pharmacological, cosmeceutical and nutraceutical potential. Molecules 21(5): 551.

Rupérez, P. 2002. Mineral content of edible marine seaweeds. Food Chemistry 79(1): 23–26.

Rupérez, P., O. Ahrazem and J.A. Leal. 2002. Potential antioxidant capacity of sulfated polysaccharides from the edible marine brown seaweed Fucus vesiculosus. Journal of Agricultural and Food Chemistry 50(4): 840–845.

Ruslan, F.S., D. Susanti, N.M. Noor and N. Iman. 2021. Bioactive compounds, cosmeceutical and nutraceutical applications of green seaweed species (Chlorophyta). Squalen Bulletin of Marine and Fisheries Postharvest and Biotechnology 16(1): 41–55.

Ryu, J., S.J. Park, I.H. Kim, Y.H. Choi and T.J. Nam. 2014. Protective effect of porphyra-334 on UVA-induced photoaging in human skin fibroblasts. International Journal of Molecular Medicine 34(3): 796–803.

Salehi, B., J. Sharifi-Rad, A.M. Seca, D.C. Pinto, I. Michalak, A. Trincone et al. 2019. Current trends on seaweeds: Looking at chemical composition, phytopharmacology, and cosmetic applications. Molecules 24(22): 4182.

Saluri, M., M. Kaldmäe, M. Rospu, H. Sirkel, T. Paalme, M. Landreh et al. 2020. Spatial variation and structural characteristics of phycobiliproteins from the red algae *Furcellaria lumbricalis* and *Coccotylus truncatus*. Algal Research 52: 102058.

Samarakoon, K. and Y.J. Jeon. 2012. Bio-functionalities of proteins derived from marine algae—A review. Food Research International 48(2): 948–960.

Sanjeewa, K.A., N. Kang, G. Ahn, Y. Jee, Y.T. Kim and Y.J. Jeon. 2018. Bioactive potentials of sulfated polysaccharides isolated from brown seaweed *Sargassum* spp. in related to human health applications: A review. Food Hydrocolloids 81: 200–208.

Sekar, S. and M. Chandramohan. 2008. Phycobiliproteins as a commodity: Trends in applied research, patents and commercialization. Journal of Applied Phycology 20: 113–136.

Senevirathne, W.S.M. and S.K. Kim. 2013. Cosmeceuticals from Algae. pp. 694–713. *In*: Domínguez, H. [Ed.]. Functional Ingredients from Algae for Foods and Nutraceuticals, 1st ed. Woodhead Publishing: Cambridge, UK.

Senni, K., F. Gueniche, A. Foucault-Bertaud, S. Igondjo-Tchen, F. Fioretti, S. Colliec-Jouault et al. 2006. Fucoidan a sulfated polysaccharide from brown algae is a potent modulator of connective tissue proteolysis. Archives of Biochemistry and Biophysics 445(1): 56–64.

Senni, K., J. Pereira, F. Gueniche, C. Delbarre-Ladrat, C. Sinquin, J. Ratiskol et al. 2011. Marine polysaccharides: A source of bioactive molecules for cell therapy and tissue engineering. Marine Drugs 9(9): 1664–1681.

Shahidi, F. and M.J. Rahman. 2018. Bioactives in seaweeds, algae, and fungi and their role in health promotion. Journal of Food Bioactives 2: 58–81.

Shannon, E., M. Conlon and M. Hayes. 2021. Seaweed components as potential modulators of the gut microbiota. Marine Drugs 19(7): 358.

Shanura Fernando, I.P., K.K. Asanka Sanjeewa, K.W. Samarakoon, H.S. Kim, U.K.D.S.S. Gunasekara, Y.J. Park et al. 2018. The potential of fucoidans from Chnoospora minima and *Sargassum polycystum* in cosmetics: Antioxidant, anti-inflammatory, skin-whitening, and antiwrinkle activities. Journal of Applied Phycology 30: 3223–3232.

Shao, P., J. Shao, L. Han, R. Lv and P. Sun. 2015. Separation, preliminary characterization, and moisture-preserving activity of polysaccharides from *Ulva fasciata*. International Journal of Biological Macromolecules 72: 924–930.

Shick, J.M.J., W.W. Dunlap and G.R. Buettner. 2000. Ultraviolet (UV) protection in marine organisms II. Biosynthesis, accumulation, and sunscreening function of mycosporine-like amino acids. Chapter 22. In Free radicals in Chemistry, Biology and Medicine-pages: 215–228. OICA International (UK).

Singh, S.P., S. Kumari, R.P. Rastogi, K.L. Singh and R.P. Sinha. 2008. Mycosporine-like amino acids (MAAs): Chemical structure, biosynthesis and significance as UV-absorbing/screening compounds. Indian J. Experim. Biol. 46: 7–17.

Skjåk-Bræk, G., H. Grasdalen and O. Smidsrød. 1989. Inhomogeneous polysaccharide ionic gels. Carbohydrate Polymers 10(1): 31–54.

Song, Y.S., M.C. Balcos, H.Y. Yun, K.J. Baek, N.S. Kwon, M.K. Kim et al. 2015. ERK activation by fucoidan leads to inhibition of melanogenesis in Mel-Ab cells. The Korean Journal of Physiology & Pharmacology: Official Journal of the Korean Physiological Society and the Korean Society of Pharmacology 19(1): 29.

Sotelo, C.G., M. Blanco, P. Ramos, J.A. Vázquez and R.I. Perez-Martin. 2021. Sustainable sources from aquatic organisms for cosmeceuticals ingredients. Cosmetics 8(2): 48.

Soulioti, I., M. Diomidous, H. Theodosopoulou, N. Violaki, H. Plessa, M. Charalambidou et al. 2013. Cosmetics: History, products, industry, legislation, regulations and implications in public health. Review of Clinical Pharmacology and Pharmacokinetics, International Edition 27(1): 5–15.

Sreekumar, K. and B. Bindhu. 2020. Alginic acid: A potential biopolymer from brown algae. Mater. Int. 2: 433–438.

Sridhar, K., B.S. Inbaraj and B.H. Chen. 2021. Recent developments on production, purification and biological activity of marine peptides. Food Research International 147: 110468.

Stengel, D.B. and S. Connan. 2015. Marine algae: A source of biomass for biotechnological applications. Natural products from marine algae: Methods and Protocols: 1–37.

Su, W., L. Wang, X. Fu, L. Ni, D. Duan, J. Xu et al. 2020. Protective effect of a fucose-rich fucoidan isolated from *Saccharina japonica* against ultraviolet B-induced photodamage *in vitro* in human keratinocytes and *in vivo* in zebrafish. Marine Drugs 18(6): 316.

Sun, Y., B. Yang, Y. Wu, Y. Liu, X. Gu, H. Zhang et al. 2015. Structural characterization and antioxidant activities of κ-carrageenan oligosaccharides degraded by different methods. Food Chemistry 178: 311–318.

Szekalska, M., K. Sosnowska, M. Tomczykowa, K. Winnicka, I. Kasacka and M. Tomczyk. 2020. *In vivo* anti-inflammatory and anti-allergic activities of cynaroside evaluated by using hydrogel formulations. Biomedicine & Pharmacotherapy 121: 109681.

Takematsu, H. and M. Seiji. 1984. Effect of macrophages on elimination of dermal melanin from the dermis. Archives of Dermatological Research 276: 96–98.

Teixeira, M.M. and P.G. Hellewell. 1997. The effect of the selectin binding polysaccharide fucoidin on eosinophil recruitment *in vivo*. British Journal of Pharmacology 120(6): 1059–1066.

Thevanayagam, H., S.M. Mohamed and W.L. Chu. 2014. Assessment of UVB-photoprotective and antioxidative activities of carrageenan in keratinocytes. Journal of Applied Phycology 26: 1813–1821.

Tiwari, B.K. and D.J. Troy. 2015. Seaweed sustainability–food and nonfood applications. In Seaweed sustainability (pp. 1–6). Academic Press.

Tønnesen, H.H. and J. Karlsen. 2002. Alginate in drug delivery systems. Drug Development and Industrial Pharmacy 28(6): 621–630.

Tümen Erden, S., C. Ekentok Atici, B. Cömez and A.D. Sezer. 2021. Preparation and in vitro characterization of laminarin based hydrogels. Journal of Research in Pharmacy, 25(2).

Usov, A.I. and N.D. Zelinsky. 2013. Chemical structures of algal polysaccharides. In Functional ingredients from algae for foods and nutraceuticals (pp. 23–86). Woodhead Publishing.

Vega, J., G. Schneider, B.R. Moreira, C. Herrera, J. Bonomi-Barufi and F.L. Figueroa. 2021. Mycosporine-like amino acids from red macroalgae: UV-photoprotectors with potential cosmeceutical applications. Applied Sciences 11(11): 5112.

Villarroel, L.H. and A.B. Zanlungo. 1981. Structural studies on the porphyran from *Porphyra columbina* (Montagne). Carbohydrate Research 88(1): 139–145.

Viskari, P.J. and C.L. Colyer. 2003. Rapid extraction of phycobiliproteins from cultured cyanobacteria samples. Analytical Biochemistry 319(2): 263–271.

Vo, T.S., D.H. Ngo, K.H. Kang, W.K. Jung and S.K. Kim. 2015. The beneficial properties of marine polysaccharides in alleviation of allergic responses. Molecular Nutrition & Food Research 59(1): 129–138.

Wahyuni, T. 2021, May. The potential and application of *Eucheuma* sp. for solid soap: A review. In IOP Conference Series: Earth and Environmental Science (Vol. 750, No. 1, p. 012048). IOP Publishing.

Wang, J., W. Jin, Y. Hou, X. Niu, H. Zhang and Q. Zhang. 2013. Chemical composition and moisture-absorption/retention ability of polysaccharides extracted from five algae. International Journal of Biological Macromolecules 57: 26–29.

Wang, J., F. Wang, Q. Zhang, Z. Zhang, X. Shi and P. Li. 2009. Synthesized different derivatives of low molecular fucoidan extracted from *Laminaria japonica* and their potential antioxidant activity *in vitro*. International Journal of Biological Macromolecules 44(5): 379–384.

Wang, J., Q. Zhang, Z. Zhang and Z. Li. 2008. Antioxidant activity of sulfated polysaccharide fractions extracted from *Laminaria japonica*. International Journal Of Biological Macromolecules 42(2): 127–132.

Wang, Y., X. Li, Y. Jiang, J. Wang, H. Hwang, X. Yang et al. 2019. Structure characterization of low molecular weight sulfate Ulva polysaccharide and the effect of its derivative on iron deficiency anemia. International Journal of Biological Macromolecules 126: 747–754.

Wang, L., T.U. Jayawardena, H.W. Yang, H.G. Lee and Y.J. Jeon. 2020. The potential of sulfated polysaccharides isolated from the brown seaweed Ecklonia maxima in cosmetics: Antioxidant, anti-melanogenesis, and photoprotective activities. Antioxidants 9(8): 724.

Wang, L., J.Y. Oh, W. Lee and Y.J. Jeon. 2021. Fucoidan isolated from Hizikia fusiforme suppresses ultraviolet B-induced photodamage by down-regulating the expressions of matrix metalloproteinases and pro-inflammatory cytokines via inhibiting NF-κB, AP-1, and MAPK signaling pathways. International Journal of Biological Macromolecules 166: 751–759.

Wells, M.L., P. Potin, J.S. Craigie, J.A. Raven, S.S. Merchant, K.E. Helliwell et al. 2017. Algae as nutritional and functional food sources: Revisiting our understanding. Journal of Applied Phycology 29: 949–982.

Wijesinghe, W.A.J.P. and Y.J. Jeon. 2012. Biological activities and potential industrial applications of fucose rich sulfated polysaccharides and fucoidans isolated from brown seaweeds: A review. Carbohydrate Polymers 88(1): 13–20.

Wong, K.H. and P.C. Cheung. 2001. Nutritional evaluation of some subtropical red and green seaweeds Part II. *In vitro* protein digestibility and amino acid profiles of protein concentrates. Food Chemistry 72(1): 11–17.

Wong, T., L. Brault, E. Gasparotto, R. Vallée, P.Y. Morvan, V. Ferrières and C. Nugier-Chauvin. 2021. Formation of amphiphilic molecules from the most common marine polysaccharides, toward a sustainable alternative? Molecules 26(15): 4445.

Xie, X.T., X. Zhang, Y. Liu, X.Q. Chen and K.L. Cheong. 2020. Quantification of 3, 6-anhydro-galactose in red seaweed polysaccharides and their potential skin-whitening activity. 3 Biotech, 10: 1–9.

Xing, M., Q. Cao, Y. Wang, H. Xiao, J. Zhao, Q. Zhang et al. 2020. Advances in research on the bioactivity of alginate oligosaccharides. Marine Drugs 18(3): 144.

Xue, C., G. Yu, T. Hirata, J. Terao and H. Lin. 1998. Antioxidative activities of several marine polysaccharides evaluated in a phosphatidylcholine-liposomal suspension and organic solvents. Bioscience, Biotechnology, and Biochemistry 62(2): 206–209.

Xue, C.H., Y. Fang, H. Lin, L. Chen, Z.J. Li, D. Deng et al. 2001. Chemical characters and antioxidative properties of sulfated polysaccharides from *Laminaria japonica*. Journal of Applied Phycology 13: 67–70.

Yaich, H., A.B. Amira, F. Abbes, M. Bouaziz, S. Besbes, A. Richel et al. 2017. Effect of extraction procedures on structural, thermal and antioxidant properties of ulvan from Ulva lactuca collected in Monastir coast. International Journal of Biological Macromolecules 105: 1430–1439.

Yang, J.H. 2012. Topical application of fucoidan improves atopic dermatitis symptoms in NC/Nga mice. Phytotherapy Research 26(12): 1898–1903.

Yang, Y., D. Liu, J. Wu, Y. Chen and S. Wang. 2011. *In vitro* antioxidant activities of sulfated polysaccharide fractions extracted from *Corallina officinalis*. International Journal of Biological Macromolecules 49(5): 1031–1037.

Yanshin, N., A. Kushnareva, V. Lemesheva, C. Birkemeyer and E. Tarakhovskaya. 2021. Chemical composition and potential practical application of 15 red algal species from the White Sea Coast (the Arctic Ocean). Molecules 26(9): 2489.

Yu, B., D. Bi, L. Yao, T. Li, L. Gu, H. Xu et al. 2020. The inhibitory activity of alginate against allergic reactions in an ovalbumin-induced mouse model. Food & Function 11(3): 2704–2713.

Yuan, Y.V., N.D. Westcott, C. Hu and D.D. Kitts. 2009. Mycosporine-like amino acid composition of the edible red alga, *Palmaria palmata* (dulse) harvested from the west and east coasts of Grand Manan Island, New Brunswick. Food Chemistry 112(2): 321–328.

Yuan, Y.V., N.D. Westcott, C. Hu and D.D. Kitts. 2009. Mycosporine-like amino acid composition of the edible red alga, *Palmaria palmata* (dulse) harvested from the west and east coasts of Grand Manan Island, New Brunswick. Food Chemistry 112(2): 321–328.

Zargarzadeh, M., A.J. Amaral, C.A. Custódio and J.F. Mano. 2020. Biomedical applications of laminarin. Carbohydrate Polymers 232: 115774.

Zhang, Q., N. Li, X. Liu, Z. Zhao, Z. Li and Z. Xu. 2004. The structure of a sulfated galactan from *Porphyra haitanensis* and its *in vivo* antioxidant activity. Carbohydrate Research 339(1): 105–111.

Zhang, Q., N. Li and J. Song. 2005. Toxicological evaluation of fucoidan extracted from *Laminaria japonica* in Wistar rats. Food and Chemical Toxicology 43(3): 421–426.

Zhang, Q., N. Li, G. Zhou, X. Lu, Z. Xu and Z. Li. 2003. *In vivo* antioxidant activity of polysaccharide fraction from *Porphyra haitanesis* (Rhodephyta) in aging mice. Pharmacological Research 48(2): 151–155.

Zhang, Y., R. Tian, H. Wu, X. Li, S. Li and L. Bian. 2020. Evaluation of acute and sub-chronic toxicity of *Lithothamnion* sp. in mice and rats. Toxicology Reports 7: 852–858.

Zhang, Z., Q. Zhang, J. Wang, H. Zhang, X. Niu and P. Li. 2009. Preparation of the different derivatives of the low-molecular-weight porphyran from *Porphyra haitanensis* and their antioxidant activities *in vitro*. International Journal of Biological Macromolecules 45(1): 22–26.

Zhao, T., Q. Zhang, H. Qi, H. Zhang, X. Niu, Z. Xu et al. 2006. Degradation of porphyran from *Porphyra haitanensis* and the antioxidant activities of the degraded porphyrans with different molecular weight. International Journal of Biological Macromolecules 38(1): 45–50.

CHAPTER 15

Phlorotannins in Cosmeceuticals

Herminia Domínguez,[1,*] *Raquel Mateos*[2] and *José R. Pérez-Correa*[3]

Cosmeceuticals

Cosmetics formulated to cleanse, beautify, and improve skin texture without altering physiological functions sit on one side of the beauty industry spectrum. On the other side, there is an expanding hybrid category called cosmeceuticals. These represent an intersection between cosmetics and pharmaceuticals, containing biologically active ingredients that provide medicinal or drug-like benefits (Vermeer and Gilchrest 1996, Edward 1997). While the regulatory landscape for cosmetics is relatively consistent worldwide, a similar formal agreement still needs to be developed for cosmeceuticals. However, this emerging category represents an area of sustained development and growth. It's expected to gain substantial market traction both presently and in the future. Cosmeceuticals have been further divided into several categories, each offering specific benefits. These include antiaging, antioxidant, antiinflammatory, antimelanogenic, anticancer, and antimicrobial properties (Murphy and Dow 2021, Borate and Mundada 2021). Despite the regulatory ambiguity, the continual advancement in this field promises a dynamic evolution of the beauty industry.

Revolutionary advancements in cosmeceuticals have recently driven the discovery and development of products with superior performance and greater market value. These advancements notably include the incorporation of natural bioactives and the replacement of potentially harmful synthetic chemicals like parabens with natural antimicrobials (Sanjeewa et al. 2016, Ahmed et al. 2022, Al-Halaseh et al. 2022, Santiesteban-Romero et al. 2022). Furthermore, the industry has embraced green extraction techniques and nanotechnology, which significantly enhance the bioavailability, safety, stability, and solubility of bioactive compounds. These innovative approaches enable the controlled release and optimized skin penetration of bioactives while also facilitating precision targeting at specific sites (Sotiropoulou et al. 2021, Goyal et al. 2022). Nanotechnology is a particularly promising tool in product development. For instance, it has been crucial in creating antiaging and antihyperpigmentation solutions (Abu Hajleh et al. 2021, Alsabeelah et al. 2021,

[1] CINBIO Universidade de Vigo, Faculty of Sciences, Department of Chemical Engineering, Campus Ourense, As Lagoas, Ourense, 32004, Spain.
[2] Institute of Food Science, Technology and Nutrition (ICTAN-CSIC), Spanish National Research Council (CSIC), José Antonio Nováis 6, Madrid, 28040, Spain.
[3] Department of Chemical and Bioprocess Engineering, Pontificia Universidad Católica de Chile, Macul, Santiago, 7810000, Chile.
* Corresponding author: herminia@uvigo.es

Tangau et al. 2022). Nanotechnology can further enhance the aesthetic properties of cosmeceuticals, such as color and transparency, especially using various nanoparticulate systems, including liposomes, solid lipid nanoparticles, and niosomes. However, nanotechnology also presents challenges. Nanoparticles can penetrate skin barriers unaltered, raising concerns over potential adverse effects on human health and the environment. Consequently, safety testing and regulatory oversight are required before such products are introduced to the market (Vaishampayan and Rane 2022).

Over the past decade, the skincare and cosmetics industry has seen a surge in demand for marine-based products, primarily attributed to the rich bioactive properties of marine organisms (Siahaan et al. 2022). Macroalgae are highly prized as a renewable, high-yield source of bioactive metabolites, capable of surviving across diverse environmental conditions (El-Chaghaby and Rashad 2021). These metabolites encompass a vast array of unique compounds, including fatty acids, fucoidan, laminarin, carrageenan, ulvan, vitamins, phenolics, carotenoids, and mycosporine-like amino acids (Sanjeewa et al. 2016, Martínez-Ruiz 2022, Santiesteban-Romero et al. 2022, López-Hortas et al. 2021). These bioactive compounds offer a wealth of benefits for skin care, ranging from sun protection, antiaging, anticellulite, and moisturizing to whitening effects (Pereira 2018). Furthermore, their inclusion in oncological cosmeceuticals provides a valuable tool in preventive strategies, as well as in pre- and post-medical treatments, through applications such as topical barrier agents, moisturizers, sunscreens, and skin healing products (Mourelle et al. 2023). In addition to their biological benefits, these algae-derived compounds provide various technological functions in cosmeceutical formulations. These include stabilizing, gelling, preserving, and providing aroma and color (Cikoš et al. 2021). Clinical studies have further validated their safety and efficacy, cementing their value in the skincare and cosmetics industry (El-Chaghaby and Rashad 2021, Murphy and Dow 2021, Pangestuti et al. 2021). Consequently, exploring and applying macroalgal bioactive compounds in this industry is a lucrative and promising avenue for continued research.

Phlorotannins

Phenolic compounds are secondary metabolites categorized into several classes according to the number of phenol rings and structural elements. In seaweeds, phenolic compounds are classified into phenolic and cinnamic acids, flavonoids, bromophenols, and phlorotannins (Subbiah et al. 2022). Phlorotannins are a class of polyphenol compounds found in brown seaweeds (Phaeophyceae). They are stored in vesicles called physodes in amounts varying among species, seaweed size, age, and tissue type, and with environmental conditions (salinity, season, nutrient, radiation, temperature) (Kamiya et al. 2010). They are dehydropolymers of phloroglucinol (1,3,5-trihydroxybenzene) with molecular weights of less than 1 to more than 100 kDa (Catarino et al. 2022). Phlorotannins are structurally highly diverse compounds, structurally different from terrestrial polyphenol compounds, such as those based on flavonoids and gallic acid) (Fernando et al. 2022, Subbiah et al. 2022). Phlorotannins have been classified according to the type of linkages: (i) fucols with phenyl bonds, (ii) fuhalols and phloroethols with ether bonds, (iii) fucophloroethols with both phenyl and ether bonds, and (iv) eckols and carmalols with dibenzodioxin bonds (Figure 1). Phlorotannins have been considered a widely available and cheap source of bioactives for the cosmeceutical industry (Sanjeewa et al. 2016). They exhibit interesting bioactive properties, including anticancer, antiinflammation, antioxidant, antiphotoaging, antiallergic, antiwrinkling, antiadipogenesis, hyaluronidase inhibitory action, and hair growth promotion capacity. Depending on their concentration, they can be highly biocompatible and nontoxic (Lopes et al. 2012, Sanjeewa et al. 2016, Sugiura et al. 2021). Some other properties, such as antimicrobial and antioxidant, make them natural preservatives for skincare and cosmetic products (Hermund et al. 2018). Phlorotannins bioactivities could be responsible for the potential of brown seaweed extracts for dermatological applications (Freitas et al. 2020). They also show bioactivities already reported for other polyphenols, such as the potential to treat

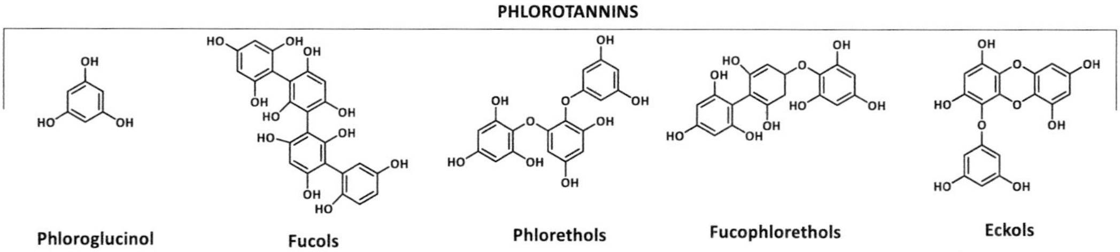

Figure 1. Examples of phlorotannins in brown seaweeds.

different types of skin and hair diseases and hair loss (Sun et al. 2022). Bak et al. (2014) reported that 7-phloroeckol increased the proliferation of human dermal papilla cells and outer root sheath cells. Furthermore, this compound promoted elongation of the hair shaft in cultured human hair follicles. In a recent review, Jayawardhana et al. (2023) compiled the activities of marine algal polyphenols on activities of cosmetic interest.

Bioactivity of phlorotannins as cosmeceutical ingredients

As mentioned above, marine phenolics exhibit many biological properties, some suitable for skin care. Below, some studies showing the potential of these compounds in the field of cosmetics and presented and a summary can be found in Table 1. This review will focus on their antioxidant, photoprotective, antiaging, whitening, antiinflammatory, and antiallergic properties.

Antioxidant properties

Oxidative stress happens when there is an imbalance between reactive oxygen species (ROS) and antioxidants. The skin is exposed to pollutants and ultraviolet (UV) radiation, which causes the production of ROS, leading to oxidative stress, damage to cellular macromolecules, and premature aging. To defend against these harmful effects, the body produces endogenous and exogenous antioxidants. Natural antioxidants, particularly those found in plants, protect against oxidative stress. Marine organisms exposed to extreme environmental conditions also accumulate polyphenols to combat stress, suppress the oxidation process, or scavenge ROS.

Various studies have been conducted to assess the antioxidant properties of marine phenolic compounds using *in vitro* tests. Balboa et al. (2013) reviewed studies that examined the antioxidant capacity of crude extracts, fractions, and components, finding that they were as effective as or even more potent than synthetic antioxidants. Shibata et al. (2007) conducted a study on phloroglucinol, eckol, phlorofucofuroeckol A, dieckol, and 8,8'-bieckol isolated from *Eisenia bicyclis*, *Ecklonia cava*, and *Ecklonia kurome*. They found these compounds to have potent inhibitory effects on phospholipid peroxidation and significant radical scavenging activity against superoxide anion and DPPH and to be more effective than α-tocopherol and ascorbic acid. Le Lann et al. (2016) established a correlation between the total phenolic compounds and antioxidant activity of an ethyl acetate fraction of phlorotannins from *Halidrys siliquosa*, with its broad-spectrum UV absorption and bactericidal capacity. Phasanasophon and Kim (2018) obtained a phlorotannin extract from *Agarum cribrosum* with a reducing power and radical scavenging activity against DPPH, ABTS, and hydrogen peroxide similar to ascorbic acid. The activity of phlorotannins in several commercially harvested French brown seaweeds was evaluated by Gager et al. (2020). They used *in vitro assays* to determine their radical scavenging activity against DPPH and FRAP to determine their reducing power. The phlorotannins obtained from *Ascophyllum nodosum*, *Halidrys siliquosa*, *Himanthalia elongata*, and *Fucus serratus* contained the most potent antioxidants.

Table 1. Bioactivity of marine phenolics as cosmeceutical ingredients.

Compounds/Marine Source	Test Model	Outcome	References
Phloroglucinol, eckol, phlorofucofuroeckol A, dieckol, and 8,8′-bieckol, isolated from *Eisenia bicyclis*, *Ecklonia cava*, and *Ecklonia kurome*	*In vitro* assay	Decreased the phospholipid peroxidation Radical-scavenging activity against the superoxide anion and DPPH	Shibata et al. 2007
Phlorotannins from *Halidrys siliquosa*	*In vitro* assay	Ethyl acetate fraction (EAF) showed higher antioxidant capacity compared to aqueous fraction EAF showed a broad-spectrum UV absorber EAF showed a strong bactericidal activity	Le Lann et al. 2016
Phlorotannin extract from *Agarum cribrosum*	*In vitro* model	Radical-scavenging activity against DPPH, ABTS, hydrogen peroxide, and reducing power Inhibited the tyrosinase, collagenase, and elastase activities	Phasanasophon and Kim 2018
Phlorotannins isolated from *Alaria esculenta*, *Ascophyllum nodosum*, *Bifurcaria bifurcata*, *Fucus serratus, Halidrys siliquosa*, *Himanthalia elongata*, and *Laminaria ochroleuca*	*In vitro* model	High radical-scavenging activity against DPPH and reducing power (FRAP), particularly *A. nodosum*, *H. elongata, H. siliquosa,* and *F. serratus* High photoprotective activity for the mentioned above brown seaweeds High anti-aging activity determined by antielastase activity of *A. nodosum* and *F. serratus* fractions	Gager et al. 2020
Dieckol, eckol, and phloroglucinol isolated from *Ecklonia cava*	*In vitro* assay: UVB-irradiated human fibroblasts	Dieckol showed the best activity to reduce cell damage induced by UV-B radiation. Dieckol showed a prominent inhibitory activity against melanogenesis	Heo et al. 2009
Fucofuroeckol-A from *Ecklonia stolonifera Okamura*	*In vitro* assay: UVB-induced allergic reaction in RBL-2H3 mast cells	Scavenging ROS production Decreased the histamine release Decreased the intracellular Ca^{2+} elevation	Vo et al. 2018
Dioxinodehydroeckol from *Ecklonia cava*	*In vitro* assay: UVB-induced apoptosis in HaCaT keratinocytes	Reduced the apoptosis by modulating pro/anti-apoptotic molecules and by preventing the release of cytochrome C	Ryu et al. 2015
Dieckol from *Ecklonia cava*	*In vitro model* *In vitro* assay: UVB-induced damage in human dermal fibroblasts	Reduced the ROS level and improved the cell viability Improved collagen synthesis and inhibit collagenase activity in cells Reduced the expression of pro-inflammatory cytokines and MMP	Wang et al. 2021

Table 1 contd. ...

...Table 1 contd.

Compounds/Marine Source	Test Model	Outcome	References
Ethyl acetate from *Polycladia myrica*	*In vitro* model *In vitro* assay: UVB-induced cytotoxicity in HaCaT keratinocytes	Scavenging DPPH radical and high antioxidant activity Reduced the UVB-induced cytotoxicity in HaCaT keratinocytes Formulated a cream enriched with 5% of this extract with promising results	Soleimani et al. 2022
Two types of extracts from *Fucus vesiculosus*	*In vitro model*	Extracts or phlorotannins in lotion (10 % w/w) showed a negligible sun protection factor.	Hermund et al. 2022
Extracts from *Ecklonia cava*, especially enriched in dieckol	*In vitro* assay: UVB-induced cytotoxicity in HaCaT keratinocytes *in vivo* assay: UVB-induced oxidative stress in a zebrafish model	Reduced the oxidative stress of HaCaT cells Reduced the oxidative stress of zebrafish model (lower ROS, nitric oxide, and cell death)	Ko et al. 2011
Phlorotannins from *E. cava*	*In vivo* assay: UVB-induced oxidative stress in a zebrafish model	Reduced ROS and nitric oxide levels	Cha et al. 2012
Brown algae polyphenols	*In vivo* assay: UVB-induced skin carcinogenesis in SKH-1 mice treated orally and topically	Reduced oxidative stress, inflammation, and cell proliferation in the skin.	Hwang et al. 2006
Eckol and dieckol from *Ecklonia stolonifera*	*In vitro* assay: human dermal fibroblasts	Inhibited the MMP-1 expression	Joe et al. 2006
An extract isolated from *Sargassum plagyophyllum*	*In vitro* assay	Inhibited the collagenase activity by 54%	Riani et al. 2018
Phlorotannin extracts from four brown seaweed (*Fucus spiralis, Cystoseira nodicaulis, Cystoseira usneoides*, and *Cystoseira tamariscifolia*)	*In vitro* assay	Inhibited the hyaluronidase activity, especially *F. spiralis* with higher molecular weight phlorotannins	Ferreres et al. 2012
A water extract from *Padina pavonica*	*In vitro* assay	Inhibited the hyaluronidase activity	Fayad et al. 2017
Phlorotannins from the blade of two *Sargassum* seaweeds and Eisenia arborea thallus	*In vitro* assay	Inhibited the hyaluronidase activity Radical-scavenging activity against DPPH and reducing power (FRAP)	Arunkumar et al. 2021
Extracts from *Sargassum muticum* (Yendo) Fensholt	*In vitro* assay	Inhibited the elastase activity	Puspita et al. 2017
An acetone extract of *Ascophyllum nodosum*	*In vitro* assay	Inhibited the tyrosinase activity	Jiménez et al. 2010

Table 1 contd. ...

...Table 1 contd.

Compounds/Marine Source	Test Model	Outcome	References
Phlorotannins isolated from *Ecklonia cava*	*In vitro* model *In vitro*: IBMX-induced melanin formation in B16F10 melanoma cells	Inhibited the tyrosinase activity Inhibited the melanin production in melanoma cells	Yoon et al. 2009
A methanolic extract of *Ecklonia stolonifera*	*In vitro* model	Inhibited the tyrosinase activity	Kang et al. 2004
Ethanolic extract from *Sargassum polycystum*	*In vitro* model *In vitro*: B16F10 murine melanoma cells	No inhibition of the tyrosinase activity was evaluated by *in vitro* model Inhibited the cellular tyrosinase activity	Chan et al. 2011
Compounds isolated from *Ecklonia stolonifera* Okamura	*In vitro* model *In vitro*: B16F10 murine melanoma cells	Reduced the cellular melanin content Reduced the tyrosinase activity Downregulated the expression of melanogenesis enzymes	Manandhar et al. 2019
Phlorotannin-rich extract from *Ecklonia maxima* containing dieckol and eckmaxol	*In vitro* model *In vitro*: alpha-melanocyte-stimulating hormone-stimulated B16F10 cells *In vitro*: UVB-induced HaCaT cell *In vitro*: UVB-irradiated HDF cells	Inhibited the tyrosinase activity Inhibited the melanogenesis Inhibited the apoptosis Inhibited the collagen degradation and MMP expression	Wang et al. 2022
Phlorofucofuroeckol-B isolated from *Eisenia arborea*	*In vitro*: rat basophile leukemia cells (RBL)-2H3 cells	Inhibited the effect on histamine release from cells	Sugiura et al. 2006
A polyphenol-rich extract isolated from *Sargassum horneri*	*In vivo:* NC/Nga	Suppressed the infiltration of inflammatory cells into the skin dermis by decreasing the infiltration of IL-13-positive cells	Mihindukulasooriya et al. 2022

Cellular models enable the assessment of different mechanisms that contribute to oxidative stress. For example, fucofuroeckol-A, derived from the brown seaweed *Ecklonia stolonifera* Okamura, protects against UVB-induced allergic reactions in RBL-2H3 mast cells. This protection was revealed by the scavenging of produced ROS, the reduced histamine release, and increased intracellular Ca2+ levels (Vo et al. 2018).

Photoprotective properties

Exposure to UV radiation from sunlight, in particular, UVB (280 – 320 nm) and UVA (320 – 400 nm) wavelengths, can damage the skin and cause sunburn, photoaging, and even cancer due to DNA damage in skin cells (Parrish et al. 1982, Brash et al. 1991). Photosynthetic organisms are subjected to UV radiation and accumulation of UV-absorbing compounds such as bromophenols and coumarins was reported in green algae, including *Ulva lactuca* (Flodin et al. 1999), *Dasycladus vermicularis* (Pérez-Rodríguez et al. 2003) and *Ulva pertusa* (Han and Han 2005), among others. The accumulation of these compounds suggests that phlorotannins may also be good candidates for photoprotective molecules.

The phlorotannin-enriched ethyl acetate fraction obtained from *Halidrys siliquosa* appeared to be a broad-spectrum UVB/UVA absorber according to in vitro assays and sun protection factor calculations (Le Lann et al. 2016). Similarly, Gager et al. (2020) confirmed the high photoprotective activity of *A. nodosum*, *H. elongata*, *H. siliquosa*, and *F. serratus* expressed as sun and UVA protection factors. An ethyl acetate fraction of the brown alga *Polycladia myrica* showed high DPPH radical scavenging and antioxidant activity and effectively reduced UVB-induced cytotoxicity in HaCaT keratinocytes. A cream containing 5% of this phenolic extract was highly protective against solar radiation (Soleimani et al. 2022). In turn, Hermund et al. (2022) observed that the sun protection of lotions (10% w/w) formulated with two *Fucus vesiculosus* extracts (7% and 14% phlorotannins/dry extract) was negligible. They concluded that further fractionation and purification of phlorotannins is required for sun protection. Ko et al. (2011) showed, using *in vitro* (HaCaT cells) and *in vivo* (zebrafish model) UVB-induced oxidative stress assays, that an ethyl acetate fraction (enriched in dieckol) of an *Ecklonia cava* extract was able to reduce ROS, nitric oxide, and cell death after treatment. Cha et al. (2012) confirmed the protective effect of phlorotannins from *E. cava* against UVB radiation in zebrafish by reducing ROS and nitric oxide levels.

Dieckol, eckol, and phloroglucinol isolated from *Ecklonia cava* showed protection against DNA damage induced by UVB radiation, with dieckol standing out. This protection was determined by measuring the damage by tail intensity and morphological changes in fibroblasts (Heo et al. 2009). Moreover, Wang et al. (2021) showed that dieckol isolated from *Ecklonia cava* effectively reduced the intracellular ROS level and improved the viability of UVB-irradiated HDF cells, suppressing skin damage. Hence, dieckol could be a valuable ingredient for the cosmeceutical industry as a photoprotective agent, especially for human dermal fibroblast (HDF) cells.

Dioxinodehydroeckol from *Ecklonia cava* effectively repaired UVB skin damage in human keratinocyte cells (Ryu et al. 2015). This phenolic compound was able to protect HaCaT cells from UVB-induced apoptosis by modulating pro/antiapoptotic molecules (Bax/Bcl-2) and preventing the release of cytochrome C into the cytosol through caspase 3, 8, and 9 and the PARP pathway.

Hwang et al. (2006) orally and topically treated UVB-induced skin carcinogenesis in SKH-1 mice with brown algae polyphenols. They observed that UVB-induced oxidative stress, inflammation, and cell proliferation in the skin were significantly prevented. Thus, brown algae polyphenols may be useful as chemopreventive agents to reduce the risk of UVB-induced skin carcinogenesis.

Antiaging properties

As we age, wrinkles become more noticeable due to skin thickness and fiber elasticity changes, according to Sanjeewa et al. (2016). However, there are various ways to manage wrinkles, such as boosting collagen and hyaluronic acid production and limiting the activity of enzymes like matrix metalloproteinase (MMP), collagenase, hyaluronidase, and elastase, as explained by Ramos-e-Silva et al. (2013).

It has been discovered that brown seaweed compounds can inhibit MMPs and collagenase. These enzymes break down the proteins and collagen in the skin matrix, reducing its elasticity, firmness, and strength, leading to wrinkles and sagging over time. Joe et al. (2006) found that eckol and dieckol from *Ecklonia* species can suppress MMP-1 expression in human dermal fibroblasts. Riani et al. (2018) showed that an extract from *Sargassum plagyophyllum* can reduce collagenase activity by 54% in an *in vitro* model. In a recent study, Wang et al. (2021) found that dieckol can decrease MMP and proinflammatory cytokine expression, increase collagen synthesis, and inhibit collagenase activity in human dermal fibroblasts.

Hyaluronidase breaks down hyaluronic acid, the main component of the extracellular matrix. Inhibition of this enzyme is therefore expected to inhibit skin aging. Ferreres et al. (2012) studied several purified phlorotannin extracts from four brown algae (*Fucus spiralis*, *Cystoseira nodicaulis*, *Cystoseira usneoides*, and *Cystoseira tamariscifolia*). They discovered that the extract with the highest molecular weight of phlorotannins from *F. spiralis* had the strongest ability to inhibit hyaluronidase. Fayad et al. (2017) reported similar antihyaluronidase activity when evaluating a water extract of *Padina pavonica*. Arunkumar et al. (2021) recently identified phlorotannins from leaf samples of two *Sargassum* seaweeds and the thallus of *Eisenia arborea* as sources of antihyaluronidase compounds.

Elastase is involved in the degeneration of elastic fibers during wrinkle formation. According to Puspita et al. (2017) a fraction enriched in phenolic compounds from *Sargassum muticum* (Yendo) Fensholt showed elastase inhibitory activity in an *in vitro* assay. In addition, Gager et al. (2020) found that fractions isolated from *A. nodosum* and *F. serratus* possessed antielastase activity.

Some brown seaweed extracts and phlorotannins have shown antiwrinkle activity through multiple simultaneous mechanisms. Phasanasophon and Kim (2018) demonstrated significant inhibitory activity of elastase, collagenase, and tyrosinase enzymes of phlorotannin extracts from *Agarum cribrosum*. Similarly, dieckol from *Ecklonia cava* was able to suppress UVB-induced skin damage in human dermal fibroblasts by inhibiting intracellular collagenase activity and reducing the expression of proinflammatory cytokines and MMPs mediated through the regulation of nuclear factor kappa B, activator protein 1, and mitogen-activated protein kinase signaling pathways (Wang et al. 2021).

Whitening properties

With aging, our skin can develop hyperpigmentation. To combat this, some studies have looked at inhibiting the enzyme tyrosinase, which is involved in melanin synthesis and contributes to skin darkening. Researchers have tested crude extracts of brown algae containing phlorotannins and found that they can inhibit tyrosinase *in vitro*, as seen in a study by Jiménez et al. (2010) using an acetone extract of *Ascophyllum nodosum*. Other studies, such as by Kang et al. (2004) have found that extracts from different seaweeds have different levels of tyrosinase inhibitory activity. However, *in vitro* tests should be followed by *in vivo* tests. For example, while an ethanolic extract of *Sargassum polycystum* did not inhibit fungal tyrosinase *in vitro*, Chan et al. (2011) found that it significantly inhibited cellular tyrosinase activity in a cellular model. On the other hand, phlorotannins isolated from *Ecklonia cava* were shown by Yoon et al. (2009) to inhibit fungal tyrosinase activity and reduce melanin production in B16F10 melanoma cells.

Certain phlorotannins effectively inhibit tyrosinase making them potentially useful in the cosmetic industry for treating hyperpigmentation. Dieckol, isolated from *Ecklonia cava*, was discovered by Heo et al. (2009) to be a powerful melanogenesis inhibitor in human fibroblasts exposed to UVB radiation. Manandhar et al. (2019) studied several compounds isolated from *Ecklonia stolonifera* Okamura, including 974-A, phlorofucofuroeckol-A, and eckol. They confirmed their ability to reduce cellular melanin content and tyrosinase activity and downregulate the expression of melanogenesis enzymes, such as tyrosinase, tyrosinase-related proteins, TRP-1 and TRP-2, in B16F10 melanoma cells. More recently, Wang et al. (2022) confirmed the antimelanogenesis and photoprotective activities of a phlorotannin extract from *Ecklonia maxima* that contains dieckol and eckmaxol. This extract inhibited fungal tyrosinase and melanogenesis in alpha-melanocyte-stimulating hormone-stimulated B16F10 cells. It also significantly suppressed UVB-induced HaCaT cell death, consistent with the inhibition of apoptosis and a reduction in intracellular ROS scavenging. Furthermore, this extract inhibited collagen degradation and MMP expression in UVB-irradiated HDF cells.

Antiinflammatory and antiallergic properties

Inflammation is a vital part of the body's defense system and can be triggered by various factors such as physical injury, UV radiation, microbial invasion, and immune responses. It manifests as redness, warmth, swelling, and pain. Prolonged inflammation is harmful and contributes to the development of several diseases. UV radiation induces various inflammatory responses in human skin, including leukocyte migration, plasma protein leakage, microvascular structural changes, and vasodilation (Talero et al. 2015).

Atopic dermatitis is a multifaceted inflammatory allergic skin disorder that poses significant challenges. It is characterized by increased eosinophilia, elevated serum immunoglobulin E (IgE) levels, and positive skin prick test responses to common environmental allergens such as food or aeroallergens (Novak and Bieber 2003). As a result, ongoing research is focused on identifying bioactive compounds from natural sources that can suppress IgE production.

Sugiura et al. (2006) found that phlorofucofuroeckol-B inhibited histamine release from RBL-2H3 cells in rats. More recently, Sugiura et al. (2021) reviewed the antiallergic effects of phlorotannins from brown algae like *Ecklonia* sp. and *Eisenia* sp. They concluded that this phenolic group could alleviate allergic inflammation, passive cutaneous anaphylaxis reaction, atopic dermatitis, and 1,2-dimethylhydrazine-induced inflammation. The explanation for these benefits involves immunomodulation, attenuation of the arachidonic acid cascade, and regulation of the MAPK/NF-κB signaling pathway. Similarly, fucofuroeckol-A from *Ecklonia stolonifera* Okamura has been shown to protect against UVB-induced allergic reactions in RBL-2H3 mast cells by decreasing histamine release (Vo et al. 2018). An interesting recent study developed by Mihindukulasooriya et al. (2022) demonstrated the activity of a polyphenol-rich extract isolated from *Sargassum horneri* to suppress atopic dermatitis-like skin lesions in NC/Nga mice by decreasing the infiltration of inflammatory IL-13 positive cells into skin dermis. Consequently, phlorotannins could help to reduce acute and chronic allergic reactions.

The following table summarizes the main cosmeceutical bioactivities reported for phlorotannins or brown seaweed extracts.

Conclusions and future trends

Current research suggests promising potential for phlorotannins as key cosmeceutical ingredients; however, comprehensive exploration is needed to evaluate the impacts of phlorotannin-rich extracts and specific compounds in their intended applications. This is imperative given that individual and synergistic effects can fluctuate and that the efficacy of the phlorotannins-rich extract from seaweeds might not remain potent during the formulation of the final cosmeceutical product, as evidenced by Hermund et al.'s (2022) research on *Fucus vesiculosus*. Furthermore, given that most studies are conducted *in vitro*, the move towards *in vivo* testing is essential to represent their efficacy in humans better. Determining skin absorption and overall bioactivity through clinical trials is a crucial next step. Despite the potential of marine phenolics, the challenges in structural elucidation can obstruct correlations between their composition and bioactivity, thus affecting commercial viability. Moving forward, emphasis should be placed on developing standardized analytical methods (Catarino et al. 2022) and delving deeper into the biological actions and underlying mechanisms (Murphy and Dow 2021, Sugiura et al. 2021, Khan et al. 2022) of these intriguing compounds.

References

Abu Hajleh, M.N., R. Abu-Huwaij, A. AL-Samydai, L.K. Al-Halaseh and E.A. 2021. Al-Dujaili. The revolution of cosmeceuticals delivery by using nanotechnology: A narrative review of advantages and side effects. J. Cosmetic Dermatol. 20(12): 3818–3828.

Ahmed, I.A., M.A. Mikail, N.H. Zamakshshari, M.R. Mustafa, N.M. Hashim and R. Othman. 2022. Trends and challenges in phytotherapy and phytocosmetics for skin aging. Saudi J. Biol. Sci. 29(8): 103363.

Al-Halaseh, L.K., S. Al-Adaileh, A. Mbaideen, M.N.A. Hajleh, A. Al-Samydai, Z.Z. Zakaraya et al. 2022. Implication of parabens in cosmetics and cosmeceuticals: Advantages and limitations. J. Cosmetic Dermatol. 21(8): 3265–3271.

Alsabeelah, N., M.F. Arshad, S. Hashmi, R.A. Khan and S. Khan. 2021. Nanocosmeceuticals for the management of ageing: Rigors and Vigors. J. Drug Deliv. Sci. Technol. 63: 102448.

Arunkumar, K., R. Raja, V.B.S. Kumar, A. Joseph, T. Shilpa and I.S. Carvalho. 2021. Antioxidant and cytotoxic activities of sulfated polysaccharides from five different edible seaweeds. J. Food Meas Charact. 15(1): 567–576. DOI: 10.1007/s11694-020-00661-4.

Bak, S.S., Y.K. Sung and S.K. Kim. 2014. 7-Phloroeckol promotes hair growth on human follicles *in vitro*. Naunyn. Schmiedebergs. Arch. Pharm. 387: 789–793. doi: 10.1007/s00210-014-0986-0.

Balboa, E.M., E. Conde, A. Moure, E. Falqué and H. Domínguez. 2013. *In vitro* antioxidant properties of crude extracts and compounds from brown algae. Food Chem. 138: 1764–1785.

Borate, S.R. and A.S. Mundada. 2021. Herbal cosmeceuticals—Intensifies health and beauty of the skin. J. Med. Pharmaceutical Allied Sci. 11(6): 40–46.

Brash, D.E., J.A. Rudolph, J.A. Simon and J. Pontén. 1991. A role for sunlight in skin cancer: UV-induced p53 mutations in squamous cell carcinoma. Proc. Natl. Acad. Sci. 88: 10124–10128. https://doi.org/10.1073/pnas.88.22.10124.

Catarino, M.D., S.M.G. Pires, S. Silva, F. Costa, S.S. Braga, D.C.G.A. Pinto et al. 2022. Overview of phlorotannins' constituents in fucales. Mar Drugs, 20(12), Article number 754.

Cha, S.H., C.I. Ko, D. Kim and Y.J. Jeon. 2012. Protective effects of phlorotannins against ultraviolet B radiation in zebrafish (*Danio rerio*). Veterinary Dermatol. 23: 51–12. https://doi.org/10.1111/j.1365-3164.2011.01009.x.

Chan, Y.Y., K.H. Kim and S.H. Cheah. 2011. Inhibitory effects of *Sargassum polycystum* on tyrosinase activity and melanin formation in B16F10 murine melanoma cells. J. Ethnopharmacol. 137: 1183–1188. https://doi.org/10.1016/j.jep.2011.07.050.

Cikoš, A.M., I. Jerković, M. Molnar, D. Šubarić and S. Jokić. 2021. New trends for macroalgal natural products applications. Nat. Prod. Res. 35(7): 1180–1191.

Edward, M.J. 1997. Lawyers, regulations, and cosmetic claims. Am. J. Contact Dermatitis 8(4): 243–246.

El-Chaghaby, G. and S. Rashad. 2021. An overview of algae prospects in cosmeceuticals. J. Egyptian Women's Dermatol Soc. 18(3): 161–166.

Fayad, S., R. Nehmé, M. Tannoury, E. Lesellier, C. Pichon and P. Morin. 2017. Macroalga Padina pavonica water extracts obtained by pressurized liquid extraction and microwave-assisted extraction inhibit hyaluronidase activity as shown by capillary electrophoresis. J. Chromatogr A 1497: 19–27. https://doi.org/10.1016/j.chroma.2017.03.033.

Fernando, I.P.S., W.W. Lee and G. Ahn. 2022. Marine algal flavonoids and phlorotannins; An intriguing frontier of biofunctional secondary metabolites. Crit. Rev. Biotechnol. 42(1): 23–45.

Ferreres, F., G. Lopes, A. Gil-Izquierdo, P.B. Andrade, C. Sousa, T. Mouga et al. 2012. Phlorotannin extracts from Fucales characterized by HPLC-DAD-ESI-MSn: approaches to hyaluronidase inhibitory capacity and antioxidant properties. Mar Drugs 10: 2766–2781. https://doi.org/10.3390/md10122766.

Flodin, C., F. Helidoniotis and F.B. 1999. Whitfield. Seasonal variation in bromophenol content and bromoperoxidase activity in *Ulva lactuca*. Phytochemistry 51: 135–138. https://doi.org/10.1016/S0031-9422(98)00668-2.

Freitas, R., A. Martins, J. Silva, C. Alves, S. Pinteus, J. Alves et al. 2020. Highlighting the biological potential of the brown seaweed *Fucus spiralis* for skin applications. Antioxidants 9: 611.

Gager, L., S. Connan, M. Molla, C. Couteau, J.F. Arbona, L. Coiffard et al. 2020. Active phlorotannins from seven brown seaweeds commercially harvested in Brittany (France) detected by 1H NMR and *in vitro* assays: Temporal variation and potential valorization in cosmetic applications. J. Appl. Phycol. 32: 2375–2386.

Goyal, A., A. Sharma, J. Kaur, S. Kumari, M. Garg, R. Sindhu et al. 2022. Bioactive-based cosmeceuticals: An update on emerging trends. Molecules 27(3): 828.

Han, Y.-S. and T. Han. 2005. UV-B induction of UV-B protection in *Ulva pertusa* (Chlorophyta). Journal of Phycology 41: 523–530. https://doi.org/10.1111/j.1529-8817.2005.00072.x.

Heo, S.J., S.C. Ko, S.H. Cha, D.H. Kang, H.S. Park, Y.U. Choi et al. 2009. Effect of phlorotannins isolated from *Ecklonia cava* on melanogenesis and their protective effect against photo-oxidative stress induced by UV-B radiation. Toxicol *in vitro* 23: 1123–1130. https://doi.org/10.1016/j.tiv.2009.05.013.

Hermund, D.B., S.Y. Heung, B.R. Thomsen, C.C. Akoh and C. Jacobsen. 2018. Improving oxidative stability of skin-care emulsions with antioxidant extracts from brown alga *Fucus vesiculosus*. J. Am. Oil Chem. Soc. 95: 1509–1520.

Hermund, DB., H. Torsteinsen, J. Vega, F.L. Figueroa and C. Jacobsen. 2022 Screening for new cosmeceuticals from brown algae *Fucus vesiculosus* with antioxidant and photo-protecting properties. Mar. Drugs 20: 687.

Hwang, H., T. Chen, R.G. Nines, H.C. Shin and G.D. Stoner. 2006. Photochemoprevention of UVB-induced skin carcinogenesis in SKH-1 mice by brown algae polyphenols. Int. J. Cancer 119: 2742–2749. https://doi.org/10.1002/ijc.22147.

Jayawardhana, H.H.A.C.K., Thilina U. Jayawardena, K.K.A. Sanjeewa, N.M. Liyanage, D.P. Nagahawatta, H.G. Lee et al. 2023. Marine algal polyphenols as skin protective agents: Current status and future prospectives. Mar. Drugs 21(5): Article number 285.

Jiménez, J.T., S. O'Connell, H. Lyons, B. Bradley and M. Hall. 2010. Antioxidant, antimicrobial, and tyrosinase inhibition activities of acetone extract of *Ascophyllum nodosum*. Chem. Papers 64: 434–442. https://doi.org/10.2478/s11696-010-0024-8.

Joe, M.J., S.N. Kim, H.Y. Choi, W.S. Shin, G.M. Park, D.W. Kang et al. 2006. The inhibitory effects of eckol and dieckol from *Ecklonia stolonifera* on the expression of matrix metalloproteinase-1 in human dermal fibroblasts. Biol. Pharmac. Bull. 29: 1735–1739. https://doi.org/10.1248/bpb.29.1735.

Kamiya, M., T. Nishio, A. Yokoyama, K. Yatsuya, T. Nishigaki, S. Yoshikawa et al. 2010. Seasonal Variation of phlorotannin in *Sargassacean* species from the coast of the sea of Japan. Phycol. Res. 58: 53–61.

Kang, H.S., H.R. Kim, D. Byun, B.W. Son, T.J. Nam and J.C. Choi. 2004. Tyrosinase inhibitors isolated from the edible brown alga *Ecklonia stolonifera*. Mar. Sci. Technol. Bull. 14: 49–56. https://doi.org/10.1007/BF02975886.

Khan, F., G.J. Jeong, M.S.A. Khan, N. Tabassum and Y.M. Kim. 2022. Seaweed-derived phlorotannins: A review of multiple biological roles and action mechanisms. Mar Drugs 20(6), Article number 384.

Ko, S.C., S.H. Cha, S.J. Heo, S.H. Lee, S.M. Kang and Y.J. Jeon. 2011. Protective effect of *Ecklonia cava* on UVB-induced oxidative stress: *in vitro* and *in vivo* zebrafish model. J. Appl. Phycol. 23: 697–708. https://doi.org/10.1007/s10811-010-9565-z.

Le Lann, K., G. Surget, C. Couteau, L. Coiffard, S. Cérantola, F. Gaillard et al. 2016. Sunscreen, antioxidant, and bactericide capacities of phlorotannins from the brown macroalga *Halidrys siliquosa*. J. Appl. Phycol. 28: 3547–3559. https://doi.org/10.1007/s10811-016-0853-0.

Lopes, G., C. Sousa, L.R. Silva, E. Pinto, P.B. Andrade, J. Bernardo et al. 2012. Can phlorotannins purified extracts constitute a novel pharmacological alternative for microbial infections with associated inflammatory conditions? PLoS ONE 7: e31145.

López-Hortas, L., N. Flórez-Fernández, M.D. Torres, T. Ferreira-Anta, M.P. Casas et al. 2021. Applying seaweed compounds in cosmetics, cosmeceuticals and nutricosmetics. Mar. Drugs 19(10): 552.

Manandhar, B., A. Wagle, S.H. Seong, P. Paudel, H.-R. Kim, H.A. Jung et al. 2019. Phlorotannins with potential anti-tyrosinase and antioxidant activity isolated from the marine seaweed *Ecklonia stolonifera*. Antioxidants 8: 240.

Martínez-Ruiz, M., C.A. Martínez-González, D.H. Kim, B. Santiesteban-Romero, H. Reyes-Pardo, K.R. Villaseñor-Zepeda et al. 2022. Microalgae bioactive compounds to topical applications products—A review. Molecules 27(11): 3512.

Mihindukulasooriya, S.P., H.J. Kim, J. Cho, K.H.I.N.M. Herath, J. Yang, D.T.T. Dinh et al. 2022. Polyphenol-rich *Sargassum horneri* alleviates atopic dermatitis-like skin lesions in NC/Nga mice by suppressing Th2-mediated cytokine IL-13. Algae 37: 331–347. https://doi.org/10.4490/algae.2022.37.11.24.

Mourelle, M.L., C.P. Gómez and J.L. Legido. 2023. Cosmeceuticals and thalassotherapy: recovering the skin and well-being after cancer therapies. Appl. Sci. 13(2): 850.

Murphy, M.J. and A.A. Dow. 2021 Clinical studies of the safety and efficacy of macroalgae extracts in cosmeceuticals. J. Clinical Aesthetic Dermatol. 14(10): 37–41.

Novak, N. and T. Bieber. 2003. Allergic and nonallergic forms of atopic diseases. J. Allergy Clin. Immunol. 112: 252–262. https://doi.org/10.1067/mai.2003.1595.

Pangestuti, R., K.H. Shin and S.K. Kim. 2021. Anti-photoaging and potential skin health benefits of seaweeds. Mar. Drugs 19(3): 172.

Parrish, J.A., K.F. Jaenicke and R.R. Anderson. 1982. Erythema and melanogenesis action spectra of normal human skin. Photochem Photobiol. 36: 187–191. https://doi.org/10.1111/j.1751-1097.1982.tb04362.x.

Pereira, L. 2018. Seaweeds as source of bioactive substances and skin care therapy-Cosmeceuticals, algotheraphy, and thalassotherapy. Cosmetics 5: 41, 68.

Pérez-Rodríguez, E., J. Aguilera, F.L. Figueroa. 2003. Tissular localization of coumarins in the green alga *Dasycladus vermicularis* (Scopoli) Krasser: A photoprotective role? J. Exp. Bot. 54: 1093–1100. https://doi.org/10.1093/jxb/erg111.

Phasanasophon, K. and S. Kim. 2018. Antioxidant and cosmeceutical activities of *Agarum cribrosum* phlorotannin extracted by ultrasound treatment. Nat. Prod. Commun. 13: 565–570

Puspita, M., M. Déniel, I. Widowati, O.K. Radjasa, P. Douzenel, C. Marty, L. Vandanjon et al. 2017. Total phenolic content and biological activities of enzymatic extracts from *Sargassum muticum* (Yendo) Fensholt. J. Appl. Phycol. 29: 2521–2537. https://doi.org/10.1007/s10811-017-1086-6.

Ramos-e-Silva, M., L.R. Celem, S. Ramos-e-Silva and A.P. Fucci-da-Costa. 2013. Anti-aging cosmetics: facts and controversies. Clinics Dermatol. 31: 750–758.

Riani, M.K.L., E. Anwar and T. Nurhayati. 2018. Antioxidant and anti-collagenase activity of *Sargassum plagyophyllum* extract as an anti-wrinkle cosmetic ingredient. Pharmacognosy 10: 932–936. https://doi.org/10.5530/pj.2018.5.157.

Ryu, B., B.-N. Ahn, K.-H. Kang, Y.-S. Kim, Y.-X. Li, C.-S. Kong et al. 2015. Dioxinodehydroeckol protects human keratinocyte cells from UVB-induced apoptosis modulated by related genes Bax/Bcl-2 and caspase pathway. J. Photochem. Photobiol. B, 153: 352–357.

Sanjeewa, K.K.A., E.A. Kim, K.T. Son and Y.J. Jeon. 2016. Bioactive properties and potentials cosmeceutical applications of phlorotannins isolated from brown seaweeds: A review. J. Photochem. Photobiol B: Biology 162: 100–105. https://doi.org/10.1016/j.jphotobiol.2016.06.027.

Santiesteban-Romero, B., M. Martínez-Ruiz, J.E. Sosa-Hernández, R. Parra-Saldívar and H.M.N. Iqbal. 2022. Microalgae photo-protectants and related bio-carriers loaded with bioactive entities for skin applications—an insight of microalgae biotechnology. Mar. Drugs 20(8): 487.

Shibata, T., K. Ishimaru, S. Kawaguchi, H. Yoshikawa and Y. Hama. 2007. Antioxidant activities of phlorotannins isolated from Japanese Laminariaceae. In Nineteenth International Seaweed Symposium; Springer: Berlin/Heidelberg, Germany, pp. 255–261.

Siahaan, E.A., R. Agusman, Pangestuti, K.H. Shin and S.K. Kim. 2022. Potential Cosmetic Active Ingredients Derived from Marine By-Products. Mar Drugs, 20, 12, Article number 734.

Soleimani, S., M. Yousefzadi, S.B.M. Nezhad, O.N. Pozharitskaya and A.N. Shikov. 2022. Evaluation of fractions extracted from *Polycladia myrica*: biological activities, UVR protective effect, and stability of cream formulation based on it. J. Appl. Phycol. 34: 1763–1777.

Sotiropoulou, G., E. Zingkou and G. Pampalakis. 2021. Redirecting drug repositioning to discover innovative cosmeceuticals. Exp. Dermatol. 30(5): 628–644.

Subbiah, V., C. Xie, F.R. Dunshea, C.J. Barrow and H.A.R. Suleria. 2022. The quest for phenolic compounds from seaweed: Nutrition, biological activities and applications. Food Rev. Int. doi: 10.1080/87559129.2022.2094406.

Sugiura, Y., H. Katsuzaki, K. Imai and H. Amano. 2021. The anti-allergic and anti-inflammatory effects of phlorotannins from the edible brown algae, *Ecklonia* sp. and *Eisenia* sp. Nat. Prod. Commun. 16, 1934578X211060924. https://doi.org/10.1177/1934578x211060924.

Sugiura, Y., K. Matsuda, Y. Yamada, M. Nishikawa, K. Shioya, H. Katsuzaki et al. 2006. Isolation of a new anti-allergic phlorotannin, phlorofucofuroeckol-b, from an edible brown alga, *Eisenia arborea*. Biosci. Biotechnol. Biochem. 70: 2807–2811. https://doi.org/10.1271/bbb.60417.

Sun, M., Y. Deng, X. Cao, L. Xiao, Q. Ding, F. Luo et al. 2022. Effects of natural polyphenols on skin and hair health: Areview. Molecules 27(22): 7832; https://doi.org/10.3390/molecules27227832.

Talero, E., S. García-Mauriño, J. Ávila-Román, A. Rodríguez-Luna, A. Alcaide and V. Motilva. 2015. Bioactive compounds isolated from microalgae in chronic inflammation and cancer. Mar. Drugs 13: 6152–6209. https://doi.org/10.3390/md13106152.

Tangau, M.J., Y.K. Chong and K.Y. Yeong. 2022. Advances in cosmeceutical nanotechnology for hyperpigmentation treatment. J. Nanoparticle Res. 24(8): 155. doi.org/10.1007/s11051-022-05534-z.

Vaishampayan, P., M. Rane. 2022. Herbal nanocosmeceuticals: A review on cosmeceutical innovation. J. Cosmet. Dermatol. 21(11): 5464–5483.

Vermeer, B.J. and B.A. Gilchrest. 1996. Cosmeceuticals: A proposal for rational definition, evaluation, and regulation. Archives of Dermatology 132(3): 337–340.

Vo, T.S., S.-K. Kim, B. Ryu, D.H. Ngo, N.-Y. Yoon, L.G. Bach et al. 2018. The suppressive activity of fucofuroeckol derived from brown alga *Ecklonia stolonifera okamura* on UVB-induced mast cell degranulation. Mar. Drugs 16: 1.

Wang, L., J.-G. Je, H.-S. Kim, K. Wang, X. Fu, J. Xu et al. 2022. Anti-melanogenesis and photoprotective effects of Ecklonia maxima extract containing dieckol and eckmaxol. Mar. Drugs 20: 557.

Wang, L., J.-G. Je, H.-W. Yang, Y.-J. Jeon and S. Lee. 2021. Dieckol, an algae-derived phenolic compound, suppresses UVB-induced skin damage in human dermal fibroblasts and its underlying mechanisms. Antioxidants 10(3): 352.

Yoon, N.Y., T.K. Eom, M.M. Kim and S.K. Kim. 2009. Inhibitory effect of phlorotannins isolated from *Ecklonia cava* on mushroom tyrosinase activity and melanin formation in mouse B16F10 melanoma cells. J. Agric. Food Chem. 57(10): 4124–4129.

Chapter 16

Recent Advancement of Phenolic Compounds from Algae

Chita Ranjan Sahoo,[1,2,*] *Sudhir Kumar Paidesetty*,[3] *Debdutta Bhattacharya*[1] and *Rabindra Nath Padhy*[2]

Introduction to algae and phenolic compounds

Algae are a group of photosynthetic organisms that can exist as either unicellular or multicellular entities. They inhabit a wide range having various sizes from macroalgae to microalgae. Presently, there is a growing global interest in the field of pharmaceutical sciences research about the production of useful metabolites by microalgae. These species possess a significant reservoir of phenolic compounds (Cichoński et al. 2022). In the realm of biotechnology algae have been shown to be both natural and effective sources of therapeutic value. There are numerous pharmacological properties of their phenolic substances and that offers a viable basis for building novel medications (Sahoo et al. 2019, Sahoo et al. 2020, Mandhata et al. 2022).

Plants produce phenolic compounds—a type of secondary metabolite during stress response. These compounds help the plant communicate with and adapt to its environment, allowing it to endure stresses like ultraviolet radiation. Vegetables, fruits, and spices all contain phenolic chemicals since higher plant organs produce them. In addition, algae are natural places to find these compounds (Swain et al. 2023). Microalgae and cyanobacteria have shown to contain various flavonoids, including flavanones, isoflavones, and dihydrochalcones. Most of the algae species possess phenolic acids such gallic, protocatechuic, caffeic, chlorogenic, vanillic and a few more bioactive compounds (Mandhata et al. 2023, Bishoyi et al. 2023). Indeed, the biological implications of macroalgae phenolic compounds, specifically phlorotannin, have been identified through research. Phlorotannin, a polyphenolic compound, is predominantly present in brown seaweed and consists of polymeric phloroglucinol as impact of biological activity (Maheswari et al. 2022).

In chemistry, phenolic compounds can take one of two forms. In the first case, we have the fundamental framework of phenolic acids and hydrolyzed tannins, a C6-C1 ring structure. The most straightforward fundamental and most crucial nonflavonoid molecules are phenolic acids.

[1] ICMR, Regional Medical Research Centre, Govt. of India, Bhubaneswar-751023, Odisha, India.
[2] Central Research Laboratory, Institute of Medical Sciences & Sum Hospital, Siksha 'O' Anusandhan (Deemed to be University), Bhubaneswar-751003, Odisha, India.
[3] Medicinal Chemistry Research Laboratory, School of Pharmaceutical Sciences, Siksha 'O' Anusandhan (Deemed to be University), Bhubaneswar-751003, Odisha, India.
* Corresponding author: chitaranjan.biotech@gmail.com

One or –OH groups and a carboxylic acid replace the phenyl in their structure. According to the number of carbon atoms between the carboxyl group and the hydroxyl group, phenolic acids can be divided into two groups: hydroxybenzoic acids and hydroxycinnamic acids. Flavonoids, condensed tannins, and phenylpropanoid acids all share the C6-C3-C6 ring as a structural motif (Sahoo et al. 2021, Kafkas et al. 2018, Kumar et al. 2019).

In this chapter, we compile the findings from several investigations on the biological activities of phenols found in algae, which together establish their biomedical significance. Moreover, we report various extraction methods, namely conventional and non-conventional, of phenolic compounds from algae.

Phenolic compounds with their classification

One of the largest classes of secondary metabolites, phenolic chemicals are found virtually everywhere. Algae and land plants rely on these substances to develop normally. The substances help them fend off pathogens, injuries, and environmental stress. The list of health benefits associated with consuming phenolic chemicals, which are included in the diets of both animals and humans, is practically limitless. Phenolic compounds are a broad and diverse class of secondary metabolites that are predominantly found in plants but can also be synthesised by other species like algae. Since these molecules have a wide variety of positive effects on human health, it is important to learn how they work. Chemically, phenolic compounds have the capacity to trap radicals that are free, reactive oxygen molecules, and chelated metallic ions due to their formation based on a structural core produced by a hydroxyl group connected directly to a phenol (Santos et al. 2019, Scania et al. 2021).

Due to their importance and contribution to organisms' growth and survival, as well as helping in the defence against pathogens and predators, phenolic compounds are present in a vast diversity of terrestrial and marine plants. These chemicals can be made via the shikimate, phenylpropanoid, or pentose phosphate pathways (Jimenez-Lopez et al. 2021). All the over 8,000 known chemical structures of phenolic compounds can be categorised because they all have a minimum of one aromatic phenol ring containing a substituted hydroxyl group which can be strongly polymerized. There are a variety of factors that can be used to complete the classification. The total number of carbon atoms in a molecule is the basis for one of these. However, other classifications can also be made based on whether the substance is soluble or insoluble, how extensively it is spread, or whether or not it is polymeric. This last categorization is especially helpful from a nutritional standpoint, as insoluble phenolic chemicals will be unable to pass through the intestinal barrier and enter the blood (Tungmunnithum et al. 2018). Since phenolic compounds have a wide range of heterogeneous structures, from the extremely simple to the highly polymerized, they fall into a wide range of categories. Indeed, phenolic substances in macroalgae range from simple molecules like phenolic and cinnamic acids or flavonoids to polymeric structures like phlorotannin. The detail phenolic compound presented in algae are presented (Figure 1–4), (Table 1) (Santos et al. 2019, Scania et al. 2021).

Extraction of phenolic compounds

There are two methods for extraction of phenolic compounds from algae. These are conventional methods and non-conventional methods (Alara et al. 2021, Jahromi et al. 2019). The detail schematic presentation is in Figure 5.

Conventional methods: Traditional extraction methods are often characterised by the employment of greater volumes of extraction solvents in addition to manual procedures that are extremely reliant on investigators and labour-intensive, leading to approaches that are not always reliable (Alara et al. 2021, Sookjitsumran et al. 2016).

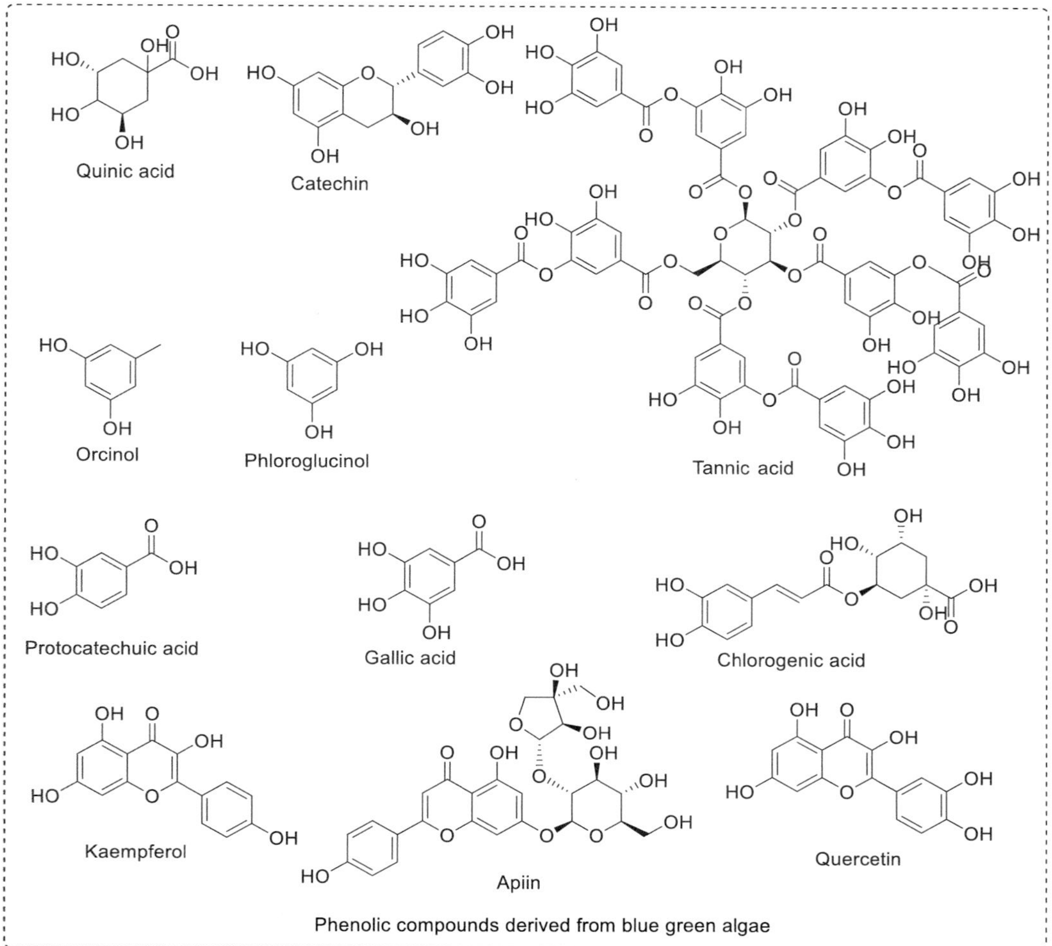

Figure 1. Phenolic compounds derived from blue green algae and structures drawn by ChemDraw Ultra.

Non-conventional: There are many reasons to favour novel extraction methods over traditional ones, including shorter extraction times, higher yields, lower amounts of harmful residues, and a higher degree of repeatability in the extraction process. There are several methods namely, pressurized extraction, microwave extraction, ultrasound extraction, water extraction, CO_2 extraction (Alara et al. 2021, Sookjitsumran et al. 2016).

The microwave is a form of electromagnetic radiation characterised by frequencies ranging from 30 to 300 MHz and the generation of heat occurs because of the activation of molecular movements, leading to the rupture of the cell wall and subsequent release of active substances within the cell. Solvents possessing a dipolar nature, such as water, have a larger dielectric constant compared to non-polar solvents. Consequently, these solvents possess the ability to absorb substantial amounts of high energy. As a result, the utilisation of dipolar solvents enhances the velocity and efficacy of phenolic compound extraction (Sookjitsumran et al. 2016).

Ultrasound waves are characterised by frequencies ranging from 20 kHz to 10 MHz, enabling them to traverse various mediums such as solids, liquids, and gases. Notably, these waves are imperceptible to the human auditory system. The current extraction technique involves the generation of cavitation bubbles in close proximity to the tissue of interest, subsequently leading to

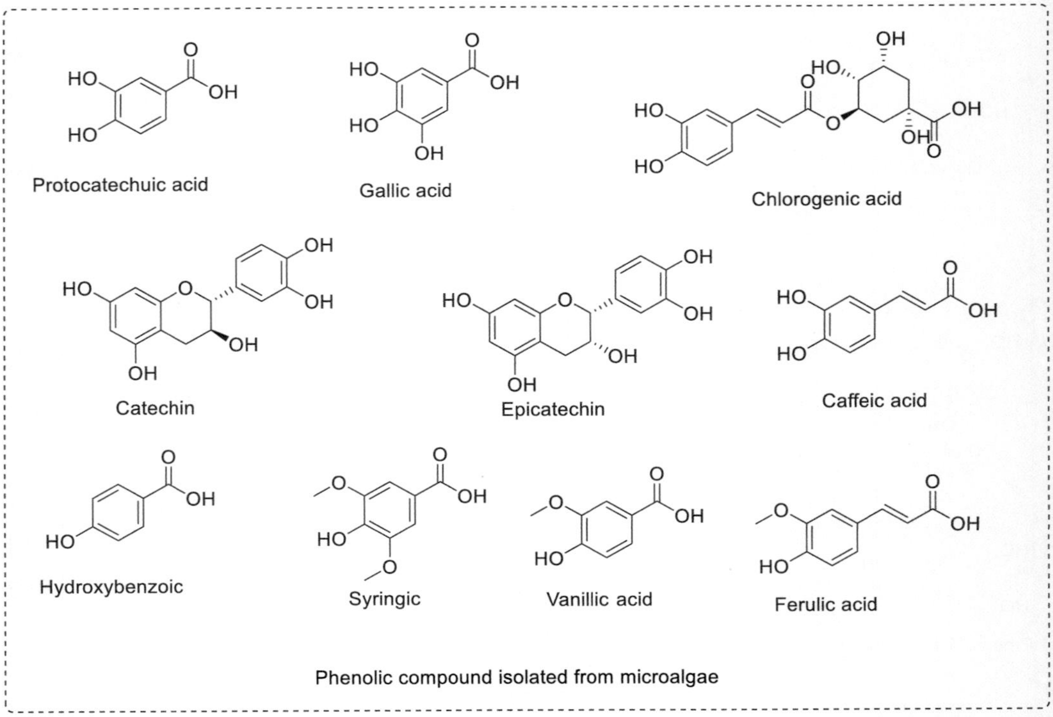

Figure 2. Phenolic compounds derived from microalgae and structures drawn by ChemDraw Ultra.

the rupture of the cellular membrane and subsequent release of intracellular contents. Ultrasound waves are employed and utilised by the process of extracting phenolic chemicals from botanical specimens. In addition to the intrinsic parameters of ultrasonic devices, such as amplitude, frequency, and wavelength, the power and intensity of these devices also exert significant influence on the extraction process, necessitating the need for optimisation. The design and configuration of a reactor, as well as the shape of the probe, can have an impact on the efficiency and effectiveness of the extraction process (Altemimi et al. 2016, Wang et al. 2018).

The utilisation of carbon dioxide (CO_2) as a supercritical fluid is necessary for the implementation of this technique.

This fluid is mostly utilised for its non-toxic nature, non-flammable properties, cost-effectiveness, and abundant availability in a high purity grade. The CO_2 extraction process has the flexibility to apply various pressure-temperature combinations, rendering it a preferred technology for producing many end products (Alara et al. 2021).

Biological activity of phenolic compounds

Increasing understanding of the biological activities and health advantages of algae has piqued the scientific community's interest in exploiting these substances. Indeed, algal bioactive compound possesses unique bioactivities like antibacterial, antifungal, antioxidant, antiproliferative, antidiabetic and a few more properties. However, phlorotannins have been extensively investigated as a class of phenolic compounds, with a significant emphasis on their biological properties compared to other phenolic compounds found in algae (Maheswari et al. 2022). The phenolic compounds found in algae have received a lot of interest for mainstream drug development cascade (Figure 6).

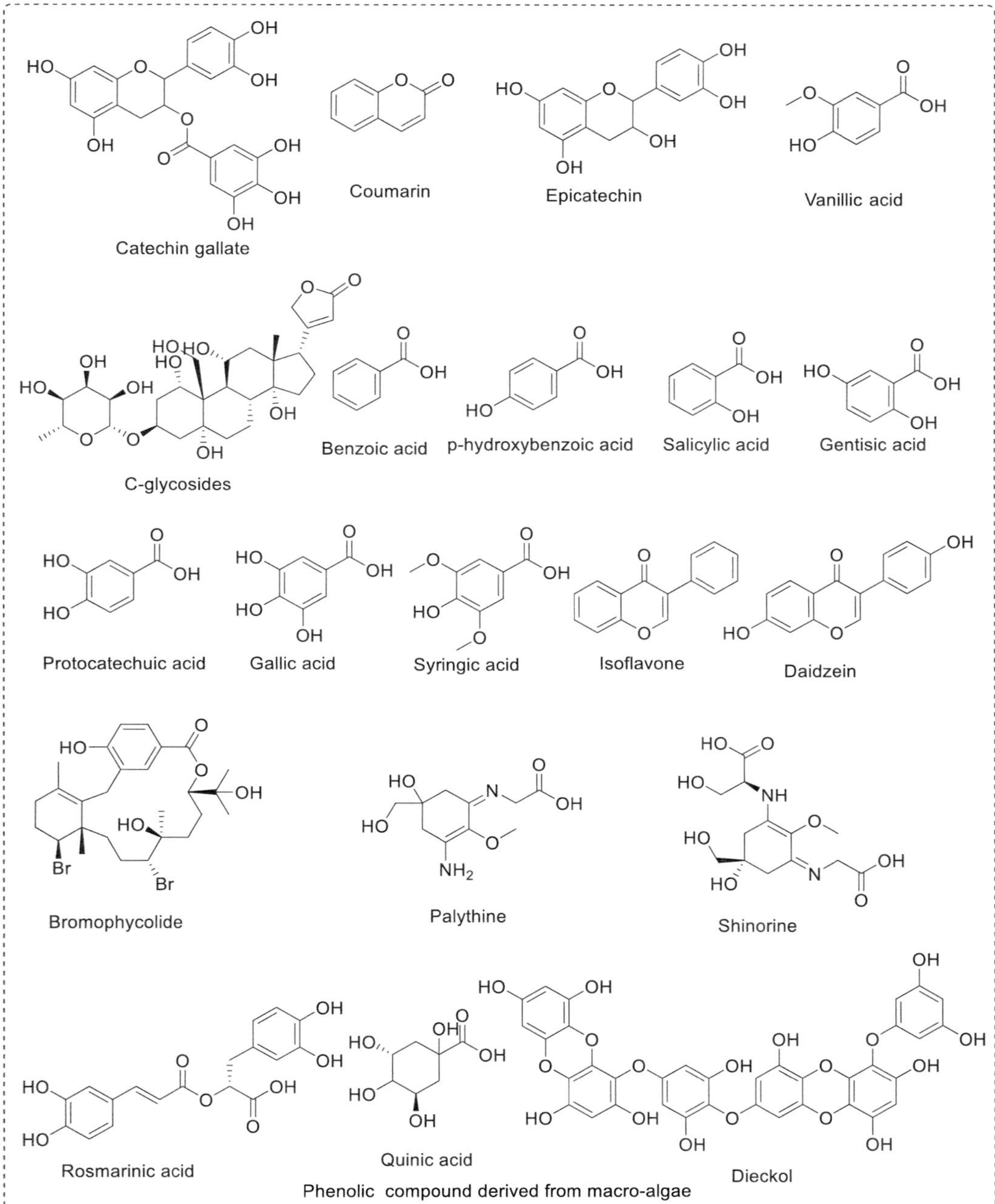

Figure 3. Phenolic compounds derived from macroalgae and structures drawn by ChemDraw Ultra.

Antimicrobial activity: The considerable antibacterial efficacy exhibited by phlorotannins against both Gram-positive and Gram-negative bacteria presents a hopeful foundation for the development of innovative pharmaceuticals intended for the management and prophylaxis of infectious ailments. These polyfunctional chemicals can be obtained from brown algae, which has a high biotechnological potential (Besednova et al. 2020).

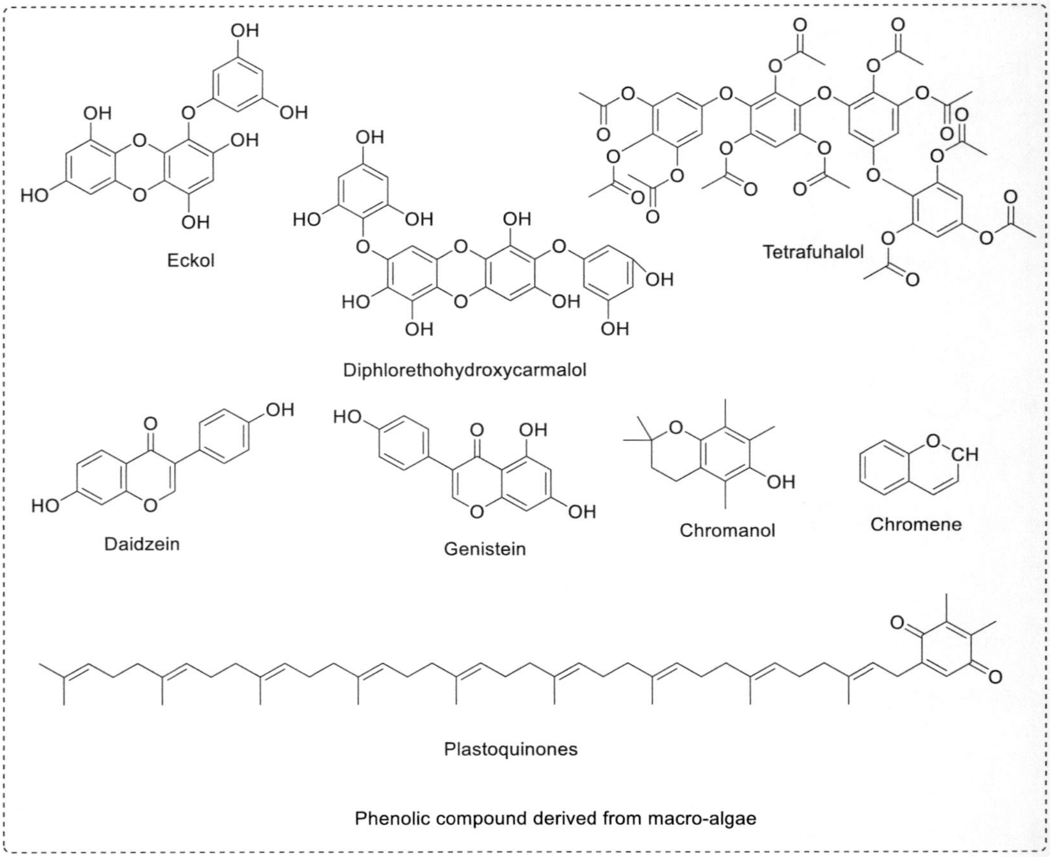

Figure 4. Phenolic compounds derived from macroalgae and structures drawn by ChemDraw Ultra.

Antioxidant activity: Polyphenols based on phloroglucinol that were isolated from brown marine algae, with a particular emphasis on *Sirophysalis trinodis* (formerly *Cystoseira trinodis*), have shown to exhibit antioxidant properties (Sathya et al. 2017).

Anticancer activity: It has been seen that several different seaweed extracts, notably those rich in phenolic compounds and also certain isolated phlorotannins, showed cytotoxic effects against a wide range of tumoral cells when tested in vitro (Rocha et al. 2018).

Anti-aging activity: The gradual breakdown of one's physiological integrity that results in the loss of one's functional capacity and an increasing risk of passing away is what we mean when we talk about ageing.

Both *Caulerpa racemosa* and *Caulerpa lentillifera* (Chlorophyta) have shown to have anti-aging and antioxidant properties, according to research by Indonesia.

Potential antimicrobials vary in chemical composition, necessitating individualised optimisation of the conditions resulting to their maximum generation, extraction, and recovery. The effects of both abiotic and biotic factors on the content of antimicrobial compounds in seaweed are reported (Pérez et al. 2016). Dieckol (Figure 7) is a phlorotannin that was discovered in the brown alga *Ecklonia cava* (Phaeophyceae). It has anti-inflammatory, anti-clotting, and antioxidant properties. Among

Table 1. Phenolic compound derived from macro-algae (Lomartire et al. 2021).

<table>
<tr><th>Name of macroalgae and class of compound</th><th>Phycompounds</th><th>Chemical profile</th></tr>
<tr><td rowspan="16">Acetabularia ryukyuensis (C) (Flavonoids),
Ascophyllum nodosum (P) (Phenolic acids),
Bifurcaria bifurcate (R) (Phenolic acids),
Cladophora socialis (C) (Phenolic acids),
Cystoseira sp. (P) (Phlorotannins),
Portieria hornemannii (formerly Chondrococcus hornemannii) (R) (Flavonoids),
Callophycus serratus (R) (Phenolic terpenoids),
Chondrus crispus (R) (Flavonoids),
Dasycladus vermicularis (C) (Flavonoids),
Durvillaea antarctica (P) (Flavonoids),
Eisenia bicyclis (P) (Flavonoids),
Ecklonia cava (P) (Flavonoids),
Fucus spiralis (P) (Phlorotannins),
Fucus vesiculosus (P) (Phenolic acids),
Gelidium elegans (R) (Flavonoids),
Gracilaria sp. (R) (Phenolic acids),
Ishige okamurae (P) (Phlorotannins),
Lessonia spicata (P) (Flavonoids),
Macrocystis pyrifera (formerly Macrocystis integrifolia) (P) (Flavonoids),
Pyropia yezoensis (R) (Flavonoids),
Nitella hookeri (C) (Flavonoids),
Padina arborescens (P) (Flavonoids),
Padina minor (P) (Flavonoids),
Palmaria palmata (C) (Aminoacids),
Sargassum muticum (P) (Flavonoids),
Stypopodium zonale (P) (Phenolic terpenoids)</td><td>Benzoic acid</td><td>Compound CID: 243
MW: 122.12g/mol</td></tr>
<tr><td>Catechin</td><td>Compound CID: 9064
MW: 290.27g/mol</td></tr>
<tr><td>C-glycosides</td><td>Compound CID: 637579
MW: 584.7g/mol</td></tr>
<tr><td>Coumarin</td><td>Compound CID: 323
MW: 146.14g/mol</td></tr>
<tr><td>Dieckol</td><td>Compound CID: 3008868
MW: 742.5g/mol</td></tr>
<tr><td>Diphlorethohydroxycarmalol</td><td>Compound CID: 16075395
MW: 512.4g/mol</td></tr>
<tr><td>Eckol</td><td>Compound CID: 145937
MW: 372.3g/mol</td></tr>
<tr><td>Epicatechin</td><td>Compound CID: 72276
MW: 290.27g/mol</td></tr>
<tr><td>Epicatechin gallate</td><td>Compound CID: 107905
MW: 442.4g/mol</td></tr>
<tr><td>Epigallocatechin</td><td>Compound CID: 72277
MW: 306.27g/mol</td></tr>
<tr><td>Epigallocatechin gallate</td><td>Compound CID: 65064
MW: 458.4g/mol</td></tr>
<tr><td>Gallate</td><td>Compound CID: 370
MW: 170.12g/mol</td></tr>
<tr><td>Quinic acid</td><td>Compound CID: 6508
MW: 192.17g/mol</td></tr>
<tr><td>Rosmarinic acid</td><td>Compound CID: 5281792
MW: 360.3g/mol</td></tr>
<tr><td>Tetrafuhalol</td><td>Compound CID: 15689631
MW: 992.8g/mol</td></tr>
<tr><td>Vanillic acid</td><td>Compound CID: 8468
MW: 168.15g/mol</td></tr>
</table>

MW: molecular weight, CID: PubChem Compound Identification, C: Chlorophyta, P: Phaeophyceae, R: Rhodophyta

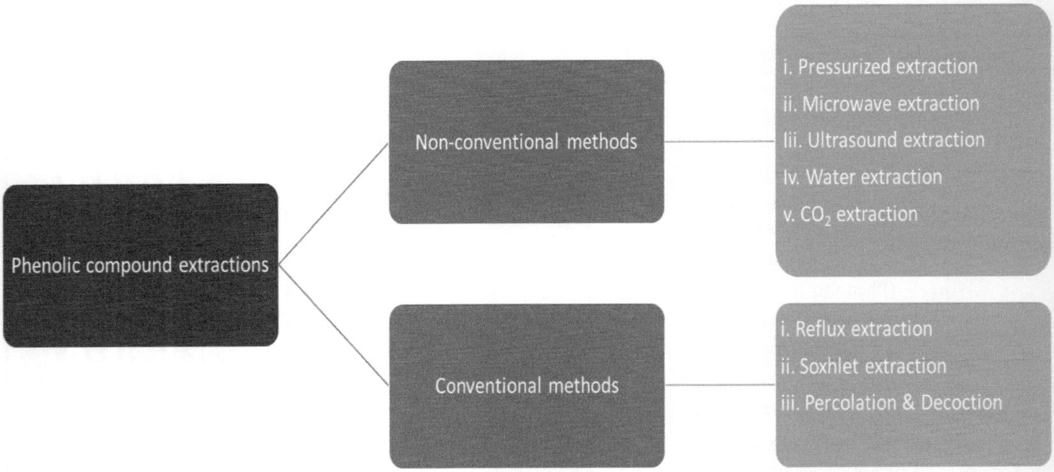

Figure 5. Graphical illustration of phenolic compound extraction methods.

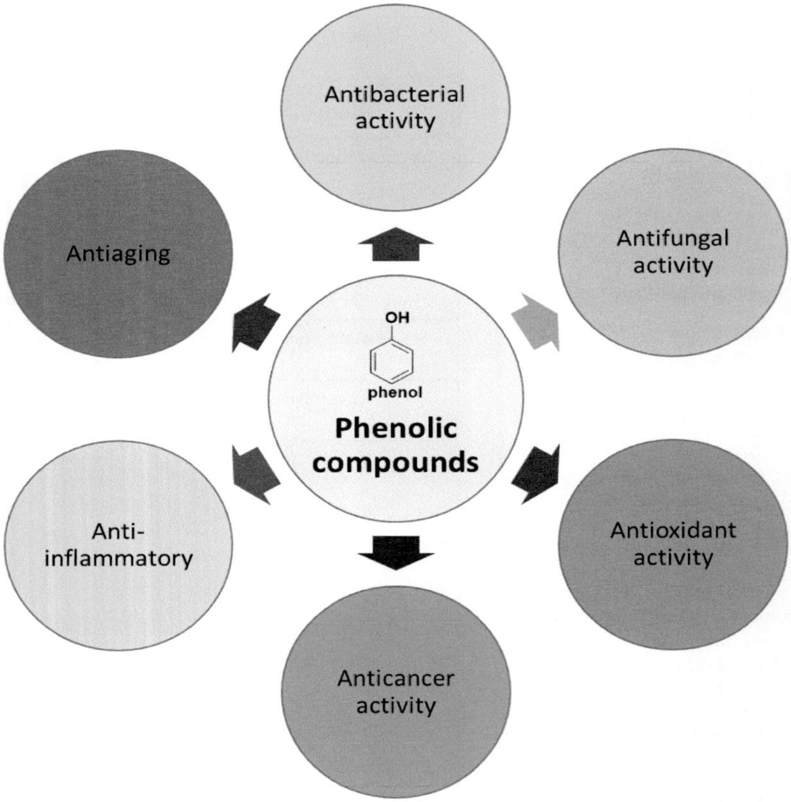

Figure 6. Schematic representation of pharmacological activities of algal phenols.

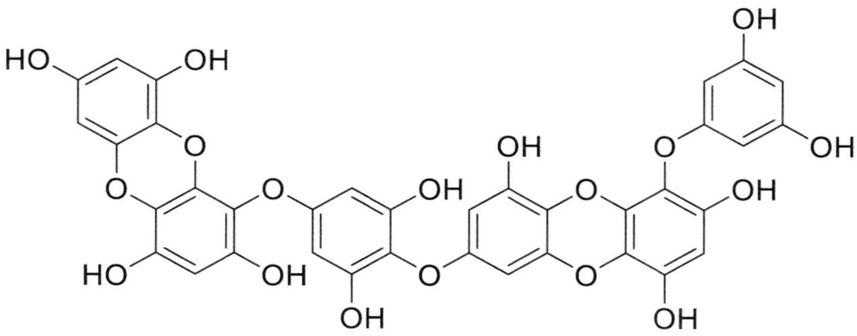

Dieckol

Figure 7. Antioxidant compound of algae.

its many functions, it acts as a metabolite, a radical scavenger, an inhibitor of (alpha-glucosidase), a hepatoprotective agent, and an anticoagulant. Phlorotannins are oxacycles that also contain aromatic ethers. It is similar to a phloroglucinol in its functionality (Maheswari et al. 2022, Besednova et al. 2020).

Concluding remarks on an application of phenolic compounds in cosmeceuticals

A broad spectrum of compounds derived from seaweed, naturally sourced from marine environments, currently have applications in different commercial industries, serving as functional food components, biomedical materials, and other products to enhance human health. An era of modern medicine began with the discovery of antibiotics. Some bacteria have become resistant to modern antibiotics, which poses a serious threat to public health in the modern world. Many nations are turning to algal products as a means of combating this issue. Phenolic compounds from algae are a rich source of phytochemicals: the isolation of which has given us a wide variety of useful compounds. Indeed, phenolic compounds from algae offer a potentially new avenue for the medicinal product development process if they are used as functional components. The significance of phenolic compounds for human well-being necessitates ongoing efforts to discover novel extraction procedures from algal materials. Such procedures may effectively recover these bioactive chemicals, hence meeting the demand for such compounds.

Acknowledgement

Authors are thankful to IMS & SUM Hospital, Siksha O Anusandhan Deemed to be University, Bhubaneswar, Odisha, and Director, ICMR-RMRC Bhubaneswar.

References

Alara, O.R., N.H. Abdurahman and C.I. Ukaegbu. 2021. Extraction of phenolic compounds: A review. Current Research in Food Science 4: 200–214.

Altemimi, A., D.G. Watson, R. Choudhary, M.R. Dasari and D.A. Lightfoot. 2016. Ultrasound assisted extraction of phenolic compounds from peaches and pumpkins. PloS one 11(2): e0148758.

Besednova, N.N., B.G. Andryukov, T.S. Zaporozhets, S.P. Kryzhanovsky, T.A. Kuznetsova, L.N. Fedyanina et al. 2020. Algae polyphenolic compounds and modern antibacterial strategies: Current achievements and immediate prospects. Biomedicines 8(9): 342.

Bishoyi, A.K., C.R. Sahoo and R.N. Padhy. 2023. Recent progression of cyanobacteria and their pharmaceutical utility: An update. Journal of Biomolecular Structure and Dynamics 41(9): 4219–4252.

Cichoński, J. and G. Chrzanowski. 2022. Microalgae as a source of valuable phenolic compounds and carotenoids. Molecules 27(24): 8852.

Jahromi, S.G. 2019. Extraction techniques of phenolic compounds from plants. Plant Physiological Aspects of Phenolic Compounds, 1–18.

Jimenez-Lopez, C., A.G. Pereira, C. Lourenço-Lopes, P. García-Oliveira, L. Cassani, M. Fraga-Corral et al. 2021. Main bioactive phenolic compounds in marine algae and their mechanisms of action supporting potential health benefits. Food Chemistry 341: 128262.

Kafkas, N.E., M. Kosar, A.T. Öz and A.E. Mitchell. 2018. Advanced analytical methods for phenolics in fruits. Journal of Food Quality, 1–6.

Kumar, N. and N. Goel. 2019. Phenolic acids: natural versatile molecules with promising therapeutic applications. Biotechnol. Rep. 24: e00370.

Lomartire, S., J. Cotas, D. Pacheco, J.C. Marques, L. Pereira and A.M. Gonçalves. 2021. Environmental impact on seaweed phenolic production and activity: An important step for compound exploitation. Marine Drugs 19(5): 245.

Maheswari, V. and P.A.S. Babu. 2022. Comprehensive reviews on phenolic compounds from Phaeophyceae as potential therapeutic agent. Journal of Applied Biology and Biotechnology 10(5): 14–21.

Mandhata, C.P., A.K. Bishoyi, C.R. Sahoo, S. Maharana and R.N. Padhy. 2023. Insight to biotechnological utility of phycochemicals from cyanobacterium *Anabaena* sp.: An overview. Fitoterapia, 105594.

Mandhata, C.P., C.R. Sahoo and R.N. Padhy. 2022. Biomedical applications of biosynthesized gold nanoparticles from cyanobacteria: An overview. Biological Trace Element Research 200(12): 5307–5327.

Pérez, M.J., E. Falqué and H. Domínguez. 2016. Antimicrobial action of compounds from marine seaweed. Marine Drugs 14(3): 52.

Rocha, D.H., A.M. Seca and D.C. Pinto. 2018. Seaweed secondary metabolites *in vitro* and *in vivo* anticancer activity. Marine Drugs 16(11): 410.

Sahoo, C.R., S. Maharana, C.P. Mandhata, A.K. Bishoyi, S.K. Paidesetty and R.N. Padhy. 2020. Biogenic silver nanoparticle synthesis with cyanobacterium *Chroococcus minutus* isolated from Baliharachandi sea-mouth, Odisha, and *in vitro* antibacterial activity. Saudi Journal of Biological Sciences 27(6): 1580–1586.

Sahoo, C.R., S.K. Paidesetty and R.N. Padhy. 2019. Nornostocine congeners as potential anticancer drugs: An overview. Drug Development Research 80(7): 878–892.

Sahoo, C.R., J. Sahoo, M. Mahapatra, D. Lenka, P.K. Sahu, B. Dehury et al. 2021. Coumarin derivatives as promising antibacterial agent (s). Arabian Journal of Chemistry 14(2): 102922.

Santos, S.A., R. Félix, A.C. Pais, S.M. Rocha and A.J. Silvestre. 2019. The quest for phenolic compounds from macroalgae: A review of extraction and identification methodologies. Biomolecules 9(12): 847.

Sathya, R., N. Kanaga, P. Sankar and S. Jeeva. 2017. Antioxidant properties of phlorotannins from brown seaweed *Cystoseira trinodis* (Forsskål) C. Agardh. Arabian Journal of Chemistry 10: S2608–S2614.

Scania, A.E. and A.R. Chasani. 2021. The anti-bacterial effect of phenolic compounds from three species of marine macroalgae. Biodiversitas Journal of Biological Diversity 22(6).

Sookjitsumran, W., S. Devahastin, A.S. Mujumdar and N. Chiewchan. 2016. Comparative evaluation of microwave-assisted extraction and preheated solvent extraction of bioactive compounds from a plant material: a case study with cabbages. International Journal of Food Science & Technology 51(11): 2440–2449.

Swain, S., S. Bej, C.P. Mandhata, A.K. Bishoyi, C.R. Sahoo and R.N. Padhy. 2023. Recent progression on phytochemistry and pharmacological activities of *Grewia asiatica* L. (Tiliaceae) and traditional uses. South African Journal of Botany 155: 274–287.

Tungmunnithum, D., A. Thongboonyou, A. Pholboon and A. Yangsabai. 2018. Flavonoids and other phenolic compounds from medicinal plants for pharmaceutical and medical aspects: An overview. Medicines 5(3): 93.

Wang, C., C. Tallian, J. Su, R. Vielnascher, C. Silva, A. Cavaco-Paulo et al. 2018. Ultrasound-assisted extraction of hemicellulose and phenolic compounds from bamboo bast fiber powder. PLoS One 13(6): e0197537.

Chapter 17
Current Status of Seaweed Utilization in Cosmeceuticals

Shreya Modi

Introduction

Cosmetics, cosmeceuticals, and nutricosmetics are becoming more and more important commercially for treating various dermatological diseases and enhancing the appearance of the skin (López-Hortas et al. 2021). The term "cosmeceuticals", which stands for "cosmetics and pharmaceuticals", denotes the presence of active substances in certain products (Thomas and Kim 2013). The cosmeceuticals industry, which uses topical cosmetic-pharmaceutical hybrids with bioactive chemicals in its makeup to promote skin health, is one of the fastest expanding in the cosmetics sector (Thiyagarasaiyar et al. 2020). Consumer's preferences towards green and eco-friendly products have drastically increased, which has led to the discovery of natural, safe and effective ingredients for the development of novel skin care products and other cosmeceuticals (Pimentel et al. 2018). Seaweeds are omnipresent, macroscopic, eukaryotic, autotrophic, benthic (at least in some species), and multicellular organisms that are found along all types of beaches, from tropical to polar. As previously said, seaweeds represent the foundation of the marine ecosystem food web since they share many characteristics with terrestrial plants (Gracia et al. 2020). Seaweeds (macroalgae) are photoautotrophic aquatic creatures that live practically everywhere on Earth, especially in coastal regions. They employ photosynthesis to metabolize the energy needed to carry out their physiological functions. They are regarded as primary producers, much like terrestrial plants, because of their tropic level. Algae are non-vascular creatures that do not have several physiological characteristics like terrestrial plants, such as a root, stem, and leaf division or conductive vessels, despite the fact that seaweeds and terrestrial plants share some characteristics, such as the existence of chlorophyll (Pereira et al. 2018). Seaweeds are aquatic, photosynthetic creatures that are taxonomically classified as algae. Rhodophyceae (red algae), Phaeophyceae (brown algae), and Chlorophyceae (green algae) are the three groups based on their pigments. The term "sea vegetables" refers to marine algae, which are also consumed. Seaweeds have been utilized as an alternative treatment for illnesses of the skin since ancient times (Jesumani et al. 2019). Seaweeds constitute a sustainable, environmentally friendly, potentially low cost and renewable resource to produce cosmetics, cosmeceuticals, and

Department of Microbiology & Biotechnology, Gandhinagar Institute of Science, Gandhinagar University, 382721.
Email: shreyamodi20@gmail.com

nutricosmetics (Pereira et al. 2018). Biomass from marine macroalgae is frequently used to make cosmetics-grade components. These ingredients can serve one of three purposes: (i) they can be additives that improve the organoleptic qualities of the product; (ii) they can be used to stabilize and preserve the product; or (iii) they can be bioactive compounds that perform actual cosmetic functions and activities (Bedoux et al. 2014).

The main objectives of this chapter are to:

- Provide an extensive overview of the importance of seaweeds in the fields of cosmeceuticals.
- Examine the function of Seaweed in cosmeceuticals with an emphasis on the advantages of these substances for skincare, such as hydration, anti-aging, skin conditioning.
- Explore the components and their applications of seaweed for the various cosmetic products and cosmeceuticals.

Seaweeds are also known as macro algae that are different form micro-algae (Cyanophyceae). Seaweeds considered as significant marine resource, which generally grow in the shallow waters of sea wherever required substrata are available (HobAllah et al. 2019). As they can be used as gelling, stabilizing and thickening agents, various ranges of sea weeds are used in different industries related to food, confectionary, paper, textile, dairy, and pharmaceuticals. Although several nations produce seaweed, the biggest producers are Chile, China, Korea, and Japan (Ferdouse et al. 2018). There are two methods for producing seaweed: (1) wild catch (from uncontrolled marine systems) and (2) aquaculture (Mac Monagail et al. 2017). The FAO reports that 33.3 million MT of aquatic plants, mostly marine macroalgae, were produced globally in 2018, with 32.4 million MT (about 97.1%) coming from aquaculture and 0.95 million MT (2.9%) through wild capture. Based on their pigments, seaweeds are divided into three groups: brown (Ochrophyta, Phaeophyceae), green (Chlorophyta), and red (Rhodophyta), which contain fucoxanthin, chlorophyll a, chlorophyll b, phycocyanin, and phycoerythrin, respectively. More than 7000 different species of red seaweed are found with brown and green seaweeds coming in second and third with 2030 and 600 species, respectively (Baweja et al. 2016). Brown seaweeds come in a variety of sizes, from large kelp with a length of over 20 m to small seaweeds with a length of 30 to 60 cm. Green seaweeds and red seaweeds both have a size range that ranges from a few centimeters to nearly a meter in length (FAO 2016). Red seaweeds are typically smaller in size (Kumar et al. 2021).

Components of seaweed and their bioactivity

Seaweeds produce both primary and secondary metabolites (Rosa et al. 2019). Secondary metabolites, which are produced under various stress conditions, such as UV radiation exposure, salinity, temperature changes, or environmental poisons, are produced by seaweeds in addition to primary metabolites, which are directly involved in normal growth, development, or reproduction conditions to perform physiological functions. In algae, polysaccharides, proteins, amino acids, and fatty acids are among the primary metabolites. Pigments, phenolic compounds, sterols, vitamins, and other useful chemicals are examples of secondary metabolites produced in seaweed tissues (Nurjanah et al. 2016, Pereira et al. 2018). Among the various classes of seaweed components, Sulphated polysaccharides, peptides (Thiviya et al. 2022), carotenoids, fatty acids, and phytohormones have shown anti-aging and antioxidant effects (Polat et al. 2023). Mycosporine-like amino acids and flavonoids have an anti-photoaging function (Fig. 1).

Polysaccharides

This research focuses on seaweed polysaccharides because of their high yields, sustainability, and unique chemical components. These include cell wall polysaccharides and mucopolysaccharides.

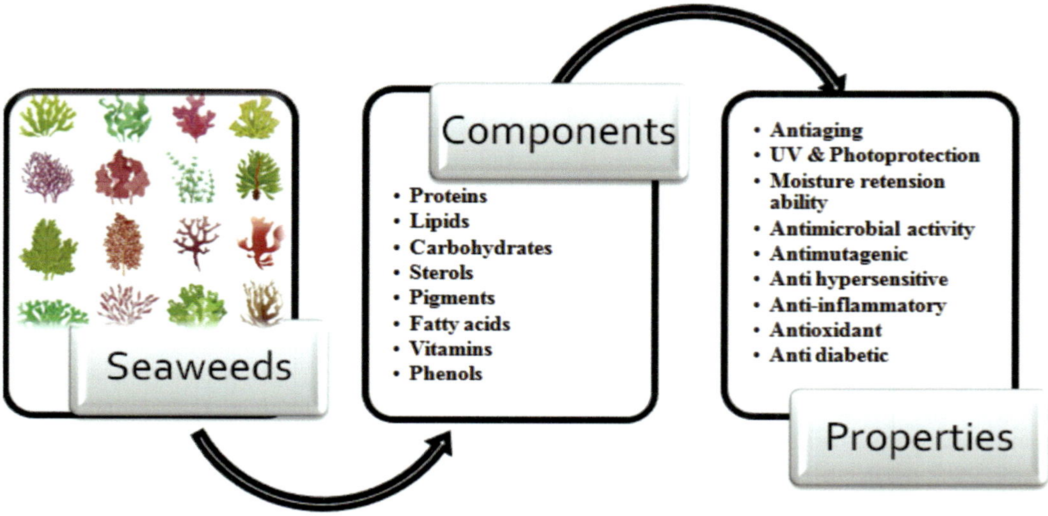

Figure 1. Components of Seaweed and their Properties.

They are particularly rich in sulfurized polysaccharides (SPS), which are widely employed in food, cosmetics, biotechnology, and other industries and have antioxidant, antiviral, and blood coagulation benefits (Zhang et al. 2022). They are unique to each type of algae and are found in brown alginate, laminaran, and fucoidan; green ulvan; and red agar, where carrageenan is the most significant. More and more research is being done on polysaccharides because of their biofunctional and physicochemical properties (Shanura Fernando et al. 2018). The main component of brown seaweeds is alginate polysaccharide. This compound is highly stable and can be used as a thickening agent and for gelling purposes in different cosmetic products. Alginates are commonly used as thickeners, protective colloids, or emulsion stabilizers in drugs and cosmetics due to their chelating properties. These substances are beneficial for hand lotions and jellies, ointment bases, pomades and other hair preparations, greaseless creams, dentifrices, and other products. In order to prevent occupational dermatitis, alginates may also be employed in the formulation of skin-protective or "barrier" lotions. Such creams make flexible films that adhere better to the skin and are ideal as an element in face packs or beauty masks (Fabrowska et al. 2015). Carrageenans generally extracted from red seaweeds, *Porphyra haitanensis, Gracilaria chouae, Graccilaria blogdgettii* serve as antioxidant (Khan et al. 2021), antitumor, antiaging, as thickening agents. They are used as radiation protectors by the different cosmeceuticals industries (Ha et al. 2022). A number of seaweed species including *Ascophyllum nodosum, Chnoospora minima, Ecklonia maxima, Hizikia fusiforme, Saccharina japonica, Sargassum hemiphyllum, Sargassum horneri, Sargassum polycystum, Sargassum vachellianum* are a good source of Fucoidans polysaccharide which has been reported as applicable for photoaging inhibition, to minimize elastage activity and as having antioxidant, anti-inflammatory properties. They are used for collagenase and elastage inhibition activities and as skin-whitening agents in the different products of cosmetics. The removal of free radicals, the decrease of inflammation, wrinkles, allergies, and the sensitive skin reaction are just a few of the ways that fucoidan has been found to protect skin and hair. The elasticity, firmness, and brightness of the skin as well as the protection, hair development, stiffness, cleanliness, and gloss have also been said to be enhanced by fucoidan (Narayanaswamy et al. 2013, Usov and Zelinsky 2013). Limanaran extracted from the *Saccharina longicruris* can be used for the reconstruction of dermis, and antioxidant and as anti-inflammation compounds.

Proteins, peptides and amino acids

On a dry weight basis, seaweeds can contain up to 47% of protein, which is comparable to the protein level of conventional protein sources including meat, egg, soybean, and milk (Naseri et al. 2020). Seaweed derived proteins, peptides and amino acids represent a number of health advantages due to their advanced antioxidant properties (Pangestuti and Kim 2018). Amino acids derived from *Ulva lactuca* have been used as antiaging, antiwrinkle and to enhance collagen formation (Leandro et al. 2019). *Pelvetia canaliculata* is a good source of amino acids which can be applied for proteoglycan' synthesis, collagen formation and also as an antioxidant (Pereira 2020, Hupel et al. 2011). Lee et al. (2015) derived bioactive peptide from pyropia yezoensis (PPY1) which can efficiently inhibit release of LPS-stimulated nitric oxide, ROS. PPY1 was found to reduce release of pro-inflammatory cytokines which significantly exhibit anti-inflammatory properties. The anti-photoaging property of porphyra-334 derived from Porphyra yezoensis was examined. Porphyra-334 supressed ROS production and the expression of metalloproteinase (MMPs) and UVA radiation. As porphyra-334 was found to be responsible for the increased levels of Extracellular matrix components such as procollagen, type I collagen and elastin, it can be also applicable to production of cosmetics (Ryu et al. 2014).

Phenolics and terpenoids

Phenolic compounds are secondary plant metabolites that have one or more aromatic rings and one or more connected hydroxyl (–OH) groups as part of their fundamental structure. They are created by plants as a component of their defense mechanisms. They are structurally less complex than terrestrial tannins (Fernando et al. 2022). Phlorotannins, which are found in seaweed, and carbohydrates such alginic acid and fucoidan are typical bioactive elements of *E. cava* ssp. *Stolonifera*. *E. cava* ssp. *Stolonifera* represented variation in their antiallergic properties with seasonal variation. Anti-allergic activity was found higher in the summer-autumn period as compared to the winter-spring period.

Vitamins

Vitamins found in algae operate as a natural anti-aging compound, activate the skin, protect it from environmental impact, and fight free radicals. While chlorophyll functions like hemoglobin in the skin's oxygen supply (Fitton et al. 2015). Many cosmeceuticals and skin care products have vitamins as prominent ingredients. These vitamins play an important role in treating skin conditions like acne, hyperpigmentation, and photoaging, protecting from UV, deactivating free radicals, and improving skin moisture retention levels of the skin (Desai and Mallya 2021). The most crucial and clinically proven vitamins for treating and preventing skin photoaging are vitamins A, C, E, K, and vitamin complex B, and vitamins A, B, C, and E are the most prevalent vitamins found in the various types of macroalgae (Pimentel et al. 2018, Peñalver et al. 2020). Ed algae and other species are good sources of vitamin B12, which has anti-aging characteristics and is necessary for healthy hair, nails, and skin (Watanabe et al 2014, Bito et al. 2018).

Pigments

A wide range of pigments found in seaweeds absorb light to facilitate photosynthesis. Chlorophylls a, b, and carotenoids, a pigment found in plants, are present in green algae. Red algae contains carotenoids, primarily—carotene, lutein, and zeaxanthin, as well as photosynthetic pigments including chlorophyll a and phycobilins like R-phycocyanin and R-phycoerythrin. The carotenoids, fucoxanthin, and chlorophylls a and c are among the pigments found in brown algae. Pigment protect the cells from ultraviolet light. Chlorophyta (green algae) contain carotenoid pigments chlorophyll a, chlorophyll b, and zeaxanthin; Rhodophyta (red algae) contain carotenoid pigments carotene, lutein, and zeaxanthin; and Phaeophyceae (brown algae) contain carotenoid pigments fucoxanthin, chlorophyll a, chlorophyll c (Kalasariya et al. 2022).

Seaweed based bioactive compounds in cosmeceuticals

Secondary metabolite with a variety of functional activities, bioactive compounds obtained from different seaweeds can be used to create cosmeceuticals. Now a days, biotechnology techniques make it possible to produce high-quality, contamination-free algae cultures, which is essential for the cosmetics sector (Michalak et al. 2020). Various kinds of seaweed extracts are incorporated as the active ingredient. They can dispense a variety of activities, such as photoprotective actions, moisturizing, antioxidant, anti-melanogenic, anti-allergic, anti-inflammatory, anti-acne, anti-wrinkling, antimicrobial, antiaging, whitening, etc. (Senevirathne and Kim 2018, Salehi et al. 2019). Compounds derived from macroalgae are becoming more and more popular in skincare products since they are renewable, abundant, natural, and biocompatible. They are becoming more widely recognized as potentially sustainable resources that are more suited for biorefinery processes because of their high biomass yields, quick growth rates, and culture procedure (Ashokkumar et al. 2022). The overproduction of free radicals and reactive oxygen species (ROS) caused by excessive sun exposure harms lipids, proteins, and deoxyribonucleic acids. The expression of collagenase, a protease that breaks down collagen, is also increased by excessive sun exposure, which can cause photoaging and skin wrinkles (López-Camarillo et al. 2011). The cosmetics industries have also placed a strong emphasis on using substances derived from plants, algae, microorganisms, and their metabolites. The marine environment has many different species with several bioactive chemicals that are in high demand. The active chemical with a wide range of uses in many fields is mostly

Table 1. List of Seaweeds having Tyrosinase inhibition activity.

Name of the seaweed	Active compound	Extract	Mode of action	References
Ecklonia stolonifera Okamura	974-A, phlorofucofuroeckol-A and eckol	Ethyl acetate	Reduce melanin content; tyrosinase activity	Manandhar et al. 2019
I.okamurae	carmalol derivative, diphlorethohydroxycarmalol (DPHC)	Ethyl acetate	Reduce ROS activity caused due to UV radiation	Heo et al. 2010
Sargassum polycystum		Ethanol & Hexane fraction (SPHF)	Inhibition of Melanogenesis	Chan et al. 2011
Chnoospora minima (CMF) and *Sargassum polycystum* (SPF)	Fucoidans	-	2,2-diphenyl-1-picrylhydrazyl (DPPH) and alkyl radical-scavenging activities, anti-inflammatory effects	Shanura Fernando et al. 2018
Leathesia difformis	Fucoxanthin	Ethanol; ethyl acetate and Water	Inhibition of Melanogenesis by the decreased gene expression of Tyr, Trp-1 snd Dct via inhibition of Mitf expression	Seo et al. 2019
Sargassum plagyophyllum & *Eucheuma cottonii*		Methanol	Oxidation inhibition of L-DOPA to DOPAAquinone	Dolorosa et al. 2020
Bangia fusco-purpurea	Phytol	95% v/v ethanol	Reduces browning index of fruits and vegetables	Wang et al. 2023

sourced from macroalgae (Berthon et al. 2017). The most often utilized skin whitening include hydroquinone, arbutin, kojic acid, ascorbic acid, and retinoic acid, but there have been some claims that these substances can be harmful and allergic to the skin. In this context, there is a growing need to create whitening agents that are risk-free, efficient, and have few adverse effects.

Collagenase and elastage inhibition activity

Reactive Oxygen Species (ROS) levels rise as a result of environmental stressors including pollution and Ultraviolet Radiation (UVR), which accelerates the ageing process of the skin and its clinical manifestations like wrinkle formation. Reductions in collagen synthesis and other enzymatic activities, such as Matrix Metalloproteinases (MMPs), which break down collagen structure in the dermis, and Tissue Inhibitor of Metalloproteinases (TIMPs), which block MMP activity, are associated with ageing of the skin (Berthon et al. 2017).

Table 2. List of Seaweeds which are involved in reducing ROS activity.

Name of the seaweed	Active compound	Mode of action	References
Hizikia fusiforme	celluclast	Reduce ROS	Wang et al. 2017
Sargassum horneri	Sargachromanol	Suppress ROS formation, collagnase expression, membrane protein expression	Kim et al. 2013
Gelidium amansii	SOD1, SOD2, Gpx, and GR enzymes	Decrease mRNA levels of ROS generator	Seo et al. 2012
Coralline pilulifera	Phenolic compound	Reduce oxidative stress	Ryu et al. 2009

Hyaluronidase inhibition

It is well known that algal sulfated polysaccharides work well as hyaluronidase inhibitors. The extracellular matrix of connective tissue, which is dispersed throughout the human body's organs and bodily fluids, is primarily made up of hyaluronic acid. Due to its high hydrophilicity, hyaluronic acid plays a crucial role in many key physiological and pathological processes, including wrinkle reduction and preserving skin smoothness and moisture (Liyanaarachchi et al. 2018). Inflammatory disorders, tumor invasiveness, metastasis, invasive breast adenocarcinoma, metastatic human melanoma, colon carcinoma, and glioblastoma cell lines are all affected by hyaluronidase, an enzyme for hydrolyzing hyaluronic acid. Hyaluronidase inhibitory activity of sulfated polysaccharide (SP) is derived from *Caulerpa lentillifera*. Results reported that SP, majorly contain Gal, Glc, Xyl, and Man, which were discovered to successfully suppress hyaluronidase activity. SP with IC_{50} of 163 µg/mL appears to allosterically inhibit hyaluronidase activity (Awanthi et al. 2023).

UV and photoprotection activity

Chronic exposure to UV radiation results in skin inflammation, photoaging, hyperpigmentation, and skin cancer by harming intracellular biomolecules (proteins, lipids, polysaccharides, and nucleic acids) (Talero et al. 2015, Berthon et al. 2017). The biological functions of marine algae photoprotective compounds include UV (ultraviolet) absorption, antioxidant, matrix-metalloproteinase inhibitor, anti-aging, and immunomodulatory actions (Pangestuti et al. 2018). Malaysian Padina australis showed protective properties against UVB-induced cytotoxicity in HaCaT human keratinocytes. The high association between phytochemical concentration and antioxidant capacity raises the possibility that phenolic or flavonoid chemicals may have had a role in the antioxidant activities (Thiyagarasaiyar et al. 2021). It has been discovered that brown algae contain phlorotannins (1,3,5-trihydroxybenzene), a special class of polyphenolic compounds that have the ability to fend against and repair UV-induced DNA damage. These phloroglucinol polymers, which are only found in marine seaweeds, are said to have photoprotective benefits because they absorb UV rays. Phlorotannin concentration and the reduction of DNA damage are clearly correlated in several species of brown algae that have been studied (Kumar et al. 2022, Erpel et al. 2020, Catarino et al. 2021). Secondary metabolites derived from seaweeds have antioxidant action and function as UV filters/sunscreen are frequently utilised in cosmetics. Brown algae, such as *Fucus vesiculosus var. alternans C. Agardh, Sargassum* sp., and *Turbinaria conoides f. laticuspidata* include polysaccharides such as fucoidan, laminarin, and alginate that have strong antioxidant properties (Wang et al. 2017).

Moisture retention ability

Moisture retention is crucial for skincare since it enhances skin texture and condition, making it appear younger and healthier. Extracts from the *Undaria pinnatifida* (Harvey) *Suringar, Codium tomentosum Stackhouse, Durvillea antarctica (Chamisso) Hariot, Cladosiphon okamuranus Tokida, A. nodosum (Linnaeus) Le Jolis, Pediastrum duplex Meyen, and Polysiphonia lanosa represented moisture retension ability* (*Linnaeus* (Jesumani et al. 2019). Seaweeds contain several molecules, such as polysaccharides, fatty acids (sophorolipids, rhamnolipids, and mannosylerythritol), and proteins, algae extract from *Laminaria, Nannochloropsis*, seaweed Undaria pinnatifida, and microalgae *Thalassiosira nannochloropsis.* Hence, they have moisturizing properties and are found to be effective in hydrating skin and restoring trans epidermal water loss (TEWL) (Mourelle et al. 2017). Wang et al. (2013) extracted seven polysaccharides from five different algae—namely brown alga *Saccharina japonica*, red alga *Porphyra haitanensis* and three different green algae *Codium fragile, Enteromorpha linza* and *Bryopsis plumose*—that have properties of moisture absorption and moisture retention. The presence of sulfated group was found to be the main active site for moisture-absorption and moisture-retention abilities.

Antimicrobial activity

A variety of phytochemicals and minerals derived from marine algae, including proteins, vitamins, polyunsaturated fatty acids, alkaloids, terpenoids, polyamides, and polyphenols, may have uses in cosmetic products and pharmaceuticals (Janarthanan and Senthil Kumar 2018). Seaweeds have the ability to produce a variety of distinctive chemical compounds with a wide range of biological functions, such as antibiotics, anticoagulants, antivirals, laxatives, anti-ulcer medication (Schwartsmann et al. 2001). *Halomonas venusta* isolated from Marine sponge *Callyspongia* sp. can produce melanin. A total of 4.92 mg/ml of melanin was produced by the improved fermented media supplemented with 1% tyrosine. Concentrations from seaweed *Gelidium spinosum* were used to manufacture and fortify the cream that included melanin. The growth of *S. aureus* (MTCC 96) and *S. pyogenes* (MTCC 442) was successfully inhibited by melanin and seaweed concentrate, which were discovered to be rich in antioxidant activities (Poulose et al. 2020). Freitas et al. 2020 assessed the dermo-cosmetic activity of different fractions derived from brown seaweed *Fucus spiralis*. The F2 and F7 fractions against *C. acnes* and *M. furfur*, respectively, exhibited the strongest antimicrobial action, which inhibited the microorganisms' growth by 50–60%. However, the proliferation of the microorganisms induced by fractions F5 (*M. furfur*), F7 and F11 (*C. acnes*), and F10 (*S. epidermidis*) was found to be stimulated. The enriched phlorotannin fraction F10 from the *F. spiralis* samples used in this study, which has a wide range of bioactivities, has shown to be suitable for additional dermo-cosmetic applications through its inclusion in skin formulations with antioxidant, anti-enzymatic, and photoprotective properties, in addition to its role in maintaining skin microbiota homeostasis. Another study, proved that the extract of three weeds namely *A. anceps, C. officinalis* and *S. filipendula* possess an effective antibacterial activity against *S. typhimurium*,

S. aureus and *E. coli* pathogenic bacteria and show antiviral activity against CVB3 and RV infections and can be used as pretreatment for their infections by attaching to the capsids to prevent the connection with the cell receptors and thereby stop their penetration into host cells. *A. anceps*, *C. officinalis*, and *S. filipendula* organic extracts each inhibit *E. coli* cells to varying degrees (57.1%, 85.7%, and 91.4%, respectively). *Staphylococcus aureus* cells are completely inhibited by the *C. officinalis* extract at the greatest concentration (100 g/mL), while *S. filipendula* and *A. anceps* extracts only inhibit 82.5% and 75% of *S. aureus*, respectively (Mofeed et al. 2022).

Seaweeds for anti-aging

By comparing the composition of products marketed in 2011 with those introduced or reformulated in 2018, the trends in the usage of marine components in a panel of anti-aging formulations commercialized in the Portuguese market were examined. While only 6% of the 190 anti-aging cosmetics in use in 2011 were from marine sources, this number increased significantly by 2018 to 33% of the 103 newly released anti-aging products under study (Resende et al. 2021). Mirata et al. (2023) concluded the potential of *E. amentacea* species derivatives having potential components for counteracting sunburn symptoms and also have anti-aging properties. The hydroalcoholic apices extracts showed the greatest therapeutic potential. They reduced oxidative stress and the release of pro-inflammatory cytokines, which are generally generated after sunburns in UV-exposed HaCaT keratinocytes and L929 fibroblasts. They also inhibited toxicity. The extracts also demonstrated anti-tyrosinase and anti-hydrolytic skin enzyme activity, which counteracted the collagenase and hyaluronidase degrading activities and may have slowed the development of wrinkles and uneven pigmentation in ageing skin.

Commercial manipulation of seaweed for the different products

According to type, the carotenoids category is anticipated to hold the biggest market share for algae-based cosmetics in 2022. The significant market share of this sector is related to the increasing popularity of algal carotenoids in cosmetic products, such as beta carotene, astaxanthin, and lutein. Due to its capacity to defend against harm brought on by UVA light, beta carotene produced from algae is utilized to prepare skincare formulas. Additionally, beta carotene-containing cosmetics improve the appearance of skin by reducing oxidative damage. Repêchage Professional Skin Care

Table 3. List of Seaweeds derived commercial products.

Seaweed	Products	Company name	References
Gracilaria sp.	Hrdogel soap Sealaria® Facial Mask Balinique®, Hydrating cream Thalasso®	Kfar Hess, Isreal Miami, FL, USA Roa Graf, Stamford, CT, USA	Torres et al. 2019
Ecklonia cava	ECKLEXT®BG	NOF group, Tokyo, Japan	Morais et al. 2021
Ulva lactuca	Chlorofiltrat®Ulva HG	CODIF, Saint-Malo, France	Morais et al. 2021
Porphyra umblicalis (red seaweed)	Sunscreen products Helionori® Helioguard® 365 Aethic Sovee®	Gelyma, Marseille, France Mibelle Biochemistry, Buchs, Switzerland AETHIC®, London, UK.	Siezen 2011 Morais et al. 2021
Chondrus crispus	Gelling, thickener and stabilizer agent Gelcarin®	Dupont Nutrition and Biosciences, Wilmington, DEL, USA	Morais et al. 2021

has introduced the Repêchage SeaSmooth Artisan Seaweed Wax line of products. The secret element in most Repechage treatments that transforms the depilation procedure is seaweed. Repêchage SeaSmooth Artisan Seaweed Wax is a smooth, soft, and creamy wax that is infused with a variety of sea plants that provide vitamins, minerals, and trace elements. The bioactive components of seaweeds are also influential in developing various hair care products. The first thalassotherapy hair line by Revlon Professionals, EksperienceTM, addresses a variety of hair-related issues, including hair loss, sensitive scalp, oily scalp, etc. The marine algae scalp treatment by Revlon Thalassotherapy combines marine algae with natural extracts to produce active mud that contains Spirulina algae and Vitamin B-stimulating essences. Some commercial products prepared using seaweeds are listed in Table 3.

Reported benefits of seaweeds for skin

A new anti-aging active component called Stemlan-173 has just been developed by the Japanese cosmetics business Shiseido. It is patented (WO2018074606A1), and has potential to enhances skin texture, rejuvenate skin, and prevent the loss of epidermal stem cells. According to Shiseido, the algae extract from green algae, red algae and brown algae, acts as an active ingredient to enhance the production of laminin-511, which decreases with aging (Mago et al. 2023). Amorepacific Corporation, in two of its recent patents (KR1972071B1, KR1901859B1), has disclosed the incorporation of marine algae extracts in cosmetic compositions. These extracts include green algae, rhodophyta, brown algae, microalgae, horsetail kelp, Spirulina, *Sargassum fulvellum*, *Codium fragile*, and *Undaria pinnatifida*. The compositions aim to improve skin tone, prevent skin aging, enhance skin elasticity, mask skin wrinkles, and protect the skin from UV rays. In its patent US9717932B2 titled "Marine extracts and biofermentations for use in cosmetics", BASF discloses that the biofermentation of marine algae, specifically *Sarcodiotheca gaudichaudi*, is useful as an anti-aging skin care active ingredient and is especially useful when combined with extracts of red algae, such as *Chondrus crispus* (Guleri and Tiwari 2020). Chung Heon (2009), created a mask pack that uses fucoidan (KR1841099B1). Washing brown algae, putting it in water, stirring to extract alginic acid and fucoidan polysaccharide, filtering the mixture to obtain purified fucoidan-containing liquid, adding acetic acid, heating to neutralize, mixing the resultant neutralized fucoidan with culture medium of *Microbacterium* sp. SC092, fermenting, drying, and pulverizing are the steps used to make fucoidan powder (Mago et al. 2023).

Challenges and future prospect

A number of seaweeds have been studied for their cosmetic properties, but still a majority of them remain unexplored. There is an urgent need for the standardization of economical and successful methods to extract bioactive substances from seaweeds with increased activity and production. (Jesumani et al. 2019). Understanding the life cycle and bioavailability of seaweed is very crucial. Seaweed needs to be thoroughly studied in order to be used as a substitute and to support its ecological use as a cosmetics, fertilizers and other purposes (de Sosa et al. 2023). It needs to be advanced approaches to create a novel seaweed species approval process that is safer, quicker, and less expensive. Apart from these, the safety problems and molecular mechanism of these compounds hold great significance for upcoming challenges in the cosmetics sector. Seasonal reliance on Indian seaweed culture is a significant obstacle that must be overcome. Sometime strong tidal fluctuations and high seawater turbulence caused by the occurrence of cyclones and typhoons affect cultivation and seed bank preservation during this period (Ganesan et al. 2019).

Table 4. List of skincare products available commercially.

Skincare Brand	Product	Key ingredient	Product link
si SKIN organics	Luminous Brightening Elixir	Red & Brown Macroalgae	si SKIN Organics
	Spot-Less Pigment Corrector	*Gelidium Cartilagineum, Laminaria Saccharina Extract*	https://www.cremedelamer.com/product/20653/30929/face/exfoliators-masks/the-intensive-revitalizing-mask#/sku/55859
Le Mer	Intensive Revitalizing mask	*Enteromorpha compressa extract*	https://la-mer.com/en/med-anti-spot/regulating-fluid
	Lifting and Firming mask	*Chlorella Vulgaris* Extract, *Corallina Officinalis*	https://www.lamer.eu/nl/en/product/9088/30156/masks/the-lifting-and-firming-mask/visibly-lifts-tightens-firms/anti-aging-face-mask#/sku/54773
Osea Malibu	Undaria Algae Body oil	*Undaria* seaweed	https://oseamalibu.com/products/undaria-algae-oil
	Undaria Body Polish	*Undaria* seaweed	https://oseamalibu.com/products/undaria-cleansing-body-polish
Skinceuticals	Daily moisture		https://www.skinceuticals.com/skincare/moisturizers
	Emollience	*Laminaria ochroleuca, Chlorella vulgaris*	https://www.skinceuticals.com/skincare/moisturizers/emollience/S26.html
PeterThomasRoth	MegaiRich Intensive Anti-Aging Cellular Crème	*Macrocystis Pyrifera*	https://www.peterthomasroth.com/mega-rich-intensive-anti-aging-creme---super-size-1806243.html
	Irish Moor Mud Mask	*Hijiki seaweed*	https://www.peterthomasroth.com/irish-moor-mud-mask-1301015.html
UNA	Day cream		https://unabiologicals.com/products/luxe-face-cream
	Ultra Rich Eye Cream		
AMA Sea Beauty	DeepSea Serum	*Mazzaella Splendens* Extract, *Macrocystis Pyrifera* Extract, *Chondrus Chrispus* Extract, *Algin*	https://www.gcimagazine.com/brands-products/skin-care/news/21867292/ama-sea-beauty-launches-marine-based-skin-care
	SeaScrub		https://www.gopicky.com/product/48070/ama-sea-beauty-seascrub
Repechage	Opti-Firm Renewable Complex Night Cream	*Ulva Lactuca, Laminaria Digitata, Chondrus Crispus (Carrageenan)* - Red Algae	https://www.repechage.com/products/opti-firm-renewal-complex
	Mineral Face Shield	*Laminaria digitata* - seaweed	https://www.repechage.com/products/face-shield
VOYA	Pearlesque hydrating Moisturiser	*Laminaria Digitata*	https://www.voya.ie/products/pearlesque-hydrating-moisturiser
	Maskerade Soothing Facial Mask	*Cystorseira amentacea*	https://la-mer.com/en/med-anti-spot/regulating-fluid

Summary of the chapter

Seaweeds are a major source of some precious components for the formulation of various range of pharmaceutical and cosmeceuticals products due to the advanced properties that they confer. It is possible to extract a variety of physiologically active substances from seaweeds, including vitamins, minerals, amino acids, carbohydrates, and lipids, and create different innovative cosmeceuticals products. Seaweed fractions can also be used for technical features, such as thickening, gelling, emulsifying, texturizing, or moistening to enhance and develop cohesive matrices. Furthermore, the availability of valorizing industrial waste streams and different varieties of algal blooms makes them an attractive, low cost, raw and renewable material. In the cosmetics industry, the active ingredients from seaweeds could be employed as an antioxidant, antibacterial whitening agent, anti-aging, anti-acne, and for moisturizing. For the commercial use of seaweed bioactive, quality control and standardization are necessary, just like with other components and uses.

References

Ashokkumar, V., S. Jayashree, G. Kumar, S.A. Sharmili, M. Gopal, S. Dharmaraj et al. 2022. Recent developments in biorefining of macroalgae metabolites and their industrial applications-A circular economy approach. Bioresource Technology 359: 127235.

Awanthi, M.G., S. Nagamoto, H. Oku, K. Kitahara and T. Konishi. 2023. Hyaluronidase-inhibiting Polysaccharide from *Caulerpa lentillifera*. Journal of Applied Glycoscience 70(1): 1–7.

Baweja, P., S. Kumar, D. Sahoo and I. Levine. 2016. Biology of seaweeds. In Seaweed in health and disease prevention (pp. 41–106). Academic Press.

Bedoux, G., K. Hardouin, A.S. Burlot and N. Bourgougnon. 2014. Bioactive components from seaweeds: Cosmetic applications and future development. In Advances in Botanical Research. Academic Press 71: 345–378.

Berthon, J.Y., R. Nachat-Kappes, M. Bey, J.P. Cadoret, I. Renimel and E. Filaire. 2017. Marine algae as attractive source to skin care. Free Radical Research 51(6): 555–567.

Bito, T., Y. Tanioka and F. Watanabe. 2018. Characterization of vitamin B 12 compounds from marine foods. Fisheries Science 84: 747–755.

Catarino, M.D., S.J. Amarante, N. Mateus, A.M. Silva and S.M. Cardoso. 2021. Brown algae phlorotannins: A marine alternative to break the oxidative stress, inflammation and cancer network. Foods 10(7): 1478.

Chan, Y.Y., K.H. Kim and S.H. Cheah. 2011. Inhibitory effects of *Sargassum polycystum* on tyrosinase activity and melanin formation in B16F10 murine melanoma cells. Journal of Ethnopharmacology 137(3): 1183–1188.

Chung Heon. 2009. Fucoidan-based mask pack (Patent No. KR1841099B1). Korean Intellectual Property Ofce.

de Sosa, L L., C.M. Navarro-Fernández, M. Panettieri, P. Madejón, A. Pérez-de-Mora and E. Madejón. 2023. Application of seaweed and pruning residue as organic fertilizer to increase soil fertility and vine productivity. Soil Use and Management 39(2): 794–804.

Desai, J. and R. Mallya. 2021. A review on novel topical formulations of vitamins. Journal of Reports in Pharmaceutical Sciences 10(2): 159.

Dolorosa, M.T., S. Purwaningsih, E. Anwar and T. Hidayat. 2019, May. Tyrosinase inhibitory activity of *Sargassum plagyophyllum* and *Eucheuma cottonii* methanol extracts. In IOP Conference Series: Earth and Environmental Science. IOP Publishing 278(1): 012020.

Erpel, F., R. Mateos, J. Perez-Jimenez and J.R. Pérez-Correa. 2020. Phlorotannins: From isolation and structural characterization, to the evaluation of their antidiabetic and anticancer potential. Food Research International 137: 109589.

Fabrowska, J., B. Łęska, G. Schroeder, B. Messyasz and M. Pikosz. 2015. Biomass and extracts of algae as material for cosmetics. pp. 681–706. In Marine algae extracts: processes, products, and applications.

FAO. 2016. Fishery and aquaculture statistics. Food and Agriculture Organization of the United Nations, 11.

Ferdouse, F., S.L. Holdt, R. Smith, P. Murúa and Z. Yang. 2018. The global status of seaweed production, trade and utilization. Globefish Research Programme, 124(I).

Fernando, I.P., W. Lee and G. Ahn. 2022. Marine algal flavonoids and phlorotannins; an intriguing frontier of biofunctional secondary metabolites. Critical Reviews in Biotechnology 42(1): 23–45.

Fitton, J.H., G. Dell'Acqua, V.A. Gardiner, S.S. Karpiniec, D.N. Stringer and E. Davis. 2015. Topical benefits of two fucoidan-rich extracts from marine macroalgae. Cosmetics 2(2): 66–81.

Freitas, R., A. Martins, J. Silva, C. Alves, S. Pinteus, J. Alves et al. 2020. Highlighting the biological potential of the brown seaweed *Fucus spiralis* for skin applications. Antioxidants 9(7): 611.

Ganesan, M., N. Trivedi, V. Gupta, S.V. Madhav, C. Radhakrishna Reddy and I.A. Levine. 2019. Seaweed resources in India– current status of diversity and cultivation: prospects and challenges. Botanica Marina 62(5): 463–482.

García-Poza, S., A. Leandro, C. Cotas, J. Cotas, J.C. Marques, L. Pereira et al. 2020. The evolution road of seaweed aquaculture: Cultivation technologies and the industry 4.0. International Journal of Environmental Research and Public Health. 2020 Sep; 17(18): 6528.

GlobeNewswire. (2016, January 4). Frutarom Expands Its Natural Ingredients Activity and Enhances Its Capabilities in Algae-Based Biotech Products. GlobeNewswire. https://www.globenewswire.com/news-release/2016/01/04/798934/0/en/Frutarom-Expands-Its-Natural-Ingredients-Activity-and-Enhances-Its-Capabilities-in-Algae-Based-Biotech-Products.html

Guleri, S. and A. Tiwari. 2020. Algae and ageing. In Microalgae biotechnology for food, health and high value products (pp. 267–293).

Ha, H.T., D.X. Cuong, L.H. Thuy, P.T. Thuan, D.T. Tuyen, V.T. Mo et al. 2022. Carrageenan of red algae Eucheuma gelatinae: Extraction, antioxidant activity, rheology characteristics, and physicochemistry characterization. Molecules 27(4): 1268.

Heo, S.J., S.C. Ko, S.M. Kang, S.H. Cha, S.H. Lee, D.H. Kang et al. 2010. Inhibitory effect of diphlorethohydroxycarmalol on melanogenesis and its protective effect against UV-B radiation-induced cell damage. Food and Chemical Toxicology 48(5): 1355–1361.

HobAllah, E., M. Saber and A. Zaghloul. 2019. Commercial bio-products from algal biomass. International Journal of Environmental Pollution and Environmental Modelling 2(2): 90–104.

Hupel, M., C. Lecointre, A. Meudec, N. Poupart and E.A. Gall. 2011. Comparison of photoprotective responses to UV radiation in the brown seaweed *Pelvetia canaliculata* and the marine angiosperm *Salicornia ramosissima*. Journal of Experimental Marine Biology and Ecology 401(1-2): 36–47.

Janarthanan, M. and M. Senthil Kumar. 2018. The properties of bioactive substances obtained from seaweeds and their applications in textile industries. Journal of Industrial Textiles 48(1): 361–401.

Jesumani, V., H. Du, M. Aslam, P. Pei and N. Huang. 2019. Potential use of seaweed bioactive compounds in skincare—A review. Marine Drugs 17(12): 688.

Kalasariya, H.S., L. Pereira and N.B. Patel. 2022. Pioneering role of marine macroalgae in cosmeceuticals. Phycology 2(1): 172–203.

Khan, B.M., L.X. Zheng, W. Khan, A.A. Shah, Y. Liu and K.L. Cheong. 2021. Antioxidant potential of physicochemically characterized *Gracilaria blodgettii* sulfated polysaccharides. Polymers 13(3): 442.

Kim, J.A., B.N. Ahn, C.S. Kong and S.K. Kim. 2013. The chromene sargachromanol E inhibits ultraviolet A-induced ageing of skin in human dermal fibroblasts. British Journal of Dermatology 168(5): 968–976.

Kumar, R.S., S. Gunaseelan, R. Sangeetha, M. Arunkumar, G. Shakambari, B. Ashokkumar et al. 2022. Bioactive phlorotannin as autophagy modulator in cervical cancer cells and advanced glycation end products inhibitor in glucotoxic C. elegans. Algal Research 67: 102820.

Kumar, Y., A. Tarafdar and P.C. Badgujar. 2021. Seaweed as a source of natural antioxidants: Therapeutic activity and food applications. Journal of Food Quality, 1–7.

Leandro, A., L. Pereira and A.M. Gonçalves. 2019. Diverse applications of marine macroalgae. Marine Drugs 18(1): 17.

Lee, H.A., I.H. Kim and T.J. Nam. 2015. Bioactive peptide from *Pyropia yezoensis* and its anti-inflammatory activities. International Journal of Molecular Medicine 36(6): 1701–1706.

Liyanaarachchi, G.D., J.K. Samarasekera, K.R. Mahanama and K.D. Hemalal. 2018. Tyrosinase, elastase, hyaluronidase, inhibitory and antioxidant activity of Sri Lankan medicinal plants for novel cosmeceuticals. Industrial Crops and Products 111: 597–605.

López-Camarillo, C., E.A. Ocampo, M.L. Casamichana, C. Pérez-Plasencia, E. Álvarez-Sánchez and L.A. Marchat. 2011. Protein kinases and transcription factors activation in response to UV-radiation of skin: Implications for carcinogenesis. International Journal of Molecular Sciences 13(1): 142–172.

López-Hortas, L., N. Flórez-Fernández, M.D. Torres, T. Ferreira-Anta, M.P. Casas, E.M. Balboa et al. 2021. Applying seaweed compounds in cosmetics, cosmeceuticals and nutricosmetics. Marine Drugs 19(10): 552.

Mac Monagail, M., L. Cornish, L. Morrison, R. Araújo and A.T. Critchley. 2017. Sustainable harvesting of wild seaweed resources. European Journal of Phycology 52(4): 371–390.

Mago, Y., Y. Sharma, Y. Thakran, A. Mishra, S. Tewari and N. Kataria. 2023. Next-generation organic beauty products obtained from algal secondary metabolites: A sustainable development in cosmeceutical industries. Molecular Biotechnology, pp. 1–21.

Mago, Y., Y. Sharma, Y. Thakran, A. Mishra, S. Tewari and N. Kataria. 2023. Next-generation organic beauty products obtained from algal secondary metabolites: A sustainable development in cosmeceutical industries. Molecular Biotechnology, 1–21.

Manandhar, B., A. Wagle, S.H. Seong, P. Paudel, H.R. Kim, H.A. Jung et al. 2019. Phlorotannins with potential anti-tyrosinase and antioxidant activity isolated from the marine seaweed *Ecklonia stolonifera*. Antioxidants 8(8): 240.

Meticulous Research. (2022, June 8). Algae Products Market for Cosmetics Worth $305.3 Million by 2029 - Exclusive Report by Meticulous Research. GlobeNewswire. https://www.globenewswire.com/en/news-release/2022/06/08/2458637/0/en/Algae-Products-Market-for-Cosmetics-Worth-305-3-Million-by-2029-Exclusive-Report-by-Meticulous-Research.html.

Michalak, I., A. Dmytryk and K. Chojnacka. 2020. Algae cosmetics. In Encyclopedia of Marine Biotechnology (pp. 65–85). Elsevier.

Mirata, S., V. Asnaghi, M. Chiantore, A. Salis, M. Benvenuti, G. Damonte et al. 2023. Photoprotective and anti-aging properties of the apical frond extracts from the mediterranean seaweed *Ericaria amentacea*. Marine Drugs 21(5): 306.

Mofeed, J., M. Deyab, A. Mohamed, M. Moustafa, S. Negm and E. El-Bilawy. 2022. Antimicrobial activities of three seaweeds extract against some human viral and bacterial pathogens. Biocell 46(1): 247.

Morais, T., J. Cotas, D. Pacheco and L. Pereira. 2021. Seaweeds compounds: An ecosustainable source of cosmetic ingredients? Cosmetics 8(1): 8.

Mourelle, M.L., C.P. Gómez and J.L. Legido. 2017. The potential use of marine microalgae and cyanobacteria in cosmetics and thalassotherapy. Cosmetics 4(4): 46.

Narayanaswamy, R., B.W. Jo, S.K. Choi and I.S. 2013. Fucoidan: Versatile cosmetic ingredient. An overview. Journal of Applied Cosmetology 31(3/4): 131–138.

Naseri, A., G.S. Marinho, S.L. Holdt, J.M. Bartela and C. Jacobsen. 2020. Enzyme-assisted extraction and characterization of protein from red seaweed *Palmaria palmata*. Algal Research 47: 101849.

Nurjanah, Nurilmala, M., T. Hidayat and F. Sudirdjo. 2016. Characteristics of seaweed as raw materials for cosmetics. Aquat. Procedia, 7: 177.

Pangestuti, R., S.K. Kim and E.A. Siahaan. 2018. Photoprotective substances derived from marine algae. Marine Drugs 16(11): 399.

PatentArt. (n.d.). Algae & Cosmetics. PatentArt Knowledge Center. https://www.patent-art.com/knowledge-center/algae-cosmetics/

Peñalver, R., J.M. Lorenzo, G. Ros, R. Amarowicz, M. Pateiro and G. Nieto. 2020. Seaweeds as a functional ingredient for a healthy diet. Marine Drugs 18(6): 301.

Pereira, L. 2018. Seaweeds as source of bioactive substances and skin care therapy—cosmeceuticals, algotheraphy, and thalassotherapy. Cosmetics 5(4): 68.

Pereira, L. 2020. Characterization of bioactive components in edible algae. Marine Drugs 18(1): 65.

Pimentel, F.B., R.C. Alves, F. Rodrigues and O.M.B. PP. 2018. Macroalgae-Derived Ingredients for Cosmetic Industry—An Update.

Polat, S., M. Trif, A. Rusu, V. Šimat, M. Čagalj, G. Alak et al. 2023. Recent advances in industrial applications of seaweeds. Critical Reviews in Food Science and Nutrition 63(21): 4979–5008.

Poulose, N., A. Sajayan, A. Ravindran, T.V. Sreechithra, V. Vardhan, J. Selvin et al. 2020. Photoprotective effect of nanomelanin-seaweed concentrate in formulated cosmetic cream: With improved antioxidant and wound healing properties. Journal of Photochemistry and Photobiology B: Biology 205: 111816.

Resende, D.I., M. Ferreira, C. Magalhães, J.S. Lobo, E. Sousa and I.F. Almeida. 2021. Trends in the use of marine ingredients in anti-aging cosmetics. Algal Research 55: 102273.

Rosa, G.P., W.R. Tavares, P.M. Sousa, A.K. Pagès, A.M. Seca and D.C. Pinto. 2019. Seaweed secondary metabolites with beneficial health effects: An overview of successes in *in vivo* studies and clinical trials. Marine Drugs 18(1): 8.

Ryu, B., Z.J. Qian, M.M. Kim, K.W. Nam and S.K. Kim. 2009. Anti-photoaging activity and inhibition of matrix metalloproteinase (MMP) by marine red alga, *Corallina pilulifera* methanol extract. Radiation Physics and Chemistry 78(2): 98–105.

Ryu, J., S.J. Park, I.H. Kim, Y.H. Choi and T.J. Nam. 2014. Protective effect of porphyra-334 on UVA-induced photoaging in human skin fibroblasts. International Journal of Molecular Medicine 34(3): 796–803.

Salehi, B., J. Sharifi-Rad, A.M. Seca, D.C. Pinto, I. Michalak, A. Trincone et al. 2019. Current trends on seaweeds: Looking at chemical composition, phytopharmacology, and cosmetic applications. Molecules 24(22): 4182.

Schwartsmann, G., A.B. da Rocha, R.G. Berlinck and J. Jimeno. 2001. Marine organisms as a source of new anticancer agents. The Lancet Oncology 2(4): 221–225.

Senevirathne, W.S. and S.K. Kim. 2013. Cosmeceuticals from algae. In Functional Ingredients from Algae for Foods and Nutraceuticals (pp. 694–713). Woodhead Publishing.

Seo, G.Y., Y. Ha, A.H. Park, O.W. Kwon and Y.J. Kim. 2019. Leathesia difformis extract inhibits α-MSH-induced melanogenesis in B16F10 cells via down-regulation of CREB signaling pathway. International Journal of Molecular Sciences 20(3): 536.

Seo, M.J., O.H. Lee, H.S. Choi and B.Y. Lee. 2012. Extract from edible red seaweed (*Gelidium amansii*) inhibits lipid accumulation and ROS production during differentiation in 3T3-L1 cells. Preventive Nutrition and Food Science 17(2): 129.

Shanura Fernando, I.P., K.K. Asanka Sanjeewa, K.W. Samarakoon, H.S. Kim, U.K. Gunasekara, Y.J. Park et al. 2018. The potential of fucoidans from Chnoospora minima and *Sargassum polycystum* in cosmetics: Antioxidant, anti-inflammatory, skin-whitening, and antiwrinkle activities. Journal of Applied Phycology 30: 3223–3232.

Shanura Fernando, I.P., K.K. Asanka Sanjeewa, K.W. Samarakoon, H.S. Kim, U.K.D.S.S. Gunasekara, Y.J. Park et al. 2018. The potential of fucoidans from Chnoospora minima and *Sargassum polycystum* in cosmetics: Antioxidant, anti-inflammatory, skin-whitening, and antiwrinkle activities. Journal of Applied Phycology 30: 3223–3232.

Siezen, R.J. 2011. Microbial sunscreens. Microbial Biotechnology 4(1): 1.

Talero, E., S. García-Mauriño, J. Ávila-Román, A. Rodríguez-Luna, A. Alcaide and V. Motilva. 2015. Bioactive compounds isolated from microalgae in chronic inflammation and cancer. Marine Drugs.

Thiviya, P., A. Gamage, N.S. Gama-Arachchige, O. Merah and T. Madhujith. 2022. Seaweeds as a source of functional proteins. Phycology 2(2): 216–243.

Thiyagarasaiyar, K., B.H. Goh, Y.J. Jeon and Y.Y. Yow. 2020. Algae metabolites in cosmeceutical: An overview of current applications and challenges. Marine Drugs 18(6): 323.

Thiyagarasaiyar, K., C.K. Mahendra, B.H. Goh, L.T. Gew and Y.Y. Yow. 2021. UVB radiation protective effect of brown Alga *Padina australis*: A potential cosmeceutical application of Malaysian Seaweed. Cosmetics 8(3): 58.

Thomas, N.V. and S.K. Kim. 2013. Beneficial effects of marine algal compounds in cosmeceuticals. Marine Drugs 11(1): 146–164.

Torres, P., J.P. Santos, F. Chow and D.Y. dos Santos. 2019. A comprehensive review of traditional uses, bioactivity potential, and chemical diversity of the genus Gracilaria (Gracilariales, Rhodophyta). Algal Research 37: 288–306.

Usov, A.I. and N.D. Zelinsky. 2013. Chemical structures of algal polysaccharides. In Functional Ingredients from Algae for Foods and Nutraceuticals (pp. 23–86). Woodhead Publishing.

Wang, H.M., X.C. Li, D.J. Lee and J.S. Chang. 2017. Potential biomedical applications of marine algae. Bioresource Technology 244: 1407–1415.

Wang, J., W. Jin, Y. Hou, X. Niu, H. Zhang and Q. Zhang. 2013. Chemical composition and moisture-absorption/retention ability of polysaccharides from five algae. International Journal of Biological Macromolecules 57: 26–29.

Wang, L., W. Lee, J.Y. Oh, Y.R. Cui, B. Ryu and Y.J. Jeon. 2018. Protective effect of sulfated polysaccharides from celluclast-assisted extract of *Hizikia fusiforme* against ultraviolet B-Induced skin damage by regulating NF-κB, AP-1, and MAPKs signaling pathways *in vitro* in human dermal fibroblasts. Marine Drugs 16(7): 239.

Wang, X., G. Chang, Y. Xu, Z. Li, X. Du, Y. Yang et al. 2023. Effect of phytol isolated from edible red alga (*Bangia fusco-purpurea*) on tyrosinase inhibition and its application on food preservation. LWT 185: 115146.

Watanabe, F., Y. Yabuta, T. Bito and F. Teng. 2014. Vitamin B12-containing plant food sources for vegetarians. Nutrients 6(5): 1861–1873.

Zhang, T., Q. Guo, Y. Xin and Y. Liu. 2022. Comprehensive review in moisture retention mechanism of polysaccharides from algae, plants, bacteria and fungus. Arabian Journal of Chemistry 15(10): 104163.

CHAPTER 18

Looking into the Future, Addressing Challenges and Unveiling Opportunities on Utilization of Seaweed in Cosmeceuticals

Taiyaba U. Vakharwala,[1] *Haresh S. Kalasariya*[2] and *João Cotas*[3,*]

Introduction to Seaweed in cosmeceuticals

Cosmeceuticals represent a dynamic and rapidly expanding sector within the skincare industry, reflecting the evolving preferences and demands of consumers. Since the early 2000s, this market has demonstrated substantial economic growth, with notable projections indicating a trajectory towards further expansion. According to Milam et al. (2016), the market value surged to approximately $9.7 billion in 2011, with estimations foreseeing sales reaching $11.7 billion by 2016. This growth trend underscores the increasing consumer inclination towards products that not only promote skin health but also align with environmental consciousness. Contemporary consumer preferences lean heavily towards green and eco-friendly formulations, emphasizing natural ingredients deemed safe for skincare applications (Pangestuti et al. 2021). Such inclinations have spurred interest in the development of cosmeceuticals, which occupy a unique niche bridging cosmetics and pharmaceuticals. Coined by Dr. Albert Kligman in the 1980s, the term "cosmeceuticals" denotes products that offer both cosmetic enhancement and therapeutic benefits. Unlike traditional cosmetics, which primarily focus on superficial improvements to appearance, cosmeceuticals contain bioactive compounds capable of modulating skin physiology and addressing specific dermatological concerns. Central to the formulation of cosmeceuticals are active ingredients sourced from various natural reservoirs, including plants, marine organisms, and microbial sources (Pereira et al. 2018).

[1] J.N.M. Patel Science College, Veer Narmad South Gujarat University, Bharthana, Vesu, Surat, 395017; vakharwalataiyaba@gmail.com

[2] Centre for Natural Products Discovery, School of Pharmacy and Biomolecular Sciences, Liverpool John Moores University, Byrom Street, Liverpool L3 3AF, UK; Faculty of Health and Life Sciences, INTI International University, Persiaran Perdana BBN, Putra Nilai, 71800 Nilai, Negeri Sembilan, Malaysia. Email: hareshahir22@gmail.com

[3] Marine Resources, Conservation and Technology, Marine Algae Lab, CFE—Centre for Functional Ecology: Science for People & Planet, Department of Life Sciences, University of Coimbra, 3000-456 Coimbra, Portugal.

* Corresponding author: jcotas@uc.pt

Among these, marine organisms stand out as a particularly rich and diverse repository of biologically active compounds. Phytochemicals, essential oils, vitamins, antioxidants, and other bioactive substances extracted from marine sources have garnered considerable attention for their potential skincare benefits. Marine organisms, with their vast biodiversity, offer a plethora of bioactive compounds that hold promise for cosmeceutical applications. From algae to sponges and corals, these organisms biosynthesize a myriad of metabolites with intriguing properties, ranging from antioxidant and anti-inflammatory to collagen-stimulating and UV-protective effects. Harnessing the therapeutic potential of marine-derived cosmeceuticals requires meticulous extraction, purification, and characterization processes to ensure efficacy and safety. Moreover, the sustainable utilization of marine resources for cosmeceutical development necessitates adherence to stringent quality control standards and regulatory frameworks. Challenges such as sourcing variability, seasonal fluctuations, and environmental sustainability further underscore the complexities inherent in harnessing marine-derived cosmeceuticals.

Algae, akin to terrestrial plants, possess specialized pigment cells essential for conducting photosynthesis, facilitating their ubiquitous distribution across a diverse array of aquatic environments (Coelho et al. 2013). Colonizing various habitats, algae exhibit an intimate association with water bodies, inhabiting fixed aquatic environments or forming symbiotic relationships with fungi, exemplified in lichens. Their geographical prevalence spans seas, rivers, lagoons, and other aqueous ecosystems, adapting to environmental niches ranging from the well-lit photic region to the dimly illuminated benthic zone, where they often anchor to substrates like rocks (macroalgae). Within the realm of macroalgae, seaweeds assume a pivotal role in the production of eco-friendly skincare products renowned for their safety and efficacy (Salehi et al. 2019). As eukaryotic macroscopic multicellular organisms, seaweeds lack traditional plant structures such as roots, stems, and leaves. Nonetheless, they possess specialized anatomical features facilitating nutrient absorption and anchorage to substrates. Notably, the presence of a stem-like structure called a stipe, alongside varied adaptations such as the absence of an air bladder or the presence of holdfast-like structures, underscores their diverse morphological repertoire.

Seaweeds exhibit remarkable taxonomic diversity, typically classified into three primary categories based on distinct characteristics: red seaweeds, brown seaweeds, and green seaweeds (El-Beltagi et al. 2022). Red seaweeds, belonging to the Rhodophyta phylum, owe their characteristic hue to the presence of pigments like phycoerythrin and phycocyanin. These algae boast a rich repertoire of photosynthetic pigments, including chlorophyll a, phycobilins, and carotenoids, with polysaccharides like carrageenan and cellulose agar constituting their cell walls. The classification of seaweed based on their cellular organization, photosynthetic pigments, reserve substances, and cell wall components is detailed in Table 1. Notable examples include coralline algae, pivotal in industries due to their calcium carbonate secretion. Brown seaweeds, classified under the Ochrophyta phylum, possess pigments such as chlorophylls a, c, and the characteristic brown pigment fucoxanthin (Misurcova et al. 2012). Fucoxanthin, renowned for its anti-inflammatory properties, is a principal marine carotenoid pigment, predominant in brown seaweeds like Ascophyllum, Laminaria, and Sargassum. The prevalence of fucoxanthin and other xanthophyll pigments imparts the characteristic brown hue to these algae. Green seaweeds, characterized by chlorophyll a and b, alongside accessory pigments like beta-carotene and xanthophylls, thrive in both aquatic and terrestrial environments (El-Beltagi et al. 2022). With origins tracing back to endosymbiotic associations, green algae like Ulva and Cladophora are prevalent in freshwater and marine habitats, while genera such as Enteromorpha and Chaetomorpha may exhibit terrestrial adaptations. Understanding the taxonomic intricacies and ecological adaptations of algae, particularly seaweeds, underscores their significance as sources of bioactive compounds for various applications, including skincare. Continued exploration of their biochemical diversity promises to unveil novel avenues for sustainable skincare innovations, rooted in the rich biodiversity of marine ecosystems.

Table 1. Seaweed classification based on their Cellular organization, Photosynthetic pigments, Reserve substances & Cell wall Components (Pereira et al. 2018, Hurd et al. 2014).

Phylum	Cellular Organization	Photosynthetic Pigments	Reserve Substances	Cell Wall Components	Ecological Importance
Chlorophyta (Green Seaweed)	Unicellular or Multicellular	Chlorophyll a; Chlorophyll b; β-carotenoids; Xanthophylls	Starch	Cellulose and Pectin	Green seaweeds play a crucial role in aquatic ecosystems as primary producers, contributing significantly to oxygen production and nutrient cycling. They provide habitats and food for various marine organisms, supporting biodiversity and ecosystem stability.
Rhodophyta (Red Seaweed)	Multicellular	Chlorophyll a; Chlorophyll d; Xanthophylls; Beta-carotenoids	Floride Starch	Cellulose, Agar, and Carrageenan	Red seaweeds contribute to reef-building processes and serve as essential components of marine habitats, providing shelter and food for numerous marine species. They also play a role in sediment stabilization and shoreline protection.
Ochrophyta (Brown Seaweed)	Multicellular	Beta-carotenoids; Chlorophyll a; Chlorophyll c	Laminarin; Starch; Mannitol	Cellulose and alginic acid	Brown seaweeds are ecologically significant as they form extensive underwater forests known as kelp forests, which provide habitats for diverse marine life and contribute to carbon sequestration. They also help mitigate coastal erosion and enhance water quality through nutrient uptake and filtration.

Importance of seaweeds in cosmeceuticals

In the contemporary landscape of consumer demand for greener and more sustainable cosmetic products, manufacturers are increasingly turning to eco-friendly and cost-effective production methods, utilizing natural components derived from renewable raw materials (Bom et al. 2019). This strategic shift is underpinned by mounting concerns over the environmental repercussions of traditional petrochemical ingredients, which not only deplete finite resources but also contribute significantly to greenhouse gas emissions, exacerbating climate change (Thiyagarasaiyar et al. 2020). In response, the cosmetic industry is witnessing a notable transition towards harnessing the rich biochemical diversity of natural resources, with algae, particularly seaweeds, emerging as pivotal raw material sources in the burgeoning field of cosmeceuticals. Seaweed's command significant attention in the cosmeceutical domain owing to their multifaceted medicinal properties, prompting a surge in demand for skincare products enriched with a plethora of bioactive compounds, including antioxidants, anti-inflammatory agents, and skin-brightening components (Pangestuti et al. 2021). Consumers today not only seek cosmetic solutions that enhance aesthetic appearance but also those that fortify skin health and combat intrinsic and extrinsic factors that contribute to aging and wrinkle formation. Seaweed extracts, renowned for their potent antimicrobial and antifungal activity, serve dual roles in cosmeceuticals, acting both as active ingredients and natural preservatives, thereby prolonging product shelf life while safeguarding against microbial spoilage.

Human skin, subjected to an ever-increasing array of environmental stressors such as ultraviolet (UV) radiation, air pollution, and oxidative stress, faces heightened vulnerability to a spectrum of dermatological maladies, including acne, hyperpigmentation, and premature aging (Pillai et al.

2005). Seaweeds, having evolved in a marine environments which is fraught with stressors, exhibit remarkable biochemical adaptations, synthesizing an array of secondary metabolites as adaptive responses. These bioactive compounds, endowed with potent antioxidant, antibacterial, and anti-inflammatory properties, confer upon seaweeds a unique therapeutic potential for skin protection and rejuvenation in cosmetic formulations (Jesumani et al. 2019). Moreover, their innate moisturizing capabilities further enhance their utility in skincare products, providing comprehensive solutions for skin health and vitality. In essence, the integration of seaweed-derived bioactive compounds into cosmeceutical formulations epitomizes a paradigm shift towards sustainable and efficacious skincare solutions. By leveraging the inherent therapeutic properties of seaweeds, cosmetic industries can effectively meet the evolving demands of consumers for environmentally conscious products, while concurrently delivering tangible benefits for skin health and rejuvenation.

Seaweeds derived ingredients used in cosmeceuticals

In contemporary cosmetic formulations, the prominence of natural ingredients, particularly seaweeds, has soared, reflecting a paradigm shift towards eco-friendly and sustainable cosmeceuticals (Amberg et al. 2019). Seaweeds, owing to their rich biochemical composition, serve as a veritable treasure trove of bioactive compounds with diverse cosmeceutical and medicinal applicability. The chemical profile of seaweeds is inherently dynamic, influenced by factors such as cultivation methods, geographical location, and harvesting periods, giving rise to a myriad of metabolites that exhibit potent bioactivity (El-Beltagi et al. 2022). Among the plethora of bioactive compounds derived from seaweeds, polyphenolic compounds, carotenoids, minerals, vitamins, phlorotannins, peptides, tocotrienols, proteins, tocopherols, and carbohydrates (polysaccharides) reign supreme, showcasing a remarkable diversity of molecular structures and functional properties (El-Beltagi et al. 2022). Polyphenolic compounds, for instance, are renowned for their antioxidant and anti-inflammatory activities, conferring upon them immense potential for combating oxidative stress and mitigating inflammatory skin conditions. Carotenoids, on the other hand, impart vibrant hues to seaweeds while serving as precursors for vitamin A synthesis, essential for skin health and rejuvenation. Minerals such as calcium, magnesium, and potassium play pivotal roles in maintaining skin integrity and functionality, promoting hydration and cellular regeneration. Vitamins, including vitamin C and vitamin E, exhibit antioxidant properties, shielding skin cells from oxidative damage and bolstering collagen synthesis. Phlorotannins, unique to brown seaweeds, possess potent antioxidant and anti-aging properties, making them coveted ingredients in anti-aging skincare formulations (Figure 1).

Peptides derived from seaweeds demonstrate remarkable skin-repairing and collagen-boosting capabilities, contributing to improved skin elasticity and firmness. Tocotrienols and tocopherols, members of the vitamin E family, offer protection against UV-induced skin damage and premature aging. Proteins extracted from seaweeds serve as valuable moisturizing agents, enhancing skin hydration and suppleness. Polysaccharides, such as alginate and carrageenan, possess film-forming properties, creating a protective barrier on the skin's surface while imparting a luxurious texture to cosmetic formulations. In essence, the utilization of seaweed-derived ingredients in cosmeceuticals underscores the immense potential of marine resources in meeting the evolving demands of consumers for efficacious, sustainable, and environmentally conscious skincare solutions. By harnessing the diverse array of bioactive compounds present in seaweeds, cosmetic industries can innovate novel formulations that not only enhance aesthetic appeal but also promote skin health and vitality.

Polysaccharides constitute a fundamental component of seaweed cell membranes, showcasing a diverse array of unique varieties across different algae types. Notably, brown algae feature alginate, which contains fucoidan, while green algae showcase ulvan, and red algae exhibit carrageenan (Figure 2). Renowned for their multifaceted biofunctional and physicochemical properties, polysaccharides emerge as prominent candidates for various applications in cosmeceuticals

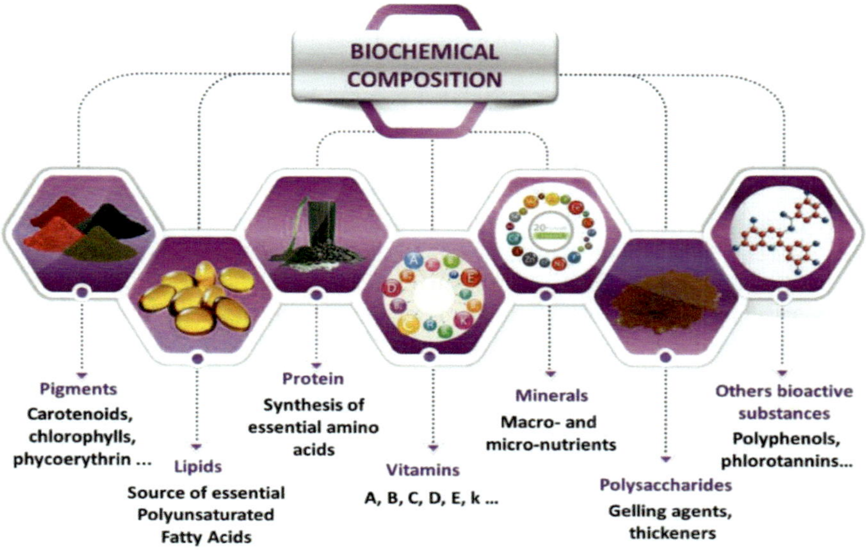

Figure 1. Illustration of Seaweed's biochemical composition.
(Source: Hentati et al. 2020) https://www.ncbi.nlm.nih.gov/pmc/articles/PMC7397078/.

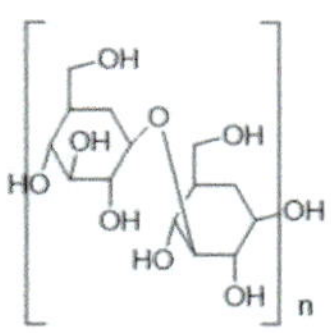
Laminarin (**1**)

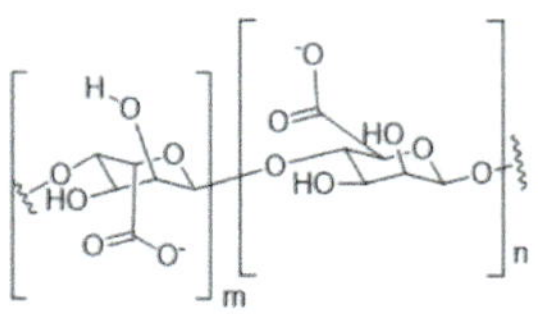
Alginate (**2**)

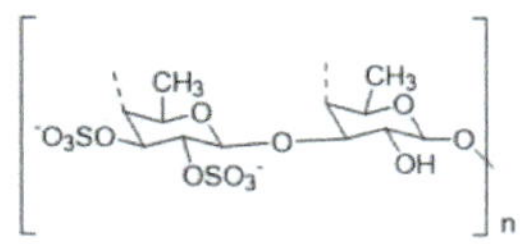
Fucoidan (**3**)

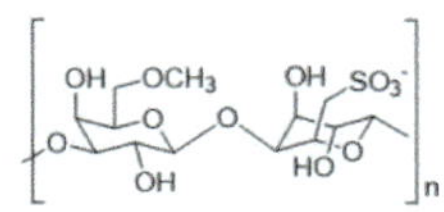
Porphyran (**4**)

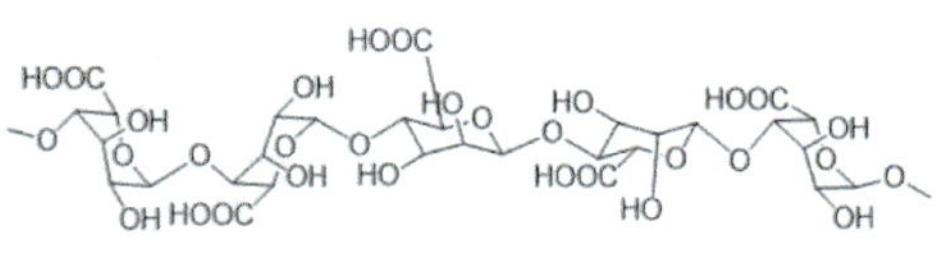
Alginic acid (**5**)

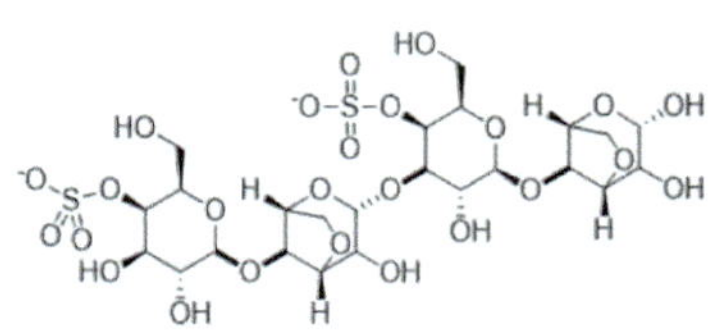
Carrageenan (**6**)

Figure 2. Chemical structures of seaweed derived polysaccharides.
(Source: Lee et al. 2021).

(Shanura Fernando et al. 2018). Comprising approximately 80% of seaweed weight, polysaccharides are classified based on their location within seaweed structures, delineating into cell membrane polysaccharides and storage polysaccharides. Predominantly, cell membrane polysaccharides are abundant in seaweed tissues (Shanura Fernando et al. 2018). The vast repertoire of bioactive polysaccharides inherent in seaweed tissues harbors moisturizing and antioxidant substances that find wide utility in cosmeceutical formulations, serving as emulsion gelling agents and stabilizers (Pereira et al. 2016). Agar, derived from seaweed polysaccharides, serves as a versatile emulsifier and stabilizer in creams, regulating moisture content in various skincare products such as hand lotions, deodorants, and foundations. Alginates, another class of seaweed polysaccharides, find application as gelling agents, thickeners, and emulsion stabilizers in cosmetics and pharmaceuticals, owing to their remarkable chelating properties. Moreover, alginates are employed as skin-protective barrier lotions to prevent dermatitis (Pereira et al. 2016). Laminarin, although less utilized for its physical properties, is prized for its bioactive attributes. Fucoidan, a sulfated polysaccharide found in seaweeds, exhibits potent anti-inflammatory properties, hastening the tissue repair process following injury or trauma. Its applications extend to muscle and joint injuries, deep wounds, and surgical procedures (Wu et al. 2016). Sulfated polysaccharides, including fucoidan, exhibit a plethora of bioactivities, ranging from anticoagulant and anti-inflammatory to antioxidative and antimicrobial properties, rendering them invaluable in cosmeceutical formulations (Saravana et al. 2018). In essence, the integration of seaweed-derived polysaccharides into cosmeceutical products epitomizes a synergistic blend of nature and science, offering consumers sustainable and effective skincare solutions enriched with bioactive compounds sourced from marine ecosystems.

Seaweeds exhibit a remarkable diversity in protein content across different species, with concentrations varying based on factors such as geographic location, cultivation methods, and environmental conditions, reaching levels as high as 45% (Peñalver et al. 2020). The proteinaceous constituents, peptides, and amino acids within seaweeds are subject to fluctuations induced by environmental perturbations and habitat characteristics. Notably, red seaweeds boast higher protein concentrations, ranging up to 47%, compared to green algae, which typically contain 9 to 26% protein, while brown algae exhibit comparatively lower levels, ranging from 3 to 15% (Peñalver et al. 2020). Seaweed-derived peptides and bioactive compounds are endowed with a myriad of health benefits, underscored by their significant antioxidant activity (Pimentel et al. 2018). These bioactive compounds, such as peptides and amino acids, play pivotal roles in cosmeceutical formulations, functioning as moisturizing agents, skin protectors against UV radiation, and stimulators of cellular proliferation (Pereira 2015). Seaweeds harbor a diverse array of amino acids, including valine, leucine, isoleucine, and taurine, which exhibit potent biological activities as antioxidants (Kadam et al. 2013, Helmi et al. 2016). These antioxidants help neutralize free radicals, reducing oxidative stress and preventing premature skin aging. Furthermore, seaweed-derived compounds like peptides can aid in collagen synthesis, improving skin elasticity and firmness, while amino acids contribute to maintaining the skin's natural moisture barrier, preventing dehydration, and enhancing hydration levels (Pereira 2015). Acidic amino acids, such as aspartic acid and glutamic acid, are prevalent in seaweed species, constituting essential components of protein structures and metabolic pathways (Laurence et al. 2004).

The multifaceted roles of proteins and amino acids derived from seaweeds extend beyond their nutritive value, encompassing diverse functionalities in cosmeceutical formulations. By harnessing the bioactive potential of these marine-derived constituents, cosmetic industries can innovate novel skincare products that promote skin health, hydration, and resilience while offering protection against environmental stressors and aging processes. These mechanisms provide a scientific basis for the efficacy of seaweed-derived compounds in cosmeceuticals, highlighting their potential in addressing various skin concerns and enhancing overall skin health. As shown in Table 2, seaweed-derived bioactive compounds exhibit a range of cosmeceutical properties and applications.

Fatty acids play a crucial role in maintaining cellular integrity and supporting metabolic processes in all living organisms (Ward et al. 2005). In seaweeds, omega-3 polyunsaturated fatty acids (PUFAs), such as eicosapentaenoic acid (EPA) and docosahexaenoic acid (DHA), are prominent constituents with significant implications for human health (Stengel et al. 2011). These omega-3 PUFAs are essential for various physiological functions, including brain function, cardiovascular health, and inflammation regulation. The multifaceted biological activities of omega-3 PUFAs make them valuable ingredients in cosmeceutical formulations. Their anti-allergic, antioxidant, and anti-inflammatory properties contribute to skin health and well-being (Ward et al. 2005). These compounds protect the skin from oxidative stress, inflammation, and allergic reactions, thereby promoting skin rejuvenation and resilience. Additionally, seaweed-derived fatty acids and lipophilic compounds act as potent moisturizers and protective agents against transepidermal water loss (TEWL) (Stengel et al. 2011). By forming a lipid barrier on the skin's surface, fatty acids help retain moisture and prevent dehydration, thereby enhancing skin hydration and suppleness. Lauric acid,

Table 2. Cosmeceutical's properties & Applications of Seaweed derived bioactive compounds.

Seaweed	Compound	Properties & Technical Functions	References
Brown Seaweed	Alginate	Emulsifier, viscosifying, moisturizing, chelating, colloids, gelling, immunostimulant, protective colloid agent	(Pimentel et al. 2018) (Tønnesen et al. 2002)
	Fucoidan	Antioxidant, anti-cellulite, antiviral, anti-inflammatory, antiaging, anti-photoaging, elastase, tyrosinase inhibitor agent	(Chizhov et al. 1999) (Wijesinghe et al. 2012)
	Laminarin	Antioxidant, anti-cellulite, anti-inflammatory agent	(Pereira et al. 2018) (Stengel et al. 2012)
	Pigments	Antioxidant, UV protector, colorant, and dye agent	(Bedoux et al. 2014), (Guillerme et al. 2017)
	Phlorotannins	Antioxidant, Collagen-producing enhancer, anti-inflammatory, antioxidants, antiaging anti-photoaging, anti-allergic, chelating, UV protection, anti-histaminic, anti-wrinkling, hair growth promoter agent	(Borowitzka et al. 2013) (Stengel et al. 2012) (Thomas et al. 2013)
Red Seaweed	Carrageenan	Thickening, viscosifying, stabilizer, sensory enhancer, moisturizing, Anticoagulant, antinociceptive and anti-inflammatory agent	(Pimentel et al. 2018) (Rocha de Souza et al. 2007)
	Agar	Emulsion stabilizer, gelling, thickening agent	(Pereira et al. 2018)
	Pigments	Antioxidant, colorant, dye, antioxidant, anti-inflammatory, radical scavenging agent	(Borowitzka et al. 2013) (Stengel et al. 2012)
	Mycosporine-like amino acids (MAAs)	UV protection agent	(Pimentel et al. 2018) (Bedoux et al. 2014)
	Bromophenols	Antioxidant, antimicrobial, antithrombotic agent	
	Fatty acids	Emollients, antimicrobial, antioxidant, regenerating compounds	(Murthy et al. 2005) (Stengel et al. 2012) (Sanghvi et al. 2010)
	Sterols	Anti-allergic, anti-inflammatory, antioxidant, radical scavenger agent	(Borowitzka et al. 2013)
Green Seaweed	Ulvan	Antioxidant, chelating, gelling, moisturizing, protective agent	(Robic et al. 2009) (Lahaye et al. 2007)
	Chlorophylls	Antibacterial, antioxidant, deodorizing, tissue growth stimulating, colorant agent	(Thomas et al. 2013) (Spears 1998) (Lanfer-Marquez et al. 2005)
	Carotenoids	Anti-inflammatory, antioxidant, tyrosinase inhibitor, anti-photoaging agents, radical scavengers, colorant agent	(Spears 1998) (Thomas et al. 2013) (Wang et al. 2013)
	Fatty acids	Antioxidant, cytoprotective agent	(Wang et al. 2013) (Stengel et al. 2012)
General Seaweed Compounds	Amino acids	Antioxidant, antiaging, moisturizing, UV protection agent	(Borowitzka et al. 2013)
	Proteins	Radical scavengers, antioxidant, chelating, moisturizing, antioxidants, UV protection agent	(Borowitzka et al. 2013) (Pimentel et al. 2018)
	Vitamins	Protects against DNA damage. Ability to promote cell regeneration	(Anyanwu et al. 2018)
	Minerals	Mineral reposition	(Pereira et al. 2018)

myristic acid, palmitic acid, and stearic acid, found in seaweeds, serve as valuable raw materials in cosmeceutical formulations, offering diverse functionalities ranging from emolliency to textural enhancement (Murphy et al. 2021).

The intricate interplay between seaweed-derived fatty acids and skin health underscores their significance in cosmeceutical applications. Through their antioxidant, anti-inflammatory, and moisturizing properties, these bioactive compounds offer a holistic approach to skincare, addressing various skin concerns and promoting overall skin well-being. Leveraging the benefits of seaweed-derived fatty acids exemplifies a synergistic fusion of scientific innovation and natural resources in developing efficacious and sustainable skincare solutions. Seaweeds exhibit a diverse array of pigments, categorizing them into three main groups based on their pigment content: Phaeophyceae (brown algae), Chlorophyceae (green algae), and Rhodophyceae (red algae). These pigments include chlorophylls and carotenoids, such as carotenes (β-carotene) and xanthophylls (fucoxanthin, violaxanthin, antheraxanthin, zeaxanthin, lutein, neoxanthin), which have been found to possess a spectrum of bioactive properties in recent studies. Carotenoids have garnered significant interest due to their antimicrobial and antioxidant attributes, which contribute to mitigating the risk of cardiovascular diseases, ophthalmologic conditions, and cancer (Francavilla et al. 2010). Various seaweed species, including *Ascophyllum nodosum*, *Cladosiphon okamuranus, Fucus serratus*, *Chaetoseros* sp., Ishige okamurae, Ecklonia stolonifera, Himanthalia elongata, and Fucus vesiculosus, are known to contain these beneficial carotenoids.

Additionally, phycobiliproteins, naturally fluorescent, water-soluble proteins, are another class of pigments present in certain seaweed species. Classified into PC (Blue pigment), PE (red pigment), and allophycocyanin (light-blue pigment), these pigments are prevalent in several red macroalgal species, with PE being the most common among them (da Vaz et al. 2016). Algal species such as Spirulina, Botryococcus, Chlorella, and Nostoc are rich sources of phycobiliproteins. Recent studies have unveiled the diverse bioactive potentials of these pigments, ranging from anti-obesity, anti-inflammatory, anti-angiogenic, antioxidant, and anti-carcinogenic properties to neuroprotective activities (Gupta et al. 2011). The intricate interplay between seaweed-derived pigments and their multifaceted bioactive properties underscores their potential applications in various industries, including cosmeceuticals. By harnessing the inherent benefits of these pigments, cosmetic formulations can offer enhanced skincare solutions, addressing diverse skin concerns and promoting overall skin health and vitality.

Seaweeds serve as a rich source of phenolic compounds, encompassing phenolic acids, tannins, flavonoids, and catechins, which hold immense potential in the cosmeceutical industry owing to their diverse array of skin benefits (Kalasariya et al. 2022, Malakar et al. 2021). These compounds exhibit multifaceted bioactivities that contribute to skin health and rejuvenation, including anti-aging, photoprotection, anti-wrinkle, antiallergic, anti-inflammatory, antioxidant, antimicrobial, antifungal, tyrosinase inhibition, anti-melanogenic, skin whitening, UVB protection, and anti-acne properties. Of particular interest are phlorotannins, a class of phenolic compounds unique to marine algae, renowned for their potent antioxidant properties and purported anti-aging effects (Thomas and Kim 2013, Nagayama et al. 2002, Hwang 2010). Phlorotannins have garnered attention for their ability to scavenge free radicals and combat oxidative stress, thereby mitigating skin damage and supporting skin resilience. Notably, their lower toxicity compared to other natural antioxidants enhances their suitability for incorporation into cosmetic formulations, offering a safer alternative for skincare products. The remarkable diversity of phenolic compounds in marine macroalgae, with more than 8000 different structures identified, underscores their vital role in the growth, survival, and defense mechanisms of seaweeds (Kalasariya et al. 2022). These compounds exhibit intricate biological activities that extend beyond mere antioxidant properties, including modulation of inflammatory pathways, regulation of melanogenesis, and enhancement of skin barrier function. As research into seaweed-derived phenolic compounds progresses, novel bioactive properties and mechanisms of action continue to be elucidated, presenting exciting opportunities for the development of innovative skincare solutions. By harnessing the multifaceted benefits of these compounds, cosmeceutical formulations can offer tailored skincare solutions that address a spectrum of skin concerns while promoting overall skin health and resilience, aligning with the growing demand for eco-conscious and efficacious skincare products.

Minerals: Seaweeds, being highly adaptable to diverse environmental conditions, boast a rich mineral profile comprising a wide array of trace elements and micronutrients (Kalasariya et al. 2022). Among these minerals are zinc, magnesium, aluminum, silica, copper, iodine, selenium, iron, manganese, calcium, sodium, phosphorus, potassium, and chlorine, each contributing distinctively to various skin health benefits. For instance, iodine, prominently present in brown seaweeds, exhibits remarkable antiseptic and anti-inflammatory properties, making it invaluable for wound healing and skincare formulations (Gupta et al. 2011). Selenium, another essential trace element abundantly found in certain seaweed species, demonstrates potent antioxidant activity, combating oxidative stress and shielding skin cells from damage induced by UV radiation and environmental pollutants (Kaushik et al. 2023). The mineral composition of seaweeds plays a pivotal role in their physiological functions and therapeutic properties, underscoring their significance in skincare applications. These minerals not only fortify the skin's protective barrier but also contribute to its overall health and resilience, offering protection against various environmental stressors and promoting skin rejuvenation.

Vitamins: Seaweeds serve as reservoirs of vitamins, encompassing both water-soluble and fat-soluble variants, each with distinct contributions to skin health and function (Peñalver et al. 2020). Water-soluble vitamins such as B1 (thiamine), B2 (riboflavin), B12 (cobalamin), and C (ascorbic acid), along with fat-soluble vitamins like E (tocopherol) and beta-carotene (provitamin A), exert multifaceted effects on the skin. Vitamin B12 plays a pivotal role in DNA synthesis and cellular division, facilitating healthy skin turnover and regeneration (Hussein et al. 2020). Vitamin C, renowned for its potent antioxidant properties, scavenges free radicals, thereby shielding skin cells from oxidative stress and promoting a youthful complexion (Pullar et al. 2017). Beta-carotene, a precursor to vitamin A, supports skin integrity and elasticity, while vitamin E aids in moisturizing and nourishing the skin, diminishing the appearance of fine lines and wrinkles (Rybak et al. 2021). The incorporation of seaweed-derived vitamins into cosmeceutical formulations offers a natural and sustainable approach to skincare, capitalizing on their therapeutic properties to enhance skin health and vitality. By harnessing the synergistic benefits of minerals and vitamins derived from seaweeds, cosmetic industries can develop innovative skincare products tailored to meet diverse skin needs while upholding principles of sustainability and eco-consciousness.

Seaweed-derived ingredients offer a diverse array of skin benefits, but they also present various biological, technical, and supply challenges, including issues related to contamination and the reproducibility of extracts obtained from raw materials. The utilization of marine biodiversity in cosmetic products and manufacturing entails both advantages and disadvantages, as the vast diversity of potential ingredients poses challenges in identifying and exploring this diversity in a sustainable and safe manner (Martins et al. 2014). Obtaining natural bioactive compounds from marine sources remains challenging, primarily due to the reliance on costly robotic and engineering techniques required to study the biodiversity of the deepest ocean regions. Consequently, most research focuses on shallow coastal waters, leading to a gap in understanding the biodiversity and chemodiversity of deeper oceanic regions (Martins et al. 2014). Moreover, the lack of correct classification and taxonomic knowledge of marine-derived products further complicates their utilization, hindering reproducibility and limiting our understanding of undiscovered biodiversity.

In addition to challenges associated with biodiversity exploration, marine organisms exhibit high biological and biochemical variability influenced by specific environmental conditions. This variability complicates the sustainable sourcing of desired compounds, as they may be present in minimal quantities within raw materials or influenced by external factors (Martins et al. 2014). Efforts to obtain large quantities of raw material have been made through optimizing molecular biology and aquaculture technologies, as well as the development of synthetic and semi-synthetic analogues. However, the intricate nature of naturally derived products poses challenges in the purification process and virtual screening of molecules.

The reproducibility of extracts from seaweed and the scale-up of cultivation from laboratory to industrial levels pose significant challenges for cosmeceutical industries. Conventional extraction methods often lack reproducibility due to manual techniques, necessitating the adoption of advanced methods such as Superficial Fluid Extraction (SFE) or electro-technologies for better reproducibility and sustainable supply (Power et al. 2020, Usman et al. 2023). Moreover, natural products can alter their metabolite levels and growth rates in response to environmental conditions, further complicating the scale-up process. Factors such as temperature, light availability, pH, oxygen, nutrient levels, risk of contamination, and biomass film formation significantly influence macroalgae growth, necessitating real-time monitoring and optimization strategies for industrial-scale production.

Contamination of raw materials and extracts derived from seaweeds poses significant challenges in the incorporation of natural components into cosmetic formulations and production processes, presenting concerns regarding both biological and chemical contaminants. Biological contamination encompasses two main categories: cell growth-affecting contamination and protein accumulation-affecting contamination (Costa et al. 2023). Sources of biological contamination include air and aquatic pollution, as well as the presence of nutrients, salts, and other organisms in bioreactors utilized for seaweed cultivation. Protein accumulation-affecting contamination refers to contaminants' ability to consume or diminish molecules produced by the organism of interest. To mitigate these issues, optimization of growth conditions, stringent controls, proper media usage, and standardized procedures are imperative to minimize contaminant levels and ensure microbiologically pure raw product units. Chemical contamination poses another significant challenge, with heavy metals such as arsenic (As), cadmium (Cd), chromium (Cr), copper (Cu), lead (Pb), nickel (Ni), and zinc (Zn) being common pollutants in marine environments (Couteau et al. 2020). Additionally, iodine, present in ocean-bound algae and marine-derived products, can contribute to chemical contamination. These chemicals undergo bioaccumulation in marine environments, with seaweeds absorbing them through the bioaccumulation process (Chang et al. 2016, Couteau et al. 2020). Incorporation of such chemicals into seaweed-derived extracts can result in potential toxicity and health risks upon incorporation into cosmetic products, necessitating adherence to EU regulations prohibiting their use in cosmetics.

Furthermore, unsustainable collection practices have led to a reduction in the genetic diversity of macroalgae, limiting their potential scientific and commercial applications (Brodie et al. 2018, Benedetti et al. 2018). Conservation strategies such as ex-situ and in-situ conservation are employed to preserve the genetic diversity of macroalgae, ensuring their sustainable utilization. The use of marine-derived products in the cosmeceutical industry necessitates further development in regulatory frameworks to ensure product safety and purity. Rigorous analysis of extract purity through specific identification tests is essential to meet regulatory standards. Sustainable utilization of marine-derived resources addresses challenges such as reducing biological and chemical contamination, monitoring growth and cultivation parameters of seaweed, ensuring extract reproducibility, scaling up processes, and ensuring sustainable supply chains. Implementation of proper techniques for collection, cultivation, and purification, alongside regulated harvesting activities and long-term safety assessments regarding skin toxicity, eye irritation, and sensitization, facilitates the extraction of valuable bioactive metabolites from seaweed for safe use in various skincare products in the cosmeceutical industry.

Future opportunities for seaweed used in the cosmeceutical industry

The quest for natural components in cosmetic formulations continues to grow as consumers increasingly prioritize skin health and seek effective solutions derived from marine organisms, particularly seaweed, renowned for its minimal cytotoxic effects on human cells. Seaweed harbors a wealth of bioactive compounds, including vitamins, polyphenols, carotenoids, and chitin, offering promising skin benefits and driving the commercialization of seaweed-based skincare products. In the realm of cosmeceuticals, the future holds tremendous potential for research and development aimed at harnessing the therapeutic properties of seaweed-derived bioactive metabolites. One key opportunity lies in the identification of novel bioactive metabolites from seaweed, fueled by the vast marine biodiversity awaiting exploration. Untapped sources such as deep oceanic realms and remote coastal areas harbor diverse algal communities, each potentially harboring unique and undiscovered metabolites with potent applications in therapeutics and cosmeceuticals. Leveraging advanced techniques like functional metagenomics and metabolomics enables researchers to delve into these uncharted territories and unearth novel metabolites with promising skincare properties (Ghosh et al. 2022, Karthikeyan et al. 2022).

Advancements in genetic engineering present another avenue for innovation in cosmeceutical product development. Biotechnological breakthroughs, including genetic engineering, offer exciting prospects for the production of unique and novel bioactive compounds on a large scale. Researchers are actively engaged in modifying and optimizing the metabolic pathways of algae to enhance the production of metabolites with desirable properties for skincare applications. By leveraging genetic engineering techniques, scientists aim to tailor algae metabolism to yield metabolites with enhanced bioactivity and efficacy, paving the way for the development of next-generation skincare formulations (Balasubramaniam et al. 2021, Hempel et al. 2023). The convergence of scientific advancements and natural resources like seaweed heralds a new era of innovation in the cosmeceutical industry. As researchers delve deeper into the biochemical intricacies of seaweed and harness cutting-edge technologies, the potential for discovering novel bioactive compounds with transformative skincare benefits grows exponentially. By capitalizing on these opportunities, the cosmeceutical industry can unlock the full potential of seaweed-derived bioactive metabolites, offering consumers innovative and effective skincare solutions rooted in nature's bounty.

The cultivation of macroalgae stands at the forefront of considerations for economically viable industrial production in recent times, driving researchers to explore various eco-friendly cultivation methods. Among these, the utilization of closed photobioreactors has garnered significant attention over traditional open pond systems. Closed systems offer greater control over environmental parameters such as temperature, light intensity, CO_2 levels, and nutrient availability, optimizing conditions for macroalgae growth and biomass generation. This controlled environment fosters the production of seaweed biomass rich in bioactive compounds, rendering the extraction process more cost-effective and efficient. Additionally, researchers are exploring mixotrophic cultivation as a strategy to enhance biomass yield and bioactive compound production. In mixotrophic cultivation, macroalgae are cultivated under conditions where they can harness both photosynthesis and heterotrophic growth, utilizing organic carbon sources in addition to CO_2. By optimizing nutrient uptake and metabolic pathways, mixotrophic cultivation offers a promising avenue for increasing the yield of bioactive metabolites from seaweed (Mourelle et al. 2017; Pereira 2018).

Synergistic Formulations: Recent efforts in cosmeceutical research have focused on leveraging the synergistic effects of combining bioactive metabolites from seaweed with other natural compounds. By formulating synergistic blends, researchers aim to enhance the efficacy and multifunctionality of skincare products, addressing a spectrum of skin-related issues. These formulations capitalize on the complementary properties of different bioactive compounds, offering comprehensive solutions for conditions such as aging, inflammation, pigmentation, and hydration. Through strategic combination and formulation optimization, synergistic blends hold the potential to revolutionize skincare treatments, providing holistic approaches to address diverse skin concerns (Mourelle et al. 2017, Pereira 2018). In conclusion, the future outlook for seaweed in the cosmeceutical industry is highly promising, driven by advancements in technology and cultivation methodologies. The application of innovative cultivation methods, including closed photobioreactors and mixotrophic cultivation, offers opportunities to enhance the production and extraction of novel bioactive compounds from seaweed. Moreover, the development of synergistic formulations capitalizes on the combined benefits of seaweed-derived metabolites with other natural ingredients, paving the way for cost-effective and eco-friendly skincare solutions. By embracing these advancements, the cosmeceutical industry can meet the growing demand for efficacious skincare products while promoting sustainability and environmental stewardship.

Concluding remarks on seaweed utilization in cosmeceuticals with its future and challenges

In conclusion, the exploration of seaweed in cosmeceuticals has unveiled a promising frontier for the skincare industry, offering a wealth of bioactive compounds with diverse applications in cosmetic formulations (González-Minero and Bravo-Díaz 2018). The comprehensive analysis presented in this chapter underscores the multifaceted potential of seaweed-derived ingredients, illuminating their role in formulating innovative cosmetic and skincare products tailored to address a spectrum of skin concerns (Ward et al. 2005). As the chapter navigates the future, it becomes increasingly apparent that seaweed-derived compounds hold the key to unlocking new possibilities in skincare innovation and efficacy (Stengel et al. 2011). However, the journey forward is not without its challenges. Efficient extraction processes, standardization of bioactive components, and sustainable sourcing practices emerge as critical areas requiring meticulous attention and interdisciplinary collaboration (Kalasariya et al. 2022). Addressing these challenges necessitates the integration of cutting-edge technologies and stringent quality control measures to ensure the reproducibility and efficacy of cosmeceutical formulations (Murphy et al. 2021). Moreover, ethical and sustainable practices in seaweed harvesting and cultivation are paramount, highlighting the importance of environmental stewardship and social responsibility in the cosmetic industry (Peñalver et al. 2020).

Looking ahead, future research endeavors should prioritize the integration of advanced technologies such as metabolomics and genomics to deepen our understanding of seaweed biochemistry (Gupta et al. 2011). By unraveling the intricate biochemical pathways within seaweeds, scientists can target the isolation of bioactive compounds with heightened efficacy and specificity, laying the groundwork for the development of next-generation cosmeceutical formulations (Malakar et al. 2021). This scientific approach not only enhances the reproducibility of results but also fosters innovation and advancement in skincare science. In essence, this chapter serves as both a summation of current knowledge and a catalyst for future scientific exploration (Thomas and Kim 2013, Nagayama et al. 2002, Hwang 2010). The uncharted waters of seaweed-based cosmeceuticals call for a collaborative and interdisciplinary approach, where researchers, industry stakeholders, and environmental advocates work hand in hand to realize the full potential of these marine wonders (Francavilla et al. 2010). This journey, marked by challenges and opportunities, underscores the imperative of ethical and sustainable practices, positioning seaweed as a cornerstone in the sustainable evolution of the cosmetic industry (Martins et al. 2014). As the chapter navigates these waters, it lays the foundation for a future where seaweed emerges as a symbol of innovation, efficacy, and sustainability in skincare solutions.

References

Amberg, N. and C. Fogarassy. 2019. Green consumer behavior in the cosmetics market. Resources 8(3): 137. https://doi.org/10.3390/resources8030137.

Anyanwu, R., C. Rodriguez, A. Durrant and A.G. Olabi. 2018. Micro-macroalgae properties and applications. In Reference module in materials science and materials engineering. Elsevier BV.

Balasubramaniam, V., R.D.-N. Gunasegavan, S. Mustar, J.C. Lee and M.F.M. Noh. 2021. Isolation of industrial important bioactive compounds from microalgae. Molecules 26(4): 943. https://doi.org/10.3390/molecules26040943.

Bedoux, A., K. Hardouin, A.-S. Burlot and N. Bourgougnon. 2014. Bioactive components from seaweeds: Cosmetic applications and future development. Advances in Botanical Research 71: 345–378. https://doi.org/10.1016/B978-0-12-408062-1.00012-3.

Benedetti, M., V. Vecchi, S. Barera and L. Dall'Osto. 2018. Biomass from microalgae: the potential of domestication towards sustainable biofactories. Microb. Cell Fact. 17: 173–218. 10.1186/s12934-018-1019-3.

Bom, S., J. Jorge, H.M. Ribeiro and J. Marto. 2019. A step forward on sustainability in the cosmetics industry: A review. Journal of Cleaner Production.

Borowitzka, M.A. 2013. High-value products from microalgae-their development and commercialisation. Environ. Boil. Fishes 25: 743–756.

Brodie, J., L.V. Ash, I. Tittley and C. Yesson. 2018. A comparison of multispectral aerial and satellite imagery for mapping intertidal seaweed communities. Aquat. Conserv. 28: 872–881. doi: 10.1002/aqc.2905.

Chang, V.S. and S.S. Teo. 2016. Evaluation of heavy metal, antioxidant and anti-tyrosinase activities of red seaweed (*Eucheuma cottonii*). International Food Research Journal 23(6): 2370.

Chizhov, A.O., A. Dell, H.R. Morris, S.M. Haslam, R.A. McDowell, A.S. Shashkov et al. 1999. A study of fucoidan from the brown seaweed *Chorda filum*. Carbohydrate Research 320(1-2): 108–119. https://doi.org/10.1016/s0008-6215(99)00148-2.

Coelho, S.M., N. Simon, S. Ahmed, J.M. Cock and F. Partensky. 2013. Ecological and evolutionary genomics of marine photosynthetic organisms. Molecular Ecology 22(3): 867–907.

Costa, J.P., L. Custódio and C.P. Reis. 2023. Exploring the potential of using marine-derived ingredients: From the extraction to cutting-edge cosmetics. Marine Drugs 21(12): 620. https://doi.org/10.3390/md21120620.

Couteau, C. and L. Coiffard. 2020. Phycocosmetics and other marine cosmetics, specific cosmetics formulated using marine resources. Marine Drugs 18(6): 322. https://doi.org/10.3390/md18060322.

da Vaz, B.S., J.B. Moreira, M.G. De Morais and J.A.V. Costa. 2016. Microalgae as a new source of bioactive compounds in food supplements. Current Opinion in Food Science 7: 73–77. [CrossRef].

El-Beltagi, H.S., A.A. Mohamed, H.I. Mohamed, K.M.A. Ramadan, A.A. Barqawi and A.T. Mansour. 2022. Phytochemical and potential properties of seaweeds and their recent applications: A review. Marine Drugs 20(6): 342. https://doi.org/10.3390/md20060342.

Francavilla, M., P. Trotta and R. Luque. 2010. Phytosterols from Dunaliella tertiolecta and Dunaliella salina: A potentially novel industrial application. Bioresource Technology 101(11): 4144–4150. https://doi.org/10.1016/j.biortech.2009.12.139.

Ghosh, S., T. Sarkar, Z.A. Kari, H.A. Edinur and R. Chakraborty. 2022. Novel bioactive compounds from marine sources as a tool for functional food development. Frontiers in Marine Science 9. https://doi.org/10.3389/fmars.2022.832957.

González-Minero, F.J. and L. Bravo-Díaz. 2018. The use of plants in skin-care products, cosmetics and fragrances: Past and present. Cosmetics 5(3): 50.

Guillerme, J.-B., C. Couteau and L. Coiffard. 2017. Applications for Marine Resources in Cosmetics. Cosmetics 4(3): 35. https://doi.org/10.3390/cosmetics4030035.

Gupta, S. and N. Abu-Ghannam. 2011. Recent developments in the application of seaweeds or seaweed extracts as a means for enhancing the safety and quality attributes of foods. Innovative Food Science and Emerging Technologies 12: 600–609.

Helmi, A. and H.I. Mohamed. 2016. Biochemical and ultrastructural changes of some tomato cultivars after infestation with Aphis gossypii Glover (Hemiptera: Aphididae) at Qalyubiyah, Egypt. Gesunde Pflanzen 68: 41–50.

Hempel, M.d.S.S., P. Colepicolo and L. Zambotti-Villela. 2023. Macroalgae biorefinery for the cosmetic industry: basic concept, green technology, and safety guidelines. Phycology 3(1): 211–241. https://doi.org/10.3390/phycology3010014.

Hentati, F., L. Tounsi, D. Djomdi, G. Pierre, C. Delattre, A.V. Ursu et al. 2020. Bioactive polysaccharides from seaweeds. Molecules 25(14): 3152. https://doi.org/10.3390/molecules25143152.

Hurd, C.L., P.J. Harrison, K. Bischof and C.S. Lobban. 2014. Seaweed Ecology and Physiology. Cambridge University Press: Cambridge, UK.

Hussain, S.A., M.A.U. Din, L.K.T. Boppana, A. Kapoor and S. Jamshed. 2020. Suspected metformin-induced cobalamin deficiency mimicking thrombotic thrombocytopenic purpura. Cureus. 12: e6921.

Hwang, H.J. 2010. Skin elasticity and sea polyphenols. Seanol Sci. Centre Rev. 1: 17.

Jesumani, V., H. Du, M. Aslam, P. Pei and N. Huang. 2019. Potential use of seaweed bioactive compounds in skincare-a review. Marine Drugs 17(12): 688. https://doi.org/10.3390/md17120688.

Kadam, S.U., B.K. Tiwari and C.P. O'Donnell. 2013. Application of novel extraction technologies for bioactives from marine algae. Journal of Agricultural and Food Chemistry 61(20): 4667–4675. https://doi.org/10.1021/jf400819p.

Kalasariya, H.S. and L. Pereira. 2022. Dermo-cosmetic benefits of marine macroalgae-derived phenolic compounds. Applied Sciences 12(23): 11954. https://doi.org/10.3390/app122311954.

Karthikeyan, A., A. Joseph and B.G. Nair. 2022. Promising bioactive compounds from the marine environment and their potential effects on various diseases. Journal of Genetic Engineering and Biotechnology 20(1): 14. https://doi.org/10.1186/s43141-021-00290-4.

Kaushik, M., U. Farooq, M.S. Ali, M.J. Ansari, Z. Iqbal and M.A. Mirza. 2023. Safety concern and regulatory status of chemicals used in cosmetics and personal care products. Dermato. 3(2): 131–157. https://doi.org/10.3390/dermato3020011.

Lahaye, M. and A. Robic. 2007. Structure and functional properties of ulvan, a polysaccharide from green seaweeds. Biomacromolecules 8(6): 1765–1774.

Lanfer-Marquez, U.M., R.M. Barros and P. Sinnecker. 2005. Antioxidant activity of chlorophylls and their derivatives. Food Research International 38(8-9): 885–891.

Laurence, J. 2004. Seaweed proteins. pp. 197–213. *In*: Yada, R. [ed.]. Proteins in Food Processing; Woodhead Publishing: Cambridge, UK.

Lee, H., B. Selvaraj and J.W. Lee. 2021. Anticancer effects of seaweed-derived bioactive compounds. Applied Sciences 11: 11261. https://doi.org/10.3390/app112311261.

Power, D.M., L. Anjos, M. Manchado, S. Letsiou, J.C.R. Cardoso, R.C. Félix et al. 2020. D7.4 Analytical Handbook for Microalgae-Based Novel High Added-Value Products for Cosmetic and Aquaculture. Project nr: 691102. Call reference: H2020-MSCA-RISE-15.

Malakar, B. and K. Mohanty. 2021. The budding potential of algae in cosmetics. *In*: Mandotra, S.K., A.K. Upadhyay and A.S. Ahluwalia [eds.]. Algae. Springer, Singapore. https://doi.org/10.1007/978-981-15-7518-1_8.

Martins, A., H. Vieira, H. Gaspar and S. Santos. 2014. Marketed marine natural products in the pharmaceutical and cosmeceutical industries: Tips for success. Marine Drugs 12(2): 1066–1101. https://doi.org/10.3390/md12021066.

Milam, E. and E. Rieder. 2016. An approach to cosmeceuticals. Journal of drugs in dermatology: JDD 15: 452–456.

Misurcova, L. 2012. Chemical Composition of Seaweeds. DOI: 10.1002/9781119977087.ch7.

Mourelle, M.L., C.P. Gómez and J.L. Legido. 2017. The potential use of marine microalgae and cyanobacteria in cosmetics and thalassotherapy. Cosmetics 4(4): 46. https://doi.org/10.3390/cosmetics4040046.

Murphy, M.J. and A.A. Dow. 2021. Clinical studies of the safety and efficacy of macroalgae extracts in cosmeceuticals. The Journal of Clinical and Aesthetic Dermatology 14(10): 37–41.

Murthy, K.C., A. Vanitha, J. Rajesha, M.M. Swamy, P. Sowmya and G.A. Ravishankar. 2005. *In vivo* antioxidant activity of carotenoids from Dunaliella salina—A green microalga. Life Science 76: 1381–1390. https://doi.org/10.1016/j.lfs.2004.08.037.

Nagayama, K., Y. Iwamura, T. Shibata, I. Hirayama and T. Nakamura. 2002. Bactericidal activity of phlorotannins from the brown alga Ecklonia kurome. The Journal of Antimicrobial Chemotherapy 50(6): 889–893. https://doi.org/10.1093/jac/dkf222.

Pangestuti, R., K.H. Shin and S.K. Kim. 2021. Anti-photoaging and potential skin health benefits of seaweeds. Marine Drugs 19(3): 172. https://doi.org/10.3390/md19030172.

Peñalver, R., J.M. Lorenzo, G. Ros, R. Amarowicz, M. Pateiro and G. Nieto. 2020. Seaweeds as a functional ingredient for a healthy diet. Marine Drugs 18(6): 301. https://doi.org/10.3390/md18060301.

Pereira, L. 2018. Seaweeds as source of bioactive substances and skin care therapy—cosmeceuticals, algotheraphy, and thalassotherapy. Cosmetics 5(4): 68. https://doi.org/10.3390/cosmetics5040068.

Pereira, L. 2015. Seaweed flora of the european north atlantic and mediterranean. *In*: Kim, S.K. [eds.] Springer Handbook of Marine Biotechnology. Springer Handbooks. Springer, Berlin, Heidelberg. https://doi.org/10.1007/978-3-642-53971-8_6.

Pereira, L. 2016. Biological and therapeutic properties of the seaweed polysaccharides. International Biological Reviews 2: 1–50.

Pereira, L. 2016. Carrageenans—Sources and Extraction Methods, Molecular Structure, Bioactive Properties and Health Effects; Nova Science Publishers: Hauppauge, NY, USA.

Pillai, S., C. Oresajo and J. Hayward. 2005. Ultraviolet radiation and skin aging: roles of reactive oxygen species, inflammation and protease activation, and strategies for prevention of inflammation-induced matrix degradation - a review. International Journal of Cosmetic Science 27(1): 17–34. https://doi.org/10.1111/j.1467-2494.2004.00241.

Pimentel, F.B., R.C. Alves, F. Rodrigues and M.B.P.P. Oliveira. 2018. Macroalgae-derived ingredients for cosmetic industry—An update. Cosmetics 5(1): 2. https://doi.org/10.3390/cosmetics5010002.

Pullar, J.M., A.C. Carr and M.C.M. Vissers. 2017. The roles of vitamin C in skin health. Nutrients 9(8): 866. https://doi.org/10.3390/nu9080866.

Regulation (EC) N° 1223/2009 on cosmetic products. Available at: https://eur-lex.europa.eu/legalcontent/EN.

Robic, A., C. Rondeau-Mouro, J.F. Sassi, Y. Lerat and M. Lahaye. 2009. Structure and interactions of ulvan in the cell wall of the marine green algae *Ulva rotundata* (Ulvales, Chlorophyceae). Carbohydrate Polymers 77(2): 206–216.

Rocha de Souza, M.C., C.T. Marques, C.M. Guerra Dore, F.R. Ferreira da Silva, H.A. Oliveira Rocha and E.L. Leite. 2007. Antioxidant activities of sulfated polysaccharides from brown and red seaweeds. Journal of Applied Phycology 19(2): 153–160. https://doi.org/10.1007/s10811-006-9121-z.

Rybak, I., A.E. Carrington, S. Dhaliwal, A. Hasan, H. Wu, W. Burney et al. 2021. Prospective randomized controlled trial on the effects of almonds on facial wrinkles and pigmentation. Nutrients 13(3): 785.

Salehi, B., J. Sharifi-Rad, A.M.L. Seca, D.C.G.A. Pinto, I. Michalak, A. Trincone et al. 2019. Current trends on seaweeds: Looking at chemical composition, phytopharmacology, and cosmetic applications. Molecules (Basel, Switzerland), 24(22): 4182. https://doi.org/10.3390/molecules24224182.

Sanghvi, A.M. and Y.M. Lo. 2010. Present and potential industrial applications of macro- and microalgae. Recent Patents on Food, Nutrition & Agriculture 2(3): 187–194.

Saravana, P.S., Y.N. Cho, M.P. Patil, Y.J. Cho, G.D. Kim, Y.B. Park et al. 2018. Hydrothermal degradation of seaweed polysaccharide: Characterization and biological activities. Food Chemistry 268: 179–187. https://doi.org/10.1016/j.foodchem.2018.06.077.

Fernando, I.P.S., K.K.A. Sanjeewa, K.W. Samarakoon, H.S. Kim, U.K.D.S. Gunasekara, Y.J. Park et al. 2018. The potential of fucoidans from Chnoospora minima and *Sargassum polycystum* in cosmetics: Antioxidant, anti-inflammatory, skin-whitening, and anti-wrinkle activities. Journal of Applied Phycology 30: 3223–3232.

Soulioti, I., M. Diomidous, H. Theodosopoulou, N. Violaki, H. Plessa, M. Charalambidou et al. 2013. Cosmetics: History, products, industry, legislation, regulations and implications in public health. Rev. Clin. Pharmacol. Pharmacokinet 27: 5–15.

Spears, K. 1988. Developments in food colourings: The natural alternatives. Trends in Biotechnology 6(11): 283–288.

Stengel, D.B., S. Connan and Z.A. Popper. 2011. Algal chemodiversity and bioactivity: Sources of natural variability and implications for commercial application. Biotechnology Advances 29(5): 483–501. https://doi.org/10.1016/j.biotechadv.2011.05.016.

Thiyagarasaiyar, K., B.H. Goh, Y.J. Jeon and Y.Y. Yow. 2020. Algae metabolites in cosmeceutical: An overview of current applications and challenges. Marine Drugs 18(6): 323. https://doi.org/10.3390/md18060323.

Thomas, N.V. and S.K. Kim. 2013. Beneficial effects of marine algal compounds in cosmeceuticals. Marine Drugs 11(1): 146–164.

Tønnesen, H.H. and J. Karlsen. 2002. Alginate in drug delivery systems. Drug Development and Industrial Pharmacy 28(6): 621–630. https://doi.org/10.1081/ddc-120003853.

Usman, M.H., M.F. Kamaroddin, M.H. Sani and N.A.N. Nik Malek. 2023. Exploring the potential of microalgae-fungi co-cultivation for sustainable bioprocessing in microalgae biorefinery. Malaysian Applied Biology 52(6): 23–46. https://doi.org/10.55230/mabjournal.v52i6.2783.

Wang, R., V.J. Paul and H. Luesch. 2013. Seaweed extracts and unsaturated fatty acid constituents from the green alga Ulva lactuca as activators of the cytoprotective Nrf2–ARE pathway. Free Radical Biology and Medicine 57: 141–153.

Ward, O.P. and A. Singh. 2005. Omega-3/6 fatty acids: Alternative sources of production. Process Biochem. 40: 3627–3652.

Wijesinghe, W. and Y.-J. Jeon. 2012. Biological activities and potential industrial applications of fucose rich sulfated polysaccharides and fucoidans isolated from brown seaweeds: A review. Carbohydr. Polym. 88: 13–20.

Wu, L., J. Sun, X. Su, Q. Yu and P. Zhang. 2016. A review about the development of fucoidan in antitumor activity: Progress and challenges. Carbohydrate Polymers 154: 96–111. https://doi.org/10.1016/j.carbpol.2016.08.005.

Index